Let's Review Regents:

Algebra I
Revised Edition

Gary M. Rubinstein
B.A. Mathematics
Tufts University
M.S. Computer Science
University of Colorado

About the Author

Gary Rubinstein has been teaching math for 25 years. He is a three-time recipient of the Math for America Master Teacher Fellowship. Gary lives with his wife Erica and his two children, Sarah and Sam. He has a YouTube channel at nymathteacher where students can find extra test tips and strategies for learning Algebra I.

Dedication

To Erica, Sarah, and Sam

Kaplan North America, LLC, d/b/a Barron's Educational Series
1515 West Cypress Creek Road
Fort Lauderdale, Florida 33309
www.barronseduc.com

ISBN: 978-1-5062-6624-4

10 9 8 7 6 5 4

Kaplan North America, LLC, d/b/a Barron's Educational Series print books are available at special quantity discounts to use for sales promotions, employee premiums, or educational purposes. For more information or to purchase books, please call the Simon & Schuster special sales department at 866-506-1949.

TABLE OF CONTENTS

PREFACE **vii**

CHAPTER 1 **SOLVING LINEAR EQUATIONS WITH ALGEBRA** **1**

1.1 Properties of Algebra 1
1.2 Solving One-Step Algebra Equations 7
1.3 Solving Multi-Step Algebra Equations 11
1.4 Isolating Variables in Equations with Multiple Variables 14

CHAPTER 2 **POLYNOMIAL ARITHMETIC** **19**

2.1 Classifying Monomials, Binomials, and Trinomials 19
2.2 Multiplying and Dividing Monomials 22
2.3 Combining Like Terms 26
2.4 Multiplying Monomials and Polynomials 29
2.5 Adding and Subtracting Polynomials 32
2.6 Multi-Step Algebra Equations Involving Polynomial Arithmetic 36
2.7 Multiplying Polynomials by Polynomials 40
2.8 Factoring Polynomials 45
2.9 More Complicated Factoring 51

CHAPTER 3 **QUADRATIC EQUATIONS** **55**

3.1 Solving Quadratic Equations by Taking the Square Root of Both Sides of the Equation 55
3.2 Solving Quadratic Equations by Guess and Check 63
3.3 Solving Quadratic Equations by Completing the Square 66
3.4 Solving Quadratic Equations by Factoring 71
3.5 The Relationship Between Factors and Roots 76
3.6 Solving Quadratic Equations with the Quadratic Formula 79
3.7 Word Problems Involving Quadratic Equations 83

CHAPTER 4 **SYSTEMS OF LINEAR EQUATIONS** **89**
4.1 Solving Systems with Guess and Check 89
4.2 Solving Simpler Systems of Equations with Algebra 95
4.3 Solving More Complicated Systems of Equations with Algebra 102
4.4 Solving Word Problems with Systems of Equations 111

CHAPTER 5 **GRAPHS OF SOLUTION SETS OF LINEAR EQUATIONS** **117**
5.1 Producing a Graph by Identifying Two or More Points 117
5.2 Calculating and Interpreting Slope 129
5.3 Slope-Intercept Form 138
5.4 Graphing Solution Sets to Linear Equations with a Graphing Calculator 146
5.5 Determining an Equation for a Given Graph 158
5.6 Word Problems Involving Finding the Equation of a Line 165

CHAPTER 6 **GRAPHING SOLUTION SETS FOR QUADRATIC EQUATIONS** **171**
6.1 Graphing Solution Sets to Quadratic Equations 171
6.2 Using the Graphing Calculator to Solve Quadratic Equations 188
6.3 Solving Linear-Quadratic Systems of Equations 196
6.4 Graphing Quadratic Equations for Real-World Applications 205

CHAPTER 7 **LINEAR INEQUALITIES** **213**
7.1 One-Variable Linear Inequalities 213
7.2 Graphing Two-Variable Linear Inequalities 218
7.3 Graphing Systems of Linear Inequalities 230

CHAPTER 8 **EXPONENTIAL EQUATIONS** **243**
8.1 Solving Exponential Equations 243
8.2 Graphing Solution Sets to Two-Variable Exponential Equations 246
8.3 Distinguishing Between Linear, Quadratic, and Exponential Equations 254
8.4 Real-World Problems Involving Exponential Equations 268

CHAPTER 9	**CREATING AND INTERPRETING EQUATIONS FROM REAL-WORLD SCENARIOS**	**273**
	9.1 Creating and Interpreting Linear Equations	273
	9.2 Creating and Interpreting Exponential Equations	279
CHAPTER 10	**FUNCTIONS**	**287**
	10.1 Describing Functions	287
	10.2 Function Graphs	292
	10.3 Defining a Function with an Equation	304
	10.4 Function Transformations	316
CHAPTER 11	**SEQUENCES**	**329**
	11.1 Types of Sequences	329
	11.2 Recursively Defined Sequences	331
	11.3 Closed Form Defined Sequences	338
CHAPTER 12	**REGRESSION CURVES**	**345**
	12.1 Line of Best Fit	345
	12.2 The Correlation Coefficient	356
	12.3 Parabolas and Exponential Curves of Best Fit	367
CHAPTER 13	**STATISTICS**	**381**
	13.1 Measures of Central Tendency	381
	13.2 Graphically Representing Data	391
CHAPTER 14	**TEST-TAKING STRATEGIES**	**399**
	14.1 Time Management	399
	14.2 Know How to Get Partial Credit	400
	14.3 Know Your Calculator	401
	14.4 Use the Reference Sheet	407
	14.5 How Many Points Do You Need to Pass?	408
ANSWERS AND SOLUTION HINTS TO PRACTICE EXERCISES		**409**
GLOSSARY OF ALGEBRA I TERMS		**439**
THE ALGEBRA I REGENTS EXAMINATION		**447**
	June 2018 Examination	450
	Answer Key	471
	June 2019 Examination	472
	Answer Key	494
INDEX		**495**

PREFACE

In June of 2014, New York State adopted a more difficult Algebra I course based on the Common Core curriculum and changed the Regents exam to reflect this new course. On the new Algebra I Regents exam, students are expected not only to solve algebra problems and answer algebra-related questions but also to explain their reasoning. This is a major shift in the testing of algebra concepts, so a thorough review is even more critical for success.

This book will introduce you to the concepts covered throughout the new Algebra I Regents course as well as to the different types of problems you will encounter on the Regents exam at the end of the year.

Who Should Use This Book?

Students and teachers alike can use this book as a resource for preparing for the Algebra I Regents exam.

Students will find this book to be a great study tool because it contains a review of all Algebra I concepts, useful examples, and practice problems of varying difficulty that can be practiced throughout the school year to reinforce what they are learning in class. The most ideal way to prepare for the Algebra Common Core Regents exam is to work through the practice problems in the review sections and then the recently administered Regents exams at the end of the book.

Teachers can use this book as a tool to help structure an Algebra I course that will culminate with the Regents exam. The topics in the book are arranged by priority, so the sections in the beginning of the book are the ones from which more of the questions on the Regents exam are drawn. There are 60 sections in the book, spread out over 14 chapters, dedicated to all topics for Algebra I, each with practice exercises and solutions.

Why Is This Book a Helpful Resource?

Becoming familiar with the specific types of questions on the Algebra I Regents exam is crucial to performing well on this test. There are questions in which the math may be fairly easy but the way in which the question is asked makes the question seem much more difficult. For example, the question "Find all zeros of the function $f(x) = 2x + 6$" is a fancy way of asking the much simpler sounding "Solve for x if $2x + 6 = 0$." Knowing exactly what the questions are asking is a big part of success on this test.

Algebra has been around for thousands of years, and its basic concepts have never changed. So fundamentally, the algebra curriculum today is not very different from the algebra taught in schools two years ago, ten years ago, or twenty years ago. But the exam that follows this Common Core-based course, with more complicated ways of asking questions and presenting problems, requires a specifically presented study plan that's more important than ever.

Gary Rubinstein

Chapter One SOLVING LINEAR EQUATIONS WITH ALGEBRA

1.1 PROPERTIES OF ALGEBRA

Algebra has a collection of tools that, when used appropriately, can be used to solve equations. These tools have names like the *addition property of equality*. Knowing the names and how to use these properties enables you to not only solve equations but also explain the logic behind your reasoning

Algebra is what you are doing when you solve equations for unknown *variables*. The most basic type of algebra problem is something like

$$x - 1 = 3$$

Of all the possible numbers that can be substituted into the variable x, the only one that makes the equation true is the number 4. Any other number would make the equation false. Another way of saying this is that the number 4 *satisfies* the equation $x - 1 = 3$ while other numbers do not satisfy that equation. All the numbers that satisfy an equation form the *solution set* of the equation. In this example, since only one number satisfies the equation, the solution set is just the set with the number 4 in it, or $\{4\}$.

If the equation is simple enough to *solve* the equation for the unknown variable x, you might be able to just see the answer, like in the $x - 1 = 3$ example. But for more involved equations, you will use the *properties of algebra*.

The Addition Property of Equality

For an equation like, $x - 1 = 3$ the way to solve this equation with algebra is to *eliminate* the -1 from the left-hand side of the equation. You do this by adding 1 to each side of the equation.

$$\begin{aligned} x - 1 &= 3 \\ +1 &= +1 \\ x &= 4 \end{aligned}$$

It is the *addition property of equality* that lets you add something to both sides of an equation to create a new equation that has the same solution set as the original equation.

The Subtraction Property of Equality

If the equation, instead, is $x + 1 = 3$ it can be solved by subtracting 1 from both sides of the equation. When you subtract something from both sides of an equation, you are using the *subtraction property of equality.*

$$\begin{aligned} x + 1 &= 3 \\ -1 &= -1 \\ x &= 2 \end{aligned}$$

The Division Property of Equality

For an equation like $2x = 10$, the 2 can be eliminated by dividing both sides of the equation by 2.

$$\begin{aligned} 2x &= 10 \\ \frac{2x}{2} &= \frac{10}{2} \\ x &= 5 \end{aligned}$$

The *division property of equality* is the reason this is allowed.

The Multiplication Property of Equality

Sometimes when fractions are involved, a way to solve the equation is to multiply both sides of the equation by some number. For example,

$$\frac{x}{2} = 10$$

By multiplying both sides by 2, the 2 in the denominator of the left-hand side of the equation will be eliminated.

$$\begin{aligned} \frac{x}{2} &= 10 \\ 2 \cdot \frac{x}{2} &= 2 \cdot 10 \\ x &= 20 \end{aligned}$$

This property is also useful for questions like $\frac{2}{3}x = 10$, which you continue by multiplying both sides of the equation by the *reciprocal* (also known as the *multiplicative inverse*) of $\frac{2}{3}$, which is $\frac{3}{2}$.

$$\frac{2}{3}x = 10$$
$$\frac{3}{2} \cdot \frac{2}{3}x = \frac{3}{2} \cdot \frac{10}{1}$$
$$x = 15$$

The Distributive Property of Multiplication Over Addition

When there is an expression like $3(2 + 5)$, there are two ways of simplifying it. If you simplify the parentheses first, it becomes $3(2 + 5) = 3 \cdot 7 = 21$. If, instead, you use *the distributive property of multiplication over addition*, it becomes

$$3(2 + 5) = 3 \cdot 2 + 3 \cdot 5 = 6 + 15 = 21$$

You get the same answer!

Even though this isn't a very practical way to simplify $3(2 + 5)$, the distributive property of multiplication over addition is useful for progressing on certain algebra problems.

$$3(2x + 5) = 27$$
$$3 \cdot 2x + 3 \cdot 5 = 27$$
$$6x + 15 = 27$$
$$-15 = -15$$
$$6x = 12$$
$$\frac{6x}{6} = \frac{12}{6}$$
$$x = 2$$

The Commutative Property of Addition or Multiplication

In both addition and multiplication, switching the order of the numbers does not change the value of the expression. So $5 + 3 = 3 + 5$ and $5 \cdot 3 = 3 \cdot 5$. There is not a commutative property for subtraction or division.

Example 1

When solving the equation $2x - 5 = 17$, Giovanny wrote $2x = 22$ as his first step. Which property justifies this step?

Solution: He must have added 5 to both sides of the equation. This is the addition property of equality.

Example 2

Brooklynn solved the equation $5(x - 3) = 20$ by writing, as her first step, $x - 3 = 4$. Which property of algebra justifies her first step?

Solution: Though this could have been done by beginning with the distributive property, she seems to have divided both sides of the equation by 5 to eliminate the 5 from the left-hand side. This is the division property of equality.

Check Your Understanding of Section 1.1

A. *Multiple-Choice*

1. Antonio started the question $2x + 1 = 11$ by writing $2x = 10$. Which property justifies this step?
(1) Commutative property of addition
(2) Distributive property of multiplication over addition
(3) Addition property of equality
(4) Subtraction property of equality

2. Mila used the multiplication property to justify the first step in solving an equation. The original equation was $\frac{x}{2} + 4 = 10$. What could the equation have been transformed into after this step?

(1) $\frac{x}{2} = 14$

(2) $\frac{x}{2} + 6 = 12$

(3) $x + 8 = 20$
(4) $x + 2 = 5$

3. To more easily simplify an expression, Michael changed $2x + 3 + 5x + 2$ into $2x + 5x + 3 + 2$. What property justifies this step?
(1) Commutative property of addition
(2) Commutative property of multiplication
(3) Addition property of equality
(4) Distributive property of multiplication over addition

4. When solving the equation $3x - 2 = 13$, Jocelyn used a property of algebra to transform the equation into $3x = 15$. Which property did she use?
(1) Subtraction property of equality
(2) Addition property of equality
(3) Multiplication property of equality
(4) Division property of equality

5. When solving the equation $\frac{x}{3} = 8$, Devin uses a property of algebra and transforms the equation into $x = 24$. Which property did he use?
(1) Subtraction property of equality
(2) Addition property of equality
(3) Multiplication property of equality
(4) Division property of equality

6. Elijah used two different properties from algebra to solve this problem. What were the two steps and in what order did he do them?

$$3x - 4 = 17$$
$$+4 = +4$$
$$3x = 21$$
$$\frac{1}{3} \cdot 3x = \frac{1}{3} \cdot 21$$
$$x = 7$$

(1) Addition property of equality followed by multiplication property of equality
(2) Addition property of equality followed by division property of equality
(3) Subtraction property of equality followed by multiplication property of equality
(4) Subtraction property of equality followed by division property of equality

Chapter One **SOLVING LINEAR EQUATIONS**

7. Which sentence uses the multiplication property of equality?
(1) $1 + 1 = 2$; therefore, $1 + 1 + 5 = 2 + 5$
(2) $1 + 1 = 2$; therefore, $\frac{1+1}{6} = \frac{2}{6}$
(3) $1 + 1 = 2$; therefore, $1 + 1 - 2 = 2 - 2$
(4) $1 + 1 = 2$; therefore, $5(1 + 1) = 5 \cdot 2$

8. Which sentence uses the subtraction property of equality?
(1) $5 + 2 = 7$; therefore, $5 + 2 - 2 = 7 - 2$
(2) $5 + 2 = 7$; therefore, $5 + 2 + 3 = 7 + 3$
(3) $5 + 2 = 7$; therefore, $2(5 + 2) = 2 \cdot 7$
(4) $5 + 2 = 7$; therefore, $\frac{5+2}{7} = \frac{7}{7}$

9. Which sentence illustrates the addition property of equality?
(1) $a - 2 = 3$; therefore, $a - 2 - 2 = 2 - 3$
(2) $a - 2 = 3$; therefore, $a - 2 + 2 = 3 + 2$
(3) $a - 2 = 3$; therefore, $2(a - 2) = 2 \cdot 3$
(4) $a - 2 = 3$; therefore, $\frac{a-2}{5} = \frac{3}{5}$

10. Which number sentence illustrates the commutative property of multiplication?
(1) $3 \cdot x \cdot 2 = 3 \cdot 2 \cdot x$
(2) $3 + x + 2 = 3 + 2 + x$
(3) $3 - x - 2 = 3 - 2 - x$
(4) $\frac{\left(\frac{3}{x}\right)}{2} = \frac{\left(\frac{3}{2}\right)}{x}$

B. *Show how you arrived at your answers.*

1. Malachi and Wesley each solved the equation $2(3x - 1) = 16$. Malachi used the distributive property of multiplication over addition to justify his first step. Wesley used the division property of equality to justify his first step. Write what equation each student would have after the first step is completed?

2. Justify each step of this solution to the equation $3(x + 3) - 5 = 22$.

$$3(x + 3) - 5 = 22$$
$$3(x + 3) = 27$$
$$3x + 9 = 27$$
$$3x = 18$$
$$x = 6$$

3. Sandra calculates $5(3 + 7)$ by writing $5 \cdot 3 + 7 = 15 + 7 = 22$. Solomon does it by writing $5(10) = 50$. Which person is correct, and why?

4. Joel rewrites $6 - x$ as $x - 6$ and justifies it with the reason "The commutative property of subtraction." What is wrong with this reasoning?

5. Natalie simplified $3(5 \cdot 2)$ as $3 \cdot 5 \cdot 3 \cdot 2$ to get 90. Lucy says the answer is 30. Who is correct, and what is wrong with the other person's logic?

1.2 SOLVING ONE-STEP ALGEBRA EQUATIONS

When an equation has a variable by itself on one side of the equals sign, we say that that variable has been *isolated*. For example, in the equation $x = 5 - 2$, the x is isolated. In order to *solve* an equation with algebra, use the properties described earlier in this chapter to isolate the variable until the *solution set* becomes clear.

Zero-Step Solutions

In an equation like $x = 3$, there is no further work to be done since the variable is already isolated. The only number you can replace the x with to make this a true equation is the number 3. Sometimes the elements of the solution set are written inside curly brackets like $\{3\}$.

One-Step Solutions

In the equation $x + 2 = 5$, the variable x is not isolated yet. In order to isolate the x, the *constant* +2 must be *eliminated* from the left-hand side of the equals sign. To eliminate the +2, use the subtraction property of equality by subtracting 2 from both sides of the equals sign.

$$\begin{aligned} x + 2 &= 5 \\ -2 &= -2 \\ x + 2 - 2 &= 5 - 2 \\ x + 0 &= 3 \\ x &= 3 \end{aligned}$$

The third and fourth lines of this solution are optional. It could have looked like this:

$$\begin{aligned} x + 2 &= 5 \\ -2 &= -2 \\ x &= 3 \end{aligned}$$

If the constant is negative, you eliminate it with the addition property of equality.

$$\begin{aligned} x - 2 &= 5 \\ +2 &= +2 \\ x &= 7 \end{aligned}$$

Some equations like $3x = 12$ have a number being multiplied by the variable. This number, which is 3 in this example, is called the *coefficient*. Coefficients can be eliminated with the division property of equality.

$$3x = 12$$

$$\frac{3x}{3} = \frac{12}{3}$$

$$1x = 4$$

$$x = 4$$

The third line of this solution is optional.

When the coefficient is a fraction, the multiplicative property of equality can be used. Multiply both sides of the equals sign by the *reciprocal* (also known as *the multiplicative inverse*) to eliminate the coefficient.

$$\frac{2}{3}x = 18$$

$$\frac{3}{2} \cdot \frac{2}{3}x = \frac{3}{2} \cdot \frac{18}{1}$$

$$1x = \frac{54}{2}$$

$$x = 27$$

The third line of this solution is optional.

MATH FACTS

When a fraction is multiplied by its reciprocal, the product is the number 1. So if the coefficient of a variable is a fraction, as in $\frac{2}{3}x = 18$, multiplying both sides of the equation by $\frac{3}{2}$ results in $1x = 27$. Then since anything multiplied by 1 is itself, $1x$ can be replaced with x.

Example 1

Solve the equation $\frac{x}{3} = 12$ for x.

Solution: Just like the equation $\frac{1}{3}x = 12$, eliminate the 3 by multiplying both sides of the equals sign by 3.

$$\frac{x}{3} = 12$$

$$3 \cdot \frac{x}{3} = 3 \cdot 12$$

$$x = 36$$

Example 2

Solve the equation $-2 + x = 7$ for x.

Solution: The constant does not have to be on the right-hand side of the variable. You still eliminate a -2 by using the addition property of equality and adding 2 to both sides of the equals sign.

$$\begin{aligned} -2 + x &= 7 \\ +2 &= +2 \\ x &= 9 \end{aligned}$$

Check Your Understanding of Section 1.2

A. Multiple-Choice

1. What value of x makes the equation $x + 4 = 12$ true?
(1) 2 (2) 4 (3) 8 (4) 16

2. What is the solution set for the equation $x - 6 = 7$?
(1) {1} (2) {6} (3) {11} (4) {13}

3. What value of x satisfies the equation $4x = 8$?
(1) 1 (2) 2 (3) 8 (4) 32

4. Solve for x in the equation $\frac{3}{4}x = 12$.

(1) 9 (2) 16 (3) 36 (4) 48

5. What value of x makes the equation $x + 8 = 3$ true?
(1) –5 (2) 5 (3) 11 (4) –11

6. What is the solution to the equation $\frac{4}{5}x = 10$?
(1) $\frac{46}{5}$ (2) 8 (3) $\frac{25}{2}$ (4) 20

7. Find the solution to $a - 5 = -2$.
(1) –7 (2) –3 (3) 3 (4) 7

8. Solve for t in the equation $-3t = -18$.
(1) –54 (2) 54 (3) –6 (4) 6

9. Find the solution for v in the equation $-\frac{3}{4}v = -12$
(1) 9 (2) –9 (3) 16 (4) –16

10. What is the solution set to the equation $x = 7$?
(1) {–7}
(2) {7}
(3) {0}
(4) It doesn't have a solution since it is not an equation.

B. *Show how you arrived at your answers.*

1. Logan decides to solve the equation $\frac{5}{6}x = 10$ by multiplying both sides of the equals sign by 6. How might he still get the correct answer?

2. Andrew says that to solve the equation $\frac{3}{5}x = 6$, you have to use the multiplication property of equality. Vanessa says that it can be done with the division property of equality. Both are correct. Explain.

3. Noah says he can solve the equation $x - 2 = 5$ by using the subtraction property of equality. He is right. How can he justify this?

4. What property of algebra could be used to solve the equation $-3 + x = 5$?

5. The equation $0x = 10$ does not have any solutions. Why?

1.3 SOLVING MULTI-STEP ALGEBRA EQUATIONS

More complicated equations require more than one step to isolate the variable. Often there is more than one possible order in which to apply the algebra properties, though some orders are quicker and easier than others.

Two-Step Algebra Equations

When both the addition property of equality and the division property of equality are used in an equation, it is simpler to use the addition property of equality first. Similarly, when the subtraction property of equality and the division property of equality are involved in the solution, use the subtraction property of equality first.

Example 1

Solve for x in the equation $2x - 4 = 6$.

Solution: Generally, it is simpler to eliminate the constant (in this case the -4) before eliminating the coefficient, which is 2.

$$2x-4=6$$
$$+4=+4$$
$$\frac{2x}{2}=\frac{10}{2}$$
$$x=5$$

It is possible to do these steps in the opposite order, as long as you do it properly.

$$\frac{2x-4}{2}=\frac{6}{2}$$
$$x-2=3$$
$$+2=+2$$
$$x=5$$

The first way is the most straightforward, but you could be asked to explain the second way for one of the part II, part III, or part IV questions.

Even if the constant is to the left of the variable, it is generally easier to eliminate the constant before eliminating the coefficient.

Example 2

Solve for x in the equation $5 - \frac{2}{3}x = -1$

Solution:

$$5 - \frac{2}{3}x = -1$$

$$-5 = -5$$

$$-\frac{2}{3}x = -6$$

$$\left(-\frac{3}{2}\right)\cdot\left(-\frac{2}{3}\right)x = \left(-\frac{3}{2}\right)\cdot\left(-\frac{6}{1}\right)$$

$$x = 9$$

Three-Step Algebra Equations

When the distributive property of multiplication over addition is involved, algebra equations usually take three steps to solve.

Example 3

What is the solution set for the equation $2(x + 3) = 14$?

Solution:

$$2(x+3) = 14$$

$$2x + 6 = 14$$

$$-6 = -6$$

$$\frac{2x}{2} = \frac{8}{2}$$

$$x = 4$$

$$\{4\}$$

Check Your Understanding of Section 1.3

A. Multiple-Choice

1. What value of x makes the equation $3x + 7 = 22$ true?
(1) 1 (2) 3 (3) 5 (4) 7

2. Solve the equation $\frac{3}{4}x - 6 = 3$ for x.
(1) 12 (2) 14 (3) 16 (4) 18

3. Find the solution set for the equation $5(x + 4) = 35$.
(1) {1} (2) {2} (3) {3} (4) {4}

4. Solve the equation $-3x + 7 = -5$ for x.
(1) 4 (2) –4 (3) 6 (4) –6

5. Solve for x. $3 + 5x = 43$.
(1) 2 (2) 4 (3) 6 (4) 8

6. What value for x makes the equation $5 - 2x = -9$ true?
(1) 7 (2) 8 (3) 9 (4) 10

7. Solve for x. $\frac{3}{5}(x - 2) = 6$.
(1) 10 (2) 12 (3) 14 (4) 16

8. Find the solution set for the equation $-9 + 7x = 19$.
(1) {4} (2) {5} (3) {6} (4) {7}

9. Solve for x in the equation $\frac{x+4}{3} = 6$.
(1) 14 (2) 16 (3) 18 (4) 20

10. Solve for x. $-\frac{2}{7}(2 - x) = -2$.
(1) –2 (2) –3 (3) –4 (4) –5

B. *Show how you arrived at your answers.*

1. Solve for x. Show all work and justify each step with a property of algebra.

$$\frac{5}{6}(x-2)=\frac{5}{2}$$

2. Solve for x. Justify each step with a property of algebra.

$$5x-4=31$$

3. Solve for x. Justify each step with a property of algebra.

$$2(x-5)=-8$$

4. Cristiano says he can solve the equation $5(x+2)=35$ with three steps, distributive property, subtraction property of equality, and division property of equality. Kayla says she can solve the same equation in two steps, division property of equality then subtraction property of equality. Are they both right? Explain.

5. To figure out what price to charge for a pair of sneakers in order to make $44,000 profit, the manufacturer must solve the equation $44=-0.3p+50$. Use properties of algebra to solve for the variable p.

1.4 ISOLATING VARIABLES IN EQUATIONS WITH MULTIPLE VARIABLES

Whether an equation has just one variable or multiple variables, the process of using the properties of algebra remain the same. The goal is to eliminate all the numbers and variables from one side of the equation except for the one that you are trying to isolate.

Algebra Equations with Multiple Variables

Solving for x in an equation like $2x + 1 = 11$ requires two steps.

$$2x + 1 = 11$$
$$-1 = -1$$
$$\frac{2x}{2} = \frac{10}{2}$$
$$x = 5$$

If the equation was, instead, to isolate the x in the equation $ax + b = c$, the process would be very similar. You have to recognize that the b is like the constant and the a is like the coefficient.

Just as you subtract 1 from each side as the first step in solving $2x + 1 = 11$, for this equation, you subtract b from both sides.

$$ax + b = c$$
$$-b = -b$$
$$ax = c - b$$

Now, eliminate the coefficient a the same way that you eliminate the 2 in the $2x + 1 = 11$ question, by dividing both sides of the equation by it.

$$\frac{ax}{a} = \frac{c-b}{a}$$
$$x = \frac{c-b}{a}$$

Example

Solve for d, in the equation $abd - e = f$

Solution: First eliminate the e, since it is like the constant in the problem. Then eliminate the a and the b, since they are like the coefficient.

$$abd - e = f$$
$$+e = +e$$
$$\frac{abd}{ab} = \frac{f+e}{ab}$$
$$d = \frac{f+e}{ab}$$

Chapter One **SOLVING LINEAR EQUATIONS**

Check Your Understanding of Section 1.4

A. Multiple-Choice

1. Solve for d in terms of c, e, and f.
$cd - e = f$

(1) $\frac{f-e}{c}$ (3) $\frac{f}{c}+e$

(2) $\frac{f+e}{c}$ (4) $\frac{f}{c}-e$

2. Solve for m in terms of a, b, and c.
$b - ma = c$

(1) $\frac{b-c}{-a}$ (3) $\frac{c-b}{-a}$

(2) $\frac{b+c}{a}$ (4) $\frac{c+b}{-a}$

3. Solve for r in terms of c and π.
$c = 2\pi r$

(1) $\frac{2c}{\pi}$ (3) $\frac{c}{2\pi}$

(2) $\frac{2\pi}{c}$ (4) $\frac{2}{c\pi}$

4. Solve for c in terms of π and d.
$\pi = \frac{c}{d}$

(1) πd (3) $\frac{d}{\pi}$

(2) $\frac{\pi}{d}$ (4) $\pi + d$

5. Solve for x in terms of y, m, and b.
$y = mx + b$

(1) $\frac{y-b}{m}$

(2) $\frac{y+b}{m}$

(3) $\frac{y}{m}-b$

(4) $\frac{y}{m}+b$

6. Solve for m in terms of x, y, and b.
$y = mx + b$

(1) $\frac{y-b}{x}$

(2) $\frac{y+b}{x}$

(3) $\frac{y}{x}+b$

(4) $\frac{y}{x}-b$

7. Solve for b in terms of x, y, and m.
$y = mx + b$

(1) $\frac{y}{mx}$

(2) ymx

(3) $y - mx$

(4) $y + mx$

8. Solve for c in terms of a, b, and d.
$abc = d$

(1) $\frac{ab}{d}$

(2) $\frac{db}{a}$

(3) $\frac{d}{ab}$

(4) $\frac{da}{b}$

9. Solve for b in terms of a, c, and d.
$abc = d$

(1) $\frac{ac}{d}$

(2) adc

(3) $\frac{da}{c}$

(4) $\frac{d}{ac}$

10. Solve for w in terms of l and p.
$2l + 2w = p$

(1) $\frac{p+2l}{2}$ (3) $\frac{2}{p+2l}$

(2) $\frac{p-2l}{2}$ (4) $\frac{2}{p-2l}$

B. *Show how you arrived at your answers.*

1. If an equation that relates profit earned (p), quantity sold (q), price (r), and fixed expenses (f), is $p = qr - f$, rewrite the equation with q solved in terms of p, r, and f.

2. The volume of a rectangular prism is $v = lwh$, where v is the volume, l is the length, w is the width, and h is the height. Rewrite this equation with w solved in terms of v, l, and h.

3. The area of a triangle is related to the base and height of the triangle by the formula $a = bh/2$. Rewrite this equation with b solved in terms of a and h.

4. The surface area of a cylinder is $s = 2\pi rh$, where s is the surface area, r is the radius, and h is the height. Rewrite this equation with r solved in terms of s, π, and h.

5. The nth term of an arithmetic sequence can be calculated with the formula $a_n = a_1 + (n - 1)d$. Rewrite this equation with n solved in terms of a_n, a_1, and d.

Chapter Two POLYNOMIAL ARITHMETIC

2.1 CLASSIFYING MONOMIALS, BINOMIALS, AND TRINOMIALS

Terms in mathematical expressions are separated by plus or minus signs. An expression like $2x + 5$ has two terms while $x^2 + 2x + 5$ has three terms. Expressions with only one term are called *monomials.* Expressions with two terms are called *binomials.* Expressions with three terms are called *trinomials.* They are all special types of *polynomials.*

Determining the Type of a Polynomial by the Number of Terms

By counting the number of terms, you can identify which type of polynomial it is. Terms are separated by either plus or minus signs.

Example 1

Is the polynomial $6x - 2$ a monomial, a binomial, or a trinomial?

Solution: There are two terms, $6x$ and 2, so this is a binomial.

Example 2

Is the polynomial $5x^2$ a monomial, a binomial, or a trinomial?

Solution: Since there are not plus or minus signs, this is just one term, and this is a monomial.

The Degree of a Polynomial

When the terms of a polynomial have exponents on the variables, the *degree* of a polynomial is the largest exponent that a variable is raised to. The binomial $2x^5 + 3x$ is degree 5 since the largest exponent that a variable is raised to is 5. The trinomial $3x^2 - 7x + 2$ is degree 2. In a monomial like 7 the degree is 0 since it is equivalent to $7x^0$.

MATH FACTS

A polynomial with a degree of 0 is called a *constant polynomial.* A polynomial with a degree of 1 is called a *linear polynomial.* A polynomial with a degree of 2 is called a *quadratic polynomial.* A polynomial with a degree of 3 is called a *cubic polynomial.*

Check Your Understanding of Section 2.1

***A.** Multiple-Choice*

1. Classify this polynomial $5x^2 + 3$.
(1) Monomial (3) Trinomial
(2) Binomial (4) None of the above

2. Classify this polynomial $7x^2 - 3x + 2$.
(1) Monomial (3) Trinomial
(2) Binomial (4) None of the above

3. Classify this polynomial $3x^3$.
(1) Monomial (3) Trinomial
(2) Binomial (4) None of the above

4. Classify this polynomial $3x^2y + 5xy^2$.
(1) Monomial (3) Trinomial
(2) Binomial (4) None of the above

5. Classify this polynomial $3x^3 + 5x^2 + 7x - 3$.
(1) Monomial (3) Trinomial
(2) Binomial (4) None of the above

6. Classify this polynomial $3x^2 - 7x + 5$.
(1) Monomial (3) Trinomial
(2) Binomial (4) None of the above

7. Which of the following is a binomial?
   (1) $2 \cdot 5x^2$
   (2) $5x^2 + 7x + 2$
   (3) $5x^2 + 2$
   (4) $5x^2 + 2 + 3x$

8. Which of the following is a trinomial?
   (1) $2x^2$
   (2) $2x^2 + 5x$
   (3) $2x^2 + 5x + 3$
   (4) $(2x^2)(5x)(3)$

9. What is the degree of the polynomial $5x^2 + 7x^3 - 2x$.
   (1) 0
   (2) 1
   (3) 2
   (4) 3

10. Which of the following is a quadratic trinomial?
    (1) $5x^2 + 8$
    (2) $5x^3 - 3x + 8$
    (3) $5x^2 - 3x + 8$
    (4) $5x^2$

B. *Show how you arrived at your answers.*

1. An expression for the height of a projectile is $-16t^2 + 72t + 5$. What kind of polynomial is this and what is its degree?

2. An equation for the amount of profit a company makes is $5q - 800$. What kind of polynomial is that and what is its degree?

3. Cross out one of the terms in this polynomial to make it into a third degree trinomial. $5x^3 + 2x^2 - 7x + 9$.

4. Create a fourth degree monomial.

5. Mark says that the expression $2 + 3$ is a binomial since it has two terms. Layla says that it is a monomial since $2 + 3 = 5$, and 5 is just one term. Who is correct?

2.2 MULTIPLYING AND DIVIDING MONOMIALS

A monomial, like $5x^3$, has a coefficient of 5 and a variable part of x^3. Just as numbers can be multiplied together, monomials can be multiplied. When multiplying monomials, you have to use the rules for exponents.

Multiplying Expressions Involving Exponents

If you multiply a number, like 5, by itself you get $5 \cdot 5 = 25$, which can also be written as 5^2. When you multiply the variable x by itself, you get $x \cdot x$, which can be written as x^2. Multiplying monomials with larger exponents can be done with a shortcut.

Example 1

Simplify $x^2 \cdot x^3$.

Solution: One way to do this is to expand each of the expressions.

$$x^2 \cdot x^3$$
$$(x \cdot x) \cdot (x \cdot x \cdot x) = x \cdot x \cdot x \cdot x \cdot x = x^5$$

A shorter way is to use the rule that when multiplying two exponential expressions with the same base, the solution has that same base and the exponent will be the sum of the two exponents.

$$x^2 \cdot x^3 = x^{2+3} = x^5$$

Multiplying Monomials That Have Coefficients and Variable Parts

To multiply two monomials, multiply the two coefficients and then multiply the two variable parts using the exponent multiplication shortcut of adding the exponents.

Example 2

Simplify $3x^2 \cdot 5x^3$.

Solution: Since $3 \cdot 5 = 15$, the coefficient of the solution is 15. Since $x^2 \cdot x^3 = x^5$, the variable part of the solution is x^5. The complete answer is $15x^5$.

Example 3

Simplify $3x^4 \cdot 4x$.

Solution: $3 \cdot 4 = 12$, which is the coefficient of the solution. $x^4 \cdot x$ is like $x^4 \cdot x^1$ so the exponent multiplication rule says that $x^4 \cdot x = x^5$. The solution is $12x^5$.

Example 4

Simplify $2x^5 \cdot 3x^4 \cdot 5x^3$.

Solution: Since $2 \cdot 3 \cdot 5 = 30$ and $x^5 \cdot x^4 \cdot x^3 = x^{12}$, the solution is $30x^{12}$.

Example 5

Simplify $4x^2y \cdot 7xy^3$.

Solution: The product of the two coefficients is 28. The product of x^2 and x is x^3. The product of y and y^3 is y^4. So the simplified expression is $28x^3y^4$.

MATH FACTS

When raising something to a negative power, like 3^{-2}, this is not the same as $-(3^2) = -9$. Instead, the rule is that $x^{-a} = \frac{1}{x^a}$ so $3^{-2} = \frac{1}{3^2} = \frac{1}{9}$. In scientific notation, there are often expressions like $5 \cdot 10^{-3}$, which is equivalent to $5 \cdot \frac{1}{10^3} = \frac{5}{1{,}000} = 0.005$.

Dividing Monomials

To divide two monomials, divide the two coefficients and subtract the exponent of the divisor from the exponent of the dividend. For example $8x^{12} \div 4x^3 = 2x^9$ because 8 divided by 4 is 2 and $12 - 3 = 9$. This is also true when the division is written as a fraction $\frac{8x^{12}}{4x^3} = 2x^9$.

Example 6

Simplify $18x^8 \div 2x^2$.

Solution: $\frac{18}{2} = 9$ and $8 - 2 = 6$ so the answer is $9x^6$.

Check Your Understanding of Section 2.2

A. *Multiple-Choice*

1. Multiply $3x^3 \cdot 4x^5$.
(1) $7x^8$
(2) $7x^{15}$
(3) $12x^8$
(4) $12x^{15}$

2. Multiply $2x^5 \cdot 5x$.
(1) $10x^6$
(2) $10x^5$
(3) $7x^6$
(4) $7x^5$

3. Which of the following would *not* simplify to $12x^6$?
(1) $2x \cdot 6x^5$
(2) $3x^2 \cdot 4x^4$
(3) $12x^3 \cdot x^3$
(4) $4x^3 \cdot 3x^2$

4. Which of the following would *not* simplify to $24x^5$?
(1) $3x^2 \cdot 8x^3$
(2) $4x \cdot 6x^4$
(3) $2x^3 \cdot 12x^2$
(4) $6x^5 \cdot 4x$

5. Multiply $x^6 \cdot x$.
(1) x^5
(2) x^6
(3) x^7
(4) x^8

6. Multiply $5 \cdot 3x^5$.
(1) $15x^4$
(2) $15x^5$
(3) $15x^6$
(4) $15x^7$

7. If $2x^3 \cdot y = 10x^6$, which expression is equivalent to y?
(1) $5x^2$
(2) $5x^3$
(3) $5x^4$
(4) $5x^5$

8. Multiply $3x^5 \cdot 2x^{-2}$.
(1) $6x^7$
(2) $6x^5$
(3) $6x^3$
(4) $6x$

9. If $5x^7 \cdot z = 20x^4$, which expression is equivalent to z?
(1) $4x^5$
(2) $4x^7$
(3) $4x^3$
(4) $4x^{-3}$

10. Simplify $(4x^3)^2$.
(1) $16x^6$
(2) $16x^9$
(3) $8x^6$
(4) $8x^9$

***B.** Show how you arrived at your answers.*

1. Alexa multiplies $4x^2 \cdot 5x^3$ and says the answer is $20x^6$ since $4 \cdot 5$ is 20 and $2 \cdot 3$ is 6. Is she correct? Explain your reasoning.

2. What does $12x^6 \div 3x^2$ simplify to?

3. Joseph says that $x^0 = 0$ since there are no xs multiplied together. Sawyer says that this is not correct. Who is right?

4. What is $5x^2 \cdot 2x^{-2}$? Explain your reasoning?

5. When multiplying a quadratic monomial by a cubic monomial, what type of polynomial will be the result?

2.3 COMBINING LIKE TERMS

Two terms are called *like terms* if they have the same variable part. For example, $3x^2$ and $5x^2$ are like terms since they both have an x^2 as the variable part. $3x^2$ and $5x^3$ are not like terms because the x^2 is different from the x^3. Like terms can be combined with addition or subtraction, whereas unlike terms cannot be combined.

Combining Two Like Terms

When you add two dogs to three dogs, you get five dogs (and a lot of dog hair!). Likewise, $2x + 3x = 5x$, $2x^2 + 3x^2 = 5x^2$, and $2x^{10} + 3x^{10} = 5x^{10}$. Unlike terms cannot be combined. For example, $2x + 3y$ cannot be simplified any further. Even though $2x$ and $3x^2$ each have a variable part containing x, they are not like terms since the variable part must be identical, Therefore, they cannot be combined.

For terms to be like terms, the variable parts must be equivalent. So $5xy^2$ cannot be combined with $5x^2y$ since the variable parts are not equivalent.

If There Is No Coefficient, It Is Really a Coefficient of 1

When a term seems to have no coefficient, it is really a coefficient of 1. To combine the like terms $5x^2 + x^2$, think of it as $5x^2 + 1x^2 = 6x^2$.

Example 1

Simplify, if it is possible, the expression $4x^3 + 6x^3$.

Solution: Since each term has a variable part of x^3, these can be combined. Just as $4 + 6 = 10$, $4x^3 + 6x^3 = 10x^3$.

Example 2

Simplify, if it is possible, the expression $6x^3 - 2x^2$.

Solution: These are not like terms since one has a variable part of x^3 and the other has a variable part of x^2, so they cannot be combined.

Example 3

Simplify, if it is possible, the expression $5x^2\,y^3 + 4x^2\,y^3$.

Solution: These are like terms so they can be combined into $9x^2\,y^3$.

Check Your Understanding of Section 2.3

A. *Multiple-Choice*

1. Which expression is equivalent to $2x^2 + 5x^2$?
(1) $10x^2$
(2) $7x^2$
(3) $7x^4$
(4) The expression cannot be simplified any further.

2. Which expression is equivalent to $3x^4 + 5x$?
(1) $8x^4$
(2) $8x^5$
(3) $8x^6$
(4) The expression cannot be simplified any further.

3. Which expression is equivalent to $-5x^3 + 2x^3$?
(1) $-7x^3$
(2) $3x^3$
(3) $-3x^3$
(4) The expression cannot be simplified any further.

4. Which expression is equivalent to $8y^2 - 10y^2 + 5y^2$?
(1) $7y^2$
(2) $-7y^2$
(3) $3y^2$
(4) The expression cannot be simplified any further.

5. Which expression is equivalent to $5xy^2 + 3xy^2 - 4x^2y$?
(1) $4x^2y$
(2) $4xy^2$
(3) $8x^2y - 4xy^2$
(4) $8xy^2 - 4x^2y$

6. Which expression is *not* equivalent to $7x^5$?
(1) $5x^2 + 2x^3$
(2) $2x^5 + 5x^5$
(3) $9x^5 - 2x^5$
(4) $7x^2 \cdot x^3$

7. Which expression is equivalent to $3x^2$?
(1) $1 + 2x^2$
(2) $x^2 + 2x^2$
(3) $x + 2x$
(4) $x + 2x^2$

8. Which expression is *not* equivalent to $7x^2y$?
(1) $2x^2y + 5x^2y$
(2) $3x^2y + 4x^2y$
(3) $7x \cdot xy$
(4) $2 + 5x^2y$

9. Which expression cannot be simplified any further?
(1) $5x + 3x$
(2) $5x^2 + 3x^2$
(3) $5x + 3x^2$
(4) $5x^2 - 3x^2$

10. Which expression is equivalent to $x^2 + x^2 + x^2 + x^2$?
(1) x^8
(2) $4x^8$
(3) $8x^4$
(4) $4x^2$

***B.** Show how you arrived at your answers.*

1. Katherine simplifies the expression $3x^2 + 5x^2$ by noticing that $x^2(3 + 5)$ would become $3x^2 + 5x^2$ if she used the distributive property of multiplication over addition, but would become $x^2(8) = 8x^2$ if she simplified the parentheses first. Was this a valid way to arrive at the correct answer $8x^2$? Explain your reasoning.

2. Chelsea says that the expression $3xy^2 + 5y^2x$ cannot be simplified any further since the terms are not like terms. Her friend Josie says that it is possible. Who is right? Explain your reasoning.

3. If $5xy^2z + a = 7xy^2z$, what expression in terms of x, y, and z must the variable a be equivalent to?

4. Combine all like terms in the expression

$$x^3 + x^2y + xy^2 + y^3 + x^2y + xy^2 + xy^2 + x^2y$$

5. Combine all like terms in the expression $6 - 3i + 2i - i^2$.

2.4 MULTIPLYING MONOMIALS AND POLYNOMIALS

KEY IDEAS

To multiply a monomial by a binomial, trinomial, or polynomial with more than three terms, use the distributive property of multiplication over addition. Just as you can use the distributive property to simplify an expression like $2 \cdot (3 + 5)$ to become $2 \cdot 3 + 2 \cdot 5 = 6 + 10 = 16$, you can multiply monomials by polynomials the same way.

Multiplying Monomials by Binomials

To multiply the monomial $2x^2$ by the binomial $3x + 5$, put the $3x + 5$ in parentheses, and write the problem as $2x^2(3x + 5)$. Now *distribute* the $2x^2$ to each of the terms inside the parentheses to get $2x^2 \cdot 3x + 2x^2 \cdot 5$. Each of these terms is now a monomial times a monomial and can be simplified separately to become $6x^3 + 10x^2$.

Multiplying Monomials by Polynomials with More than Two Terms

The distributive property applies even when there are more than two terms in the parentheses. So $5(1 + 2 + 3) = 5 \cdot 1 + 5 \cdot 2 + 5 \cdot 3 = 30$. The same thing happens when multiplying a monomial by a polynomial with more than two terms.

To multiply $5x$ by $2x^2 + 7x - 3$, multiply the $5x$ by each of the three terms in the second expression and then simplify the terms if possible.

$$5x(2x^2 + 7x - 3) = 5x \cdot 2x^2 + 5x \cdot 7x + 5x(-3) = 10x^3 + 35x^2 - 15x$$

Example 1

Multiply $4x^2$ by $5x^3 + 7x^5$.

Solution: Write as

$$4x^2(5x^3 + 7x^5)$$

Distribute the $4x^2$ through each term in the parentheses.

$$4x^2 \cdot 5x^3 + 4x^2 \cdot 7x^5$$
$$20x^5 + 28x^7$$

Example 2

Multiply -3 by $2x - 3$.

Solution: First, write as $-3(2x - 3)$. Then distribute the -3 through each term in the parentheses.

$$-3(2x) = -6x$$
$$-3(-3) = +9$$

So the answer is $-6x + 9$. Because $-3(-3) = 9$, you put a plus sign between the two terms in the answer.

Example 3

Multiply -4 by $3x + 5$.

Solution:

$$-4(3x + 5)$$
$$-4(3x) = -12x$$
$$-4(5) = -20$$

The answer is $-12x - 20$. Since $-4\ (5) = -20$, put a minus sign between the two terms in the answer.

Example 4

Multiply $2x^2 - 5x + 2$ by $3x$.

Solution: It is simpler if you write the $3x$ on the left of the parentheses.

$$3x(2x^2 - 5x + 2)$$

Distribute the $3x$ through each term in the parentheses.

$$6x^3 - 15x^2 + 6x$$

Check Your Understanding of Section 2.4

A. Multiple-Choice

1. Multiply 5 by $2x + 3$.

(1) $10x + 3$
(2) $10x + 15$
(3) $2x + 15$
(4) $7x + 15$

2. Multiply –2 by $5x^2 - 2x + 6$.
(1) $-10x^2 - 4x - 12$
(2) $-10x^2 - 2x + 6$
(3) $-10x^2 - 4x + 12$
(4) $-10x^2 + 4x - 12$

3. Multiply $3x^2 + 7x - 5$ by $4x^2$.
(1) $-17x^2 + 7x$
(2) $12x^4 + 7x - 5$
(3) $12x^4 + 28x^3 - 20x^2$
(4) $12x^4 + 28x^3 + 20x^2$

4. Simplify $6x(2x + 3)$.
(1) $8x^2 + 18x$
(2) $12x^2 + 18x$
(3) $12x^2 + 3$
(4) $20x$

5. Simplify $-2x(3x - 5)$.
(1) $-6x - 5$
(2) $-6x^2 - 5$
(3) $-6x^2 - 10x$
(4) $-6x^2 + 10x$

6. Simplify $-4x(-2x - 7)$.
(1) $-8x^2 - 28x$
(2) $-8x^2 + 28x$
(3) $8x^2 - 28x$
(4) $8x^2 + 28x$

7. Simplify $3x(5x^2 - 2x + 3)$.
(1) $15x^3 - 2x + 3$
(2) $15x^3 - 6x^2 + 9x$
(3) $5x^2 + x + 3$
(4) $15x^3 + 6x^2 - 9x$

8. Simplify $4x^2y\,(3xy - 2xy^2)$.
(1) $12x^2y^2 - 8x^2y^3$
(2) $12x^3y^2 - 6x^3y^3$
(3) $12x^3y^2 - 8x^3y^3$
(4) $12x^3y^3 - 6x^2y^3$

9. If $a(2x + 5) = 8x^2 + 20x$, which expression is a equivalent to?
(1) $4x$
(2) $6x$
(3) $4x^2$
(4) $6x^2$

10. Which expression is *not* equivalent to $2x^3 + 10x^2 - 6x$?
(1) $2(x^3 + 5x^2 - 3x)$
(2) $2x^2(x^2 + 5x - 6)$
(3) $x(2x^2 + 10x - 6)$
(4) $2x(x^2 + 5x - 3)$

B. Show how you arrived at your answers.

1. Tucker does the question $5(2x + 3x)$ by first combining the like terms and gets $5(5x) = 25x$. Alexander does the same question by applying the distributive property. Show how Tucker completed the question, and determine if Alexander gets the same answer as Tucker.

2. Jack says that $(3x^2 + 2x + 1)5x = 15x^3 + 10x^2 + 5x$. His friend Izabella says that the monomial has to be on the left of the parentheses to do it this way. Who is correct and why?

3. The polynomial $10x^3 - 6x^2 + 14x$ can be written as $a(5x^2 - 3x + 7)$. What expression must the variable a represent? Explain your reasoning.

4. Simplify and write as a trinomial $4x(2x + 3) - 2(2x + 3)$.

5. Simplify and write as a trinomial $2x(3x - 5) - 6(3x - 5)$.

2.5 ADDING AND SUBTRACTING POLYNOMIALS

Polynomials are composed of one or more monomials. When you add or subtract polynomials, you locate the like terms and combine them separately. Trinomials often have a constant term, an x term, and an x^2 term. If two trinomials like this are combined, the result is also a trinomial.

Adding Polynomials

When two polynomials have like terms, those terms can be combined the way you combined monomials.

Example 1

Simplify the expression $(2x + 3) + (3x + 5)$.

Solution: When adding polynomials, you can disregard the parentheses and get $2x + 3 + 3x + 5$. The $2x$ and the $3x$ are like terms that can be combined to become $+5x$. The 3 and the 5 are also like terms that can be combined to become $+8$, so the answer is $5x + 8$.

Example 2

Simplify the expression $(x^2 + 3x - 2) + (x^2 - 5x + 7)$.

Solution:

$$x^2 + 3x - 2 + x^2 - 5x + 7$$
$$x^2 + x^2 = 2x^2$$
$$+3x - 5x = -2x$$
$$-2 + 7 = +5$$

The answer is $2x^2 - 2x + 5$.

Subtracting Polynomials

When subtracting polynomials, you have to beware of one of the most common errors in all of high school math: not distributing the negative sign through the parentheses.

Example 3

Simplify the expression $(8x + 5) - 3(3x + 2)$.

Solution: You can disregard the parentheses of the expression on the left side of the minus sign.

$$8x + 5 - 3(3x + 2)$$

Now be sure to distribute the -3 through, as if the $8x + 5$ were not there. Since $-3(3x + 2)$ is $-9x - 6$, this becomes

$$8x + 5 - 9x - 6$$

Now the like terms can be combined.

$$8x - 9x = -1x = -x$$
$$+5 - 6 = -1$$

The simplified answer is $-x - 1$.

MATH FACTS

One of the most common mistakes in algebra is to not distribute the minus sign through all the terms in a polynomial subtraction problem. If the problem is $(2x + 5) - (3x + 2)$, the correct thing to do is write it as $2x + 5 - 3x - 2$, not as $2x + 5 - 3x + 2$.

Example 4

Simplify the expression $(5x - 2) - (2x + 7)$.

Solution: Think of this as $(5x - 2) -1(2x + 7)$.
Remove the parentheses from the first expression and distribute the -1 through the second expression. Since $-1(2x + 7)$ is $-2x - 7$, this becomes

$$5x - 2 - 1(2x + 7)$$
$$5x - 2 - 2x - 7$$
$$5x - 2x = 3x$$
$$-2 - 7 = -9$$

So the answer is $3x - 9$.

Warning! The most common mistake students make is to just remove both sets of parentheses and get $5x - 2 - 2x + 7$ with a $+7$ instead of the correct -7 before combining like terms.

Example 5

Simplify the expression $(7x + 3) - (4x - 5)$.

Solution:

$$7x + 3 - 1(4x - 5)$$
$$7x + 3 - 4x + 5 \text{ (notice the +5 is correct, not -5.)}$$
$$3x + 8$$

Check Your Understanding of Section 2.5

A. *Multiple-Choice*

1. Simplify $(2x + 3) + (4x + 5)$.
(1) $6x + 8$ (3) $6x + 15$
(2) $8x + 15$ (4) $8x + 8$

2. Simplify $(3x - 2) + (6x + 5)$.
(1) $9x + 3$ (3) $9x + 7$
(2) $9x - 7$ (4) $9x - 3$

3. Simplify $(x^2 + 3x + 5) + (x^2 + 4x - 2)$.
(1) $2x^4 + 7x^2 + 3$ (3) $2x^2 + 7x - 7$
(2) $2x^2 + 7x + 3$ (4) $2x^2 + 7x + 7$

4. Simplify $(x^2 - 6x + 4) + 3(x^2 + 2x - 5)$.
(1) $4x^4 - 4x - 1$
(2) $4x^4 - 12x - 11$
(3) $4x^2 - 11$
(4) $4x^2 - 12x + 11$

5. Simplify $(5x + 3) - (3x + 2)$.
(1) $2x - 5$
(2) $2x - 1$
(3) $2x + 5$
(4) $2x + 1$

6. Simplify $(3x - 4) - (5x - 3)$.
(1) $-2x - 7$
(2) $-2x - 1$
(3) $2x - 1$
(4) $2x - 7$

7. Simplify $2x(3x + 4) - 4x(2x - 5)$.
(1) $-2x^2 + 28x$
(2) $-2x^2 + 8x - 5$
(3) $-2x^2 - 12x$
(4) $-2x^2 - 20x$

8. If $C = 2x^2 + 3x - 2$ and $D = -4x^2 - 3x + 5$, then $C - D$ equals
(1) $-2x^2 + 3$
(2) $6x^2 + 3$
(3) $6x^2 + 6x - 7$
(4) $6x^2 - 7$

9. Simplify $(3x^2 - 4x + 2) - 2(x^2 - 2x + 1)$.
(1) x^2
(2) $x^2 - 2x + 1$
(3) $x^2 + 2x - 1$
(4) $x^2 - 4x + 5$

10. Simplify $(x^2 - 5x + 6) - (x^2 - 2x + 1)$.
(1) $-7x + 7$
(2) $-7x + 5$
(3) $-3x + 5$
(4) $-3x + 7$

B. *Show how you arrived at your answers.*

1. Nathan tried to simplify the expression $(2x + 3) - (4x + 2)$ by first writing $2x + 3 - 4x + 2$ and then simplifying to $-2x + 5$. Did he do this correctly? Explain your reasoning.

2. If $C = 3x + 5$ and $D = 2x - 5$, determine $C - 3D$ and simplify as much as possible.

3. To solve $49 - 2(13)$, Kaleigh changed it into $(40 + 9) - 2(10 + 3)$ and then simplified to $40 + 9 - 20 - 6 = 20 + 3 = 23$. Explain why this produced the right answer.

4. If $F = -2x + 5$ and $G = 3x - 7$, express $2F - 5G$ in terms of x and y simplified as much as possible.

5. If $P = x^2 + 5x + 2$, $Q = x^2 - 3x + 4$, and $R = x^2 - 2x + 5$, what is $P + Q - R$ in terms of x simplified as much as possible?

2.6 MULTI-STEP ALGEBRA EQUATIONS INVOLVING POLYNOMIAL ARITHMETIC

More complicated algebra equations often require combining like terms before solving. These equations can take three or more steps. When variables are on both sides of the equals sign, they can be eliminated from one side, which will require combining like terms on the other side.

Simplifying One Side of the Equation Before Solving

Linear equations of the form $ax + b = c$ are the simplest to solve. If there is a linear equation that is not in this form, often it can first be converted into an equation of this form and then solved. This could require using the distributive property and/or combining like terms.

Example 1

Solve for x in the equation $5(x + 3) - 2x = 27$.

Solution: First, simplify the left-hand side, using the distributive property and then combining like terms.

$$5x + 15 - 2x = 27$$
$$3x + 15 = 27$$

Then solve the two-step equation.

$$3x + 15 = 27$$
$$-15 = -15$$
$$3x = 12$$
$$\frac{3x}{3} = \frac{12}{3}$$
$$x = 4$$

Solving an Equation with Variables on Both Sides of the Equals Sign

Variables can be eliminated from one side of an equation by using the addition property of equality or the subtraction property of equality.

Example 2

Solve for x in the equation $5x + 3 = 3x + 15$.

Solution: In order to make this equation so that there are only variables on one side of the equation, it is simplest to subtract $3x$ from both sides of the equation

$$\begin{aligned} 5x + 3 &= 3x + 15 \\ -3x &= -3x \end{aligned}$$

Now combine the like terms $5x$ and $-3x$, and complete the two-step equation.

$$\begin{aligned} 2x + 3 &= 15 \\ -3 &= -3 \\ 2x &= 12 \\ \frac{2x}{2} &= \frac{12}{2} \\ x &= 6 \end{aligned}$$

Word Problems Involving Combining Like Terms

Some word problems can be converted into algebra equations where there are like terms that have to be combined or variables on both sides of the equation. It is important to set these equation up properly.

Example 3

Angelina has x dollars. Alexander has five more than three times the amount of money that Angelina has. Together they have 41 dollars. How much does each have?

Solution: Alexander has $3x + 5$ dollars. The equation to solve this problem, then, is

$$x + (3x + 5) = 41$$
$$x + 3x + 5 = 41$$
$$4x + 5 = 41$$
$$-5 = -5$$
$$\frac{4x}{4} = \frac{36}{4}$$
$$x = 9$$
$$3x + 5 = 3 \cdot 9 + 5 = 27 + 5 = 32$$

So Angelina has \$9 and Alexander has \$32.

Example 4

In five years Gabriel will be three years less than twice his current age. How old is he now?

Solution: Call his current age x. In five years, his age will be $x + 5$. The equation becomes

$$x + 5 = 2x - 3$$
$$-x = -x$$
$$5 = x - 3$$
$$+3 = +3$$
$$8 = x$$

Check Your Understanding of Section 2.6

A. Multiple-Choice

1. Solve for x: $(2x + 3) + (4x + 5) = 32$.
(1) 4 (2) 5 (3) 6 (4) 7

2. Solve for x: $(5x + 7) - (2x + 3) = 25$.
(1) 5 (2) 6 (3) 7 (4) 8

3. Solve for x: $x + 3x - 5 = 195$.
(1) 30 (2) 40 (3) 50 (4) 60

4. Solve for y: $y + (y + 1) + (y + 2) = 45$.
(1) 14 (2) 15 (3) 16 (4) 17

5. Solve for x: $5x + 50 = 7x$.
(1) 22 (2) 23 (3) 24 (4) 25

6. Solve for z: $30z = 300 - 20z$.
(1) 3 (2) 4 (3) 5 (4) 6

7. Solve for a: $2a + 5 = 4a - 1$.
(1) 2 (2) 3 (3) 4 (4) 5

8. Solve for x. $x^2 + 4x + 6 = x^2 - 2x + 18$
(1) 0 (2) 1 (3) 2 (4) 3

9. The width of a rectangle is 5 inches more than its length. The perimeter of the rectangle is 58 inches. Which equation could be used to find the length of the rectangle?
(1) $2x + 2(x + 5) = 58$
(2) $x + 2(x - 5) = 58$
(3) $x + (x - 5) = 58$
(4) $x + (x + 5) = 58$

10. Jonathan weighs thirty pounds more than three times his son Aidan's weight. If their combined weight is 210 pounds, which equation could be used to find the weight of Aidan?
(1) $x + 3x + 30 = 210$
(2) $x + 3x - 30 = 210$
(3) $3x - 30 - x = 210$
(4) $x\,(3x + 30) = 210$

***B.** Show how you arrived at your answers.*

1. In 20 years Lillian will be three years older than twice her current age. Set up an equation that can be used to solve for x, Lillian's current age.

2. The equation $2x + 5 = 6x - 7$ can be solved several ways. Makayla solves it by first subtracting $2x$ from both sides of the equation. David solves it by first subtracting $6x$ from both sides. Which method would you choose and why?

3. Solve for x in $2(x + 3) - 5(x - 2) = 4$.

4. Two trains are 300 miles away from each other on the same set of tracks. The first train goes east at a speed of 30 miles per hour. The second train goes west at a speed of 20 miles per hour. An equation that can be used to determine how long it will take for the trains to pass is $30x = 300 - 20x$. Explain how this equation was formed, and solve for x to determine when the trains will pass.

5. On Tuesday an electronics store sells five items less than double the amount it sold on Monday. On Wednesday the store sells ten items more than triple the amount it sold on Monday. If it sold 125 items altogether on the three days, create an equation and use it to determine the number of items sold on each day.

2.7 MULTIPLYING POLYNOMIALS BY POLYNOMIALS

Multiplying polynomials can be done by applying the distributive property several times and then combining like terms. When multiplying a binomial by another binomial, some teachers teach the FOIL shortcut.

Multiplying Binomials with the Distributive Property

Just as $2\,(x + 5)$ can be first simplified with the distributive property into $2 \cdot x + 2 \cdot 5$, a more complicated multiplication like $(x + 2)\,(x + 5)$ can be first simplified to

$$(x + 2) \cdot x + (x + 2) \cdot 5.$$

To make this more familiar, change this to $x \cdot (x + 2) + 5 \cdot (x + 2)$. Now each of the terms can be simplified with the distributive property.

$$x^2 + 2x + 5x + 10$$

After combining like terms, this becomes $x^2 + 7x + 10$.

Example 1

Multiply $(x + 4) \cdot (x - 3)$.

Solution:

$$(x + 4)\,x + (x + 4)(-3)$$
$$x\,(x + 4) - 3(x + 4)$$
$$x^2 + 4x - 3x - 12$$
$$x^2 + 1x - 12$$
$$x^2 + x - 12$$

Multiplying Binomials with the FOIL Shortcut

Four different multiplications occur when multiplying binomials with the distributive property method. The FOIL shortcut is a way to skip two steps of the longer way and accomplish all four of these multiplications before combining like terms.

$$
\begin{aligned}
(x-6)(x+2) &= \overbrace{x \cdot x}^{F} + \overbrace{2 \cdot x}^{O} + \overbrace{(-6)(x)}^{I} + \overbrace{(-6)(+2)}^{L} \\
&= x^2 + \quad [2x \ -6x] \quad -12 \\
&= \mathbf{x^2 - 4x - 12}
\end{aligned}
$$

Example 2

Multiply $(x + 2)(x + 5)$.

Solution: The F stands for "Firsts." The first term in the first binomial is x. The first term in the second binomial is also x. When you multiply these firsts together, you get x^2.

$$x^2$$

The O stands for "Outers." The x term in the first binomial and the +5 in the second binomial are the outermost terms. Multiply them together to get $+5x$.

$$x^2 + 5x$$

The I stands for "Inners." The +2 in the first binomial and the x in the second binomial are the two innermost terms. Multiply them together to get $+2x$.

$$x^2 + 5x + 2x$$

The L stands for "Lasts." The last term in the first binomial is +2. The last term in the second binomial is +5. Multiply them together to get +10.

$$x^2 + 5x + 2x + 10$$

Now combine like terms to get $x^2 + 7x + 10$, which is the product.

Example 3

Multiply $(x + 4)(x - 2)$.

Solution: Using the FOIL shortcut, the product of the Firsts is x^2, the product of the Outers is $-2x$, the product of the Inners is $+4x$, and the product of the Lasts is -8. $x^2 - 2x + 4x - 8$ becomes $x^2 + 2x - 8$ after combining like terms.

Example 4

Multiply $(2x - 3)(5x + 2)$.

Solution: Using FOIL, the four terms become $10x^2$, $+4x$, $-15x$, and -6. Combine the like terms of $10x^2 + 4x - 15x - 6$ to get $10x^2 - 11x - 6$.

Example 5

Multiply $(x + 5)(x + 5)$.

Solution: The four terms you get with FOIL are x^2, $+5x$, $+5x$, and $+25$, which combine to get $x^2 + 10x + 25$.

Squaring Binomials

To square a binomial, like $(x + 5)^2$, write it as $(x + 5)(x + 5)$ and then use FOIL to get $x^2 + 10x + 25$. The answer is *not* $x^2 + 25$ as there is no distributive property for exponents. In general, the square of a binomial $(x + a)$ is $x^2 + 2ax + a^2$.

Example 6

Simplify $(x + 6)^2$.

Solution: Rewrite as $(x + 6)(x + 6)$ and use FOIL.

$$(x + 6)(x + 6) = x^2 + 6x + 6x + 36 = x^2 + 12x + 36.$$

Note that it did not become $x^2 + 36$, as some students get when they try to distribute the exponent through the parentheses, which is not a valid way to calculate this.

Multiplying the $(a - b)(a + b)$ Pattern

When two binomials are the same except that one has a minus sign between the two terms and the other has a plus sign between the two terms, like $(a - b)(a + b)$, the answer will be $a^2 - b^2$. If you do it with FOIL, the middle two terms, the OI, will cancel each other out.

Example 7

Multiply $(x + 5)(x - 5)$.

Solution: The four terms you get with FOIL are x^2, $-5x$, $+5x$, and -25. When you combine them you get $x^2 - 5x + 5x - 25 = x^2 + 0x - 25 = x^2 - 25$, which is a binomial.

This can also be done with the $(a - b)(a + b)$ shortcut. In this case the a is the x and the b is the 5 so the solution is $x^2 - 5^2 = x^2 - 25$.

Multiplying Polynomials by Polynomials

If one of the polynomials has more than two terms, the FOIL shortcut cannot be used and the problem must be done the long way with the distributive property.

Example 8

Multiply $(x + 3)(x^2 + 5x + 6)$.

Solution: Using the distributive property, first get

$$(x + 3)x^2 + (x + 3)5x + (x + 3)6$$

Using the distributive property three more times, get

$$x^3 + 3x^2 + 5x^2 + 15x + 6x + 18$$

Combine like terms

$$x^3 + 8x^2 + 21x + 18$$

Check Your Understanding of Section 2.7

A. *Multiple-Choice*

1. $(x + 2)(x + 7) =$
(1) $x^2 + 14$
(2) $x^2 + 5x + 14$
(3) $x^2 + 9x + 14$
(4) $x^2 + 14x + 9$

2. $(x + 2)(x - 7) =$
(1) $x^2 - 5x - 14$
(2) $x^2 - 17$
(3) $x^2 + 5x + 14$
(4) $x^2 - 5x + 14$

3. $(2x + 3)(3x - 1) =$
(1) $6x^2 - 3$
(2) $6x^2 + 11x - 3$
(3) $6x^2 + 7x - 3$
(4) $6x^2 - 7x - 3$

4. $(2a + 5b)(4a + 3b) =$
(1) $8a^2 + 15b^2$
(2) $8a^2 + 30ab + 15b^2$
(3) $8a^2 + 28ab + 15b^2$
(4) $8a^2 + 26ab + 15b^2$

5. $(x + 5)^2 =$
(1) $x^2 + 25$
(2) $x^2 - 25$
(3) $x^2 + 10x - 25$
(4) $x^2 + 10x + 25$

6. $(x - 7)(x + 7) =$
(1) $x^2 + 49$
(2) $x^2 - 49$
(3) $x^2 - 14x - 49$
(4) $x^2 + 14x - 49$

7. $(x^2 - 2)(x^2 + 6) =$
(1) $x^4 + 4x^2 - 12$
(2) $x^4 - 12$
(3) $x^2 + 4x - 12$
(4) $x^4 - 4x^2 - 12$

8. $(x^2 - 4)(x^2 - 9) =$
(1) $x^4 + 36$
(2) $x^4 - 13x^2 + 36$
(3) $x^4 + 13x^2 + 36$
(4) $x^4 - 13x^2 - 36$

9. If $(x + 4)(x + a) = x^2 + 7x + 12$. What is the value of a?
(1) 1
(2) 2
(3) 3
(4) 4

10. A 4 foot by 6 foot patio has a 2 foot border around it. The area of the combined patio and border can be expressed as $(2x + 4)(2x + 6)$. Which expression is equivalent to this?
(1) $4x^2 - 20x + 24$
(2) $4x^2 + 20x + 24$
(3) $4x^2 + 20x - 24$
(4) $4x^2 + 24$

B. *Show how you arrived at your answers.*

1. If $C = x + 4$ and $D = x - 1$, express $CD + 2C$ as a trinomial.

2. Express $(2 + 3i)^2 + (1 + 2i)$ as a trinomial.

3. Maria says that $(x + 4)^2 = x^2 + 16$ because of the distributive property. Is she correct? Explain your reasoning.

4. What is $(x + 3)(x^2 + 7x + 10)$ expressed as a four-term polynomial?

5. Noah says that he has a shortcut for multiplying $48 \cdot 52$ in his head. He starts by writing 48 as $50 - 2$ and 52 as $50 + 2$. Explain how he can complete this question.

2.8 FACTORING POLYNOMIALS

Factoring a number is when you find two numbers whose product is that number. For example, the number 15 can be *factored* into $3 \cdot 5$. Some polynomials can also be factored by reversing the multiplication processes from the previous sections.

Greatest Common Factor Factoring

When you use the distributive property to multiply an expression like $4(3x + 5)$, you get $12x + 20$. *Factoring* $12x + 20$ requires undoing the distributive property. Examine the terms $12x$ and 20 and determine if they have a common factor. In this case since 4 is a factor of both $12x$ and of 20, it is a *common factor*. There are other common factors, such as 2, but of all the common factors, 4 is the largest one, or the *greatest common factor*. To factor $12x + 20$, first write the greatest common factor, 4, and then put an empty set of parentheses, which will be filled in later. Since the goal is to find what must be multiplied by 4 to get $12x + 20$, put the $12x + 20$ on the right-hand side of the equals sign.

$$4(\qquad) = 12x + 20$$

Now think, "Four times what is $12x$?" The answer is $3x$. Then think, "Four times what is 20?" The answer is +5. These are the two terms that go inside the parentheses.

$$4(3x + 5)$$

You can check your answer by using the distributive property.

$$4(3x + 5) = 4 \cdot 3x + 4 \cdot 5 = 12x + 20$$

Example 1

Factor the trinomial $2x^2 - 8x + 10$.

Solution: The greatest common factor is 2.

$$2(\qquad) = 2x^2 - 8x + 10$$
$$2(x^2 - 4x + 5) = 2x^2 - 8x + 10$$

Example 2

Factor the trinomial $12x^3 + 18x^2 - 24x$.

Solution: In this case, the greatest common factor is not just 6, but $6x$ since all of the terms have xs in them.

$$6x(\qquad) = 12x^3 + 18x^2 - 24x$$
$$6x(2x^2 + 3x - 4) = 12x^3 + 18x^2 - 24x$$

Factoring a Trinomial into the Product of Two Binomials

When you multiply the binomials $(x + 2)$ and $(x + 5)$, you get $x^2 + 7x + 10$. The answer has three terms, an x^2 term, an x term, and a constant. The coefficient of the x term is 7 and the constant is 10. Notice that 5 and 2 are the two constants of the binomials and that $7 = 5 + 2$ and $10 = 5 \cdot 2$. This is the key to factoring trinomials that have the form $x^2 + bx + c$. You will need to find two numbers that add up to the b coefficient and whose product is the c constant as the first step.

Example 3

Factor $x^2 + 8x + 15$ into the product of two binomials.

Solution: You need to find two numbers whose product is +15 and whose sum is +8. There are a lot of pairs of numbers whose sum is +8, but not as many whose product is +15. The list of numbers whose product is +15 is $+1 \cdot +15$, $+3 \cdot +5$, $-1 \cdot -15$, and $-3 \cdot -5$.

Of these four pairs, +3 and +5 are the two numbers that also have a sum of +8. The solution, then, is $(x + 3)(x + 5)$.

Check your answer by multiplying with FOIL.

$$(x + 3)(x + 5) = x^2 + 5x + 3x + 15 = x^2 + 8x + 15$$

Example 4

Factor $x^2 - 3x - 18$ into the product of two binomials.

Solution: The six factor pairs of –18 are –1 · +18, +1 · –18, –2 · +9, +2 · 9, –3 · +6, and +3 · –6. From this list, the pair whose sum is –3 is +3 and –6, so the solution is $(x + 3)(x - 6)$. Multiply to check that the product of these is $x^2 - 3x - 18$.

Example 5

Factor $3x^2 + 30x + 63$.

Solution: If possible, first look for a common factor. In this case all three terms have a factor of 3.

$$3x^2 + 30x + 63 = 3(x^2 + 10x + 21)$$

Now factor the $x^2 + 10x + 21$ by finding two numbers with a product of 21 and a sum of 10. The two numbers are +3 and +7, so it factors into $(x + 3)(x + 7)$. The answer is $3(x + 3)(x + 7)$.

MATH FACTS

There are four different combinations of plus and minus signs that can separate the terms in a quadratic trinomial. $x^2 + bx + c$, $x^2 - bx + c$, $x^2 + bx - c$, and $x^2 - bx - c$. The signs between the terms give hints of what the factors could be.

Case 1: $x^2 + bx + c$. The factors will be of the form $(x + p)(x + q)$. For example, $x^2 + 7x + 10 = (x + 2)(x + 5)$.

Case 2: $x^2 - bx + c$. The factors will be of the form $(x - p)(x - q)$. For example $x^2 - 7x + 10 = (x - 2)(x - 5)$.

For both cases 1 and 2, the rule is that when the *c* term is positive, the signs between the terms in both factors will be the same as the sign of the *b* term.

Case 3: $x^2 + bx - c$. The factors will be of the form $(x - p)(x + q)$ where *q* is the larger factor. For example, $x^2 + 3x - 10 = (x - 2)(x + 5)$.

Case 4: $x^2 - bx - c$. The factors will be of the form $(x - p)(x + q)$ where *p* is the larger factor. For example, $x^2 - 3x - 10 = (x - 5)(x + 2)$.

For both cases 3 and 4, the rule is that when the *c* term is negative, one of the factors will have a plus sign between the two terms and the other will have a minus sign between the two terms. The larger number will go after whatever the sign of the *b* term is.

Factoring a Perfect Square Trinomial into the Square of a Binomial

When you multiply a number by itself, like 3 times 3, the result is 9, which is known as a *square* number because it can be factored into 3^2. When you use FOIL to square a binomial, like $(x + 5)^2$, you get $x^2 + 10x + 25$. This is known as a *perfect square trinomial* because it can be factored back into $(x + 5)^2$. By recognizing that a trinomial is a perfect square trinomial, it is easier to factor. All perfect square trinomials $x^2 + bx + c$ have the property that $c = \left(\frac{b}{2}\right)^2$, and factor into $\left(x+\frac{b}{2}\right)^2$.

Example 6

Is $x^2 + 12x + 36$ a perfect square trinomial?

Solution: Yes. If you calculate $\left(\frac{12}{2}\right)^2$ you get 36, which is equal to the constant term.

Example 7

Is $x^2 - 6x + 8$ a perfect square trinomial?

Solution: No. $\left(-\frac{6}{2}\right)^2 = 9$, which is not equal to the constant term, 8.

Example 8

If $x^2 - 14x + 49$ is a perfect square trinomial, factor it.

Solution: Since $\left(-\frac{14}{2}\right)^2$ does equal 49, it is a perfect square trinomial.

It factors into $(x - 7)^2$. You can check by calculating $(x - 7)^2$. Note that $(x - 7)^2$ is NOT $x^2 - 49$. Instead, write $(x - 7)$ twice and use FOIL.

$$(x - 7)^2 = (x - 7)(x - 7) = x^2 - 7x - 7x + 49 = x^2 - 14x + 49.$$

Difference of Perfect Squares Factoring

When you multiply the binomials $(x - 5)$ and $(x + 5)$ with FOIL, you get $x^2 + 5x - 5x - 25$. When combining like terms, the $+5x$ and the $-5x$ become $0x$, which is the same thing as 0. So $(x - 5)(x + 5) = x^2 - 25$. Notice that x^2 and 25 are both perfect squares. When you have a binomial like $x^2 - 25$ where there is a perfect square subtracted from another perfect square, there is a shortcut for factoring it. The rule is that $(x^2 - b^2) = (x - b)(x + b)$.

Example 9

If possible, factor $x^2 - 16$.

Solution: Since both x^2 and 16 are perfect squares and one is being subtracted from the other, this difference of perfect squares pattern applies. Rewrite as $x^2 - 4^2$ and follow the pattern.

$$(x^2 - b^2) = (x - b)(x + b)$$
$$(x^2 - 4^2) = (x - 4)(x + 4)$$

Example 10

If possible, factor $x^2 + 49$.

Solution: Though this does resemble the $x^2 - b^2$ pattern, there is a plus sign between the two terms. The difference between perfect squares rule cannot be applied here.

Example 11

If possible, factor $3x^2 - 48$.

Solution: Though this is not a difference of perfect squares, you can first factor out a 3.

$$3x^2 - 48 = 3(x^2 - 16)$$

Now the $x^2 - 16$ can be written as $x^2 - 4^2$, which can be factored with the difference of perfect squares rule.

$$3(x^2 - 4^2) = 3(x - 4)(x + 4)$$

Test-taking Strategy for Multiple-Choice Factoring Questions

If a factoring question is a part A multiple-choice question, the correct answer can be found by multiplying all the answer choices and seeing which one is equivalent to the expression that needs to be factored.

Example 12

Which choice is a factorization of $x^2 - 3x - 180$?

(1) $(x - 18)(x + 10)$
(2) $(x + 12)(x - 15)$
(3) $(x - 12)(x + 15)$
(4) $(x - 20)(x + 9)$

Solution: Multiply the answer choices until one matches the question.

Testing choice 1: $(x - 18)(x + 10) = x^2 + 10x - 18x - 180 = x^2 - 8x - 180$. No.

Testing choice 2: $(x + 12)(x - 15) = x^2 - 15x + 12x - 180 = x^2 - 3x - 180$. Yes!

Check Your Understanding of Section 2.8

A. Multiple-Choice

1. Factor $3x^2 + 6x + 15$.
(1) $3(x + 5)(x + 1)$
(2) $3(x^2 + 6x + 15)$
(3) $(3x + 5)(x + 3)$
(4) $3(x^2 + 2x + 5)$

2. Factor completely $2x^3 - 8x^2 + 14x$.
(1) $2(x^3 - 4x^2 + 7x)$
(2) $x(2x^2 - 8x + 14)$
(3) $2x(x^2 - 4x + 7)$
(4) $(2x^2 + 7)(x + 2)$

3. Factor $x^2 + 10x + 16$.
(1) $(x + 2)(x + 8)$
(2) $(x + 4)(x + 4)$
(3) $(x + 16)(x + 1)$
(4) $(x + 3)(x + 7)$

4. Factor $x^2 - 2x - 15$.
(1) $(x - 3)(x - 5)$
(2) $(x + 3)(x - 5)$
(3) $(x - 15)(x + 1)$
(4) $(x + 15)(x - 1)$

5. Factor $x^2 + x - 42$.
(1) $(x - 21)(x + 2)$
(2) $(x - 6)(x + 7)$
(3) $(x - 14)(x + 3)$
(4) $(x + 6)(x - 7)$

6. Factor $x^2 + 8x + 16$.
(1) $(x + 4)^2$
(2) $(x + 8)(x + 2)$
(3) $(x - 4)^2$
(4) $(x - 4)(x + 4)$

7. Factor $x^2 - 12x + 36$.
(1) $(x - 6)^2$
(2) $(x + 6)^2$
(3) $(x - 6)(x + 6)$
(4) $(x + 12)(x + 3)$

8. Factor $x^2 - 100$.
(1) $(x - 10)(x + 10)$
(2) $(x - 10)^2$
(3) $(x + 10)^2$
(4) $(x - 20)(x + 5)$

9. Factor completely $5x^2 - 20$.

(1) $5(x^2 - 4)$ (3) $5(x + 2)^2$
(2) $5(x - 2)^2$ (4) $5(x - 2)(x + 2)$

10. Factor completely $6x^2 + 12x - 144$.

(1) $6(x^2 + 2x - 24)$ (3) $6(x + 4)(x - 6)$
(2) $6(x - 4)(x + 6)$ (4) $(3x - 4)(2x + 36)$

B. *Show how you arrived at your answers.*

1. Kenzie says that $x^2 + 25$ can be factored into $(x + 5)^2$. Ariana says that it cannot. Who is correct? Explain your answer.

2. Factor completely $x^3 + 4x^2 - 21x$.

3. Factor completely $5x^2 + 20x + 20$.

4. Korbin says that there are two different ways to factor $x^2 - 5x + 6$. One is $(x - 2)(x - 3)$ and the other is $(x - 6)(x - 1)$. Is he correct? Explain your answer.

5. The polynomial $2x^2 + 13x + 15$ can be factored into $(2x + 3)$ multiplied by another factor. What is the other factor? Explain your reasoning.

2.9 MORE COMPLICATED FACTORING

Some polynomials don't, at first, appear to match any of the factoring patterns. When this happens, it is sometimes possible to rewrite it as something that does match one of the patterns. The three most frequent patterns are the factoring of a trinomial into two binomials, the difference of perfect squares, and the perfect square trinomial pattern.

Recognizing the Difference of Perfect Squares Pattern

If there are only two terms and they are separated by a minus sign, there is a chance that it will be possible to eventually use the difference of perfect squares pattern.

Example 1

Factor $9x^2 - 4$.

Solution: First, look for a common factor. In this case there is no number other than 1 that is a factor of both $9x^2$ and 4. Since there are two terms separated by a minus sign, see if each of the terms can be written as the square of other terms.

Since $9x^2 = (3x)^2$ and $4 = 2^2$,

$$9x^2 - 4 = (3x)^2 - 2^2 = (3x - 2)(3x + 2)$$

Example 2

Factor $x^4 - 9$.

Solution: x^4 can be written as $(x^2)^2$, and 9 can be written as 3^2. This is another difference of perfect squares situation.

$$x^4 - 9 = (x^2)^2 - 3^2 = (x^2 - 3)(x^2 + 3)$$

Example 3

Completely factor $x^4 - 81$.

Solution: This is like the previous question with an additional step.

$$(x^2)^2 - 9^2 = (x^2 - 9)\ (x^2 + 9)$$

Notice that the $x^2 - 9$ factor is, itself, a difference of perfect squares and can be factored further.

$$(x^2 - 9)(x^2 + 9) = (x - 3)(x + 3)(x^2 + 9)$$

Recognizing the Factoring a Trinomial into Two Binomials Pattern

If a polynomial has three terms where the exponent of one of the terms is half the largest exponent and one of the terms is a constant, it can sometimes be factored.

Example 4

Factor $x^4 + 4x^2 - 12$.

Solution: This can be written as $(x^2)^2 + 4(x^2) - 12$. Notice that this resembles the equation $y^2 + 4y - 12$. Just as $y^2 + 4y - 12$ would factor to $(y + 6)(y - 2)$, this polynomial can be factored into $(x^2 + 6)(x^2 - 2)$.

Multiply to check your answer.

Example 5

Factor the expression $x^4 + 8x^2 - 9$ completely.

Solution: This is a trinomial that can be rewritten as $(x^2)^2 + 8(x^2) - 9$. It has the same structure as the trinomial $y^2 + 8y - 9$, which would factor as $(y - 1)(y + 9)$. So this factors as $(x^2 - 1)(x^2 + 9)$.

Since $(x^2 - 1)$ can be written as $(x^2 - 1^2)$ the first factor can be further factored by the difference of perfect squares rule.
The answer, then, is $(x - 1)(x + 1)(x^2 + 9)$.

Recognizing the Perfect Square Trinomial Pattern

If there is a trinomial where one of the exponents is half of another exponent, and one term is a constant, it will be a perfect square trinomial when the square of half of the coefficient of the term with the smaller exponent is equal to the constant.

Example 6

Factor $x^6 + 8x^3 + 16$.

Solution: This can be written as $(x^3)^2 + 8(x^3) + 16$, which is similar in structure to $y^2 + 8y + 16$. Since $\left(\frac{8}{2}\right)^2 = 16$, this is a perfect square and can be factored into $(x^3 + 4)^2$.

Check Your Understanding of Section 2.9

A. Multiple-Choice

1. Factor $(2x)^2 - 25$.
(1) $(2x - 5)(2x - 5)$
(2) $(2x + 5)(2x + 5)$
(3) $(2x - 5)(2x + 5)$
(4) This cannot be factored.

2. Factor $x^4 - 49$.
(1) $(x^2 - 7)(x^2 + 7)$
(2) $(x^2 - 7)(x^2 - 7)$
(3) $(x^2 + 7)(x^2 + 7)$
(4) This cannot be factored.

3. Factor $x^4 + 7x^2 + 12$.
(1) $(x^2 + 6)(x^2 + 2)$
(2) $(x^2 - 3)(x^2 - 4)$
(3) $(x^2 + 3)(x^2 + 4)$
(4) This cannot be factored.

4. Factor $x^4 - x^2 - 20$.
(1) $(x^2 - 5)(x^2 + 4)$
(2) $(x^2 + 5)(x^2 + 4)$
(3) $(x^2 - 5)(x^2 - 4)$
(4) This cannot be factored.

5. Factor $9x^2 - 16y^2$.
(1) $(3x - 4y)(3x - 4y)$
(2) $(3x - 4y)(3x + 4y)$
(3) $(3x + 4y)(3x + 4y)$
(4) This cannot be factored.

6. Factor $a^4 + 10a^2 + 25$.
(1) $(a^2 - 5)(a^2 + 5)$
(2) $(a^2 - 5)^2$
(3) $(a^2 + 5)^2$
(4) This cannot be factored.

7. Factor $x^6 - 4$.
(1) $(x^3 - 2)^2$
(2) $(x^3 + 2)^2$
(3) $(x^3 - 2)(x^3 + 2)$
(4) This cannot be factored.

8. Factor $(3x)^2 + 8(3x) + 12$.
(1) $(3x + 4)(3x + 3)$
(2) $(3x + 6)(3x + 2)$
(3) $(3x + 1)(3x + 12)$
(4) This cannot be factored.

9. Factor $x^4 - 10x^2 + 16$.
(1) $(x^2 - 8)(x^2 - 2)$
(2) $(x^2 - 8)(x^2 + 2)$
(3) $(x^2 + 8)(x^2 + 2)$
(4) This cannot be factored.

10. Factor $(2x + 1)^2 - 9$.
(1) $(2x + 1 + 3)^2$
(2) $(2x + 1 - 3)(2x + 1 + 3)$
(3) $(2x + 1 - 3)^2$
(4) This cannot be factored.

B. *Show how you arrived at your answers.*

1. Factor completely $x^4 - 81$.

2. Factor completely $x^4 + 4x^2 - 5$.

3. Factor completely $x^4 - 13x^2 + 36$.

4. Factor $(3x - 5)^2 + 2(3x - 5) - 3$.

5. The polynomial $a^3 - b^3$ can factor into $(a - b)(a^2 + ab + b^2)$. Use this factoring pattern to factor the polynomial $x^6 - 8$.

Chapter Three

QUADRATIC EQUATIONS

3.1 SOLVING QUADRATIC EQUATIONS BY TAKING THE SQUARE ROOT OF BOTH SIDES OF THE EQUATION

An equation that has a variable raised to the second power is called a quadratic equation. Examples of quadratic equations are $x^2 = 9$, $x^2 + 5x = 24$, and $x^2 + 5x + 6 = 0$. Quadratic equations require more advanced techniques to solve. Most quadratic equations have two solutions. The simplest quadratic equations to solve are the ones where there is no x term, like $x^2 = 25$ or $x^2 + 2 = 38$.

One-Step Quadratic Equations

If a quadratic equation is in the form $x^2 = c$, the exponent can be eliminated by taking the square root of both sides. There are two answers, generally, since positive times positive equals a positive, whereas negative times negative also equals a positive.

Example 1

Solve for all values of x that satisfy the equation $x^2 = 25$.

Solution: By looking at the equation and thinking through the multiplication tables, you may notice that 5 is one of the solutions since $5^2 = 5 \cdot 5 = 25$. Less obvious is the other answer, –5, since $-5 \cdot -5$ also equals positive 25.

This problem can also be solved with algebra. Using algebra, you could take the square root of both sides of the equation.

$$x^2 = 25$$
$$\sqrt{x^2} = \pm\sqrt{25}$$
$$x = \pm 5$$
$$x = 5 \text{ or } x = -5$$

Since +5 or –5 can be squared to become +25, when you take the square root of both sides, put the ± symbol in front of the square root sign on the right side of the equation. You can leave out the second line of this solution and still get full credit.

MATH FACTS

Just as $\sqrt{3^2}$ is 3 and $\sqrt{5^2}$ is 5 (Quick! What is $\sqrt{173^2}$?), when x is a positive number $\sqrt{x^2}$ is x. This is the property that allows you to do the algebraic process of eliminating the exponent by taking the square root of both sides of the equation.

Example 2

Solve for all values of x that satisfy the equation $x^2 = 26$.

Solution:

$$x^2 = 26$$
$$\sqrt{x^2} = \pm\sqrt{26}$$
$$x = \pm\sqrt{26}$$
$$x = \sqrt{26} \text{ or } x = -\sqrt{26}$$

Since 26 is not a perfect square, the answer should be left in this form unless the question says to round the answer. In that case, you would use your calculator to find $\sqrt{26} \approx 5.09901951$ and round appropriately.

Two-Step Quadratic Equations with Constants

More complicated quadratic equations sometimes require the addition property of equality or the subtraction property of equality in order to solve.

Example 3

Solve for all values of x that satisfy the equation $x^2 + 5 = 54$.

Solution: First isolate the radical by subtracting 5 from both sides of the equation. After that, the equation will be just like the one step quadratic equations.

$$x^2 + 5 = 54$$
$$-5 = -5$$
$$x^2 = 49$$
$$x = \pm 7$$

Example 4

Solve for all values of x that satisfy the equation $x^2 + 9 = 36$.

Solution: It is tempting to try to take the square root of both sides and get $x + 3 = 6$, but it is not true that $\sqrt{x^2+9} = x + 3$. Instead, solve the same way as in the previous example.

$$\begin{aligned} x^2 + 9 &= 36 \\ -9 &= -9 \\ x^2 &= 27 \\ x &= \pm\sqrt{27} \end{aligned}$$

Example 5

Solve for all values of x that satisfy the equation $(x + 1)^2 = 25$.

Solution: Do not multiply out the left-hand side. Instead, look at how similar this question is to the question $x^2 = 25$, which had two solutions +5 and –5. Begin by taking the square root of both sides of the equation.

$$\begin{aligned} (x + 1)^2 &= 25 \\ \sqrt{(x+1)^2} &= \pm\sqrt{25} \\ x + 1 &= \pm 5 \end{aligned}$$

Complete the question by subtracting 1 from each side.

$$\begin{aligned} x + 1 &= \pm 5 \\ -1 &= -1 \\ x &= \pm 5 - 1 \\ x &= +5 - 1 = +4 \text{ or } x = -5 - 1 = -6 \end{aligned}$$

You can check that these two answers are correct by substituting them back into the original equation.

$$\begin{aligned} (4 + 1)^2 &\stackrel{?}{=} 25 \\ 5^2 &\stackrel{?}{=} 25 \\ 25 &\stackrel{\checkmark}{=} 25 \end{aligned} \qquad \begin{aligned} (-6 + 1)^2 &\stackrel{?}{=} 25 \\ (-5)^2 &\stackrel{?}{=} 25 \\ 25 &\stackrel{\checkmark}{=} 25 \end{aligned}$$

Example 6

Solve for all values of x that satisfy the equation $(x + 3)^2 = 17$.

Solution:

$$(x + 3)^2 = 17$$
$$\sqrt{(x+3)^2} = \pm\sqrt{17}$$
$$x + 3 = \pm\sqrt{17}$$
$$-3 = -3$$
$$x = -3 \pm \sqrt{17}$$
$$x = -3 + \sqrt{17} \text{ or } x = -3 - \sqrt{17}$$

MATH FACTS

When combining something like –3 with $\sqrt{17}$, it is customary to put the $\sqrt{17}$ on the right as $-3 + \sqrt{17}$ rather than as $\sqrt{17} - 3$ which might get confused with $\sqrt{17-3} = \sqrt{14}$.

Example 7

Solve for all values of x that satisfy the equation $x^2 + 6x + 9 = 17$.

Solution: There are some more advanced techniques for a problem like this, but if you notice that $x^2 + 6x + 9$ is a perfect square trinomial and can be factored into $(x + 3)^2$, the equation becomes $(x + 3)^2 = 17$, which is the same as in Example 6.

Two-Step Quadratic Equations with Coefficients

To solve the equation $2x^2 = 50$ with a coefficient of 2 in front of the x^2 term, first divide both sides of the equation by 2.

$$2x^2 = 50$$
$$\frac{2x^2}{2} = \frac{50}{2}$$
$$x^2 = 25$$
$$x = \pm\sqrt{25} = \pm 5$$

Two-Step Quadratic Equations with Multiple Variables

If an equation has more than one variable and the goal is to isolate one of the variables, follow the same steps you would if the other variables were numbers.

Example 8

Solve for x in terms of a and b in the equation $ax^2 = b$.

Solution:

$$ax^2 = b$$

$$\frac{ax^2}{a} = \frac{b}{a}$$

$$x^2 = \frac{b}{a}$$

$$x = \pm\sqrt{\frac{b}{a}}$$

Example 9

The formula for the volume of a cylinder is $V = \pi r^2 h$. Solve this equation for r in terms of V, h, and π.

Solution: First eliminate the π and the h from the right-hand side.

$$\frac{V}{\pi h} = \frac{\pi r^2 h}{\pi h}$$

$$\frac{V}{\pi h} = r^2$$

Eliminate the exponent by taking the square root of both sides of the equation:

$$\frac{V}{\pi h} = \sqrt{r^2}$$

$$\sqrt{\frac{V}{\pi h}} = r$$

Since the r represents the radius of the cylinder, the negative answer can be disregarded.

Solving Square Root Equations

Related to quadratic equations are equations like $\sqrt{x} = 3$ and $\sqrt{x-2} = 5$. To eliminate a square root sign, raise each side of the equation to the second power.

$$\sqrt{x} = 3$$
$$(\sqrt{x})^2 = 3^2$$
$$x = 9$$

$$\sqrt{x-2} = 5$$
$$(\sqrt{x-2})^2 = 5$$
$$x - 2 = 25$$
$$+2 = +2$$
$$x = 27$$

If the square root sign is not already isolated, first isolate it with algebra.

$$\sqrt{x} + 3 = 8$$
$$-3 = -3$$
$$\sqrt{x} = 5$$
$$(\sqrt{x})^2 = 5^2$$
$$x = 25$$

Check Your Understanding of Section 3.1

A. Multiple-Choice

1. Find all solutions to $x^2 = 36$.
(1) 6 (2) –6 (3) 36, –36 (4) 6, –6

2. Find all solutions to $x^2 = 37$.
(1) $\sqrt{37}, -\sqrt{37}$
(2) $\sqrt{37}$
(3) $-\sqrt{37}$
(4) There are no solutions since 37 is not a perfect square.

3. Find all solutions to $x^2 - 9 = 40$.

(1) 7
(2) –7
(3) 7, –7
(4) $3 + \sqrt{40}$, $3 - \sqrt{40}$

4. Find all solutions to $(x + 2)^2 = 64$.

(1) $\sqrt{62}$, $-\sqrt{62}$
(2) 6, –10
(3) 6
(4) –10

5. Find all solutions to $(x - 3)^2 = 13$.

(1) $3 + \sqrt{13}$
(2) $3 - \sqrt{13}$
(3) $3 + \sqrt{13}$, $3 - \sqrt{13}$
(4) $\sqrt{10}$, $-\sqrt{10}$

6. Find all solutions to $(x + 2)^2 = 17$.

(1) $-2 + \sqrt{17}$
(2) $-2 - \sqrt{17}$
(3) $\sqrt{15}$, $-\sqrt{15}$
(4) $-2 + \sqrt{17}$, $-2 - \sqrt{17}$

7. Find all solutions to $3x^2 = 108$.

(1) $\sqrt{\frac{108}{3}}, -\sqrt{\frac{108}{3}}$
(2) 6
(3) –6
(4) 6, –6

8. Find all solutions to $4x^2 + 3 = 103$.

(1) 5
(2) –5
(3) 5, –5
(4) 10, –10

9. Solve for x in terms of c and d.
$cx^2 = d$.

(1) $\frac{d}{c}$
(2) $\sqrt{\frac{c}{d}}, -\sqrt{\frac{c}{d}}$
(3) $\sqrt{\frac{d}{c}}, -\sqrt{\frac{d}{c}}$
(4) $\frac{\sqrt{d}}{c}, -\frac{\sqrt{d}}{c}$

10. Solve for x in terms of g and h.
$(x + g)^2 = h$

(1) $-g + \sqrt{h}$, $-g - \sqrt{h}$
(2) $-g + \sqrt{h}$
(3) $-g - \sqrt{h}$
(4) $\sqrt{h-g}$, $-\sqrt{h-g}$

Chapter Three **QUADRATIC EQUATIONS**

B. *Show how you arrived at your answers.*

1. Find all solutions to the equation $x^2 = 15$, rounded to the nearest hundredth.

2. The left-hand side of the equation $x^2 + 10x + 25 = 64$ can be factored into $(x + 5)^2$. Show how you can find the two solutions to this equation by first factoring the left-hand side of the equation.

3. If the area of the largest square in this diagram (composed of the two smaller squares and the two rectangles) is 49 square units, what is the length of segment x?

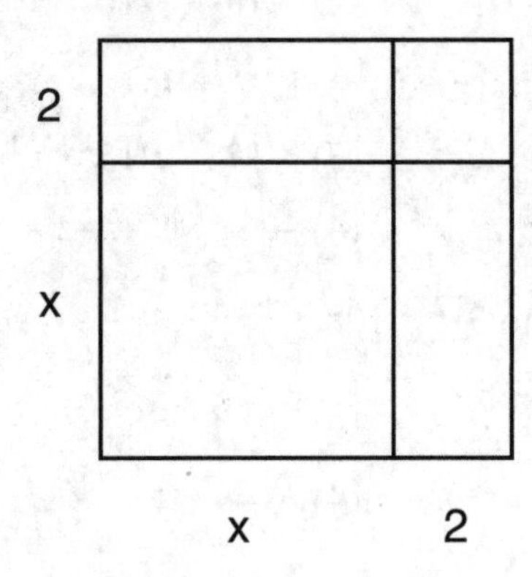

4. Diana solves the equation $(x + 4)^2 = 97$ as follows:

$$(x + 4)^2 = 97$$
$$x^2 + 16 = 97$$
$$-16 = -16$$
$$x^2 = 81$$
$$x = 9, -9$$

What is wrong with this reasoning?

5. Find the solution to the equation $(x + 1)^3 = 8$.

3.2 SOLVING QUADRATIC EQUATIONS BY GUESS AND CHECK

KEY IDEAS

There are several ways to solve a quadratic equation like $x^2 + 10x = 39$, which has the $10x$ term involved. For some examples, when the answer is an integer, it is possible to find the answer through guess and check.

Guess = 1	Guess = 2	Guess = 3
$1^2 + 10 \cdot 1 \overset{?}{=} 39$	$2^2 + 10 \cdot 2 \overset{?}{=} 39$	$3^2 + 10 \cdot 3 \overset{?}{=} 39$
$1 + 10 \overset{?}{=} 39$	$4 + 20 \overset{?}{=} 39$	$9 + 30 \overset{?}{=} 39$
$11 \neq 39$	$24 \neq 39$	$39 \overset{\checkmark}{=} 39$

Guess and check is useful if the quadratic equation is a multiple-choice question so there are only four things to check.

Example 1

Which of the four choices is a solution to the equation $x^2 + 5x - 3 = 33$?
(1) 2 (2) 3 (3) 4 (4) 5

Solution: Since $4^2 + 5 \cdot 4 - 3 = 16 + 20 - 3 = 33$, 4 is a solution to the equation, which is choice (3).

MATH FACTS

The solutions to a quadratic equation are sometimes called the *roots* or the *zeros* of the equation.

Example 2

Which of the four choices is a root of the equation $x^2 - 6x + 4 = 0$?

(1) $3 + \sqrt{2}$ (2) $3 + \sqrt{3}$ (3) $3 + \sqrt{5}$ (4) $3 + \sqrt{6}$

Solution: To test the four answer choices, there is a useful feature on the graphing calculator.

For the TI–84:

Store the value from choice (1) into the x variable by pressing [3], [+], [2ND], [x^2], [2], [ )], [STO>], [x]. The [STO>] key is right above the [ON] key and the [x] key is next to the [ALPHA] key in the second row.

Then check the answer by entering [x], [x^2], [–], [6], [x], [+], [4], and [ENTER].

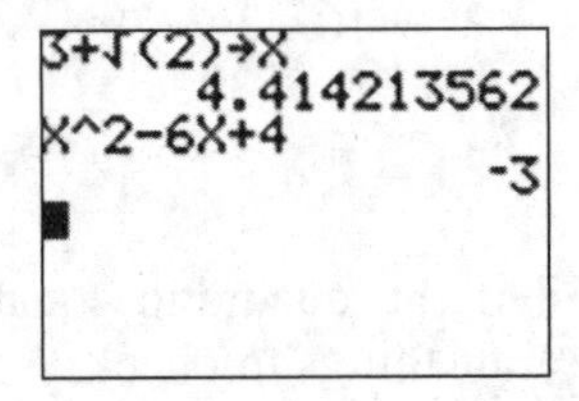

Since it did not evaluate to 0, choice (1) is not the answer.

When using this method to check choice (3), it becomes

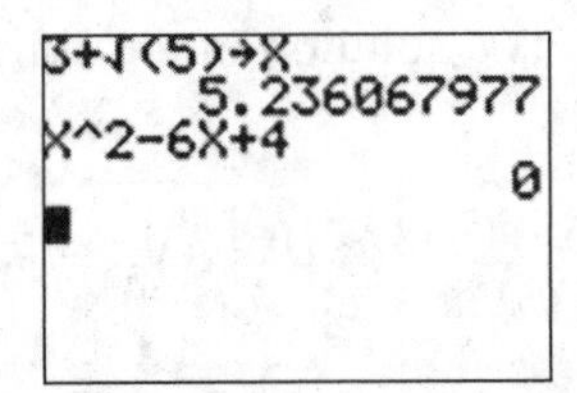

For the TI-Nspire:

Store the value from choice (1) into the x variable by pressing [3], [+], [ctrl], [x^2], [2], [right arrow], [ctrl], [var], [x].

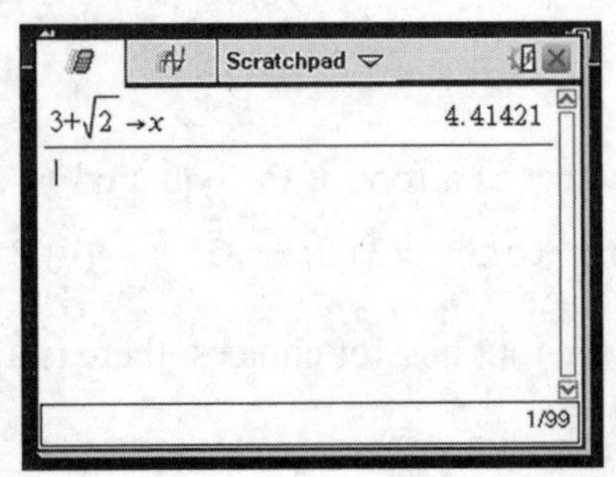

Check if this solves the quadratic by entering [x], [x^2], [–], [6], [x], [+], [4], and [enter].

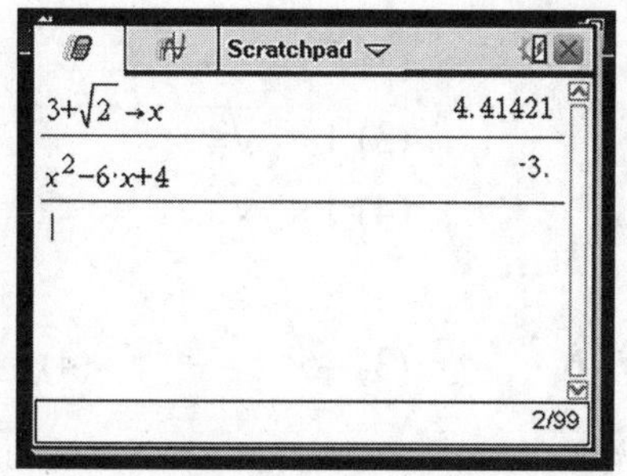

When checking choice (3), it will look like this:

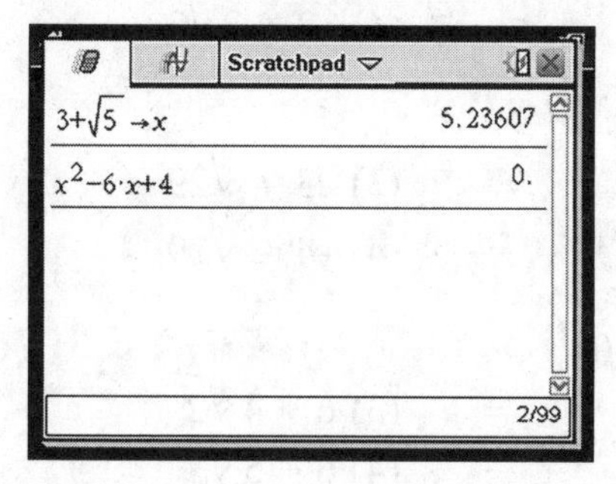

Either calculator will confirm that the answer is choice (3).

Check Your Understanding of Section 3.2

A. Multiple-Choice

For each question, use your calculator to check which answer satisfies the quadratic equation.

1. $x^2 - 8x + 15 = 0$
(1) 3 (2) 4 (3) 6 (4) 7

2. $x^2 - 2x - 8 = 0$
(1) 1 (2) 2 (3) 3 (4) 4

3. $x^2 + 5x - 6 = 0$
(1) 6 (2) –6 (3) –4 (4) –2

4. $2x^2 + 5x - 3 = 0$
(1) $\frac{1}{4}$ (2) $\frac{1}{3}$ (3) $\frac{1}{2}$ (4) 1

5. $x^2 + 7x = 30$

(1) 1 (2) 3 (3) 5 (4) 7

6. $x^2 - 2x - 2 = 0$

(1) $1 + \sqrt{6}$ (3) $1 + \sqrt{4}$

(2) $1 + \sqrt{5}$ (4) $1 + \sqrt{3}$

7. $x^2 + 5x = 6$

(1) -7 (2) -6 (3) -5 (4) -4

8. $x^2 + 10x + 13 = 0$

(1) $-5 + 2\sqrt{3}$ (3) $-3 + 2\sqrt{3}$

(2) $-4 + 2\sqrt{3}$ (4) $-2 + 2\sqrt{3}$

9. $x^2 + 4x = 25$

(1) $-1 - \sqrt{29}$ (3) $-3 - \sqrt{29}$

(2) $-2 - \sqrt{29}$ (4) $-4 - \sqrt{29}$

10. $x^2 - 12x - 14 = 0$

(1) $6 + 2\sqrt{2}$ (3) $6 + 4\sqrt{2}$

(2) $6 + 3\sqrt{2}$ (4) $6 + 5\sqrt{2}$

3.3 SOLVING QUADRATIC EQUATIONS BY COMPLETING THE SQUARE

Completing the square is a technique for solving quadratic equations that relies on the concept of a perfect square trinomial. It is not the quickest method to solve certain quadratic equations, but it can be used for all quadratic equations.

A perfect square trinomial, like $x^2 + 6x + 9$ is one where the constant term, the 9 in this case, is equal to the square of half the coefficient of the x, in this case the 6. Since $\left(\frac{6}{2}\right)^2$ does equal 9, this is a perfect square trinomial and is equal to $\left(x+\frac{6}{2}\right)^2 = (x+3)^2$.

If the left-hand side of a quadratic equation is already a perfect square, you can factor it and solve using the technique in Section 3.1, example 7.

If the left-hand side is not a perfect square trinomial, it is possible to make it into one by adding to both sides of the equation. For the equation $x^2 + 10x = 39$, you look at the left-hand side and ask "what constant would make the left-hand side of the equation into a perfect square trinomial." Since $\left(\frac{10}{2}\right)^2$ is 25, $x^2 + 10x + 25$ would be a perfect square trinomial.

Add 25 to both sides and complete the question:

$$
\begin{aligned}
x^2 + 10x &= 39 \\
+25 &= +25 \\
x^2 + 10x + 25 &= 64 \\
(x+5)^2 &= 64 \\
x + 5 &= \pm 8 \\
-5 &= -5 \\
x &= -5 \pm 8 \\
x &= -5 + 8 = 3 \text{ or } x = -5 - 8 = -13
\end{aligned}
$$

You can check these answers by first substituting 3 for both *x*s in the equation and then by substituting –13 for both *x*s in the equation and making sure that they evaluate to 39.

Example 1

What are the roots of the equation $x^2 + 12x = 28$.

Solution:

$$
\begin{aligned}
x^2 + 12x &= 28 \\
x^2 + 12x + 36 &= 28 + 36 \\
(x+6)^2 &= 64 \\
x + 6 &= \pm 8 \\
x &= -6 + 8 = 2 \text{ or } x = -6 - 8 = -14
\end{aligned}
$$

Example 2

What are the roots of the equation $x^2 + 4x - 21 = 0$.

Solution: Completing the square is simplest when there is no constant term on the left-hand side of the equation. Eliminate the –21 by adding 21 to both sides. Then continue as in the previous example.

$$
\begin{aligned}
x^2 + 4x - 21 &= 0 \\
+21 &= +21 \\
x^2 + 4x &= 21 \\
x^2 + 4x + 4 &= 21 + 4 \\
(x+2)^2 &= 25 \\
x + 2 &= \pm 5 \\
x &= -2 + 5 = 3 \text{ or } x = -2 - 5 = -7
\end{aligned}
$$

Example 3

What are the roots of the equation $x^2 + 6x - 3 = 0$?

Solution: At the end of the completing the square process, something unusual happens.

$$x^2 + 6x - 3 = 0$$
$$+3 = +3$$
$$x^2 + 6x = 3$$
$$x^2 + 6x + 9 = 3 + 9$$
$$(x + 3)^2 = 12$$
$$x + 3 = \pm\sqrt{12}$$
$$x = -3 \pm\sqrt{12}$$

Since 12 is not a perfect square, it does not become a whole number as it did in previous examples. There is something further you can do, though. By factoring 12 into $4 \cdot 3$, $\sqrt{12}$ can be simplified to $\sqrt{4 \cdot 3} = \sqrt{4} \cdot \sqrt{3} = 2\sqrt{3}$. If the question asks for *simplest radical form*, the solution would be written as $x = -3 \pm 2\sqrt{3}$.

Example 4

Which equation has the same roots as $x^2 - 8x - 19 = 0$.

(1) $(x + 4)^2 = 35$ (3) $(x + 4)^2 = 3$
(2) $(x - 4)^2 = 35$ (4) $(x - 4)^2 = 3$

Solution: Notice that the answer choices all look like one of the steps of the completing the square process. Begin the question as if you are solving by completing the square, but you can stop when the equation resembles the structure of the answer choices.

$$x^2 - 8x - 19 = 0$$
$$+19 = +19$$
$$x^2 - 8x = 19$$
$$x^2 - 8x + 16 = 19 + 16$$
$$(x - 4)^2 = 35$$

You do not have to go any further. The answer is choice (2).

The Nature of the Roots

The quadratic equation $x^2 = 5$ has two solutions $x = \sqrt{5}$ and $x = -\sqrt{5}$. Since 5 is not a perfect square, these two solutions are known as *irrational* numbers. The equation $x^2 = 5$ has two irrational solutions.

Since 4 is a perfect square, the quadratic equation $x^2 = 4$ has two solutions, $x = 2$ and $x = -2$. These are known as *rational* numbers. A rational number is any number that can be written as a fraction with integers in both the numerator and the denominator. Numbers like $\frac{2}{5}$, $\frac{-7}{3}$, and $\frac{8}{1}$ are all rational. Integers like 2 and –2 are rational because they can be expressed as $\frac{2}{1}$ and $\frac{-2}{1}$. The equation $x^2 = 4$ has two rational solutions.

The quadratic equation $x^2 - 4x - 1 = 0$ has solutions $x = 2 + \sqrt{5}$ and $x = 2 - \sqrt{5}$. When a rational number is added to an irrational number, the result is an irrational number. The equation $x^2 - 4x - 1 = 0$ has two irrational solutions.

MATH FACTS

When a rational number is added to or subtracted from a rational number, the result is always a rational number. When a rational number is added to or subtracted from an irrational number, the result is always an irrational number. When an irrational number is added to or subtracted from another irrational number, the result is often irrational, but not always. For example, $3 - \sqrt{5}$ is an irrational number, but $\sqrt{5} + (3 - \sqrt{5}) = 3$, which is rational. The product of a rational and an irrational number is always irrational, but the product of two irrationals can be either irrational or rational. For example, $\sqrt{3} \cdot \sqrt{2} = \sqrt{6}$ which is irrational, but $\sqrt{5} \cdot \sqrt{5} = 5$ is rational.

Example 5

Which of the following expressions is a rational number?

(1) $\sqrt{7}$
(2) $\sqrt{9} + \sqrt{7}$
(3) $\sqrt{9} + \sqrt{4}$
(4) $2\sqrt{5}$

Solution: The answer is (3) because $\sqrt{9} + \sqrt{4} = 3 + 2 = 5$

Check Your Understanding of Section 3.3

A. *Multiple-Choice*

1. $x^2 + 12x + c$ is a perfect square trinomial. What is the value of c?
(1) 36 (2) 18 (3) 12 (4) 6

2. $x^2 + bx + 49$ is a perfect square trinomial. What is one possible value of b?
(1) 7 (2) 14 (3) 21 (4) 49

3. Use completing the square to find both solutions for x in the equation $x^2 + 8x + 16 = 9$.
(1) –1, –7 (2) –1, –2 (3) –2, –3 (4) –3, –4

4. Use completing the square to find both solutions for x in the equation $x^2 + 10x = 24$.
(1) 2, –10 (2) 4, –12 (3) 3, –8 (4) 2, –12

5. Use completing the square to find both solutions for x in the equation $x^2 - 18x = 40$.
(1) –2, 10 (2) –2, 20 (3) –4, 20 (4) –4, 10

6. What is one solution to the equation $x^2 - 6x = 8$?
(1) $3 + \sqrt{9}$ (3) $3 + \sqrt{17}$
(2) $3 + \sqrt{13}$ (4) $3 + \sqrt{21}$

7. The quadratic equation $x^2 + 2x - 24 = 0$ has the same solutions as which of the following equations?
(1) $(x + 1)^2 = 24$ (3) $(x - 1)^2 = 24$
(2) $(x + 1)^2 = 25$ (4) $(x - 1)^2 = 25$

8. Find the solutions using completing the square $x^2 + 4x - 5 = 0$.
(1) 1, –5 (2) 1, –6 (3) 2, –5 (4) 2, –6

9. Solve by completing the square $x^2 - 12x + 32 = 0$.
(1) 4, 8 (2) 4, 6 (3) 3, 8 (4) 3, 6

10. Which of the following is not a rational number?
(1) $\frac{7}{5}$ (3) $\sqrt{7} \cdot \sqrt{7}$
(2) $\sqrt{7} - \sqrt{7}$ (4) $7 + \sqrt{7}$

***B.** Show how you arrived at your answers.*

1. What are all possible integer solutions for b and p in the equation $x^2 + bx + 64 = (x + p)^2$?

2. Use the completing the square method to find the two solutions to the equation $x^2 - 16x - 7 = 0$.

3. Ramon says that π is irrational and that $\pi = \frac{22}{7}$; therefore, $\frac{22}{7}$ is irrational also. What is wrong with his reasoning process? Explain.

4. Use the completing the square method to find the two solutions to the equation $x^2 - 5x = 14$.

5. Use the completing the square method to find the two solutions to the equation $x^2 + 2ax = b$ in terms of a and b.

3.4 SOLVING QUADRATIC EQUATIONS BY FACTORING

When the quadratic equation contains a trinomial that can be factored, the equation can be solved very quickly. This requires first getting the equation in a form with all the terms on one side of the equation and a zero on the other side of the equation.

Solving a Quadratic Equation That Is Already in Factored Form

The only way for the product of two numbers to be zero is if one, or both, is the number zero. In the equation $(x - 2)(x - 3) = 0$, there are two expressions $(x - 2)$ and $(x - 3)$ that have a product of zero. This means that you need to find x values that make either of these expressions equal to zero.

$$(x - 2)(x - 3) = 0$$

This equation will be satisfied if either $(x - 2)$ or $(x - 3)$ is equal to 0.
This becomes two separate questions

$$x - 2 = 0 \quad \text{or} \quad x - 3 = 0$$
$$x - 2 + 2 = 0 + 2 \qquad x - 3 + 3 = 0 + 3$$
$$x = +2 \quad \text{or} \quad x = +3$$

Either of these two values, 2 or 3, would make the expression $(x - 2)(x - 3)$ equal to 0.

$$(2 - 2)(2 - 3) = 0 \cdot -1 = 0$$
$$(3 - 2)(3 - 3) = 1 \cdot 0 = 0$$

Example 1

Solve for both values of x that satisfy the equation $(x - 4)(x - 5) = 0$.

Solution: This is true for x values that make either $x - 4$ or $x - 5$ equal to 0.

$$x - 4 = 0 \quad \text{or} \quad x - 5 = 0$$
$$x - 4 + 4 = 0 + 4 \qquad x - 5 + 5 = 0 + 5$$
$$x = +4 \quad \text{or} \quad x = +5$$

So the solutions are 4 and 5.

Example 2

Determine the two roots of the equation $(x - 2)(x + 6) = 0$.

Solution:

$$x - 2 = 0 \quad \text{or} \quad x + 6 = 0$$
$$x - 2 + 2 = 0 + 2 \qquad x + 6 - 6 = 0 - 6$$
$$x = +2 \quad \text{or} \quad x = -6$$

The solutions are 2 and -6.

Solving Equations by First Factoring the Quadratic Trinomial

One of the most common types of quadratic equations in the course is when you have to first factor and then solve the equation. This technique is likely to be applied several times on each test.

Example 3

Determine the roots of the equation $x^2 + 8x + 15 = 0$.

Solution: Though this question can be solved by completing the square, it is much quicker if you can first factor the left-hand side and use the techniques from this section to complete the question. In Section 2.8, the trinomial $x^2 + 8x + 15$ was factored into $(x + 3)(x + 5)$.

$$x^2 + 8x + 15 = 0$$
$$(x + 3)(x + 5) = 0$$

$$x + 3 = 0 \quad \text{or} \quad x + 5 = 0$$
$$x + 3 - 3 = 0 - 3 \qquad x + 5 - 5 = 0 - 5$$
$$x = -3 \quad \text{or} \quad x = -5$$

Example 4

Determine the roots of the equation $x^2 + 3x - 10 = 0$.

Solution:

$$x^2 + 3x - 10 = 0$$
$$(x + 5)(x - 2) = 0$$

$$x + 5 = 0 \quad \text{or} \quad x - 2 = 0$$
$$x + 5 - 5 = 0 - 5 \qquad x - 2 + 2 = 0 + 2$$
$$x = -5 \quad \text{or} \quad x = 2$$

Example 5

Determine the roots of the equation $x^2 - x - 6 = 14$.

Solution: Do not factor the left-hand side of the equation until you have arranged the equation so that a zero is on the right-hand side.

$$x^2 - x - 6 = 14$$
$$-14 = -14$$
$$x^2 - x - 20 = 0$$
$$x^2 - 1x - 20 = 0$$
$$(x - 5)(x + 4) = 0$$

$$x - 5 = 0 \quad \text{or} \quad x + 4 = 0$$
$$x - 5 + 5 = 0 + 5 \qquad x + 4 - 4 = 0 - 4$$
$$x = 5 \quad \text{or} \quad x = -4$$

Example 6

Find the values of x that satisfy the equation $x^2 - 7x = 0$.

Solution: Since this polynomial does not have a constant term, it can be factored with the greatest common factor method.

$$x(x - 7) = 0$$

$$x = 0 \quad \text{or} \quad x - 7 = 0$$
$$x = 0 \quad \text{or} \quad x = 7$$

Example 7

Find the values of x that satisfy the equation $2x^2 + 12x + 16 = 0$.

Solution: Start by factoring out the common factor of 2.

$$2(x^2 + 6x + 8) = 0$$
$$2(x + 4)(x + 2) = 0$$

The only way this expression can be 0 is if either $x + 4 = 0$ or $x + 2 = 0$.

$$x + 4 = 0 \quad \text{or} \quad x + 2 = 0$$
$$x = -4 \quad \text{or} \quad x = -2$$

Example 8

Find all values of x that are solutions to the equation $x^2 - 25 = 0$.

Solution: One way to do this is to factor the left-hand side.

$$x^2 - 25 = 0$$
$$(x - 5)(x + 5) = 0$$

$$x - 5 = 0 \quad \text{or} \quad x + 5 = 0$$
$$x = 5 \quad \text{or} \quad x = -5$$

Another way is to add 25 to both sides of the equation.

$$x^2 - 25 = 0$$
$$+25 = +25$$
$$x^2 = 25$$
$$x = \pm\sqrt{25}$$
$$x = \pm 5$$

Check Your Understanding of Section 3.4

A. Multiple-Choice

For each equation, find all values of x that satisfy it.

1. $(x - 2)(x - 3) = 0$
(1) 2, 3 (2) –2, –3 (3) –2, 3 (4) 2, –3

2. $(x - 3)(x + 4) = 0$
(1) 3, –4 (2) –3, 4 (3) –3, –4 (4) 3, 4

3. $x(x - 5) = 0$
(1) 5 (2) 0 (3) –5, 0 (4) 5, 0

4. $x^2 + 10x + 24 = 0$
(1) 4, –6 (2) –4, –6 (3) –4, 6 (4) 4, 6

5. $x^2 + 2x = 15$
(1) –3, 5 (2) 3, –5 (3) 3, 5 (4) –3, –5

6. $x^2 + 6x = 0$
(1) 0 (2) –6 (3) 0, –6 (4) 0, 6

7. $x^2 - 6x + 9 = 0$
(1) 3 (2) 3, –3 (3) –3 (4) 3, 0

8. $2x^2 - 16x + 14 = 0$
(1) 1 (2) 7 (3) 1, 7 (4) –1, –7

9. $3x^2 + 15x - 108 = 0$
(1) –4, 9 (2) 4, –9 (3) 4, 9 (4) –4, –9

10. $x^2 - 9 = 0$
(1) 3 (2) –3 (3) 3, –3 (4) 3, 0

***B.** Show how you arrived at your answers.*

1. Faith tries to solve the equation $x^2 - 3x = 0$ by first dividing both sides by x to get $x - 3 = 0$. She concludes that the only solution is $x = 3$. Is she correct? Explain.

2. The cubic polynomial $x^3 - 6x^2 + 11x - 6$ can be factored into $(x - 1)(x - 2)(x - 3)$. How can you use this fact to find all solutions to the equation $x^3 - 6x^2 + 11x - 6 = 0$?

3. If $ab = 0$ and it is known that $a \neq 0$, what conclusion can you make about b? Explain.

4. The equation $x^4 - 13x^2 + 36 = 0$ has four solutions. What are they?

5. Stephanie tries to solve the equation $x^2 + 6x = 40$ by first factoring the left-hand side into $x(x + 6) = 40$ and then concludes that either $x = 40$ or $x + 6 = 40$ giving two solutions 40 and 34. What is wrong with this reasoning?

3.5 THE RELATIONSHIP BETWEEN FACTORS AND ROOTS

When a quadratic trinomial $x^2 + bx + c = 0$ factors into $(x + r)(x + s) = 0$, the roots of the polynomial $x^2 + bx + c$ are $x = -r$ and $x = -s$. These are also sometimes called the roots of the equation $x^2 + bx + c = 0$. The roots, $-r$ and $-s$, are very closely related to the factors of the quadratic expression $(x + r)(x + s)$. It is possible to very quickly translate between the factors and the roots.

The quadratic trinomial $x^2 + 8x + 15$ has *factors* of $(x + 3)$ and $(x + 5)$. The quadratic equation $x^2 + 8x + 15 = 0$ has two solutions $x = -3$ and $x = -5$. These two solutions are called the *roots* of the polynomial $x^2 + 8x + 15$. In general, the two roots are the opposites of the constants in the two factors.

In the equation $x^2 + 3x - 10 = 0$, the factors of the trinomial on the left-hand side of the equation are $(x + 5)$ and $(x - 2)$, and the roots of the polynomial are $x = -5$ and $x = 2$.

Example 1

What are the roots of the equation $(x + 6)(x - 4) = 0$?

Solution: The first root is the opposite of the constant of the first factor. That constant is +6 so the root is $x = -6$. The second root is the opposite of the constant of the second factor. That constant is -4 so the root is $x = -(-4) = +4$.

Example 2

If the roots of a quadratic equation are +2 and +7, what could the equation be?

Solution: Since the roots are the opposites of the constants of the factors, the constants of the factors must be $(x - 2)$ and $(x - 7)$. So the equation is $(x - 2)(x - 7) = 0$.

Example 3

If the roots of a quadratic equation are –5 and 3, what could the equation be?

Solution: Since the opposite of –5 is +5 and the opposite of +3 is –3, the factors are $(x + 5)$ and $(x - 3)$ so the equation could be $(x + 5)(x - 3) = 0$.

Example 4

If the roots of a quadratic equation are +2 and +8, what could the equation be?

(1) $x^2 + 10x + 16 = 0$
(2) $x^2 - 10x + 16 = 0$
(3) $x^2 + 10x - 16 = 0$
(4) $x^2 - 10x - 16 = 0$

Solution: Since the roots are +2 and +8, the factors are $(x - 2)$ and $(x - 8)$ and the equation is $(x - 2)(x - 8) = 0$. Use FOIL to multiply the left-hand side of the equation to make it look more like the answer choices and get $(x - 2)(x - 8) = x^2 - 8x - 2x + 16 = x^2 - 10x + 16$, which is choice (2).

Check Your Understanding of Section 3.5

A. Multiple-Choice

1. What are the roots of the equation $(x - 2)(x + 5) = 0$?
(1) 2 and 5
(2) 2 and –5
(3) –2 and –5
(4) –2 and 5

2. What are the roots of the polynomial $(x + 4)(x - 7)$?
(1) 4 and 7
(2) 4 and –7
(3) –4 and 7
(4) –4 and –7

3. If the roots of an equation are 3 and –6, what could the equation be?
(1) $(x - 3)(x - 6) = 0$
(2) $(x - 3)(x + 6) = 0$
(3) $(x + 3)(x - 6) = 0$
(4) $(x + 3)(x + 6) = 0$

4. If the roots of a polynomial are 4 and –2, what could the polynomial be?
(1) $(x - 4)(x - 2)$
(2) $(x + 4)(x + 2)$
(3) $(x - 4)(x + 2)$
(4) $(x + 4)(x - 2)$

5. If the factors of a polynomial are $(x - 5)$ and $(x - 2)$, what are the roots of that polynomial?
(1) 5 and 2
(2) 5 and –2
(3) –5 and 2
(4) –5 and –2

6. If the roots of a polynomial are 1 and –8, what could be the factors?
(1) $(x - 1)$ and $(x + 8)$
(2) $(x - 1)$ and $(x - 8)$
(3) $(x + 1)$ and $(x + 8)$
(4) $(x + 1)$ and $(x - 8)$

7. If a polynomial has factors of $(x - p)$ and $(x + q)$, what are the roots of the polynomial?
(1) p and q
(2) $-p$ and q
(3) $-p$ and $-q$
(4) p and $-q$

8. If a cubic polynomial has factors of $(x + 2)$, $(x - 3)$, and $(x - 7)$, what are the roots of the polynomial?
(1) 2, –3, and –7
(2) –2, 3, and 7
(3) 2, 3, and 7
(4) –2, –3, and –7

9. If a cubic polynomial has roots of 5, 3, and –1, what could the factors of the polynomial be?
(1) $(x + 5)$, $(x + 3)$, and $(x - 1)$
(2) $(x + 5)$, $(x + 3)$, and $(x + 1)$
(3) $(x - 5)$, $(x - 3)$, and $(x + 1)$
(4) $(x - 5)$, $(x - 3)$, and $(x - 1)$

10. Which is a root of the equation $x^3 - 6x^2 + 13x - 120 = 0$?
(1) 1
(2) 3
(3) 5
(4) 7

B. *Show how you arrived at your answers.*

1. A cubic equation has three roots, 2, –2, and 4. What could the equation be?

2. If an equation has two roots, –4 and –7, what could the equation be?

3. Camila says the roots of a polynomial are just the factors with the signs changed. Is this accurate? Explain.

4. If a polynomial has factors $(2x + 3)$ and $(2x + 5)$, what are the two roots?

5. The polynomial $x^2 - 4x - 3$ does not seem to factor into $(x - p)(x - q)$ with p and q as integers, but it might factor if p and q don't have to be integers. But by solving the equation $x^2 - 4x - 3 = 0$ by completing the square, it is possible to find the roots. Find the roots, and then use them to find the factors.

3.6 SOLVING QUADRATIC EQUATIONS WITH THE QUADRATIC FORMULA

KEY IDEAS

Many quadratic polynomials do not factor. Though completing the square is a technique that will work when factoring is not possible, it is a lengthy process, with many opportunities for careless errors. An efficient way to solve for the roots of a quadratic equation is to use the quadratic formula.

MATH FACTS

The two solutions to the quadratic equation $ax^2 + bx + c = 0$ are $x = \frac{-b \pm \sqrt{b^2 - 4ac}}{2a}$. These are the same two answers you would get if you did the completing the square process with the equation $x^2 + \frac{b}{a}x + \frac{c}{a} = 0$. If the terms of the quadratic are in a different order, like $bx + c + ax^2$, the a value is still the coefficient of the x^2, the b value is still the coefficient of the x, and the c value is the constant.

Applying the Quadratic Formula

The equation $x^2 + 5x + 2 = 0$ cannot be solved with factoring since $x^2 + 5x + 2$ does not factor easily. It would also be difficult to use completing the square for this equation since the 5 is odd so the constant would need to be $\left(\frac{5}{2}\right)^2 = \frac{25}{4}$. This is a job for the quadratic formula!

Since the equation could be written as $1x^2 + 5x + 2 = 0$, the a coefficient is 1, the b coefficient is 5, and the c is 2.

Substituting these three values into the quadratic formula lead to

$$x = \frac{-5 \pm \sqrt{5^2 - 4 \cdot 1 \cdot 2}}{2 \cdot 1}$$

Rather than type this entire expression into the calculator, evaluate the number inside the radical sign, and also the denominator to get the expression

$$x=\frac{-5\pm\sqrt{25-8}}{2}=\frac{-5\pm\sqrt{17}}{2}$$

If you get a decimal approximation of this, it is safest to first calculate $-5+\sqrt{17}$ and then divide by 2 to get approximately –0.44. Then calculate $-5-\sqrt{17}$ and divide that by 2 to get approximately –4.56.

If this is a multiple-choice question, the answer might also look like $-\frac{5}{2}\pm\frac{\sqrt{17}}{2}$.

Applying the Quadratic Formula When Some of the Coefficients Are Negative

Be careful when you see negatives in the quadratic equation. When b is negative, then the $-b$ becomes positive while the b^2 is still positive. If c is negative while a is positive, the expression $-4ac$ becomes positive when the product ac is negative.

For the equation $x^2+5x-2=0$, the quadratic equation becomes

$$x=\frac{-5\pm\sqrt{5^2-4\cdot1\cdot(-2)}}{2\cdot1}=\frac{-5\pm\sqrt{25+8}}{2}=\frac{-5\pm\sqrt{33}}{2}$$

Notice that the 8 got added to the 25 because the expression became positive.

Example 1

Use the quadratic formula to solve for all values of x that satisfy the equation $x^2+8x+15=0$.

Solution: Though this one can be done by factoring, the quadratic formula can be used on any equation like this.

$$a=1,\ b=8,\text{ and } c=15.$$

$$x=\frac{-8\pm\sqrt{8^2-4\cdot1\cdot15}}{2\cdot1}=\frac{-8\pm\sqrt{64-60}}{2}=\frac{-8\pm\sqrt{4}}{2}=\frac{-8\pm2}{2}$$

$$x=\frac{-8+2}{2}=\frac{-6}{2}=-3\text{ or }x=\frac{-8-2}{2}=\frac{-10}{2}=-5$$

Example 2

Which is a solution to the quadratic equation $x^2 + 6x + 2 = 0$?

(1) $3 + \sqrt{7}$ (2) $3 - \sqrt{7}$ (3) $-3 + \sqrt{7}$ (4) $-3 + \sqrt{5}$

Solution:

$$a = 1, b = 6, c = 2$$

$$x = \frac{-6 \pm \sqrt{6^2 - 4 \cdot 1 \cdot 2}}{2 \cdot 1} = \frac{-6 \pm \sqrt{36 - 8}}{2} = \frac{-6 \pm \sqrt{28}}{2}$$

This doesn't exactly match any of the four answer choices. But $\sqrt{28}$ can be written as $\sqrt{4 \cdot 7} = \sqrt{4} \cdot \sqrt{7} = 2\sqrt{7}$, which allows you to simplify the answer to $\frac{-6 \pm 2\sqrt{7}}{2} = -3 \pm \sqrt{7}$ so the answer is choice (3).

Example 3

Use the quadratic formula to solve for all values of x that satisfy the equation $3x^2 - 4x - 5 = 0$

Solution: $a = 3$, $b = -4$, and $c = -5$.

$$x = \frac{-(-4) \pm \sqrt{(-4)^2 - 4 \cdot 3 \cdot (-5)}}{2 \cdot 3} = \frac{4 \pm \sqrt{16 + 60}}{6} = \frac{4 \pm \sqrt{76}}{6}$$

The $\sqrt{76}$ can be further simplified into $\sqrt{4 \cdot 19} = 2\sqrt{19}$, if needed.

Check Your Understanding of Section 3.6

A. Multiple-Choice

For each question, find all solutions for the variable using the quadratic formula.

1. $x^2 - 6x + 8 = 0$
(1) 2, 4 (2) 3 (3) –2, –4 (4) 2, –4

2. $x^2 + 2x - 15 = 0$
(1) –3, 5 (2) –5, 3 (3) –1, 15 (4) 3, 5

3. $x^2 + 6x - 16 = 0$
(1) –2, 8 (2) 2, –8 (3) 4, –4 (4) –1, 16

4. $-x^2 - 9x - 20 = 0$
(1) 4, –5 (2) –4, 5 (3) 4,5 (4) –4, –5

5. $2x^2 + 7x - 4 = 0$
(1) $\frac{1}{2}, -4$ (2) 1, –4 (3) $-\frac{1}{2}, 4$ (4) –1, 4

6. $x^2 + 4x - 7 = 0$
(1) 1.3 (2) $2 \pm \sqrt{11}$ (3) $2 \pm \sqrt{12}$ (4) $-2 \pm \sqrt{11}$

7. $6x^2 - 13x + 6 = 0$
(1) $-\frac{2}{3}, -\frac{3}{2}$ (2) $\frac{2}{3}, -\frac{3}{2}$ (3) $\frac{2}{3}, \frac{3}{2}$ (4) $-\frac{2}{3}, \frac{3}{2}$

8. $x^2 - 4 - 2x = 0$
(1) $1 \pm \sqrt{5}$ (2) $2 \pm 2\sqrt{6}$ (3) $-1 \pm \sqrt{5}$ (4) $-2 - 2\sqrt{6}$

9. $x^2 = x + 1$
(1) $\frac{1 \pm \sqrt{5}}{2}$ (2) $\frac{-1 \pm \sqrt{5}}{2}$ (3) $\frac{1 \pm \sqrt{-3}}{2}$ (4) $\frac{-1 \pm \sqrt{-3}}{2}$

10. $-16t^2 + 64t + 80 = 0$
(1) 1, –5 (2) –1, 5 (3) –1, –5 (4) 1, 5

***B.** Show how you arrived at your answers.*

1. Solve for all values of x that satisfy the equation $x^2 - 10x + 22 = 0$. Round answers to the nearest hundredth.

2. If the solution to a quadratic equation, $ax^2 + bx + c = 0$ is $x = \frac{-5 \pm \sqrt{17}}{2}$, what could the values of a, b, and c be?

3. Use the quadratic formula to solve for both values of x in terms of c. $x^2 + 4x + c = 0$.

4. Are the solutions to the quadratic equation $x^2 + 3x - 2 = 0$ rational or irrational? Explain.

5. What are the two solutions to the equation $x^2 + 3 = 4x + 6$?

3.7 WORD PROBLEMS INVOLVING QUADRATIC EQUATIONS

Many real-world situations, especially ones that involve products, can be modeled as quadratic equations. After the equation is set up, the techniques from the previous sections in this chapter can be used to complete the question.

Questions Involving the Area of a Rectangle

Since the area of a rectangle is equal to the length times the width, questions about areas of rectangle often become quadratic equations.

Example 1

The length of a rectangle is 4 feet more than its width. The area of the rectangle is 140 square feet. If the width is represented by x, the equation that could be used to solve for x is $x(x + 4) = 140$. Solve this equation for x to find the width of the rectangle.

Solution:

$$x(x + 4) = 140$$
$$x^2 + 4x = 140$$
$$x^2 + 4x - 140 = 0$$
$$(x + 14)(x - 10) = 0$$
$$x + 14 = 0 \text{ or } x - 10 = 0$$
$$x = -14 \text{ or } x = 10$$

Since x represents the width, it must be a positive number, so the answer is 10. The quadratic equation $x^2 + 4x - 140 = 0$ could have also been solved with completing the square or with the quadratic formula.

Questions Involving the Height of a Projectile

When a projectile is shot into the air, the relationship between the time in the air and the projectile's height above the ground can be represented with a quadratic equation.

Example 2

A projectile is shot into the air such that its height can be calculated with the expression $-16x^2 + 48x + 64$, where x is the number of seconds since the projectile has been shot. The equation $-16x^2 + 48x + 64 = 0$ can be used to find the amount of time it will take for the projectile to land on the ground. Solve for x to find this answer.

Solution:

$$-16x^2 + 48x + 64 = 0$$
$$-16(x^2 - 3x - 4) = 0$$
$$-16(x - 4)(x + 1) = 0$$
$$x = 4 \text{ or } x = -1$$

Since x represents the amount of time since the projectile was shot, only the answer $x = 4$ is possible.

Questions Involving the Price of an Item

Usually when the cost of something increases, the number of things sold decreases. Another type of word problem that becomes a quadratic equation is one that describes the effect on sales of raising the price of an item.

The price to rent a bicycle is \$10 an hour, and a store rents 40 bicycles in a day. Each time the store raises the rental price \$2, the number of bicycles the store rents decreases by 5. When they charge \$10 and rent 40 bicycles, their total revenue per day is $\$10 \cdot 40 = \$400$. If they increase the price to \$12 an hour, they only rent 35 bicycles a day, and their revenue becomes $\$12 \cdot 35 = \$420$.

In general, the equation for revenue is $R = (10 + 2x)(40 - 5x)$ where x is the number of \$2 increases.

To find out what they should charge to make a revenue of $300, solve the equation.

$$
\begin{aligned}
300 &= (10 + 2x)(40 - 5x) \\
300 &= 400 - 50x + 80x - 10x^2 \\
300 &= -10x^2 + 30x + 400 \\
-\,300 &= -300 \\
0 &= -10x^2 + 30x + 100 \\
0 &= -10(x^2 - 3x - 10) \\
0 &= -10(x - 5)(x + 2) \\
x - 5 = 0 &\text{ or } x + 2 = 0 \\
x = 5 &\text{ or } x = -2
\end{aligned}
$$

When $x = 5$, they charge $\$10 + \$2 \cdot 5 = \$20$.
When $x = -2$, they charge $\$10 + \$2 \cdot (-2) = \$6$.

So if they charge $20 or $6, they will make $300 in revenue.

Check Your Understanding of Section 3.7

A. Multiple-Choice

1. A printer wants to put 54 square inches of text into a rectangle on an 8 by 11 sheet of paper. She wants the text to be surrounded by a border of constant width. Which equation could be used to find the width of the border (x)?

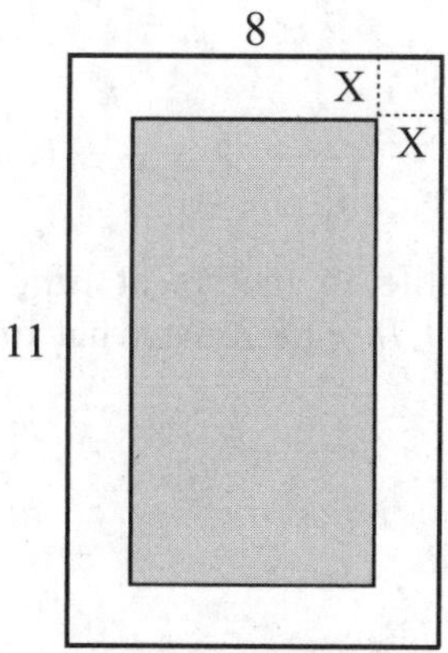

(1) $2(8 - 2x) + 2(11 - 2x) = 54$
(2) $(8 - 2x) + (11 - 2x) = 54$
(3) $(8 - x)(11 - x) = 54$
(4) $(8 - 2x)(11 - 2x) = 54$

2. The width of a rectangle is 10 inches longer than its length. If the area of the rectangle is 56 square inches, which equation could be used to determine its length (l)?
(1) $l(l + 10) = 56$
(2) $l(l - 10) = 56$
(3) $2l + 2(l + 10) = 56$
(4) $2l - 2(l + 10) = 56$

3. A 100-foot fence is used to enclose a rectangular land plot. If the area of the land plot is 456 square feet, which equation can be used to determine the length (x) of the land plot?
(1) $x(100 - x) = 456$
(2) $2x + 2(100 - x) = 456$
(3) $x(50 - x) = 456$
(4) $2x + 2(50 - x) = 456$

4. The height of a projectile in feet at time t is determined by the equation $h = -16t^2 + 128t + 320$. At what time will the projectile be 560 feet high?
(1) 4 seconds
(2) 5 seconds
(3) 6 seconds
(4) 7 seconds

5. The height of a projectile in feet at time t is determined by the equation $h = -16t^2 + 112t + 128$. At what time will the projectile be 0 feet high?
(1) 5 seconds
(2) 6 seconds
(3) 7 seconds
(4) 8 seconds

6. The height of a projectile in meters at time t is determined by the equation $h = -4.9t^2 + 14.7t + 88.2$. At what time will the projectile be 68.6 meters high?
(1) 3 seconds
(2) 4 seconds
(3) 5 seconds
(4) 6 seconds

7. When a textbook sells for \$20, 70 copies are purchased. Every time the price is increased by \$3, the number of people purchasing the book decreases by 8. Which equation can be used to determine the number of \$3 increases needed to make the total revenue from the books equal \$1,426?
(1) $(20 - 3x)(70 - 8x) = 1{,}426$
(2) $(20 - 3x)(70 + 8x) = 1{,}426$
(3) $(20 + 3x)(70 + 8x) = 1{,}426$
(4) $(20 + 3x)(70 - 8x) = 1{,}426$

8. When the cost of a movie is \$10, 200 people buy tickets. Each time the price of the movie is increased by \$3, the number of people buying tickets decreases by 20. Which equation can be used to determine the number of \$3 increases needed to make the total revenue from the movie tickets equal to \$2,660?
(1) $(10 - 3x)(200 - 20x) = 2{,}660$
(2) $(10 - 3x)(200 + 20x) = 2{,}660$
(3) $(10 + 3x)(200 - 20x) = 2{,}660$
(4) $(10 + 3x)(200 + 20x) = 2{,}660$

9. When a candy bar costs \$0.50, 500 kids buy it. Each time the price increases by \$0.05, the number of kids buying it decreases by 10. Which equation could be used to determine the number of \$0.05 increases needed to make the total revenue from the candy equal to \$450?
(1) $(0.5 + 0.05x)(500 - 10x) = 450$
(2) $(0.5 - 0.05x)(500 - 10x) = 450$
(3) $(0.5 + 0.05x)(500 + 10x) = 450$
(4) $(0.5 - .005x)(500 + 10x) = 450$

10. The surface area of a box with a square base is $S=2l^2 + 4hl$ where l is the length of the side of the square base and h is the height of the box. If the height of the box is 5 feet, which equation could be used to find l to make the surface area 238 square feet?
(1) $l^2h = 238$
(2) $2l^22 + 5l = 238$
(3) $2l^2 + 4l = 238$
(4) $2l^2 + 20l = 238$

B. *Show how you arrived at your answers.*

1. The height, in feet, of a projectile after t seconds is determined by the equation $h = -16t^2 + 128t + 144$. At what two times is the projectile exactly 384 feet high?

2. A 4 inch by 6 inch photo is put into a picture frame with a border of constant width. If the area of the frame, including the picture, is 80 square inches, find an equation for determining the width of the border, and use that equation to solve for it.

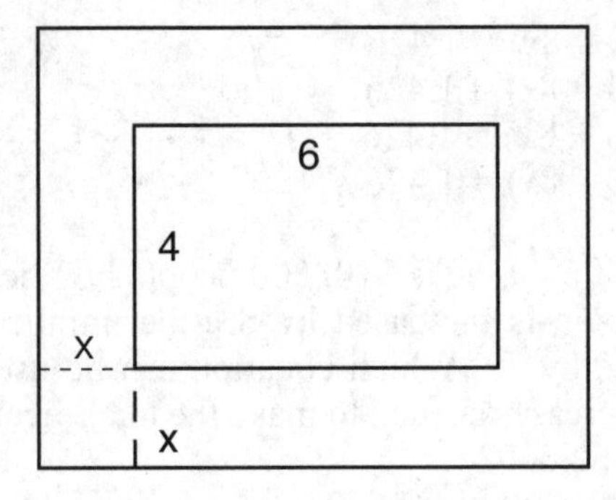

3. When a song costs $0.99 it gets downloaded 200,000 times. Each time the price is increased by $0.10, the number of downloads decreases by 10,000. How may $0.10 increases need to happen for the total revenue to be $223,500? Create an equation and solve.

4. A man is 35 years older than his son. The product of their ages is 884. Write an equation that can be used to solve for the man's age. How old is the man?

5. The perimeter of a rectangle is 44 inches. The area of the rectangle is 120 square inches. If the width of the rectangle is x inches, determine the length of the rectangle using an algebraic solution.

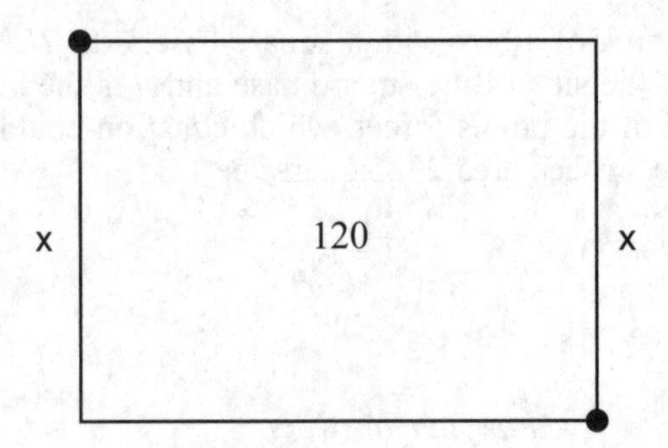

Chapter Four SYSTEMS OF LINEAR EQUATIONS

4.1 SOLVING SYSTEMS WITH GUESS AND CHECK

An equation like $x + 3 = 10$ has just one number in its solution set, $x = 7$. When an equation has two variables, for instance $x + y = 10$, there is no longer just one solution. There are many pairs of numbers that satisfy the equation $x + y = 10$. Since $4 + 6 = 10$, one solution is $x = 4$, $y = 6$. This can be written as the *ordered pair* (4, 6) where 4 is called the *x-coordinate* and 6 is the *y-coordinate*. Other ordered pairs that satisfy the equation are (3, 7), (1, 9), (9, 1), and (–4, 14).

Finding an Ordered Pair That Satisfies an Equation with Two Variables

An equation with two variables has an infinite number of ordered pairs that satisfy it. To find one solution, try picking any number you want for the x-coordinate. Then substitute that into the equation and solve for y to find the other number in the ordered pair.

If the equation is $2x + y = 10$, you could pick 3 for the x-value and substitute that into the equation.

$$\begin{aligned} 2(3) + y &= 10 \\ 6 + y &= 10 \\ -6 &= -6 \\ y &= 4 \end{aligned}$$

The ordered pair, then, is (3, 4). You can check by substituting 3 for x and 4 for y in the equation.

$$\begin{aligned} 2(3) + 4 &\stackrel{?}{=} 10 \\ 6 + 4 &\stackrel{?}{=} 10 \\ 10 &\stackrel{\checkmark}{=} 10 \end{aligned}$$

Checking to See If an Ordered Pair Is Part of the Solution Set of a Two-Variable Equation

By substituting the x-coordinate of an ordered pair in for the x-variable and the y-coordinate of the ordered pair in for the y-variable in an equation, you can determine if the point is in the solution set for that equation. If the resulting equation is true, then it is. If the resulting equation is not true, then the ordered pair is not part of the solution set.

Example 1

Which of the following choices is part of the solution set of the equation $2x + 3y = 14$?

(1) (1, 5)　　(2) (2, 3)　　(3) (3, 3)　　(4) (4, 2)

Solution: There are an infinite number of solutions to the equation $2x + 3y = 14$. One of them is among the four choices. Check each choice by substituting the first coordinate into x and the second coordinate into y, and see which make the equation true.

Check (1, 5)

$$\begin{aligned} 2(1) + 3(5) &\stackrel{?}{=} 14 \\ 2 + 15 &\stackrel{?}{=} 14 \\ 17 &\neq 14 \end{aligned}$$

Check (2, 3)

$$\begin{aligned} 2(2) + 3(3) &\stackrel{?}{=} 14 \\ 4 + 9 &\stackrel{?}{=} 14 \\ 13 &\neq 14 \end{aligned}$$

Check (3, 3)

$$\begin{aligned} 2(3) + 3(3) &\stackrel{?}{=} 14 \\ 6 + 9 &\stackrel{?}{=} 14 \\ 15 &\neq 14 \end{aligned}$$

Check (4, 2)

$$\begin{aligned} 2(4) + 3(2) &\stackrel{?}{=} 14 \\ 8 + 6 &\stackrel{?}{=} 14 \\ 14 &\stackrel{\checkmark}{=} 14 \end{aligned}$$

So the answer is (4, 2).

Identifying Equations That Have Equivalent Solution Sets

The equation $x + y = 10$ has an infinite number of ordered pairs in its solution set. A few of them are (2, 8), (5, 5), and (7, 3) since $2 + 8 = 10$, $5 + 5 = 10$, and $7 + 3 = 10$. If you multiply both sides of this equation by the same number, it becomes a new equation with the same solution set as the original.

If you multiply both sides of $x + y = 10$ by 2, it becomes

$$2(x + y) = 2(10)$$
$$2x + 2y = 20$$

If you test the points (2, 8), (5, 5), and (7, 3), you will see that they are solutions of $x + y = 10$ and of $2x + 2y = 20$.

$$2(2) + 2(8) = 4 + 16 = 20$$
$$2(5) + 2(5) = 10 + 10 = 20$$
$$2(7) + 2(3) = 14 + 6 = 20$$

Every solution to $x + y = 10$ will also be a solution of $2x + 2y = 10$. When both sides of an equation are multiplied by the same number, the new equation has the same solution set as the original equation.

Example 2

Which of the following equations has the same solution set as $x + 3y = 12$?

(1) $4x + 12y = 12$
(2) $3x + 9y = 48$
(3) $4x + 12y = 48$
(4) $2x + 9y = 36$

Solution: If you multiply both sides of the equation $x + 3y = 12$ by 4, it becomes $4x + 12y = 48$, choice (3).

If you multiply both sides by 3, it becomes $3x + 9y = 36$, not $3x + 96 = 48$ so choice (2) is incorrect. The other two choices are also not possible to obtain by multiplying both sides of $x + 3y = 12$ by the same number.

Solving Systems of Equations by Guess and Check

A system of equations is a set of two equations that each has an x and a y. The goal is to find the ordered pairs that satisfy both equations at the same time.

An example is the system of equations:

$$x + y = 10$$
$$x - y = 2$$

By going through some of the ordered pairs that satisfy the first equation (9, 1), (8, 2), (7, 3), (6, 4), and (5, 5), you might notice that the ordered pair

(6, 4) also satisfies the second equation so the solution to the system of equations is (6, 4).

$$6 + 4 \stackrel{?}{=} 10$$
$$10 \stackrel{\checkmark}{=} 10$$

$$6 - 4 \stackrel{?}{=} 2$$
$$2 \stackrel{\checkmark}{=} 2$$

This guess and check method is not very practical unless it is a multiple-choice question where there are at most four ordered pairs to check.

Example 3

Which ordered pair is a solution to the system of equations?

$$2x + 3y = 11$$
$$4x - y = 15$$

(1) (1, 3) (2) (2, –7) (3) (3, 4) (4) (4, 1)

Solution: Only if the ordered pair is a solution to both equations, is it a solution to the system. Check each of the choices in both equations.

Check (1, 3)

$$2(1) + 3(3) \stackrel{?}{=} 11$$
$$2 + 9 \stackrel{?}{=} 11$$
$$11 \stackrel{\checkmark}{=} 11$$

$$4(1) - 3 \stackrel{?}{=} 15$$
$$4 - 3 \stackrel{?}{=} 15$$
$$1 \neq 15$$

Since it satisfies only the first equation, choice (1) is not a solution to the system of equations.

Choice (2) doesn't satisfy the first equation, and choice (3) doesn't satisfy either of the equations. But when you test choice (4), it becomes

$$2(4) + 3(1) \stackrel{?}{=} 11$$
$$8 + 3 \stackrel{?}{=} 11$$
$$11 \stackrel{\checkmark}{=} 11$$

$$4(4) - 1 \stackrel{?}{=} 15$$
$$16 - 1 \stackrel{?}{=} 15$$
$$15 \stackrel{\checkmark}{=} 15$$

Choice (4) is a solution to both equations so it is a solution to the system of equations.

Check Your Understanding of Section 4.1

A. Multiple-Choice

1. (2, 5) is a solution to which equation?
(1) $x + 2y = 9$
(2) $2x + y = 9$
(3) $8x - y = 9$
(4) $3x + 4y = 25$

2. Which ordered pair is a solution to the equation $3y - 4x = 3$?
(1) (3, 5) (2) (5, 3) (3) (1, 4) (4) (8, 12)

3. Which ordered pair is *not* a solution to the equation $3y - 2x = 3$?
(1) (6, 5) (2) (9, 7) (3) (12, 9) (4) (10, 8)

4. Which equation has the same solution set as the equation $2x + 3y = 5$?
(1) $8x + 12y = 20$
(2) $8x + 12y = 15$
(3) $6x + 9y = 12$
(4) $4x + 6y = 8$

5. Which equation does *not* have the same solution set as the equation $4x - 8y = 12$?
(1) $6x - 7y = 20$
(2) $x - 2y = 3$
(3) $2x - 4y = 6$
(4) $8x - 16y = 24$

6. Which is the solution set to this system of equations?

$$2x + 3y = 20$$
$$5x - 2y = 31$$

(1) (4, 4) (2) (1, 6) (3) (–2, 8) (4) (7, 2)

7. Which is the solution set to this system of equations?

$$-4x + 2y = 18$$
$$3x - 6y = -36$$

(1) (–2, 5) (2) (–1, 7) (3) (–3, 3) (4) (0, 9)

8. The ordered pair (3, –7) is a solution to which system of equations?
(1) $2x - 3y = 27$
$5x - 3y = 38$
(2) $4x + 2y = -2$
$3x - 7y = 59$
(3) $2x - 3y = 27$
$4x + 2y = -2$
(4) $5x - 3y = 38$
$3x + 7y = -40$

Chapter Four **SYSTEMS OF LINEAR EQUATIONS**

9. If $(a, 5)$ is a solution to the equation $3x + 6y = 42$, what is the value of a?
(1) 1 (2) 2 (3) 3 (4) 4

10. Which is *not* a solution to the system of equations?

$$x + y = 12$$
$$2x + 2y = 24$$

(1) (8, 4) (2) (3, 10) (3) (7, 5) (4) (5, 7)

B. *Show how you arrived at your answers.*

1. The equation $2x + 3y = 11$ has two solutions in which both coordinates are positive integers less than 6. What are those two solutions?

2. Use guess and check to find the solution to the system of equations.

$$x + y = 8$$
$$x - y = 6$$

3. The system of equations

$$2x + 5y = 25$$
$$2x + 5y = 26$$

has no solutions. Explain why?

4. The equation $x + 3y = c$ has the point (4, 7) in its solution set. What is c?

5. $(6, -2)$ is a solution to the equation $5x - 6y = 42$. Find a solution to $15x - 18y = 126$.

4.2 SOLVING SIMPLER SYSTEMS OF EQUATIONS WITH ALGEBRA

Certain systems of equations can be solved by combining the two equations such that one of the variables gets eliminated. Three ways of doing this are addition, when the two equations are added to each other; subtraction, when one equation is subtracted from the other; and substitution, when one of the variables in one equation is replaced by an expression involving the other variable.

Combining Two Equations to Form a New Equation

If two equations have an ordered pair that makes each of them true, then the equation formed by adding the two equations will also have that ordered pair as one of its solutions.

The equations $2x + 5y = 26$ and $3x - 2y = 1$ both have the solution (3, 4). If you add the two equations together, you get

$$\begin{array}{rrcrcr} & 2x & + & 5y & = & 26 \\ + & & & & & \\ & 3x & - & 2y & = & 1 \\ \hline & 5x & + & 3y & = & 27 \end{array}$$

This new equation also is satisfied by the ordered pair (3, 4).

$$5(3) + 3(4) \stackrel{?}{=} 27$$
$$15 + 12 \stackrel{?}{=} 27$$
$$27 \stackrel{\checkmark}{=} 27$$

MATH FACTS

If in a system of equations, the coefficient of one of the variables is the opposite sign of the coefficient of the same variable in the other equation, the system can be quickly solved by adding the two equations together to form an equation with just one variable.

Example 1

Which equation has, as one of its solutions, the solution of the system of equations?

$$\begin{aligned} 3x - 2y &= 16 \\ 4x + 3y &= -7 \end{aligned}$$

(1) $7x + y = 9$
(2) $7x + y = 8$
(3) $7x + y = 7$
(4) $7x + y = 6$

Solution: If you add the two equations, the new equation has as one of its solutions, the solution to the system of equations. In this case,

$$\begin{array}{rrcrcr} & 3x & - & 2y & = & 16 \\ + & 4x & + & 3y & = & -7 \\ \hline & 7x & + & y & = & 9 \end{array}$$

choice (1)

Adding Two Equations to Eliminate a Variable

The equation $5x + 2y = 36$ has an infinite number of solutions. The equation $3x - 2y = 12$ also has an infinite number of solutions. Finding the ordered pair that makes both equations true can be done quickly in this case by adding the two equations.

$$\begin{array}{rrcrcr} & 5x & + & 2y & = & 36 \\ + & 3x & - & 2y & = & 12 \\ \hline & \frac{8x}{8} & & & = & \frac{48}{8} \\ & x & & & = & 6 \end{array}$$

In this example, because the $+2y$ and the $-2y$ have a sum of $0y$, which is 0, the new equation you get by adding the two equations together is an equation with only one variable. This equation, $8x = 48$ is solved by dividing both sides of the equation by 8, to get $x = 6$.

6 is the x-coordinate of the ordered pair that solves the original system of equations. To find the y-coordinate, substitute the 6 in for x in either of the original two equations.

$$\begin{aligned} 5(6) + 2y &= 36 \\ 30 + 2y &= 36 \\ -30 &= -30 \\ \frac{2y}{2} &= \frac{6}{2} \\ y &= 3 \end{aligned}$$

The solution to the original system of equations is (6, 3).

To check, substitute 6 for x and 3 for y into both original equations.

$$5(6) + 2(3) = 30 + 6 = 36$$
$$3(6) - 2(3) = 18 - 6 = 12$$

Example 2

Find the ordered pair that solves the system of equations.

$$4x - 3y = 22$$
$$2x + 3y = 2$$

Solution: Since the $-3y$ in the top equation will cancel out the $+3y$ in the bottom equation, this system can be solved by adding the two equations.

$$\begin{array}{rrcrcr} & 4x & - & 3y & = & 22 \\ + & 2x & + & 3y & = & 2 \\ \hline & \frac{6x}{6} & & & = & \frac{24}{6} \\ & x & & & = & 4 \end{array}$$

Substitute $x = 4$ into either of the original two equations.

$$4(4) - 3y = 22$$
$$16 - 3y = 22$$
$$-16 = -16$$
$$\frac{-3y}{3} = \frac{6}{-3}$$
$$y = -2$$

The solution is $(4, -2)$.

Example 3

Find the solution set for the system of equations.

$$-3x + 7y = 16$$
$$3x + 4y = -5$$

Solution: Since the coefficients of the x in the first equation is the opposite sign of the coefficient of the x in the second equation, this system can be solved by adding the two equations together.

$$\begin{array}{rrcrcr} & -3x & + & 7y & = & 16 \\ + & 3x & + & 4y & = & -5 \\ \hline & & & \frac{11y}{11} & = & \frac{11}{11} \\ & & & y & = & 1 \end{array}$$

$$-3x + 7(1) = 16$$
$$-3x + 7 = 16$$
$$-7 = -7$$
$$\frac{-3x}{-3} = \frac{9}{-3}$$
$$x = -3$$

The ordered pair that satisfies the system is (–3, 1).

Subtracting Two Equations to Eliminate a Variable

If in the two equations in a system, one of the variables has the same coefficient as the same variable has in the other equation, that variable can be eliminated by subtracting the two equations.

In this system

$$4x + 5y = 37$$
$$2x + 5y = 31$$

the y variables in both equations have a coefficient of +5. Subtract the two equations

$$\begin{array}{rcrcrcr} & 4x & + & 5y & = & 37 \\ - & 2x & + & 5y & = & 31 \\ \hline & \frac{2x}{2} & & & = & \frac{6}{2} \\ & x & & & = & 3 \end{array}$$

Substitute $x = 3$ into either equation and solve for y.

$$4(3) + 5y = 37$$
$$12 + 5y = 37$$
$$-12 = -12$$
$$\frac{5y}{5} = \frac{25}{5}$$
$$y = 5$$

The solution is (3, 5).

Solving Systems of Equations with the Substitution Method

If one of the two equations is in the form $y = mx + b$, the system of equations can be solved by replacing the y in the other equation by the expression $mx + b$ and then solving the new one-variable equation.

For the system

$$y = 2x + 1$$
$$3x + 2y = 16$$

one of the equations is in the form $y = mx + b$.

Replace the y in the second equation with the expression $2x + 1$.

$$3x + 2(2x + 1) = 16$$
$$3x + 4x + 2 = 16$$
$$7x + 2 = 16$$
$$-2 = -2$$
$$\frac{7x}{7} = \frac{14}{7}$$
$$x = 2$$

Substitute this x into either equation.

$$y = 2(2) + 1$$
$$y = 4 + 1$$
$$y = 5$$

The solution is (2, 5).

Example 4

Solve this system of equations with the substitution method.

$$y = 5x - 7$$
$$3x - 4y = -6$$

Solution: Replace the y in the second equation with the expression $5x - 7$.

$$3x - 4(5x - 7) = -6$$

Distribute the -4 through the parentheses.

$$3x - 20x + 28 = -6$$
$$-17x + 28 = -6$$
$$-28 = -28$$
$$\frac{-17x}{-17} = \frac{-34}{-17}$$
$$x = 2$$

Substitute the $x = 2$ into the first equation

$$y = 5x - 7 = 5(2) - 7 = 10 - 7 = 3$$

The solution is (2, 3).

Check Your Understanding of Section 4.2

A. Multiple-Choice

1. Which equation has as one of its solutions, the solution to the system of equations?

$$4x + 2y = 10$$
$$3x - 5y = 14$$

(1) $7x - 3y = 21$ (3) $7x - 3y = 23$
(2) $7x - 3y = 22$ (4) $7x - 3y = 24$

2. Solve the system of equations.

$$3x + 2y = 17$$
$$4x - 2y = 4$$

(1) (4, 3) (2) (5, 1) (3) (3, 4) (4) (1, 6)

3. Solve the system of equations.

$$y = 5x + 3$$
$$2x + 6y = 50$$

(1) (1, 8) (2) (8, 1) (3) (–1, –8) (4) (–8, –1)

4. Solve the system of equations.

$$6x - 5y = 19$$
$$-3x + 5y = -7$$

(1) (–4, –1) (2) (–4, 1) (3) (4, –1) (4) (4, 1)

5. Solve the system of equations.

$$2x + 3y = -1$$
$$-2x + 5y = -23$$

(1) (–4, 3) (2) (4, 3) (3) (4, –3) (4) (–4, –3)

6. Solve the system of equations.

$$-3x + 7y = 1$$
$$3x + 3y = -21$$

(1) (5, 2) (2) (–5, 2) (3) (–5, –2) (4) (5, –2)

7. Solve the system of equations.

$$3x + 7y = -2$$
$$x - 7y = -10$$

(1) (3, 1) (2) (–3, –1) (3) (3, –1) (4) (–3, 1)

8. Solve the system of equations.

$$5x + 3y = 19$$
$$2x + 3y = 4$$

(1) (5, –2) (2) (5, 2) (3) (–5, –2) (4) (–5, 2)

9. Solve the system of equations.

$$y = 3x - 2$$
$$4x - 2y = -4$$

(1) (10, 4) (2) (4, 10) (3) (–10, –4) (4) (–4, –10)

10. Solve the system of equations.

$$4x + 8y = 36$$
$$4x + 5y = 33$$

(1) (7, 1) (2) (7, –1) (3) (–7, 1) (4) (–7, –1)

***B.** Show how you arrived at your answers.*

1. Find an equation that has as one of its solutions, the solution to the system of equations.

$$3x + 4y = 11$$
$$2x + 6y = 14$$

2. A student likes to use the substitution method for systems of equation. How can he use it with a system that is not in the proper form for substitution? Show with this system.

$$-2x + y = 4$$
$$3x + 4y = 49$$

3. The system

$$y = 2x - 9$$
$$y = -3x + 16$$

can be solved several ways.

Student 1 wants to use subtraction. Student 2 wants to use substitution. Show how each student would do this.

4. Two numbers, x and y, have a sum of 17 but a difference of 11. Write a system of equations that can be used to solve this and then use it to find the solution.

5. The system of equations

$$\begin{aligned} 5x + 2y &= 8 \\ -5x + 2y &= -32 \end{aligned}$$

can be solved several ways.

Student 1 wants to use addition. Student 2 wants to use subtraction. Who is right? Explain your answer.

4.3 SOLVING MORE COMPLICATED SYSTEMS OF EQUATIONS WITH ALGEBRA

Often when solving systems of equations, we do not get so lucky that the coefficient of one variable in the top equation is the opposite of, or the same as, the coefficient of the same variable in the bottom equation. When this happens, questions like this involve more steps.

Solving a System by Changing One of the Equations

Look at this system of equations

$$\begin{aligned} 3x + 4y &= 31 \\ 5x - 2y &= -9 \end{aligned}$$

Adding or subtracting the two equations will not eliminate one of the variables.

$$\begin{array}{rrcrcr} & 3x & + & 4y & = & 31 \\ + & 5x & - & 2y & = & -9 \\ \hline & 8x & + & 2y & = & 22 \end{array}$$

$$\begin{array}{rrcrcr} & 3x & + & 4y & = & 31 \\ - & 5x & - & 2y & = & -9 \\ \hline & -2x & - & 6y & = & 40 \end{array}$$

Fortunately there is a way to change the equations so that a variable will get eliminated when you add the changed equations together. You are permitted to multiply both sides of either equation by whatever number you want.

In the above example, If the $3x$ were a $-5x$, it could cancel out the $+5x$, but it isn't easy to turn a $3x$ into a $-5x$ by multiplying. If the $5x$ were a $-3x$, that would also work but, again, it isn't easy to do this. Now look at the ys. If the $-2y$ in the bottom equation were a $-4y$, then it would be the opposite sign of the $+4y$ in the top equation. And best of all, it is possible to make that $-2y$ into a $-4y$ by just multiplying both sides of the bottom equation by 2. Multiplying both sides of an equation by a number does not change the solution set for that equation, or the solution set for the system of equations.

$$3x + 4y = 31$$
$$2(5x - 2y) = 2 \cdot -9$$

The system becomes

$$3x + 4y = 31$$
$$10x - 4y = -18$$

This equation is now the type that can be solved by adding the two equations together to get $13x = 13$, which leads to $x = 1$.

Now you can substitute $x = 1$ into either of the original two equations.

$$3(1) + 4y = 31$$
$$3 + 4y = 31$$
$$-3 = -3$$
$$4y = 28$$
$$y = 7$$

Example 1

Solve this system of equations by first multiplying both sides of one of the equations by a number.

$$2x + 6y = 2$$
$$3x + 2y = 10$$

Solution: Since the $6y$ is a multiple of the $2y$, multiply both sides of the bottom equation by -3 so that the $2y$ becomes $-6y$.

$$2x + 6y = 2$$
$$-3(3x + 2y) = -3(10)$$

$$2x + 6y = 2$$
$$-9x - 6y = -30$$

Now add the equations together to eliminate the *y*s.

$$\begin{array}{rrcrcr} & 2x & + & 6y & = & 2 \\ + & -9x & - & 6y & = & -30 \\ \hline & \frac{-7x}{-7} & & & = & \frac{-28}{-7} \\ & x & & & = & 4 \end{array}$$

$$\begin{aligned} 2(4) + 6y &= 2 \\ 8 + 6y &= 2 \\ -8 &= -8 \\ \frac{6y}{6} &= \frac{-6}{6} \\ y &= -1 \end{aligned}$$

The solution is (4, –1).

Solving a System by Changing Both Equations

Sometimes the only way to change the equations so that a variable will drop out when adding them is to change both equations. This is needed when none of the coefficients is evenly divisible by the corresponding coefficient in the other equation.

An example of a system where both equations need to be changed is

$$\begin{aligned} 2x + 3y &= 23 \\ 3x + 4y &= 31 \end{aligned}$$

If you want to make it so the y variable drops out after adding, first take the coefficient of the y in the bottom equation, change the sign, and multiply both sides of the first equation by it.

Since 4 is the coefficient of the y in the bottom equation, multiply both sides of the top equation by –4.

$$\begin{aligned} -4(2x + 3y) &= -4(23) \\ -8x - 12y &= -92 \end{aligned}$$

For the bottom equation, multiply both sides of it by the coefficient of the y in the top equation. In this case, multiply both sides of the bottom equation by +3.

$$\begin{aligned} +3(3x + 4y) &= 3(31) \\ 9x + 12y &= 93 \end{aligned}$$

Multiplying one or both equations in a system by numbers does not change the solution. The new system of equations has the same solution set as the original system of equations but is easier to solve.

Now add the two new equations and complete the process

$$\begin{array}{rrcrcr} & -8x & - & 12y & = & -92 \\ + & 9x & + & 12y & = & 93 \\ \hline & 1x & & & = & 1 \\ & x & & & = & 1 \end{array}$$

To get the y value, substitute the x value of 1 into one of the original two equations.

$$\begin{aligned} 2(1) + 3y &= 23 \\ 2 + 3y &= 23 \\ -2 &= -2 \\ \frac{3y}{3} &= \frac{21}{3} \\ y &= 7 \end{aligned}$$

And the solution is (1, 7).

Example 2

Determine the solution set of the system of equations.

$$\begin{aligned} 2x + 3y &= -11 \\ 4x - 7y &= 43 \end{aligned}$$

Since $-(-7) = +7$, multiply the top equation by 7 and the bottom equation by 3.

$$\begin{aligned} 7(2x + 3y) &= 7(-11) \\ 3(4x - 7y) &= 3(43) \end{aligned}$$

$$\begin{array}{rrcrcr} & 14x & + & 21y & = & -77 \\ + & 12x & - & 21y & = & 129 \\ \hline & \frac{26x}{26} & & & = & \frac{52}{26} \\ & x & & & = & 2 \end{array}$$

$$\begin{aligned} 2(2) + 3y &= -11 \\ 4 + 3y &= -11 \\ -4 &= -4 \\ \frac{3y}{3} &= \frac{-15}{3} \\ y &= -5 \end{aligned}$$

So, the solution is (2, –5).

Example 3

Find the solution to the system of equations.

$$y = 2x - 1$$
$$y = 3x - 4$$

Solution: Rearrange the equations so they are in standard form ($ax + by = c$).

$$-2x + y = -1$$
$$-3x + y = -4$$

Multiply both sides of the top equation by –1.

$$-1(-2x + y) = -1(-1) \text{ becomes } 2x - y = 1$$

$$\begin{array}{rrcrcr} & 2x & - & y & = & 1 \\ + & -3x & + & y & = & -4 \\ \hline & -x & & & = & -3 \\ & x & & & = & 3 \end{array}$$

$$y = 2(3) - 1$$
$$y = 5$$

The solution is (3, 5).

Example 4

Which system of equations has the same solution set as the following system?

$$3x - 4y = 7$$
$$4x + y = 22$$

(1) $3x - 4y = 7$
$16x + 4y = 22$

(2) $6x - 8y = 7$
$4x + y = 22$

(3) $3x - 4y = 7$
$16x + 4y = 22$

(4) $3x - 4y = 7$
$16x + 4y = 88$

Solution: The system of equations in choice (4) can be obtained by multiplying both sides of the bottom equation by 4. This new system will have the same solution set as the original. This would also be true if both equations were multiplied by numbers. The system

$$6x - 8y = 14$$
$$16x + 4y = 88$$

would also be correct if it were a choice since it is obtained by multiplying both sides of the top equation by 2 and both sides of the bottom equation by 4.

Notice that the system in choice (4) would be the logical first step in solving the original system.

Systems of Equations That Have No Solutions or Infinite Number of Solutions

For some systems of equations, there are no ordered pairs that satisfy both equations. An example of this is

$$x + y = 10$$
$$x + y = 11$$

Any ordered pair that satisfies the first equation, like (7, 3), will not satisfy the second equation, and any ordered pair that satisfies the second equation, like (7, 4),will not satisfy the first equation.

To determine that a system of equations has no solution, do the steps for elimination.

$$x + y = 10$$
$$-1(x + y) = -1(11)$$

$$\begin{array}{rrcrcr} & x & + & y & = & 10 \\ + & -x & - & y & = & -11 \\ \hline & 0 & & & = & -1 \end{array}$$

Since 0 does not equal –1, the original system had no solution.

The system of equations

$$x + y = 10$$
$$2x + 2y = 20$$

also has something unusual occur.

$$-2(x + y) = -2(10)$$
$$2x + 2y = 20$$

$$\begin{array}{rrcrcr} & -2x & - & 2y & = & -20 \\ + & 2x & + & 2y & = & 20 \\ \hline & 0 & & & = & 0 \end{array}$$

Since 0 does equal 0, the original system has an infinite number of solutions. Any ordered pair that satisfies $x + y = 10$, like (2, 8) will also satisfy $2x + 2y = 20$, $2(2) + 2(8) = 4 + 16 = 20$.

MATH FACTS

If when doing the process to eliminate a variable, both variables get eliminated, there are either no solutions or an infinite number of solutions. If the equation that is left is false saying that 0 is equal to something other than zero, then there are no solutions. If the equation that is left is 0 = 0, there are an infinite number of solutions—any pair that satisfies the top equation will satisfy the bottom equation too.

Check Your Understanding of Section 4.3

A. Multiple-Choice

1. The system of equations

$$5x - 4y = -3$$
$$3x + 2y = 7$$

has the same solution as the system

(1) $5x - 4y = -3$
$6x + 4y = 7$

(2) $5x - 4y = -3$
$6x + 2y = 14$

(3) $5x - 4y = -3$
$6x + 4y = 14$

(4) $5x - 4y = -3$
$9x + 6y = 14$

2. In order to eliminate the y from this system of equations,

$$3x + 12y = 21$$
$$6x + 4y = 22$$

you could
(1) Multiply both sides of the first equation by –2.
(2) Multiply both sides of the second equation by –3.
(3) Multiply both sides of the first equation by –3.
(4) Multiply both sides of the second equation by 1/2.

3. In order to eliminate the x from this system of equations,

$$12x - 3y = 21$$
$$-2x + 6y = 2$$

you could
(1) Multiply both sides of the first equation by 2.
(2) Multiply both sides of the second equation by 6.
(3) Multiply both sides of the first equation by –2.
(4) Multiply both sides of the second equation by 1/2.

4. Which step would not cause a variable to be eliminated after adding the new equations together on this system?

$$3x - 6y = 6$$
$$9x + 2y = 38$$

(1) Multiply both sides of the first equation by –3.
(2) Multiply both sides of the second equation by +3.
(3) Multiply both sides of the first equation by +1/3.
(4) Multiply both sides of the first equation by 4.

5. One way to eliminate the y from this system of equations

$$3x + 4y = 16$$
$$2x - 5y = 3$$

is to
(1) Multiply both sides of the first equation by –2 and both sides of the second equation by +3.
(2) Multiply both sides of the first equation by +4 and both sides of the second equation by –5.
(3) Multiply both sides of the first equation by +5 and both sides of the second equation by +4.
(4) Multiply both sides of the first equation by +2 and both sides of the second equation by –3.

6. Solve the system of equations:

$$8x - 2y = 28$$
$$4x + 3y = 6$$

(1) (3, 2) (2) (–3, 2) (3) (–3, –2) (4) (3, –2)

7. Solve the system of equations:

$$4x - 3y = 31$$
$$3x + 5y = 16$$

(1) (7, –1) (2) (7, 1) (3) (–7, –1) (4) (–7, 1)

8. Solve the system of equations:

$$x - 3y = 14$$
$$4x + 5y = 22$$

(1) (8, 2) (2) (–8, –2) (3) (–8, 2) (4) (8, –2)

9. Solve the system of equations:

$$3x - 2y = -4$$
$$5x - 4y = -12$$

(1) (4, 8) (2) (4, –8) (3) (–4, 8) (4) (–4, –8)

10. The system of equations

$$2x - 7y = 2$$
$$6x - 21y = 7$$

has
(1) Exactly one solution.
(2) Every ordered pair possible satisfies this equation.
(3) Every ordered pair that satisfies the first equation also satisfies the whole system.
(4) No ordered pairs that solve the system.

B. *Show how you arrived at your answers.*

1. For the system

$$2x + 5y = 17$$
$$6x + 7y = 27$$

Samuel wants to eliminate the y, but Georgia wants to eliminate the x. Which student's choice will require less work?

2. Solve for x and y in this system

$$x + y = 40$$
$$2x + 4y = 136$$

3. Aubree solves the system of equations

$$2x + 3y = 21$$
$$4x + 6y = 42$$

and gets the solution

$$0 = 0$$

She says this means that every ordered pair is a solution to the system of equations. Is this accurate? Explain.

4. If it is possible, find an ordered pair that is a solution to the system

$$5x - 2y = 5$$
$$10x - 4y = 11$$

If not, explain why.

5. For the system of equations

$$4x - 3y = 18$$
$$2x + 12y = 36$$

Victoria starts by multiplying both sides of the second equation by –2. Porter starts by multiplying both sides of the second equation by +2. How is it that either of these methods will lead to the correct solution?

4.4 SOLVING WORD PROBLEMS WITH SYSTEMS OF EQUATIONS

KEY IDEAS

Word problems requiring this simultaneous equation process are common in the free response parts of this test. This is generally a two-part process: (1) setting up the system of equations and (2) solving the system of equations.

Setting Up the System of Equations

A typical word problem that can be solved with systems of equations is: A restaurant sells hamburgers and hot dogs. One customer purchases 3 hot dogs and 5 hamburgers and the total bill is \$37. Another customer purchases 5 hot dogs and 2 hamburgers and the total bill is \$30. What is the cost of one hot dog and the cost of one hamburger?

Let x be the price of one hot dog and y be the price of one hamburger. The cost of 3 hotdogs is $3x$ dollars. The cost of 5 hamburgers is $5y$ dollars. So the cost of 3 hot dogs and 5 hamburgers is $3x + 5y$ dollars. The cost of 5 hot dogs and 2 hamburgers is $5x + 2y$ dollars.

Since the question says that 3 hot dogs and 5 hamburgers cost \$37, the first equation is

$$3x + 5y = 37$$

Since the question says that 5 hot dogs and 2 hamburgers cost \$30, the second equation is

$$5x + 2y = 30$$

Solve the system of equations.

$$\begin{aligned} 3x + 5y &= 37 \\ 5x + 2y &= 30 \end{aligned}$$

$$\begin{aligned} -2(3x + 5y) &= -2(37) \\ 5(5x + 2y) &= 5(30) \end{aligned}$$

$$\begin{array}{rrcrcr} & -6x & - & 10y & = & -74 \\ + & 25x & + & 10y & = & 150 \\ \hline & \frac{19x}{19} & & & = & \frac{76}{19} \\ & x & & & = & 4 \end{array}$$

$$\begin{aligned} 3(4) + 5y &= 37 \\ 12 + 5y &= 37 \\ -12 &= -12 \\ 5y &= 25 \\ y &= 5 \end{aligned}$$

So the cost of a hot dog is \$4, and the cost of a hamburger is \$5.

Example

The cost of a movie ticket for a child is \$7, while the cost of a movie ticket for an adult is \$11. A group of 23 people, some children and some adults, go to the movies. The total cost of all the tickets is \$197. How many child tickets and how many adult tickets were purchased?

Solution: If there are x children and y adults, the total number of people is $x + y$. The cost for x child tickets is $7x$ dollars. The cost for y adult tickets is $11y$ dollars. The total cost is $7x + 11y$ dollars.

The two equations are

$$\begin{aligned} x + y &= 23 \\ 7x + 11y &= 197 \end{aligned}$$

Multiply both sides of the top equation by -11

$$\begin{aligned} -11(x + y) &= -11(23) \\ 7x + 11y &= 197 \end{aligned}$$

$$\begin{array}{rrcrcr} & -11x & - & 11y & = & -253 \\ + & 7x & + & 11y & = & 197 \\ \hline & \frac{-4x}{-4} & & & = & \frac{-56}{-4} \\ & x & & & = & 14 \end{array}$$

$$14 + y = 23$$
$$-14 = -14$$
$$y = 9$$

There were 14 child tickets and 9 adult tickets purchased.

Check Your Understanding of Section 4.4

A. Multiple-Choice

1. Which system of equations can be used to model the following scenario? There are 50 animals. Some of the animals have 2 legs and the rest of them have 4 legs. In total there are 172 legs.
   (1) $x + y = 172$; $2x + 4y = 50$
   (2) $x + 50 = y$; $2x + 172 = 4y$
   (3) $y + 50 = x$; $4y + 172 = 2x$
   (4) $x + y = 50$; $2x + 4y = 172$

2. Which system of equations could be used to model the following scenario? There are 8 people in an elevator. Some are adults and the rest are children. Each adult weighs 150 pounds. Each child weighs 50 pounds. The total weight of the 8 people is 800 pounds.
   (1) $x + y = 8$; $150x + 50y = 800$
   (2) $x + y = 800$; $150x + 50y = 50$
   (3) $x + 50 = y$; $150x + 800 = 50y$
   (4) $y + 50 = x$; $50y + 800 = 150x$

3. Which system of equations could be used to model the following scenario? There are 20 coins. Some are quarters and the rest are dimes. The quarters are worth 25 cents each and the dimes are worth 10 cents each. The total value of the 20 coins is \$2.90.
   (1) $x + y = 20$; $25x + 10y = 290$
   (2) $x + y = 20$; $25x + 10y = 2.90$
   (3) $x + y = 290$; $25x + 10y = 20$
   (4) $x + y = 2.90$; $25x + 10y = 20$

4. A pet store has 30 animals. Some are cats and the rest are dogs. The cats cost \$50 each. The dogs cost \$100 each. If the total cost for all 30 animals is \$1900, how many cats are there?
   (1) 8 (2) 20 (3) 22 (4) 24

5. A restaurant sells only two desserts, pie and cake. A piece of pie costs \$4. A piece of cake costs \$5. The restaurant sells 100 desserts, which cost a total of \$473. How many pieces of pie did they sell?
(1) 73 (2) 27 (3) 18 (4) 11

6. A fence is put around a rectangular plot of land. The perimeter of the fence is 28 feet. Two of the opposite sides of the fence cost \$10 per foot. The other two sides cost \$12 per foot. If the total cost of the fence is \$148, what are the dimensions of the fence?
(1) 8 by 20
(2) 4 by 10
(3) 3 by 11
(4) 2 by 12

7. One number is five bigger than another number. When three times the larger number is added to twice the smaller number, the result is 60. What are the two numbers?
(1) 9 and 14
(2) 10 and 15
(3) 11 and 16
(4) 12 and 17

8. Peanuts cost \$7 a pound. Cashews cost \$9 a pound. There is a mixture of peanuts and cashews that weighs 10 pounds and costs \$84. How many pounds of peanuts are there in the mixture?
(1) 1 (2) 2 (3) 3 (4) 4

9. The current in a river makes boats going upstream slower by y miles per hour and boats going downstream faster by y miles per hour. Upstream boats go 3 miles per hour and downstream boats go 13 miles per hour. Which system of equations can be used to find the speed of the boat if it were in still water?

(1) $x + y = 13$
$x - y = 3$

(2) $x + y = 3$
$x - y = 13$

(3) $xy = 13$
$\frac{x}{y} = 3$

(4) $x + 2y = 13$
$x - 3y = 3$

10. A basketball player made some shots worth two points and the rest worth three points. If she made 16 shots for a total of 38 points, how many of each type did she make?
(1) 5 three pointers and 11 two pointers
(2) 6 three pointers and 10 two pointers
(3) 7 three pointers and 9 two pointers
(4) 8 three pointers and 8 two pointers

B. *Show how you arrived at your answers.*

1. Red roses cost \$3 each. Pink roses cost \$2 each. A man buys 24 flowers for his wife with some pink roses and the rest red roses. The total cost of the flowers is \$68. Write a system of equations to model this situation and use the equation to determine how many red and how many pink roses he bought.

2. Ice cream cones cost \$4 each. Milkshakes cost \$6 each. If ten items are purchased, either ice cream cones or milkshakes: (a) What is the greatest amount of money that can be spent? (b) What is the least amount of money that can be spent? (c) How much would it cost for 5 cones and 5 shakes? (d) If \$52 is spent on ten items, how many cones were purchased and how many shakes? Explain how you got your answer to part d.

3. Marshmallows cost \$2 an ounce and have 500 calories in an ounce. Cookies cost \$3 an ounce and have 400 calories in an ounce. A mixture of marshmallows and cookies costs \$29 and has 4,800 calories. Create a system of equations to model this scenario. How many ounces of marshmallows are there in the mixture?

4. Three burgers and two orders of French fries cost \$24. Five burgers and one order of French fries cost \$33. What is the cost of one burger? What is the cost of one order of French fries?

5. A cycling store sells two-wheel bicycles and three-wheel tricycles. It sells 58 cycles that had a total of 134 wheels. How many of each type of cycle did they sell?

Chapter Five GRAPHS OF SOLUTION SETS OF LINEAR EQUATIONS

5.1 PRODUCING A GRAPH BY IDENTIFYING TWO OR MORE POINTS

A linear equation, like $x + y = 10$, is one where neither of the variables has an exponent greater than or equal to 2. The set of ordered pairs that makes this equation true include (2, 8), (3, 7), and (9, 1). If each ordered pair that satisfies an equation is plotted as a point on the *coordinate plane*, the result is the *graph of the solution set of the equation.* For linear equations, the graph is always a line so getting two points in the solution set and drawing the line that passes through those two points will create the graph of the solution set for that line.

Graphing the Solution Set of a Linear Equation by Making a Table of Values

The equation $x + y = 10$ has an infinite number of ordered pairs that satisfy it. One way to organize the information before creating a graph is to make a table of values.

This is a chart with three ordered pairs satisfying the equation $x + y = 10$. For a linear equation, only two ordered pairs are needed, but it is wise to do an extra ordered pair in case one of your first two is incorrect.

x	y
2	8
3	7
9	1

Plot the ordered pair (2, 8) on the coordinate plane by locating the point that is two units to the right of the y-axis and 8 units above the x-axis. One way to do this is to start at the origin point where the two axes intersect and move to the right two units from there and then up 8 units.

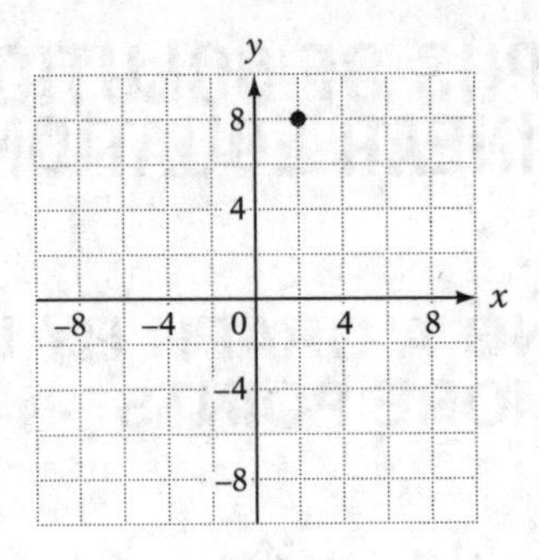

Do the same for the other two ordered pairs on the chart (3, 7) and (9, 1).

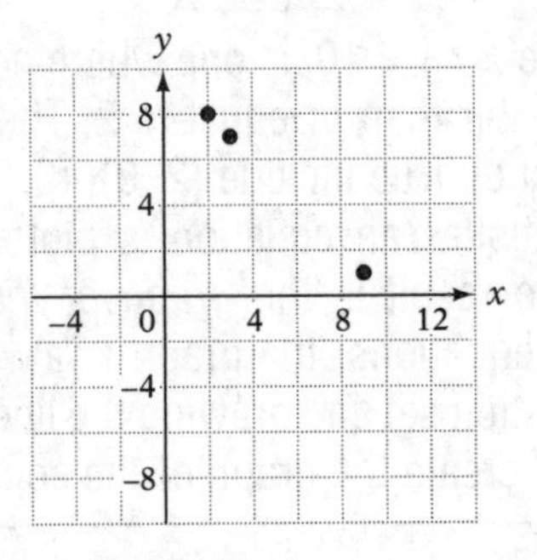

Draw a line through the three points. If the three points do not all lie on the same line, one of your ordered pairs is incorrect.

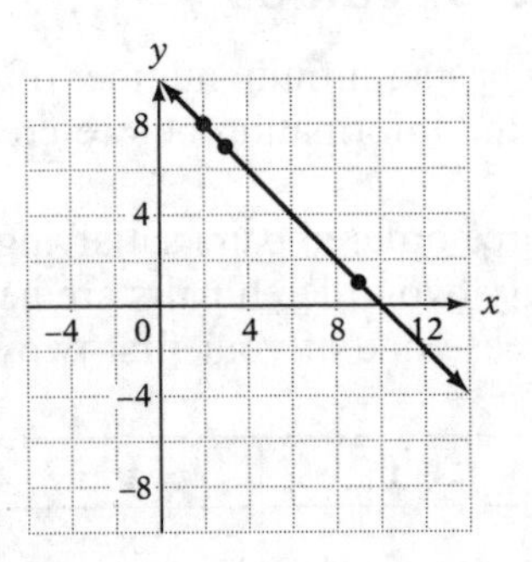

MATH FACTS

The line on a graph contains an infinite number of points. Each point corresponds to an ordered pair that is part of the solution set for the equation, and each ordered pair that is part of the solution set for the equation corresponds to a point on the line.

Example 1

Fill in this chart with three ordered pairs that satisfy the equation $2x + y = 10$. Then graph the three points that correspond to those three ordered pairs and use them to create the graph of the solution set of $2x + y = 10$.

x	*y*

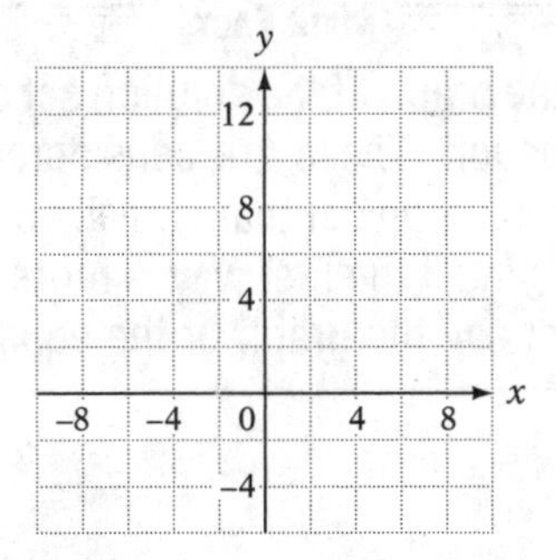

Solution: One way to determine ordered pairs that satisfy an equation is to pick small values for x and then substitute those values into the equation and use algebra to solve for the corresponding y value that would make the equation true.

For $x = 0$, the equation becomes $2(0) + y = 10$. When you solve for y, it becomes $y = 10$. One ordered pair is (0, 10).

For $x = 1$, the equation becomes $2(1) + y = 10$. When you solve for y, it becomes $y = 8$. Another ordered pair is (1, 8).

For $x = 2$, the equation becomes $2(2) + y = 10$. When you solve for y, it becomes $y = 6$. A third ordered pair is (2, 6).

x	*y*
0	10
1	8
2	6

Plot the three points and draw a line through them to complete the graph.

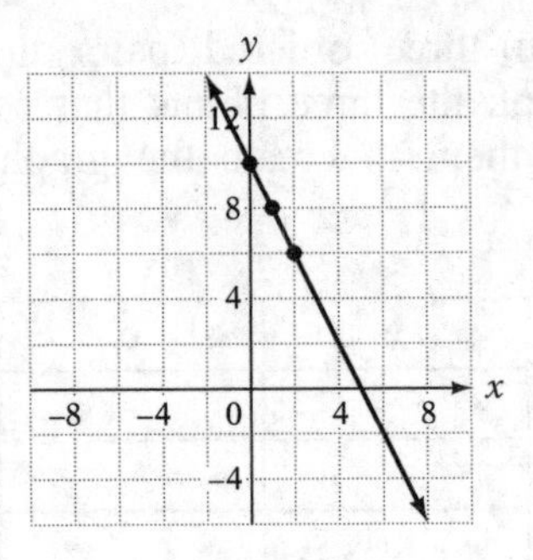

MATH FACTS

It is common to call the graph of the solution set of an equation simply the graph of the equation. These are considered to mean the same thing though it is the solution set that is really being graphed and not the equation itself. Saying it the first way is more accurate, but usually on a test it would just say the graph of the equation.

Example 2

Will the point (9, 2) be on the line that is the graph of the solution set of the equation $x - 2y = 5$?

Solution: If an ordered pair is a solution to the equation, then the point that represents that ordered pair will be on the graph of the solution set. By substituting 9 for x and 2 for y into the equation, it becomes

$$9 - 2(2) \stackrel{?}{=} 5$$
$$9 - 4 \stackrel{?}{=} 5$$
$$5 \stackrel{\checkmark}{=} 5$$

So, (9, 2) is in the solution set for $x - 2y = 5$ and the point (9, 2) will be on the graph of the solution set.

Example 3

Below is the graph of the solution set for an equation. Based on this graph, is the ordered pair (4, 7) a solution to the equation?

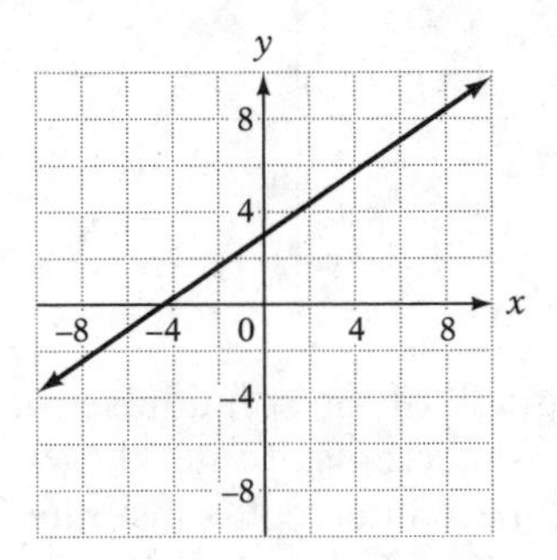

Solution: If you plot the point (4, 7) on the same coordinate plane, it is close to being on the line, which is the graph of the solution set. If it were on the line, (4, 7) would be a solution to the equation, but since it is not, (4, 7) is not a solution to the equation.

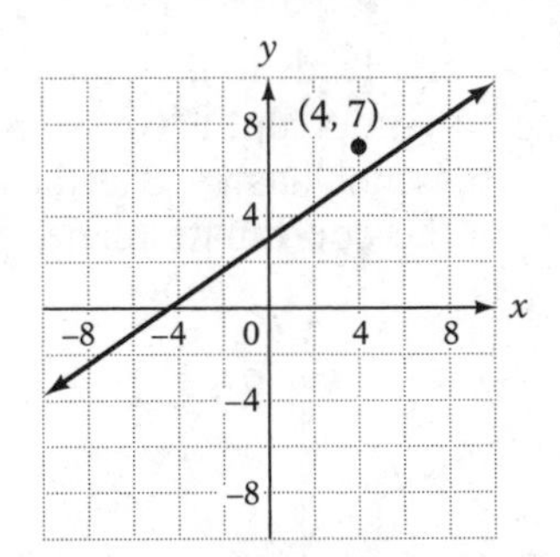

Graphing the Solution Set of Linear Equations Using the Two-Intercept Method

MATH FACTS

Unless a line is vertical or horizontal, it will cross both the x-axis and the y-axis. The point where the line crosses the x-axis is called the *x-intercept* and the point where it crosses the y-axis is called the *y-intercept.* Any point on the y-axis has an x-coordinate of 0 and any point on the x-axis has a y-coordinate of 0. In the graph below, the x-intercept of the line is (5, 0) and the y-intercept is (0, 8).

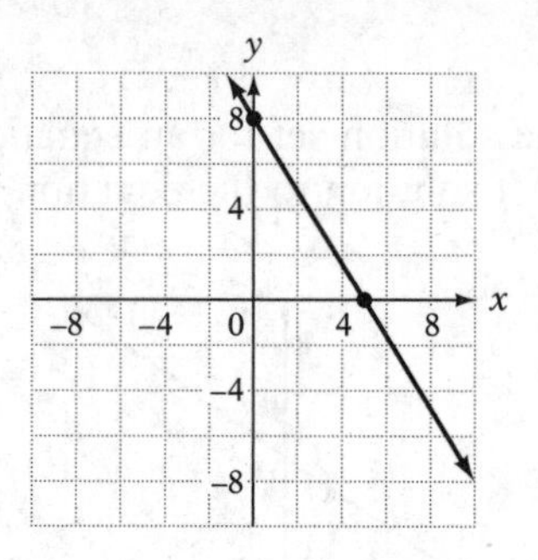

A quick way to make a graph of the solution set of a two variable equation is to calculate the x and y-intercepts. To get the x-intercept, substitute 0 for y and solve for x. To get the y-intercept, substitute 0 for x and solve for y.

Example 4

Find the x-intercept and y-intercept of the graph of the solution set of $2x + 4y = 16$. Use the intercepts to sketch the graph.

Solution: Substituting 0 for x leads to the equation $2(0) + 4y = 16$, which has $y = 4$ as the solution. The y-intercept is (0, 4). Substituting 0 for y leads to the equation $2x + 4(0) = 16$, which has $x = 8$ as the solution. The x-intercept is (8, 0). Plot both points on the coordinate plane, and draw a line through them.

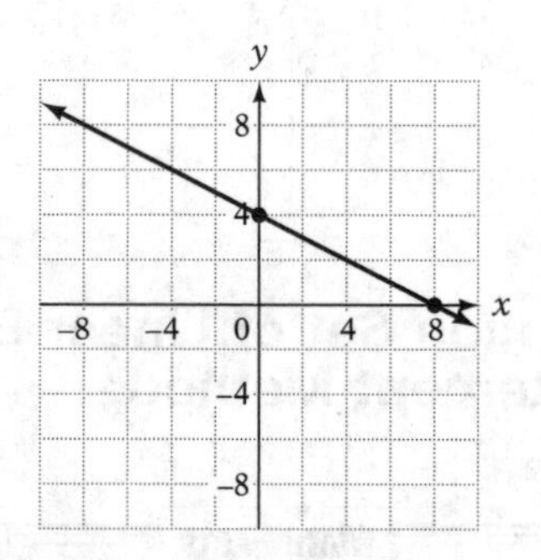

Equations for Horizontal or Vertical Lines

An equation with one variable can also have a solution set of ordered pairs. One ordered pair that satisfies the equation $y = 3$ is (0, 3) since it is true that $3 = 3$ if you substitute the 3 for the y (there is no x to substitute with the 0!). Other ordered pairs that satisfy this equation are (1, 3), (7, 3), and (29, 3). When all the ordered pairs that satisfy the equation $y = 3$ are graphed, it becomes a horizontal line with y-intercept of (0, 3).

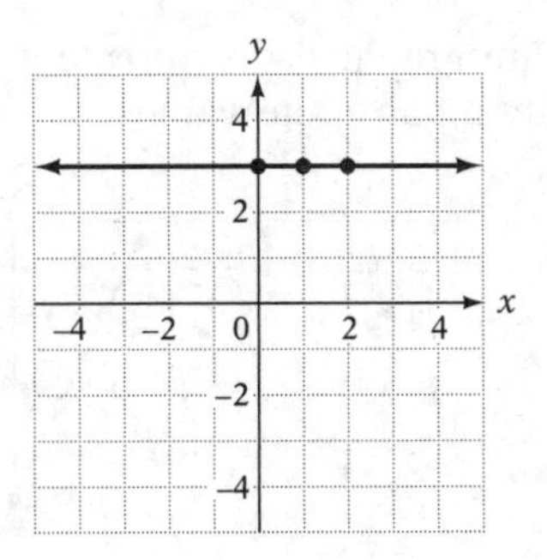

For an equation with only an x variable, like $x = 5$, some ordered pairs that satisfy it are (5, 0), (5, 1), and (5, 2). When these are graphed, they make a vertical line.

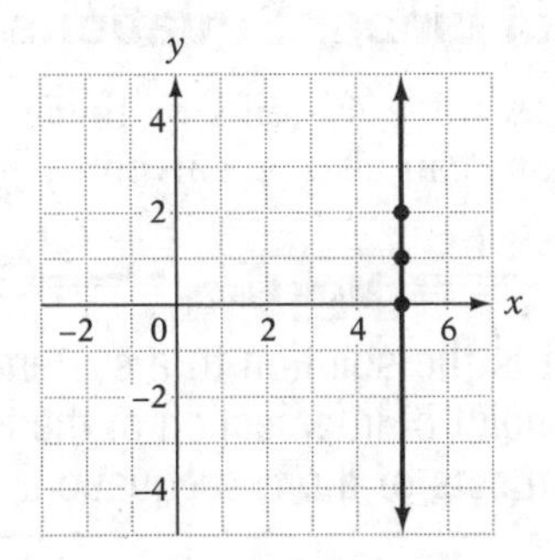

Graphing Linear Equations Involving Absolute Value

The *absolute value* of a number is defined as the distance that number is from zero on the number line. The symbol for absolute value of a is $|a|$. The absolute value of 5 is 5. The absolute value of –5 is also 5.

A two variable equation can have absolute values like $y = |2x - 2|$. One way to graph the solution set is to create a chart.

x	$\|2x - 2\|$
–2	$\|2(-2) - 2\| = \|-4 - 2\| = \|-6\| = 6$
–1	$\|2(-1) - 2\| = \|-2 - 2\| = \|-4\| = 4$
0	$\|2(0) - 2\| = \|0 - 2\| = \|-2\| = 2$
1	$\|2(1) - 2\| = \|2 - 2\| = \|0\| = 0$
2	$\|2(2) - 2\| = \|4 - 2\| = \|2\| = 2$
3	$\|2(3) - 2\| = \|6 - 2\| = \|4\| = 4$
4	$\|2(4) - 2\| = \|8 - 2\| = \|6\| = 6$

When these five points are graphed, they create a V shape. This is typical for linear equations involving absolute values.

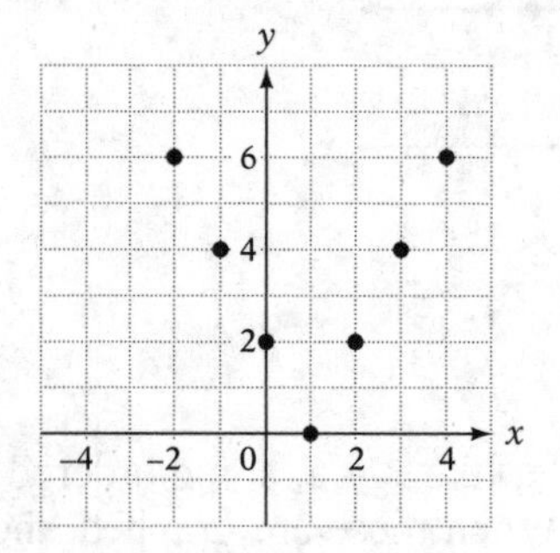

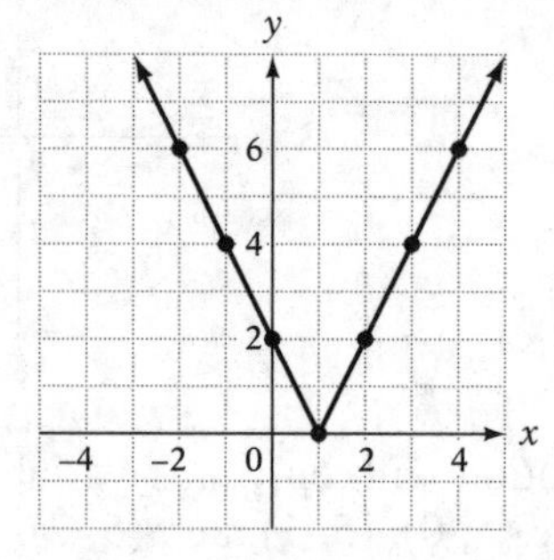

Solving Systems of Linear Equations by Graphing

Chapter 4 showed how to solve systems of linear equations with algebra. Systems of linear equations can also be solved by graphing.

MATH FACTS

The ordered pair that is the solution to a system of linear equations will be related to the point of intersection of the two lines that are the graphs of the solution sets of the two equations.

In Chapter 4, there was an example where you had to solve the system of equations.

$$x + y = 10$$
$$x - y = 2$$

Using algebra, the solution was (6, 4).

This system can also be solved by graphing the solution sets of each equation and locating the intersection point.

Find the intercepts:

$$x + y = 10$$
$$x\text{-intercept: } x + 0 = 10, x = 10$$
$$y\text{-intercept: } 0 + y = 10, y = 10$$

So the intercepts for the top equation are (10, 0) and (0, 10).

$$x - y = 2$$
$$x\text{-intercept: } x - 0 = 2, x = 2$$
$$y\text{-intercept: } 0 - y = 2, y = -2$$

So, the intercepts for the bottom equation are (2, 0) and (0, –2).

Plot the intercepts on graph paper and draw the lines carefully with a ruler.

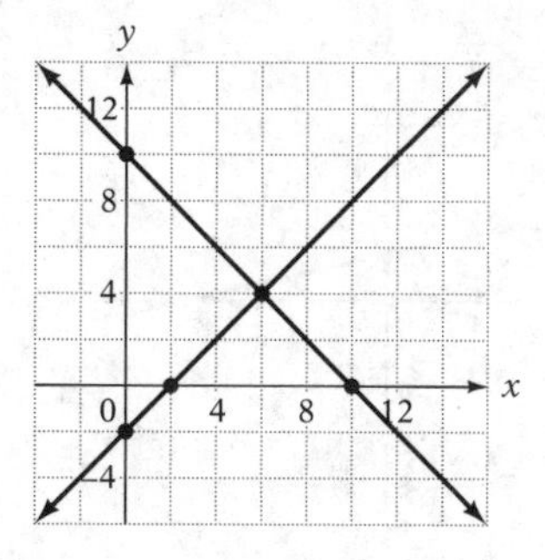

The lines intersect at the point (6, 4) so the solution to the system of equations is $x = 6$ and $y = 4$.

MATH FACTS

Solving with algebra is generally the faster and more accurate method for solving a system of linear equations. But this is a technique you need for systems of linear inequalities, coming later.

Check Your Understanding of Section 5.1

A. *Multiple-Choice*

1. Which are the coordinates of a point that will be on the line that represents the solution set of the equation $2x + 3y = 12$?
(1) (4, 1) (2) (2, 3) (3) (5, 0) (4) (3, 2)

2. What is the x-intercept of the graph of the solution set of the equation $2x + 5y = 20$?
(1) (10, 0) (2) (4, 0) (3) (0, 10) (4) (0, 4)

3. What is the y-intercept of the graph of the solution set of the equation $3x - 8y = 24$?
(1) (0, –3) (2) (0, 3) (3) (8, 0) (4) (0, 8)

4. The point (2, k) is on the graph of the solution set of the equation $3x + y = 15$. What is the value of k?
(1) 7 (2) 8 (3) 9 (4) 10

5. Based on the x-intercept and y-intercept, this is the graph of the solution set of which equation?

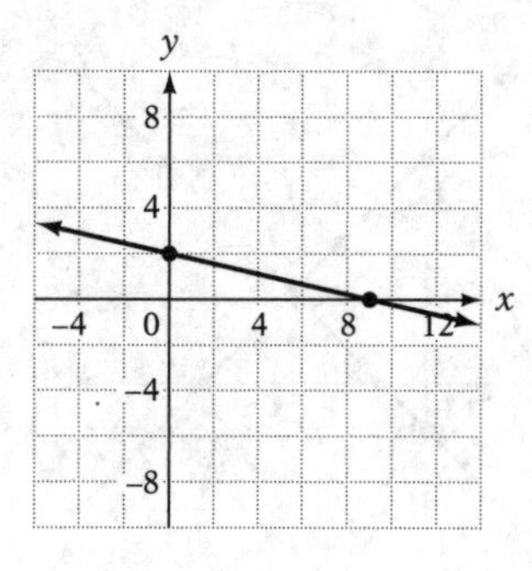

(1) $9x + 2y = 18$
(2) $2x + 9y = 18$
(3) $9x - 2y = 18$
(4) $2x - 9y = 18$

6. Which is a graph of the solution set of the equation $y = 5$?

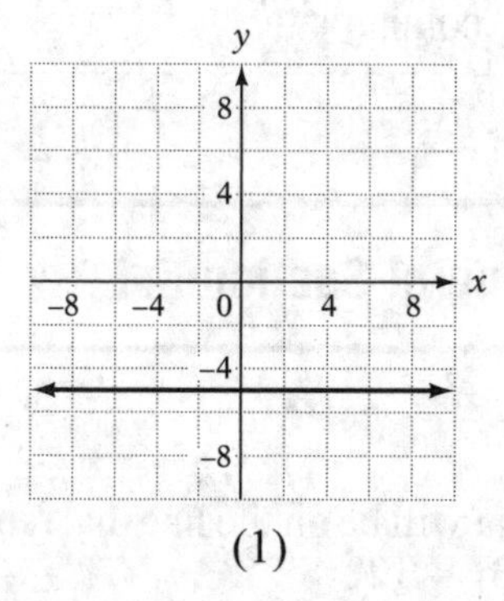

(1)

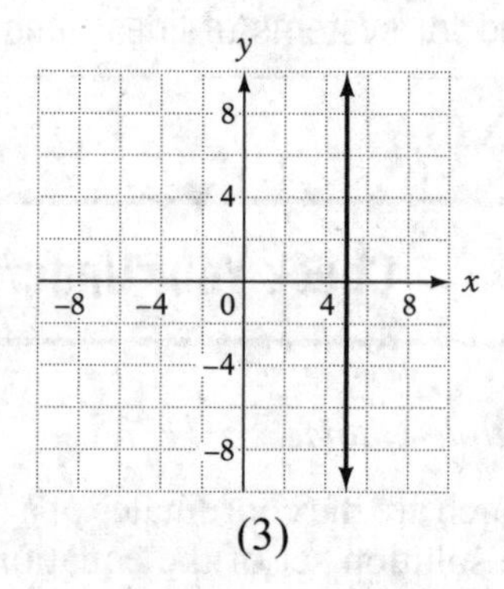

(3)

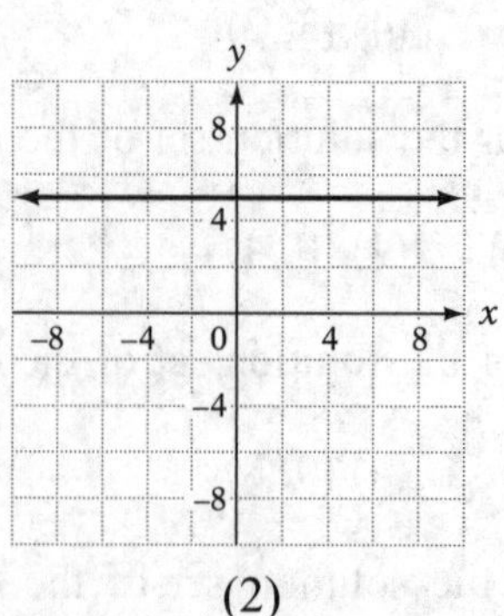

(2)

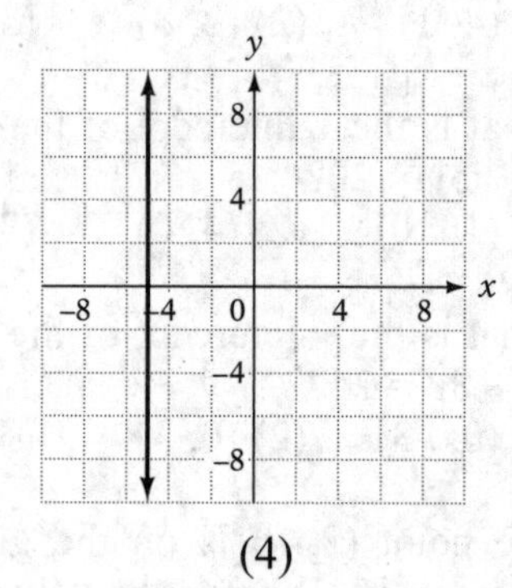

(4)

7. This is a graph of the solution set of which equation?

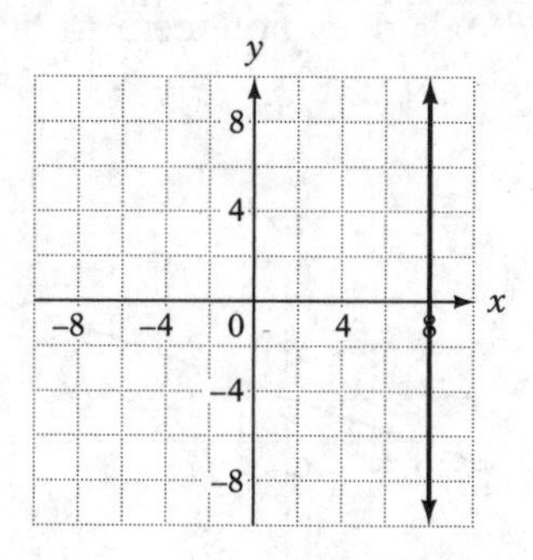

(1) $x = 8$
(2) $x = -8$
(3) $y = 8$
(4) $y = -8$

8. The equations $2x - 3y = 9$ and $3x + 2y = 20$ are graphed below. What is the solution to the system of equations?

$$2x - 3y = 9$$
$$3x + 2y = 20$$

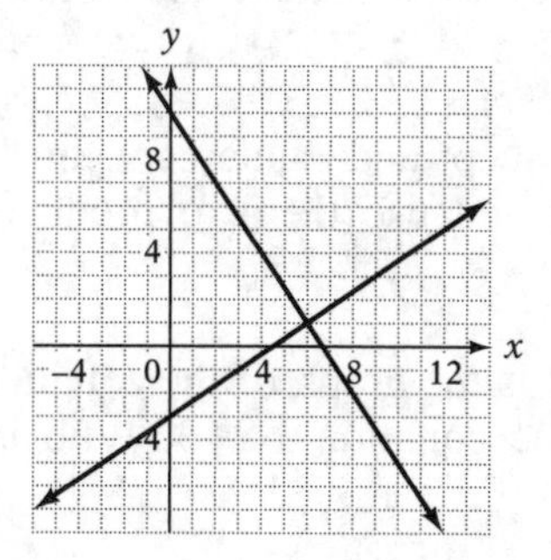

(1) (5, 2)
(2) (2, 5)
(3) (1, 6)
(4) (6, 1)

9. Below is the graph of the solution set of an equation. Based on this graph, which ordered pair does not seem to be part of the solution set of the equation $y = \frac{1}{3}x + 2$?

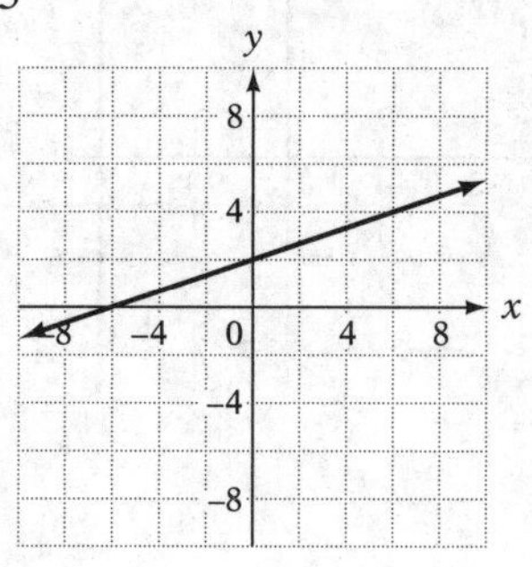

(1) (3, 3) (2) (6, 4) (3) (6, 8) (4) (9, 5)

10. What is the equation of the x-axis?
(1) $x = 0$ (2) $y = 0$ (3) $x + y = 0$ (4) $x - y = 0$

B. *Show how you arrived at your answers.*

1. Make a table of values to graph the solution set of the equation $y + 2x = 8$.

2. Determine the x-intercept and y-intercept for the graph of the solution set of this equation and use them to produce a sketch of the graph $4x - 6y = 24$.

3. Tenley says that this is the graph of the equation $y = 3$. Ingrid says that it is the graph of the equation $x = 3$. Who is correct? Explain your reasoning.

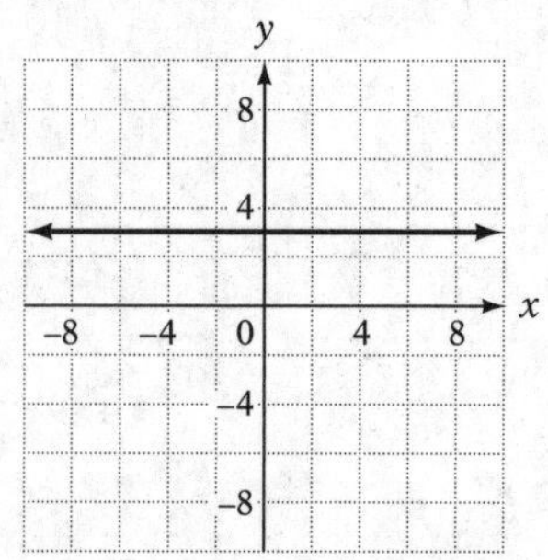

4. Graphically solve the system of equations

$$x - y = 4$$
$$3x + 5y = 20$$

5. Below is the graph of $2x + y = k$ with intercepts at $(0, 8)$ and $(4, 0)$. What must the value of k be?

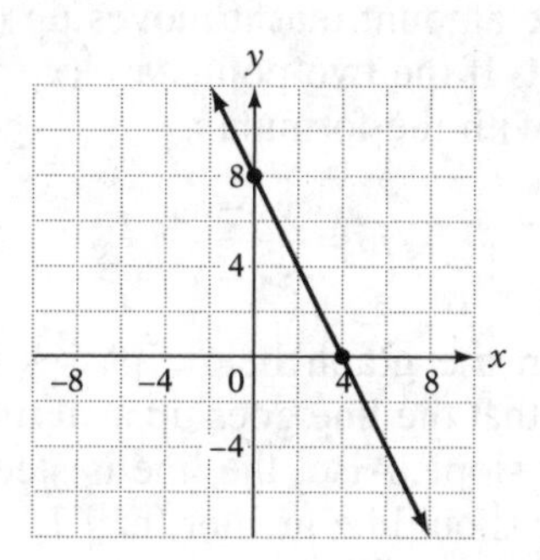

5.2 CALCULATING AND INTERPRETING SLOPE

The *slope* of a line is a number that measures how steep it is. A horizontal line has a slope of 0. A line with a positive slope goes up as it goes to the right. A line with a negative slope goes down as it goes to the right. The variable used for slope is the letter *m*.

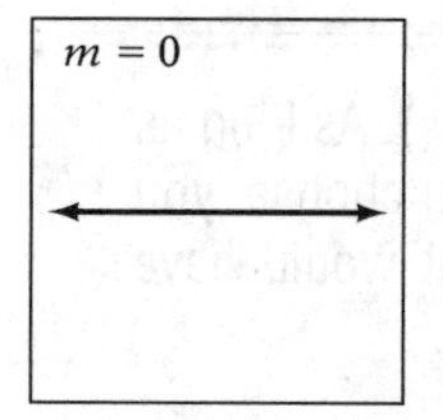

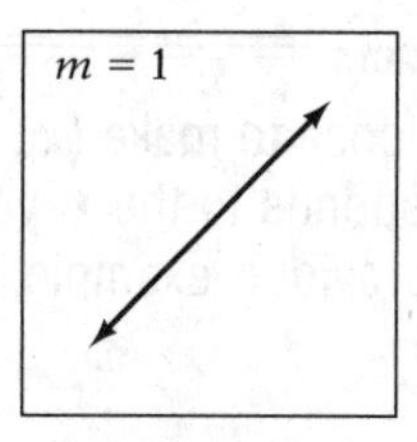

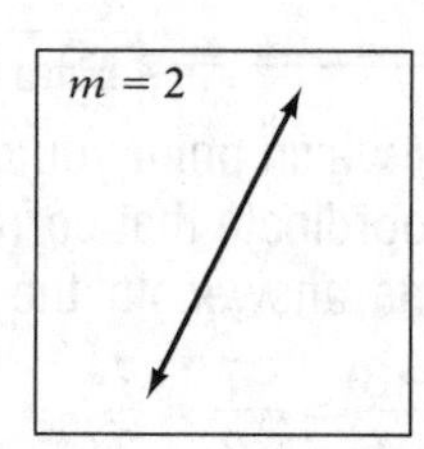

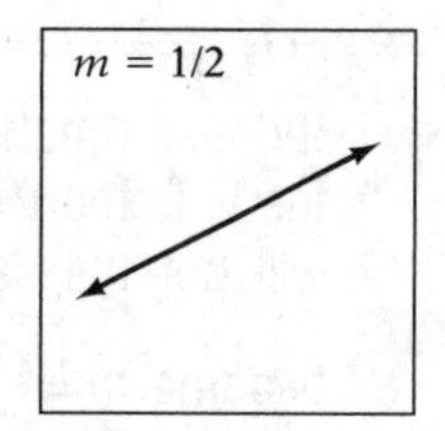

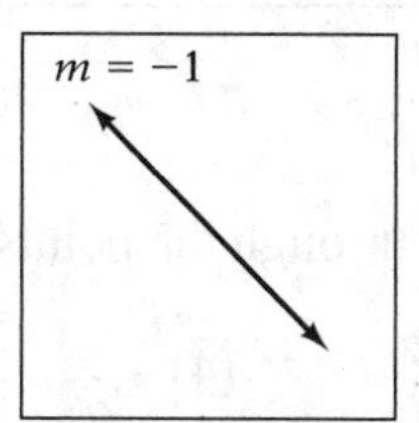

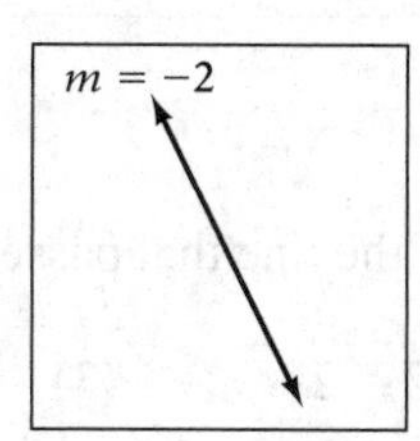

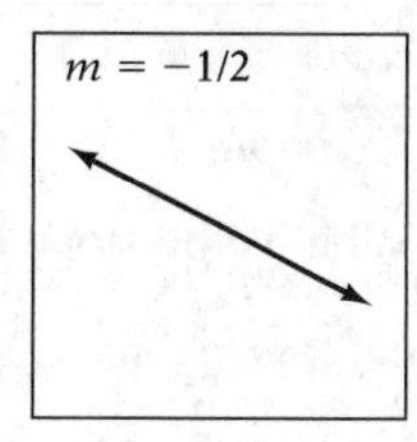

Calculating the Slope of a Line

The slope of a line is the amount that it moves up (or down) for every one unit it moves to the right. If the two points are (x_1, y_1) and (x_2, y_2), then the slope can be calculated with the formula

$$m = \frac{y_2 - y_1}{x_2 - x_1}$$

For example, the line in the graph below passes through the two points (3, 1) and (6, 8). Notice that the line goes up as it moves to the right, so you should expect a positive slope. Also, the line is steeper than the line with a slope of 1 so the answer should be greater than 1.

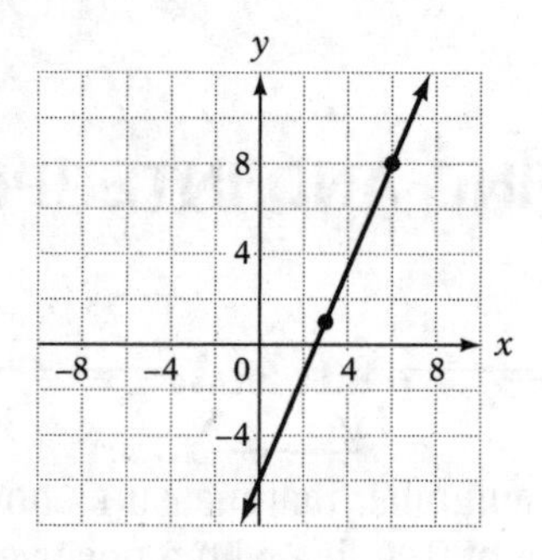

Substituting into the slope formula $x_1 = 3, y_1 = 1, x_2 = 6$, and $y_2 = 8$, you get $m = \frac{8-1}{6-3} = \frac{7}{3}$ so the slope is $\frac{7}{3}$.

MATH FACTS

It doesn't matter which point you choose to make (x_1, y_1). As long as the y_1 is the y-coordinate that corresponds to the x_1 you choose, you will get the same answer. In the previous example, it would have become $m = \frac{1-8}{3-6} = \frac{-7}{-3} = \frac{7}{3}$.

Example 1

What is the slope of the line that passes through the points (1, 2) and (9, –2)?

(1) $\frac{1}{2}$ (2) –2 (3) $-\frac{2}{3}$ (4) $-\frac{1}{2}$

Solution: A sketch of the graph indicates that the solution should be negative.

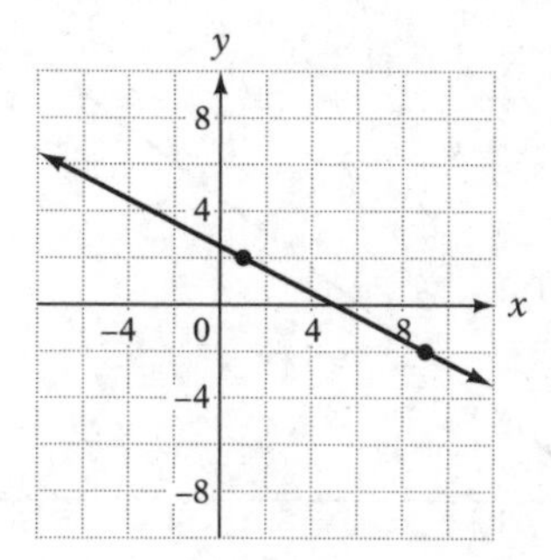

Using the formula, you get $m = \frac{-2-2}{9-1} = \frac{-4}{8} = -\frac{1}{2}$, choice (4).

MATH FACTS

When two lines have the same slope they are parallel. If the lines have different slopes, then they are not parallel. A special case of nonparallel lines are lines whose slopes have a product of −1. When this happens, the lines are perpendicular. When two fractions have a product of −1, one is the *negative reciprocal* of the other. For example, the negative reciprocal of $\frac{2}{3}$ is $-\frac{3}{2}$ and $\left(\frac{2}{3}\right)\left(-\frac{3}{2}\right) = -1$.

MATH FACTS

A horizontal line has a slope of 0, whereas the slope of a vertical line is said to be *undefined*. When using the slope formula for two points on a vertical line, like (5, 2) and (5, 3), it becomes $m = \frac{(3-2)}{(5-5)} = \frac{1}{0}$. Since division by 0 is undefined, the slope of the vertical line is undefined too.

Interpreting Slope in a Time/Distance Graph

If a car is traveling 30 miles per hour, it will have traveled 30 miles after 1 hour, 60 miles after 2 hours, and 90 miles after 3 hours. If you graph the points (1, 30), (2, 60), and (3, 90) on a coordinate axis with elapsed time as the horizontal axis and distance traveled as the vertical axis, it will look like this:

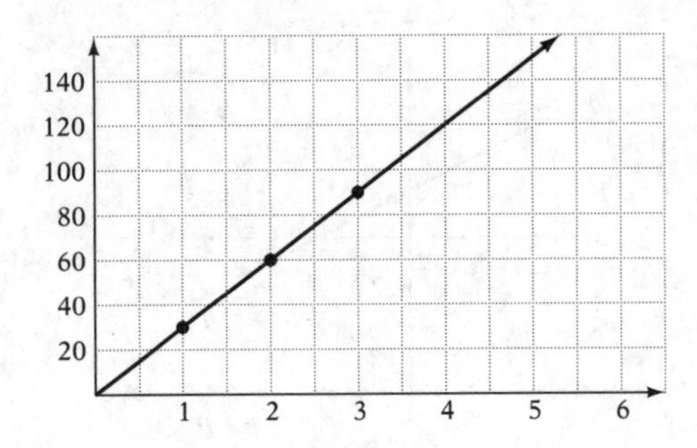

The slope of this line is positive since it goes up as it goes to the right. It isn't easy to estimate the slope since the unit size on the time axis is larger than the unit size on the distance axis. But the slope can still be calculated using any two points, like (1, 30) and (2, 60).

$m = \frac{60-30}{2-1} = \frac{30}{1} = 30$. The slope is 30 and the speed is 30 miles per hour.

MATH FACTS

In a distance time graph, the slope of a line segment corresponds to the speed of the thing that is moving.

Distance Time Graphs for Things That Are Changing Speeds

When a car is traveling at a constant speed, the distance time graph will be a line with the same slope on *every* interval. If the car changes speed, the slope of the distance time graph will also change.

In the graph below, the car has traveled 30 miles after 1 hour, 60 miles after 2 hours, and 140 miles after 4 hours.

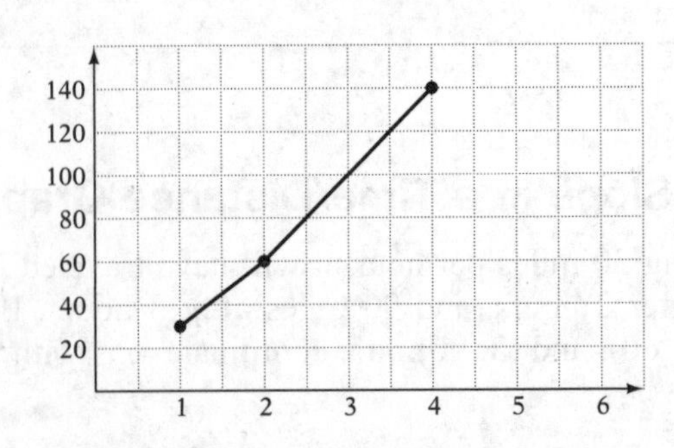

Since the line segment between (2, 60) and (4, 140) is steeper than the line segment between (1, 30) and (2, 60), the car was moving faster between hours 2 and 4 than it was between hours 1 and 2.

If necessary, you can also calculate the two different speeds by finding the slopes of the two segments.

$$m = \frac{60 - 30}{2 - 1} = \frac{30}{1} = 30 \text{ mph between hours 1 and 2.}$$

$$m = \frac{140 - 60}{4 - 2} = \frac{80}{2} = 40 \text{ mph between hours 2 and 4.}$$

Example 2

Below is the distance time graph for a 120 mile bicycle trip. During which interval was the average speed of the bicycle the least?

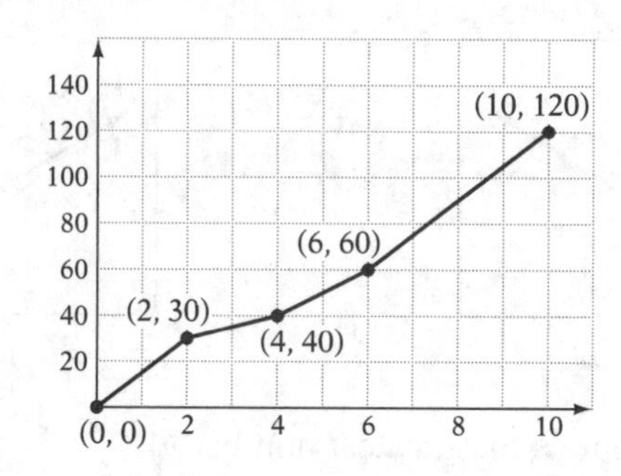

(1) hours 0 to 2
(2) hours 2 to 4
(3) hours 4 to 6
(4) hours 6 to 10

Solution: Between hours 2 and 4, the slope appears to be the least steep, which means the bicycle is going slowest in that interval. To make sure, you can calculate the slopes for each interval separately. Between hours 0 and 2, the speed was 15 mph, between hours 2 and 4 the speed was 5 mph, between hours 4 and 6 the speed was 10 mph, and between hours 6 and 10 the speed was 15 mph. The answer is choice (2).

Check Your Understanding of Section 5.2

A. Multiple-Choice

1. Which line has a positive slope?

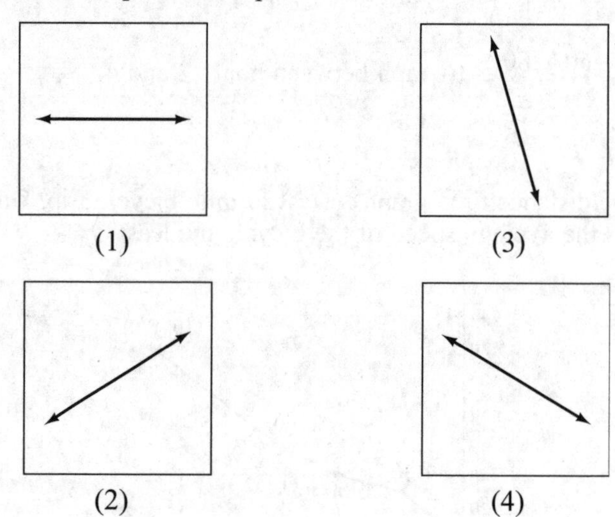

2. Which line's slope is the greatest number?

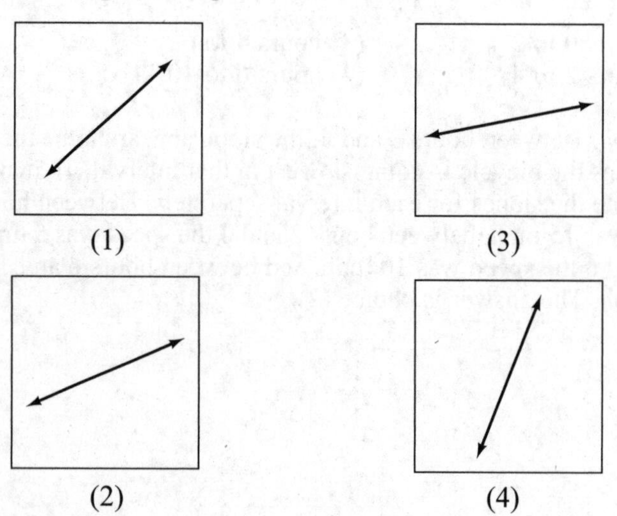

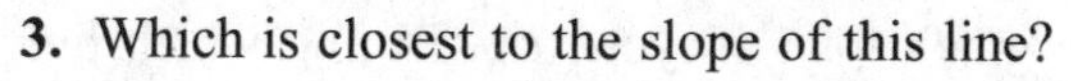

3. Which is closest to the slope of this line?

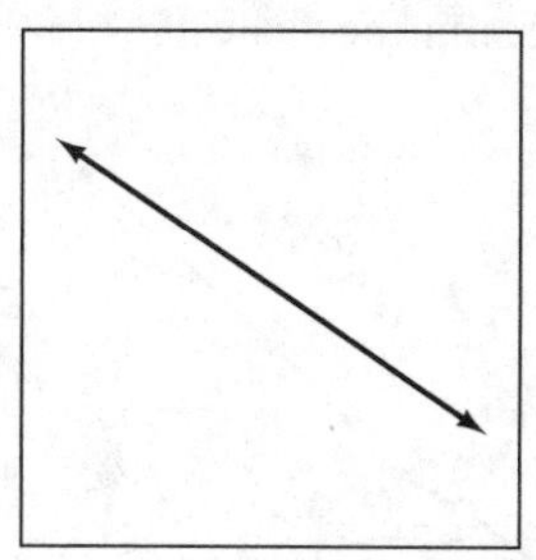

(1) $-\frac{2}{3}$ (2) -3 (3) $\frac{2}{3}$ (4) 3

4. What is the slope of the line that passes through $(-2, 1)$ and $(8, 5)$?

(1) $\frac{2}{5}$ (2) $-\frac{2}{5}$ (3) $\frac{5}{2}$ (4) $-\frac{5}{2}$

5. What is the slope of the line that passes through $(2, 8)$ and $(5, -1)$?

(1) $-\frac{1}{3}$ (2) $\frac{1}{3}$ (3) -3 (4) 3

6. A line with a slope of $\frac{3}{4}$ passes through the point $(2,1)$ and the point $(10, a)$. What must the value of a be?

(1) 7 (2) 8 (3) 9 (4) 10

7. Line 1 passes through the points $(-3, -2)$ and $(3, 8)$. Line 2 is parallel to line 1. What is the slope of line 2?

(1) $\frac{3}{5}$ (2) $\frac{5}{3}$ (3) $-\frac{3}{5}$ (4) $-\frac{5}{3}$

8. Below is a distance–time graph for a bicycle trip. During which time interval is the cyclist going the fastest?

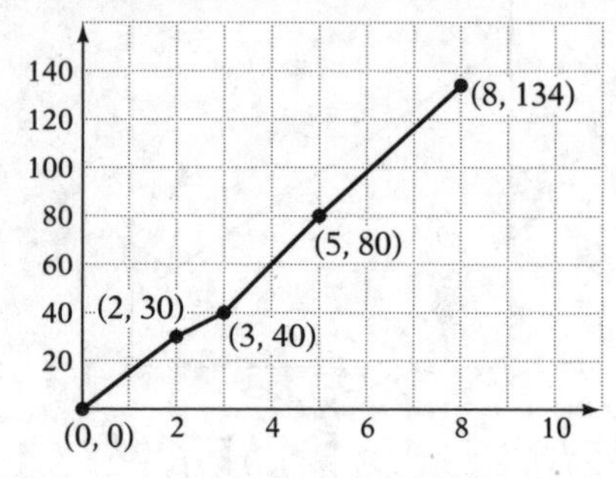

(1) 0 to 2 hours (3) 3 to 5 hours
(2) 2 to 3 hours (4) 5 to 8 hours

9. A line with a slope of 3 passes through the point (4, 1). Which point will this line *not* pass through?
(1) (5, 4) (2) (6, 7) (3) (7, 10) (4) (7, 11)

10. What is the slope of the hypotenuse of this right triangle?

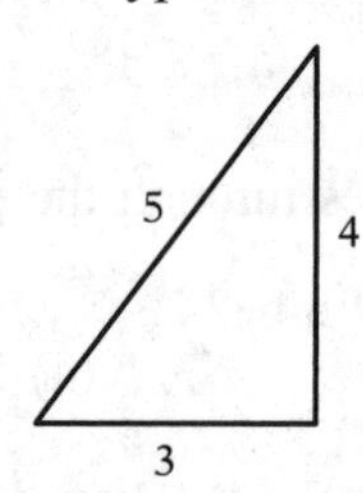

(1) $\frac{3}{4}$ (2) $\frac{4}{5}$ (3) $\frac{4}{3}$ (4) $\frac{3}{5}$

B. *Show how you arrived at your answers.*

1. Line 1 passes through the points (–3, 7) and (7, 1). Line 2 is perpendicular to line 1. What is the product of the slopes of line 1 and line 2?

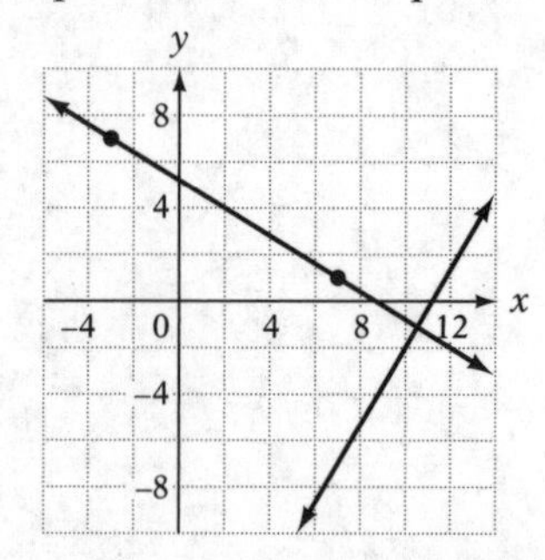

2. A triangle has vertex A at (0, 0), vertex B at (2, 5), and vertex C at (4, 5). Which side of the triangle has the greatest slope?

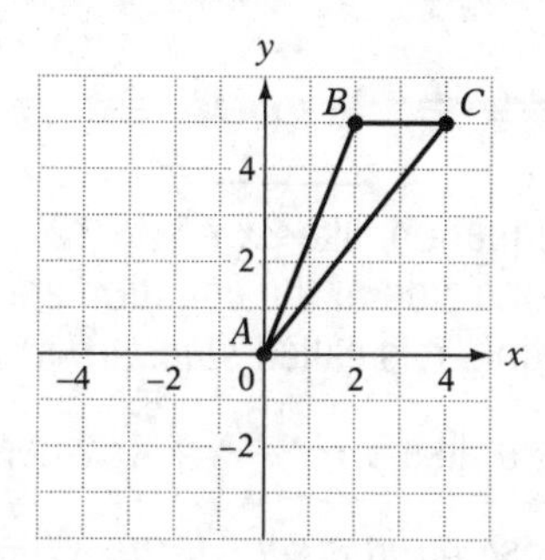

3. The slope of a line is $\frac{2}{3}$, and it passes through the point (–5, 2). What are three other points on this line that have coordinates that are integers?

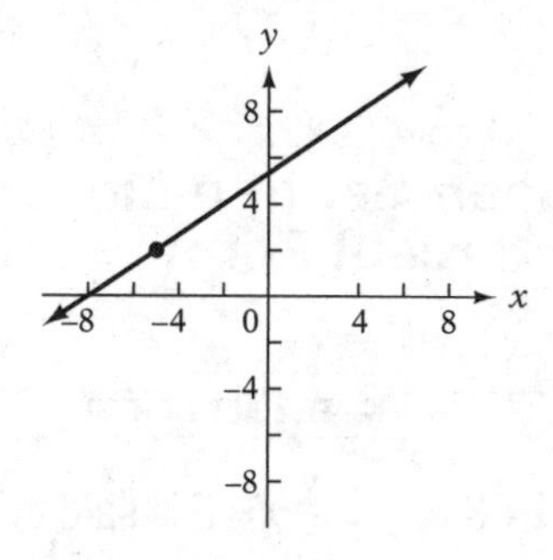

4. A line passes through the points (–7, 2) and (–3, –1). It also passes through the point (a, –7). What is the value of a?

5. Find the x-intercept and y-intercept of the graph of the solution set of $4x - 6y = 12$, and use them to find the slope of the line defined by that equation.

5.3 SLOPE-INTERCEPT FORM

When a two variable equation, like $2x + 3y = 12$, is written in this form with the variables on one side of the equation and the constant on the other side of the equation, it is called *standard form*. When it is written in the form $y = mx + b$, like $y = -\frac{2}{3}x + 4$, where m and b could be fractions, it is called *slope-intercept form*. When an equation is in slope-intercept form, there is a fast way to graph the solution set. Also, when an equation is in slope intercept form, it is often quicker to solve for x or y when the other variable's value is known.

Graphing the Solution Set of a Linear Equation That Is in Slope-Intercept Form

MATH FACTS

An equation like $y = 2x + 3$ or $y = \frac{2}{3}x - 5$ is said to be in *slope-intercept form*. In general, an equation of the form $y = mx + b$ is in slope-intercept form with the m representing the slope and the b representing the y-intercept.

To graph the solution set of a linear equation that is already in slope–intercept form:

1. Plot the point $(0, b)$, which is on the y-axis. If the equation is $y = 2x + 3$, plot the point $(0, 3)$. This is known as the y-intercept of the graph.

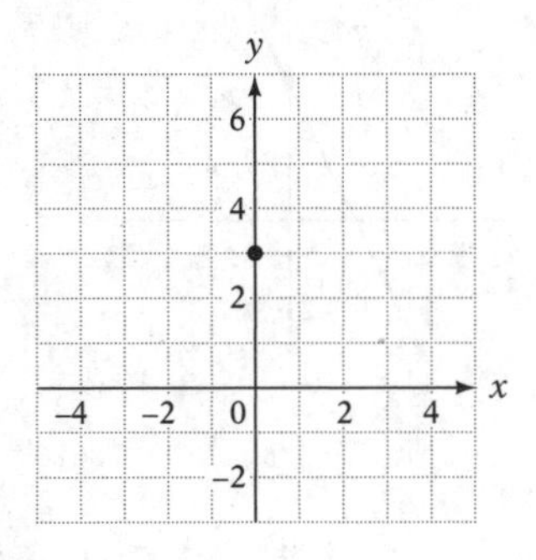

2. If the coefficient of the x term is not already a fraction, turn it into a fraction by putting the coefficient in the numerator of a fraction and a 1 in the denominator. If the coefficient is a negative fraction, make the numerator negative and the denominator positive. For the example, $y = 2x + 3$, the slope, denoted by m, is 2, which gets changed into $\frac{2}{1}$.

3. Starting at the y-intercept you already plotted, move right the number in the denominator of the slope. Then, from where you stopped, move up (down if it is negative) the number in the numerator of the slope. For the $y = 2x + 3$ example, the slope is $m = \frac{2}{1}$ so from $(0, 3)$, you move to the right 1 unit and then up 2 units to get to the point $(1, 5)$.

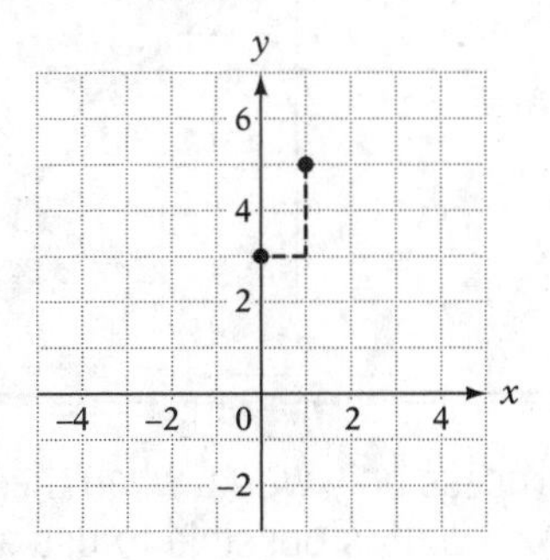

4. Draw a line through the y-intercept and the new point. Put arrows on both sides of the line to indicate that it continues forever on both sides.

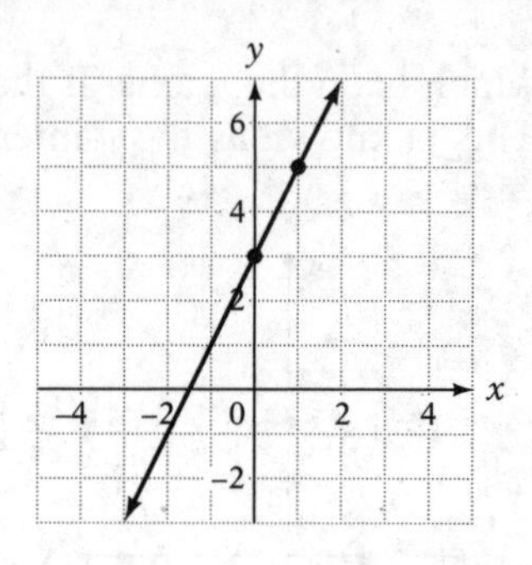

Example 1

Make a sketch of the solution set of the graph $y = -\frac{2}{3}x + 5$ using the slope-intercept process.

Solution: Since the constant is 5, the y-intercept is (0, 5). Since the coefficient of the x term is $-\frac{2}{3}$, $m = \frac{-2}{3}$ Plot the point (0, 5). From that point, move 3 to the right and 2 down (because it is negative) and then plot a new point. That's all you need for your line, but you can make another point by again moving 3 to the right and 2 down to get a third point.

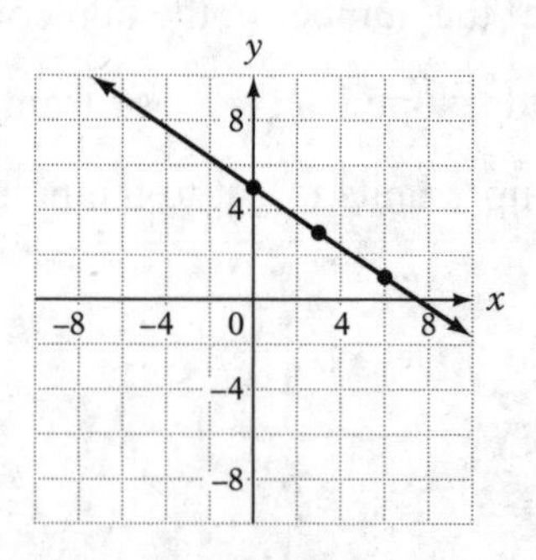

MATH FACTS

The slope-intercept process only works when the equation is in slope-intercept form, $y = mx + b$. It is possible to use algebra to change an equation that is not in slope-intercept form into one that is by isolating the y term.

Example 2

Change $2y + 6x = 4$ into slope-intercept form.

Solution: In order to isolate the y term, first eliminate the $6x$ term from the left-hand side of the equation by subtracting $6x$ from both sides.

$$2y + 6x = 4$$
$$-6x = -6x$$

Usually we put the $-6x$ in front of the $+4$, rather than have $4 - 6x$, as we are trying to get this to look like $y = mx + b$.

$$2y = -6x + 4$$

Now, eliminate the coefficient on the y by dividing both sides by 2.

$$\frac{2y}{2} = \frac{-6x}{2} + \frac{4}{2}$$
$$y = -3x + 2$$

Finding the Equation in Slope-Intercept Form When the *y*-Intercept and Another Point Are Known

The two things needed before creating the equation are the slope and the y-intercept. If the y-intercept is given, half the work is done. To calculate the slope, use the slope formula $m = \frac{y_2 - y_1}{x_2 - x_1}$.

Example 3

The graph of the solution set of a two variable equation is a line that passes through (0, 3) and (5, 6). What could the equation be?

Solution: A sketch of the graph looks like this:

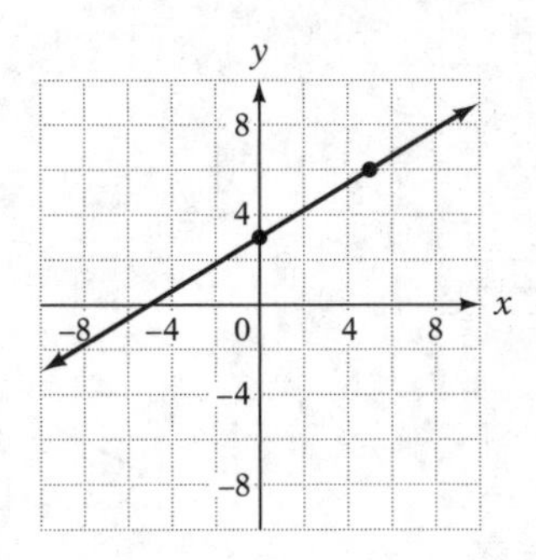

The b value is 3 since the y-intercept is (0, 3). To calculate the slope, use the slope formula with the points (0, 3) and (5, 6).

$m = \frac{6-3}{5-0} = \frac{3}{5}$. The equation is $y = \frac{3}{5}x + 3$.

MATH FACTS

The y-intercept can be described by the point $(0, b)$ or as just the number b. For example, the y-intercept of the equation $y = 5x + 7$ could be described as (0, 7) or as just 7.

Check Your Understanding of Section 5.3

A. Multiple-Choice

1. Which is the graph of $y = 2x - 5$?

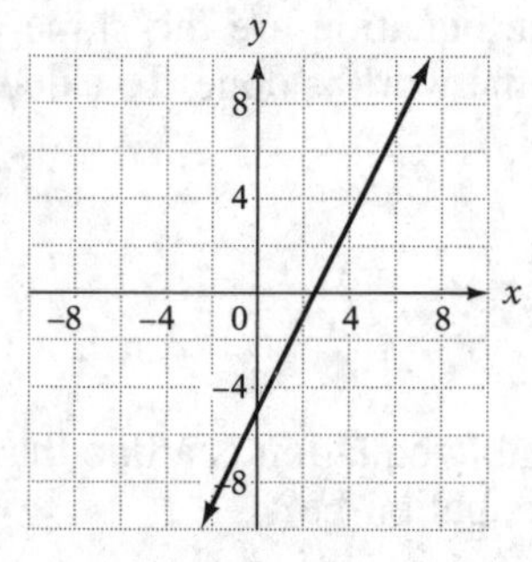

(1)

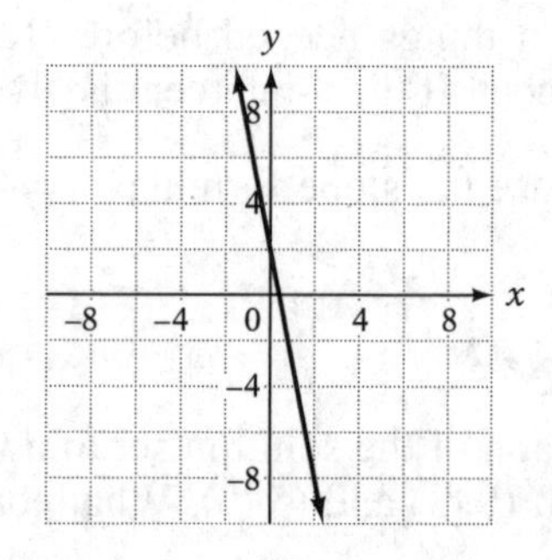

(3)

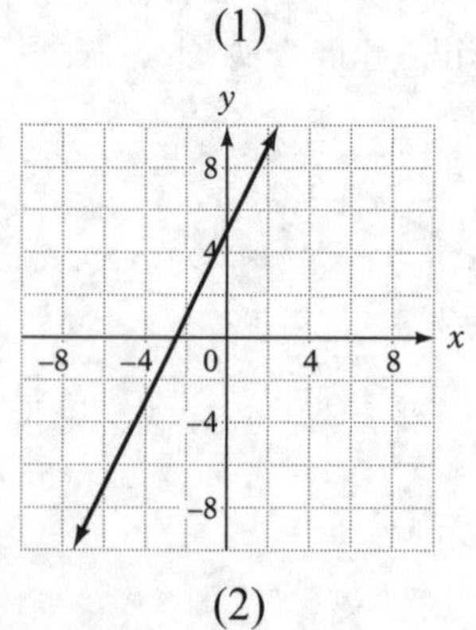

(2)

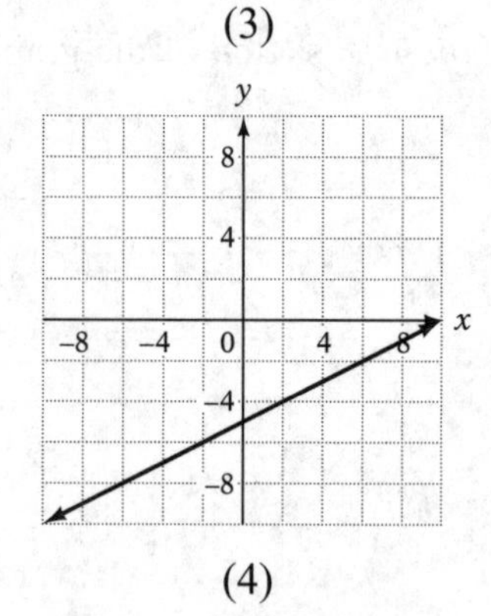

(4)

2. Which is the graph of $y = -\frac{5}{3}x + 4$?

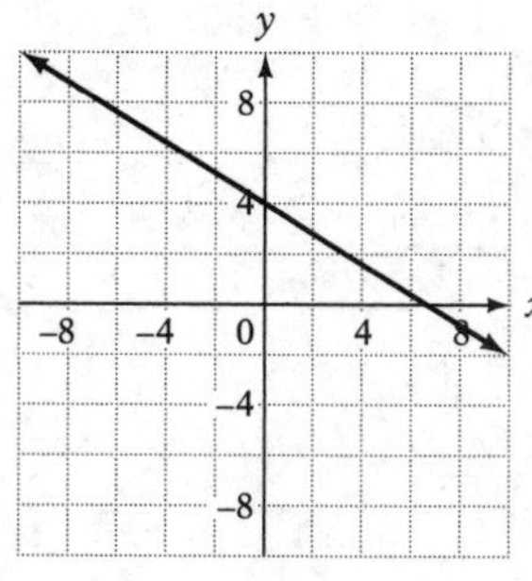

(1)

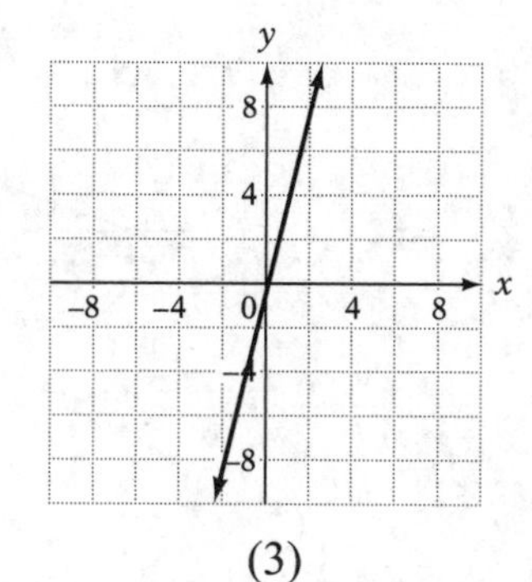

(3)

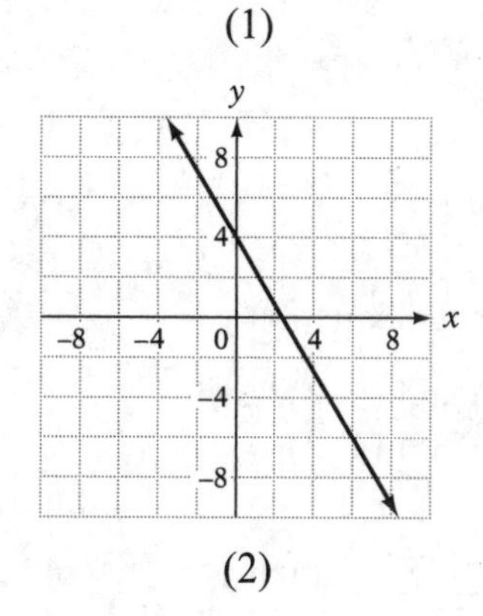

(2)

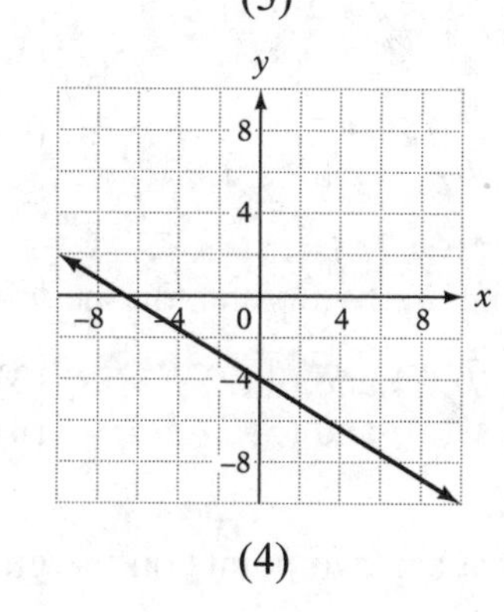

(4)

3. What is the slope of the line defined by the equation $y = -3x + 4$?
(1) 3 (2) –3 (3) 4 (4) –4

4. What is the y-intercept of the line defined by the equation $y = 5x + 2$?
(1) (2, 0) (2) (0, 2) (3) (5, 0) (4) (0, 5)

5. In the equation $y = 5$, the 5 is the
(1) y-intercept
(2) x-intercept
(3) slope
(4) length

6. The line defined by the equation $y = -\frac{4}{7}x + 5$ is perpendicular to the line defined by which equation?

(1) $y = \frac{4}{7}x + 3$ (3) $y = -\frac{7}{4}x + 3$

(2) $y = -\frac{4}{7}x + 3$ (4) $y = \frac{7}{4}x + 3$

7. This is the graph of the solution set of which equation?

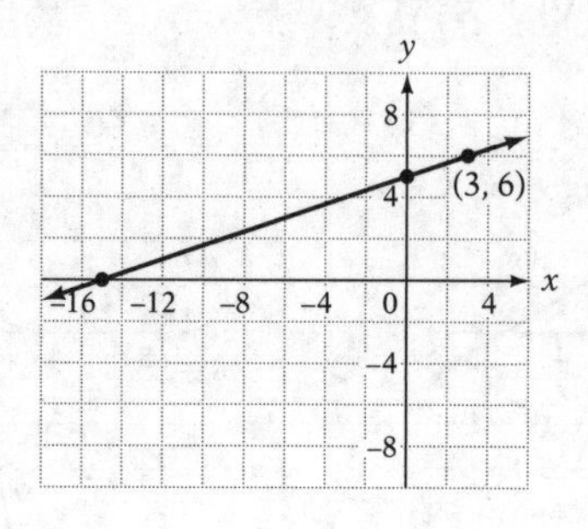

(1) $y = 5x + \frac{1}{3}$ (3) $y = 3x + 5$

(2) $y = 5x + 3$ (4) $y = \frac{1}{3}x + 5$

8. What are two points on the graph for the equation $y = -\frac{4}{5}x + 7$?

(1) (0, 7) and (4, 2) (3) (0, 7) and (5, 3)
(2) (7, 0) and (5, 3) (4) (7, 0) and (4, 2)

9. What are two points on the graph for the equation $y = \frac{5}{6}x - 4$?

(1) (0, 4) and (6, 1) (3) (0, –4) and (6, 1)
(2) (4, 0) and (6, 1) (4) (0, –4) and (5, 2)

10. What is the slope of the graph of the solution set of the equation $y = 3 + 2x$?
(1) –3 (2) 3 (3) 2 (4) –2

***B.** Show how you arrived at your answers.*

1. Identify the slope and the y-intercept of the graph for the equation $y = 3x - 9$, and sketch the graph on graph paper.

2. Jaiden says the graph for the equation $2y = 3x - 8$ has a slope of 3 and a y-intercept of –8. Edwin says that this is not correct and that the slope is $\frac{3}{2}$ and the y-intercept is –4. Which student is correct?

3. A portion of a line that passes through (2, 4) and (4, 5) is shown below. What is the y-intercept of this line?

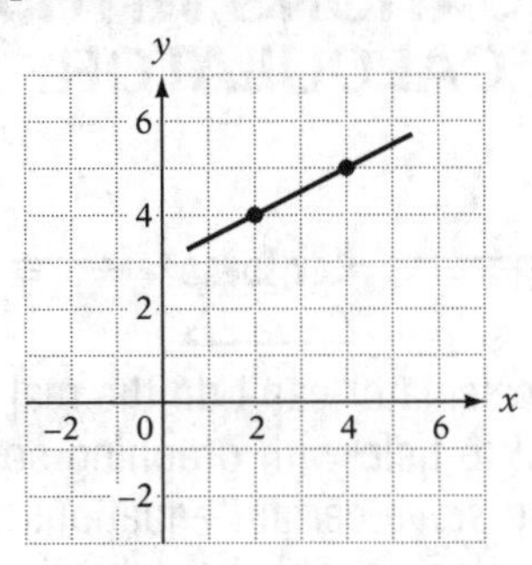

4. Solve the system of equations graphically. An algebraic solution will not be accepted.

$$y = -2x + 8$$
$$y = -\frac{2}{3}x + 4$$

5. What do the equations from which these five lines were determined have in common?

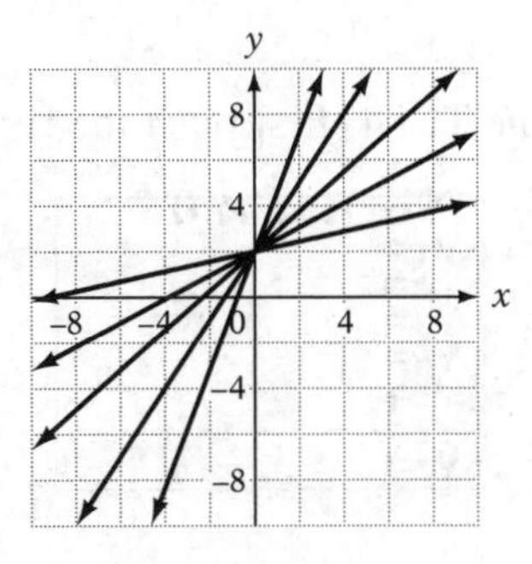

Chapter Five GRAPHS OF SOLUTION SETS OF LINEAR EQUATIONS

5.4 GRAPHING SOLUTION SETS TO LINEAR EQUATIONS WITH A GRAPHING CALCULATOR

Though the graphing calculator can't do the real "thinking" for you, it can be a powerful tool to help with graphing solution sets. It is also possible to *graphically solve* certain equations to find solution sets without needing to use algebra. This book has instructions for the two most popular calculators, the TI-84 Plus and the TI-Nspire.

Graphing Solution Sets for Linear Equations

Any graphing calculator can graph solution sets to two variable equations. Both the TI-84 and the TI-Nspire require that the y-variable is first isolated, however. Here are instructions for graphing $y = 2x - 5$.

For the TI-84:

To graph $y = 2x - 5$ on the TI-84, first push the [y=] key.

```
Plot1 Plot2 Plot3
\Y1=
\Y2=
\Y3=
\Y4=
\Y5=
\Y6=
\Y7=
```

Then type $2x - 5$. Be sure to use the minus sign that is two keys above [ENTER] and not the (–), which is the negative sign, and which is one key to the left of [ENTER].

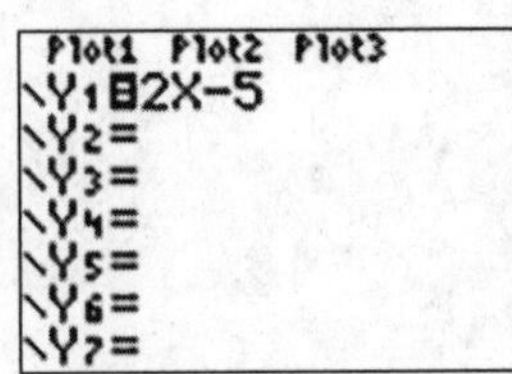

Push [ZOOM] and then [6] for ZStandard. This will make the graph on a grid that has a minimum value of –10 for both x and y and a maximum value of +10 for both x and y.

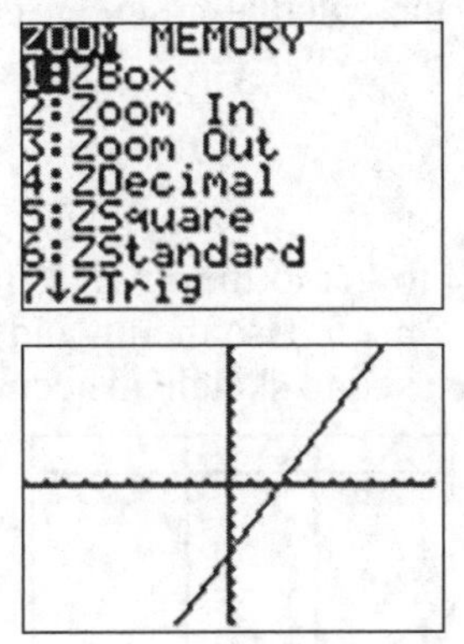

For the TI-Nspire:

Press [home] to get to the home screen. Then select [B] for Graph.

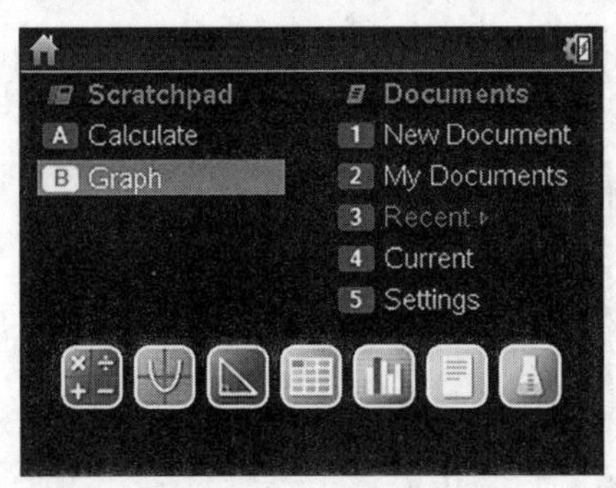

At the bottom of the screen, type $2x–5$ on the entry line after $f1(x)$= and then press [enter].

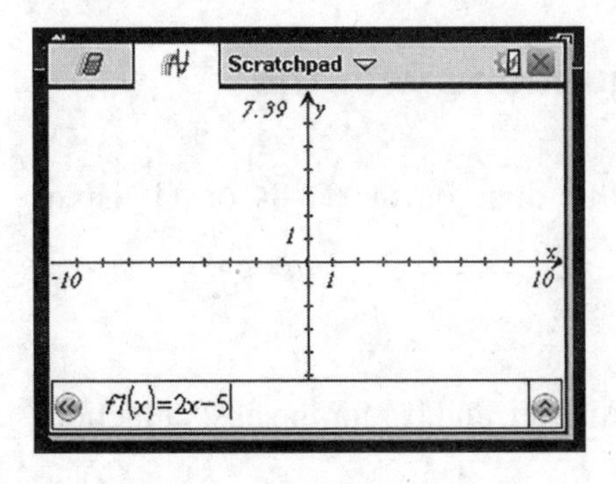

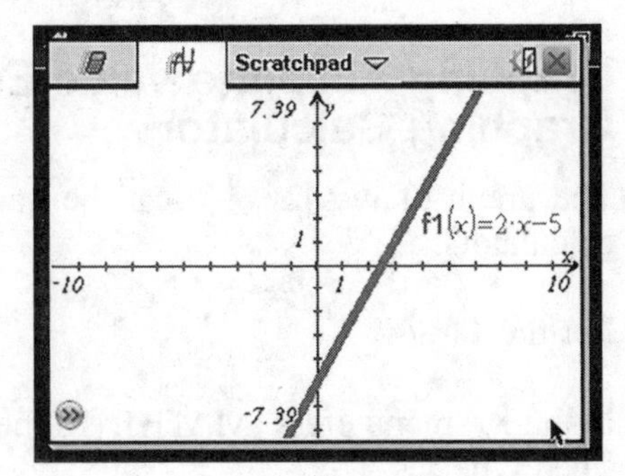

To see the entry line, press [tab].

Making a Table of Values with the Calculator

The graph is useful, but since you will likely have to draw the graph in your test booklet, the image on the calculator screen is not as helpful as a table of values.

For the TI-84:

Press [2ND] and [GRAPH] to get to the TABLE function. This is a table of values for the equation $y = 2x - 5$. Use the up and down arrows to see more values. These values can be used to sketch an accurate graph on graph paper.

X	Y_1
0	-5
1	-3
2	-1
3	1
4	3
5	5
6	7

X=0

For the TI-Nspire:

If you are on the graphing screen, press and hold down [ctrl] and then press [t]. To hide the table, click on the graph and push [ctrl] and [t] again.

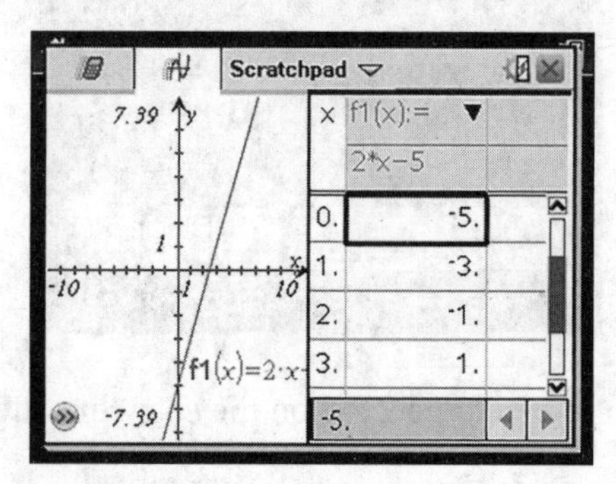

Graphing Absolute Value Equations with the Graphing Calculator

The graph of $y = |2x - 2|$ can be quickly done on the TI-84 or TI-Nspire calculators.

For the TI-84:

In the $Y=$ menu press [MATH], [Right Arrow], and [1] for the abs() function. Then type $2x - 2$.

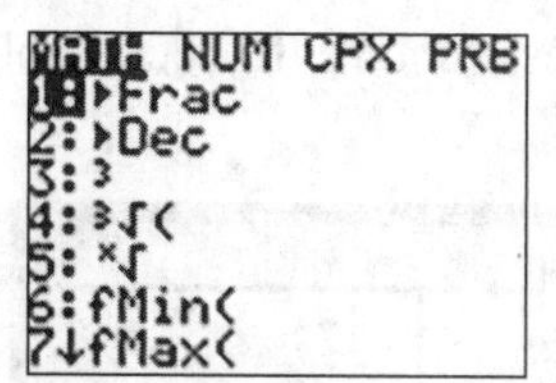

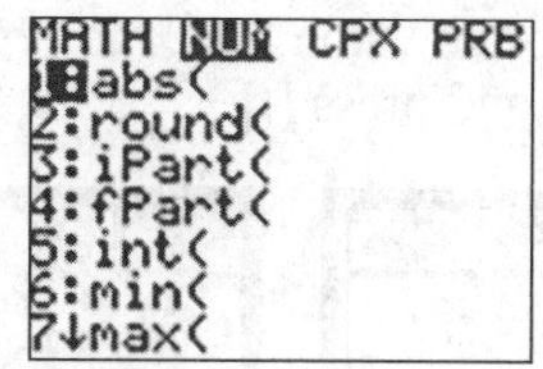

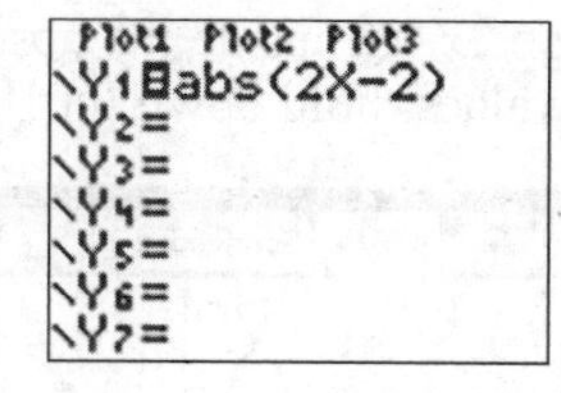

Press [ZOOM] and [6] to graph or [2nd] and [GRAPH] to see a table of values.

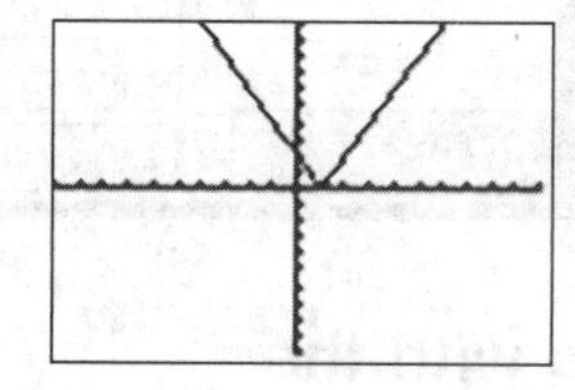

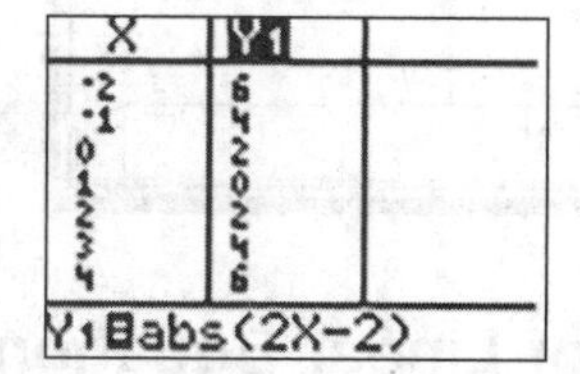

For the TI-Nspire:

From the home screen, press [B] to get to the Graph scratchpad.

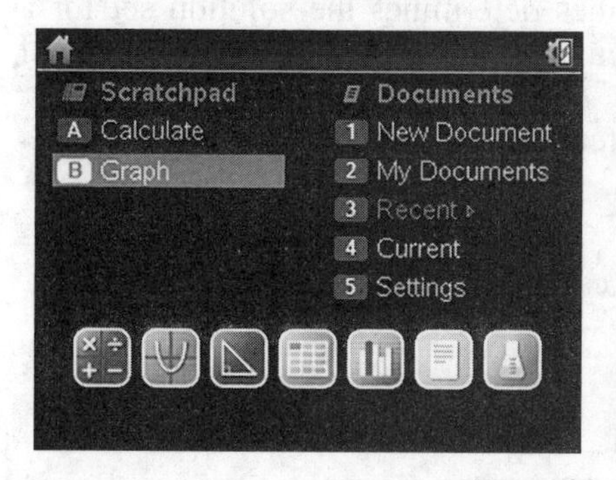

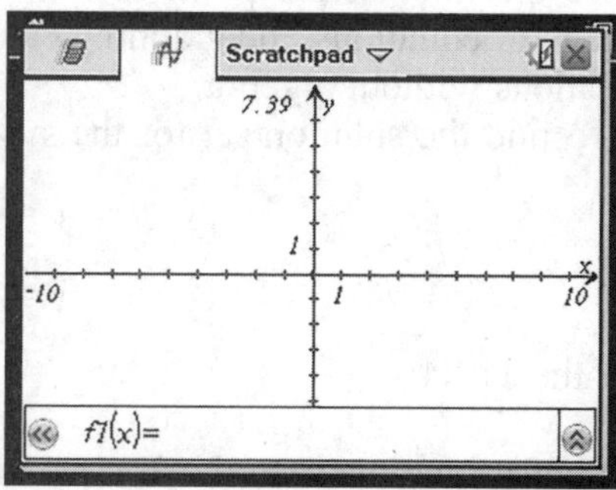

Press [templates] button next to the [9] key and select the absolute value template in the second row on the far left.

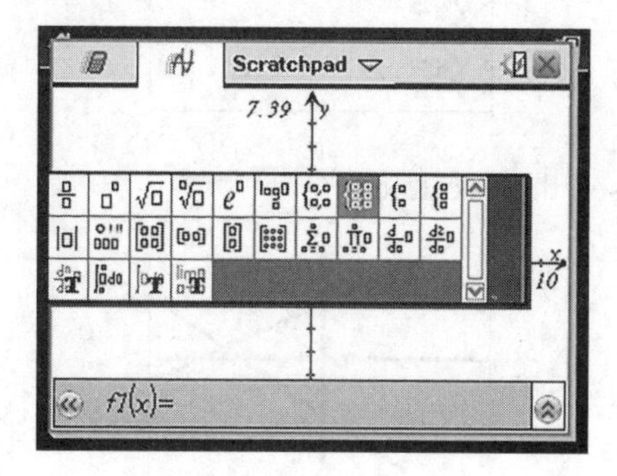

Type $2x - 2$ inside the absolute value bars and press [enter]. To see a table of values, hold down [ctrl] and press [t].

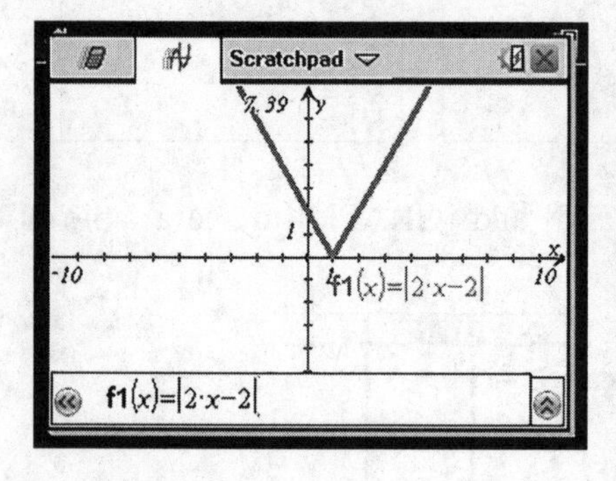

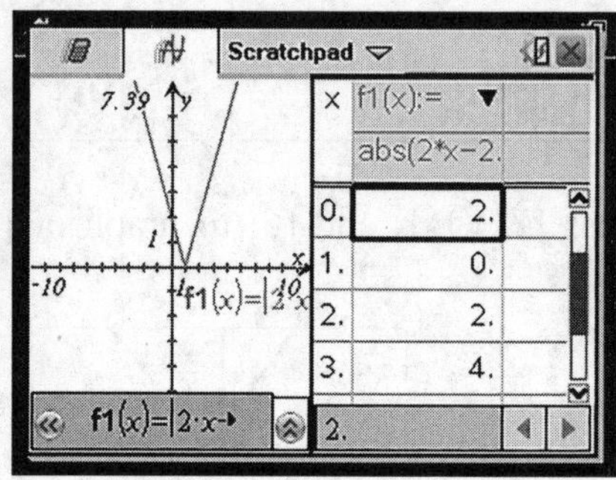

Systems of Linear Equations with the Intersect Feature

When two solution sets are graphed on the graphing calculator, there is a feature called intersect that determines the intersection point of two graphs. Since the intersection point of two lines determines the solution set for a system of equations, the graphing calculator can be used to solve systems of equations without algebra.

To find the solution set for the system of equations

$$y = -\frac{2}{3}x + 5$$
$$y = x - 5$$

For the TI-84:

Put both equations into the $Y =$ menu.

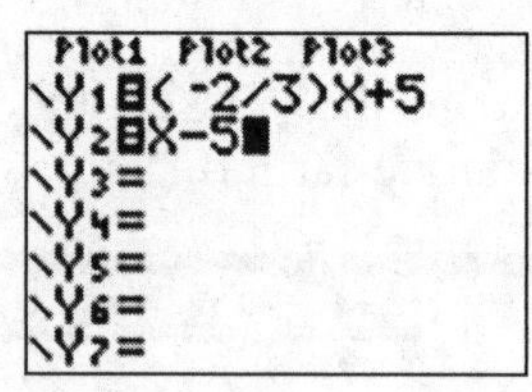

Press[GRAPH] to see the two lines.

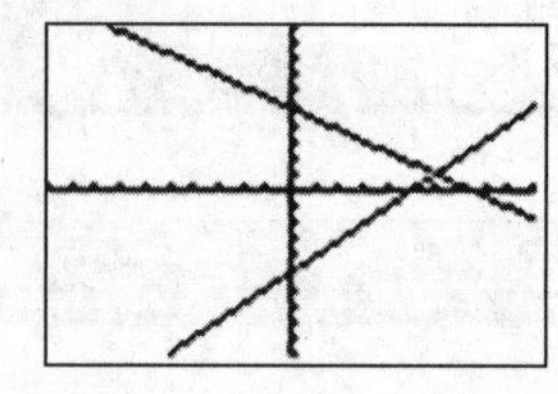

Using the arrow keys, you can move the cursor as close as you can to the intersection.

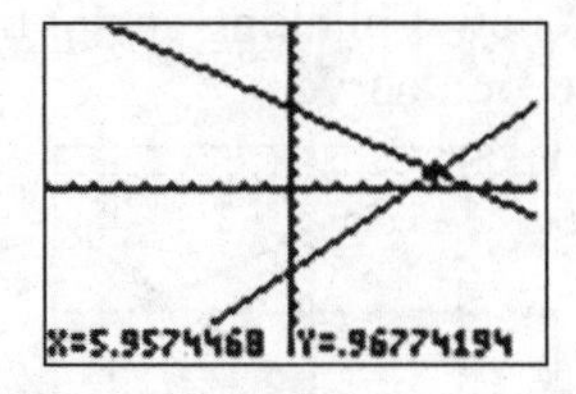

Because the answer is likely to have integer coordinates, the solution set appears to be (6, 1).

For a more exact answer, press [2ND] and [TRACE], which displays the CALCULATE menu.

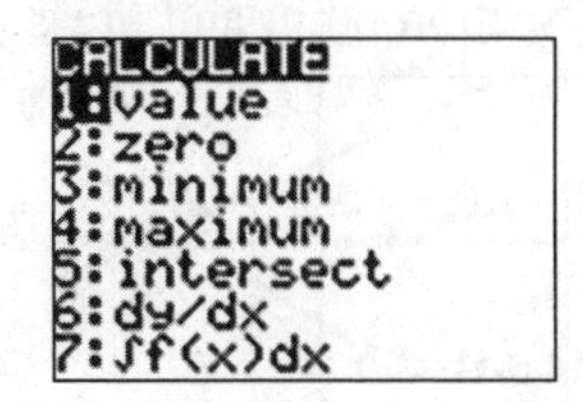

Select [5], intersect.

Because there may be more than two curves on the screen, you have to inform the calculator which two things you are trying to find the intersection for.

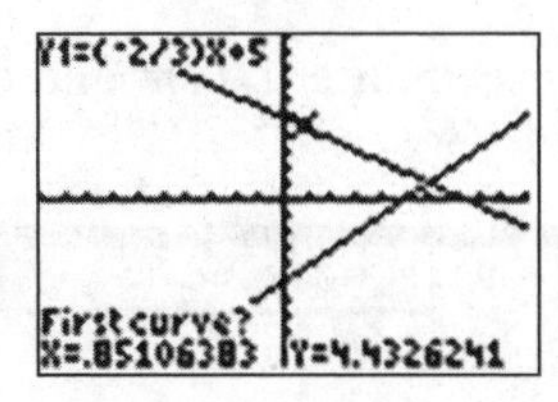

There is a cursor on the line for $y = -\frac{2}{3}x + 5$. Press [ENTER] to select the first line.

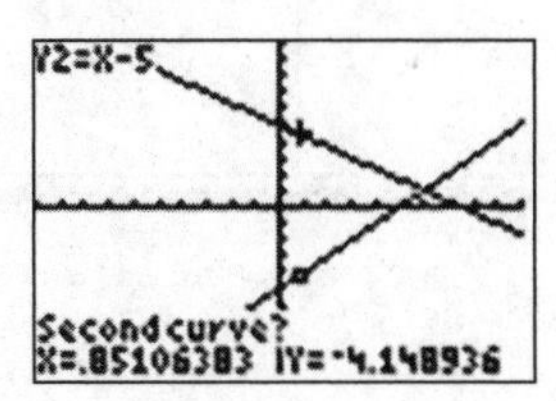

A + is now on the $y = -\frac{2}{3}x + 5$ line, and the calculator is asking what the second curve is. As there is now a blinking cursor on the $y = x - 5$ line, press [ENTER] to set it as the second curve.

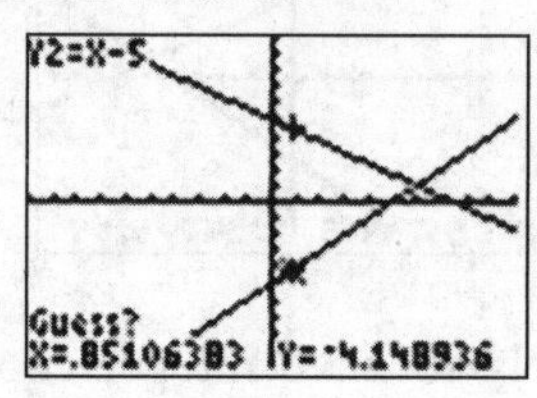

As there may be several intersection points, the calculator is asking for a "guess" near the intersection point you are looking for. Use the right and left arrows to get near the intersection point and press [ENTER] again.

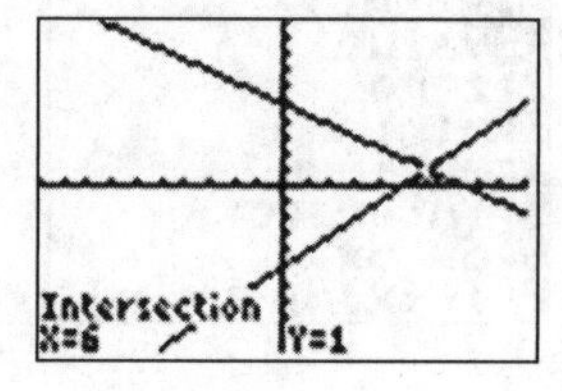

The accurate answer is (6, 1).

For the TI-Nspire:

In the entry field, enter $-\frac{2}{3}x + 5$ after *f1*(*x*)= and press [enter].

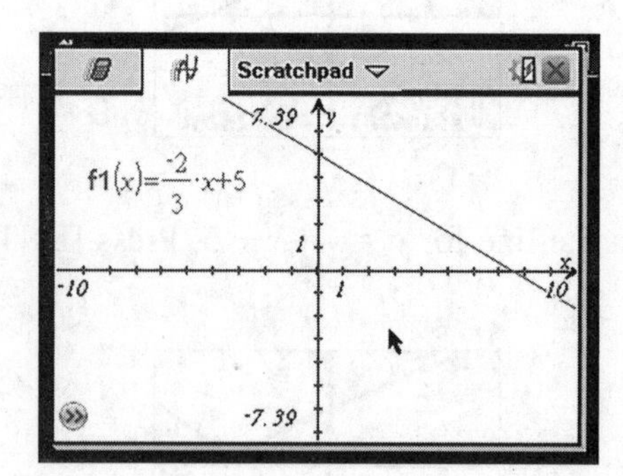

Press [tab] to see the entry field again and enter $x - 5$ after $f2(x)$= and press [enter].

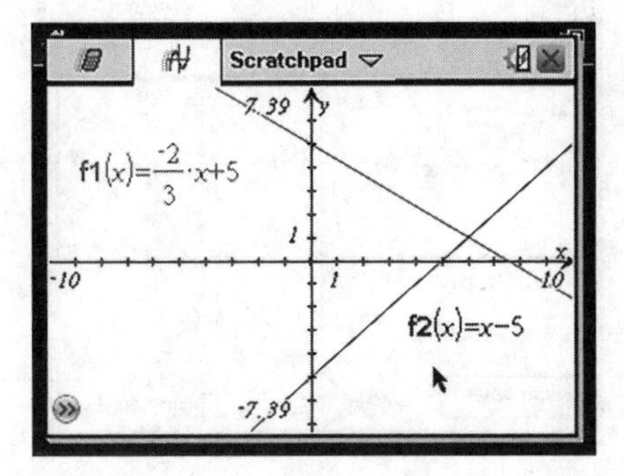

Press [menu][6][4] for Intersection.

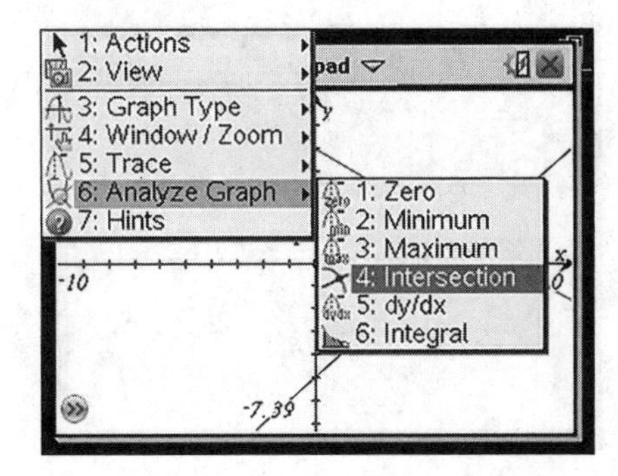

Move the pointer to the right of the intersection point and press [enter] to set the upper bound.

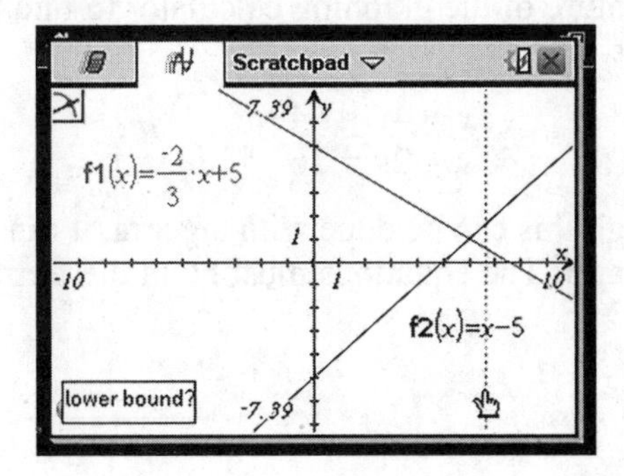

Move the pointer to the left of the intersection and press [enter] to find the intersection point.

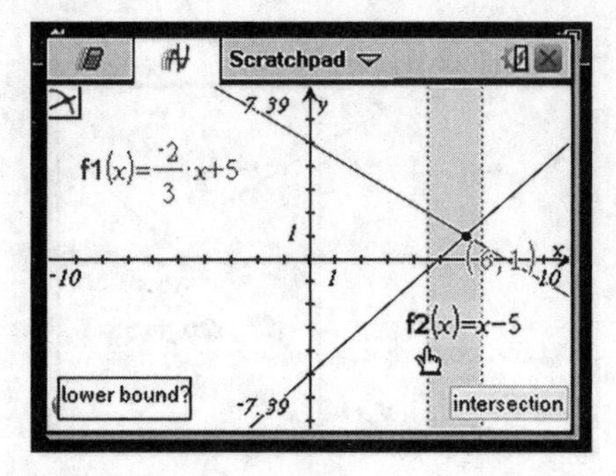

The intersection point (6, 1) is labeled.

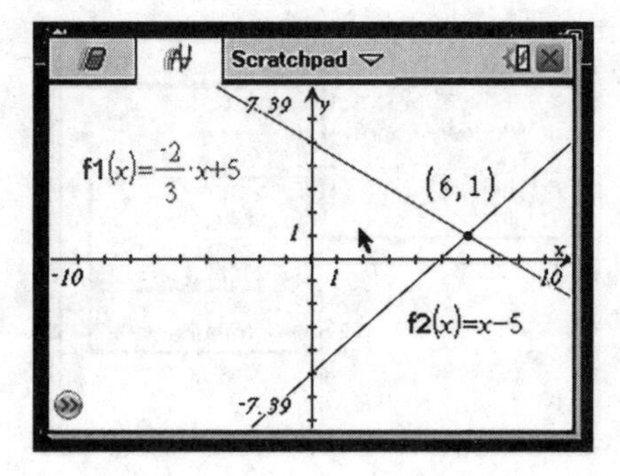

Example

Use the intersect feature of the graphing calculator to find the solution to the system of equations

$$-3x + 4y = -4$$
$$3x + 2y = 16$$

Solution: Though this can be done with algebra, it can also be done with the graphing calculator. The equations must be in the form of $y=$ in order to graph it.

Isolate the y in each equation.

$$\begin{aligned} -3x+4y&=-4\\ +3x&=+3x\\ \frac{4y}{4}&=\frac{3x-4}{4}\\ y&=\frac{3}{4}x-1 \end{aligned}$$

$$\begin{aligned} 3x+2y&=16\\ -3x&=-3x\\ \frac{2y}{2}&=\frac{-3x+16}{2}\\ y&=-\frac{3}{2}x+8 \end{aligned}$$

For the TI-84:

Enter these equations on the *Y*= menu.

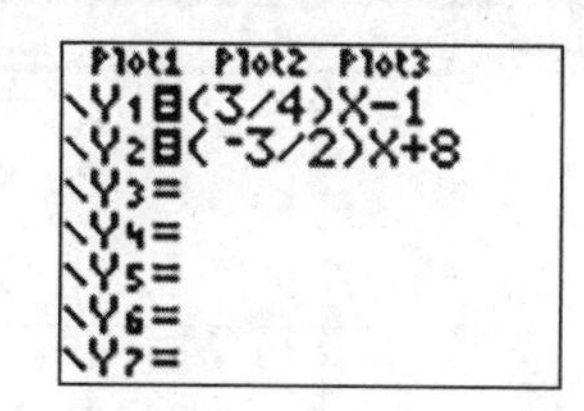

Graph and find the intersection point.

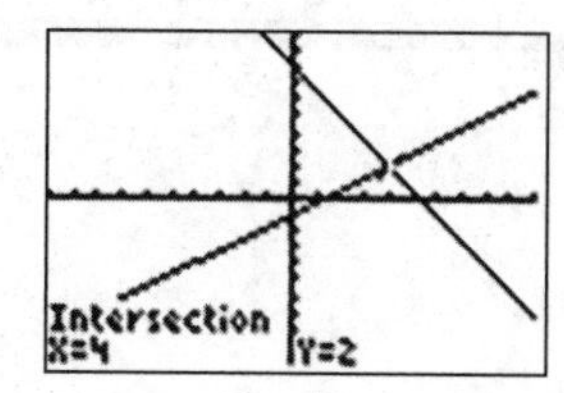

The solution is $x = 4$, $y = 2$.

For the TI-Nspire:

Enter the two equations in the entry line and press [enter].

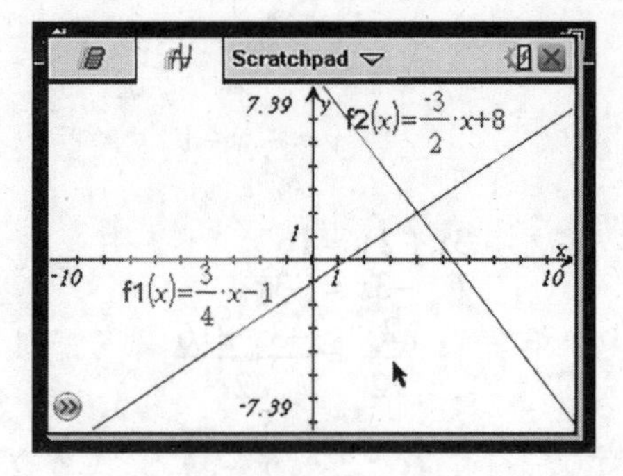

Press [menu], [6], and [4] and click to the right and then to the left of the intersection.

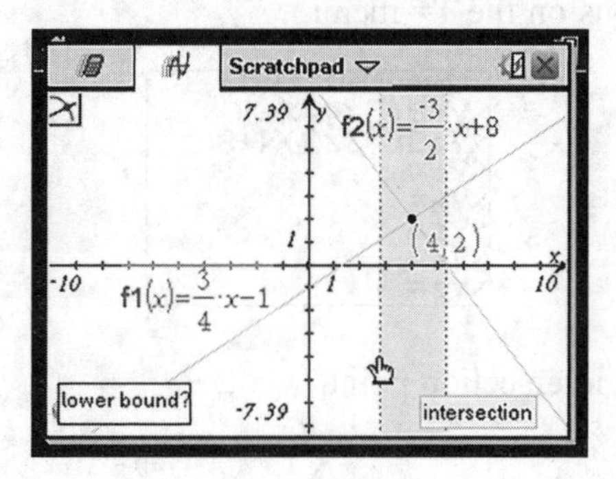

The solution is (4, 2).

Check Your Understanding of Section 5.4

B. Show how you arrived at your answers.

1. Use your graphing calculator to graph the equation $y = x - 3$.

2. Use your graphing calculator to graph the equation $y = -\frac{3}{4}x + 2$.

3. An algebra equation like $2x - 3 = 5$ can be solved graphically by intersecting the two graphs $y = 2x - 3$ and $y = 5$. Graph these two equations and find the x-coordinate of the intersection to solve the algebra problem $2x - 3 = 5$.

4. Use your graphing calculator to solve the system of equations.

$$y = -\frac{4}{5}x + 2$$
$$y = 2x + 16$$

5. Use your graphing calculator to solve the system of equations.

$$y = \frac{1}{4}x + 1$$
$$y = -x + 11$$

5.5 DETERMINING AN EQUATION FOR A GIVEN GRAPH

Just as it is possible to create a graph of a solution set when given a two-variable equation, it is also possible to find the two-variable equation for which a graph is the solution set. Often this equation is then used to answer other questions about the graph.

Finding the Equation When the Slope and *y*-Intercept Are Known

In the equation $y = mx + b$, the m represents the slope, and the b is the y-intercept. If these two values are given, substitute them into the equation.

Example 1

What is the equation for a line that has a y-intercept of (0, –2) and a slope of 5?

Solution: Since m is the slope, $m = 5$. And since $(0, b)$ is the y-intercept, b is –2. The answer is $y = 5x - 2$.

Finding the Equation When the *y*-Intercept and Another Point Are Known

If the y-intercept is known, then the b value in the $y = mx + b$ equation is the same number. With one other point, the slope formula can be used with that point and with the y–intercept to calculate the value for m.

Example 2

If a line passes through the points (0, 4) and (6, 8), what is the equation for that line?

Solution: Since one of the points has an x-coordinate of 0, it is the y-intercept and b is 4. Then to calculate m, use the slope formula with points (0, 4) and (6, 8).

$$m = \frac{8-4}{6-0} = \frac{4}{6} = \frac{2}{3}$$

The equation is $y = \frac{2}{3}x + 4$.

Finding the Equation When Two Points Are Known

If two points in the solution set for a linear equation are known, neither of which is the y-intercept, there is a two-part process for finding the equation. First, use the two points to calculate m, the slope of the line. Then substitute that slope and also the x and y values from one of the points into the equation $y = mx + b$ or $b = y - mx$ to solve for b.

Example 3

The graph of the solution set of a linear equation contains the points (4, –1) and (8, –4). What could be the equation?

Solution: Making a sketch of the graph shows that the y-intercept is something positive and that the slope is negative.

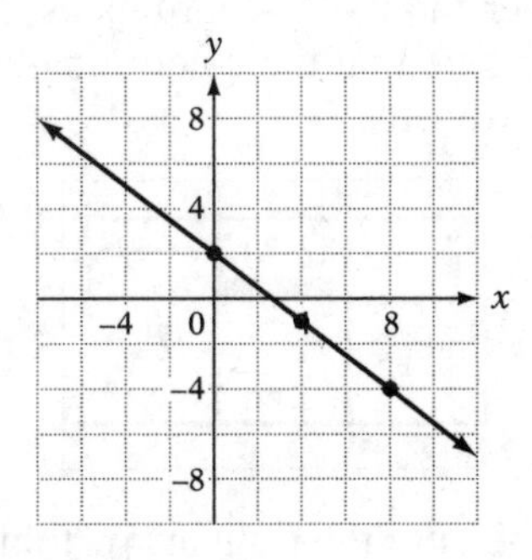

To calculate m, use the slope formula

$$m = \frac{-4-(-1)}{8-4} = \frac{-3}{4} = -\frac{3}{4}$$

Now choose just one of the known points, like (4, –1) and substitute 4 for x, –1 for y, and $-\frac{3}{4}$ for m into $y = mx + b$.

$$-1 = -\frac{3}{4}\cdot 4 + b$$
$$-1 = -3 + b$$
$$+3 = +3$$
$$2 = b$$

So, the equation is $y = -\frac{3}{4}x + 2$.

Using the Calculator to Find the Equation When Two Points Are Known

If the question is a multiple-choice question about finding the linear equation when two ordered pairs in the solution set are known, the TI calculator has the ability to do this.

Here is how you could do the last example with the calculator.

The points on the line are (4, –1) and (8, –4).

For the TI-84:

Press [STAT] and then choose [1], Edit.

In the L1 column, enter the two x-coordinates. In the L2 column, enter the two y-coordinates. Be sure that the y-coordinates are in the same row as their corresponding x-coordinates.

L1	L2	L3 2
4	-1	------
8	-4	

L2(3) =

Press [STAT] again and then use the right arrow to move to the CALC menu and press [4] to select LinReg and then [ENTER].

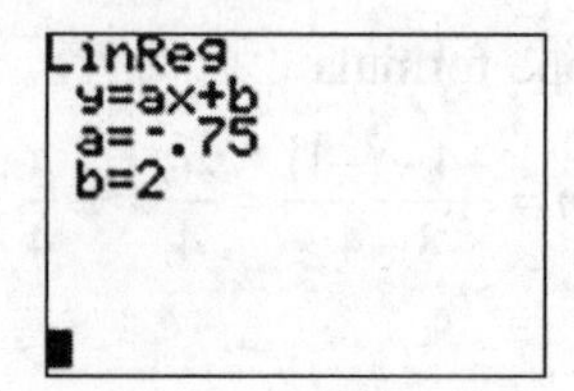

The calculator uses the variable a for slope instead of m. The solution is $y = -0.75x + 2$, as above.

For the TI–Nspire:

Press [home] to get to the home screen, and then select the Add Lists & Spreadsheet icon

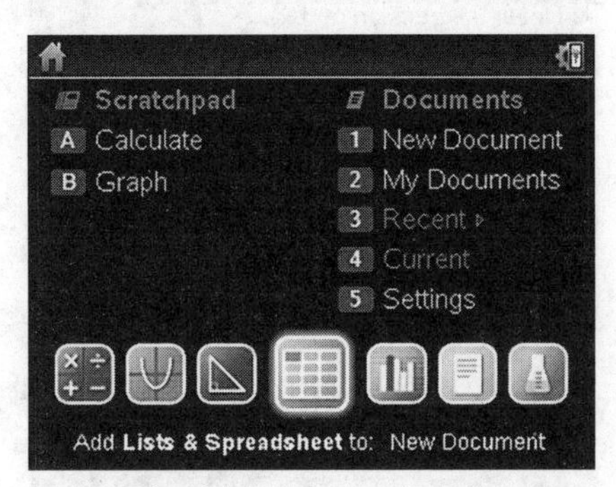

On the spreadsheet, enter *x* to name the first column and *y* to name the second column, and then enter 4 into cell A1, –1 into cell B1, 8 into cell A2, and –4 into cell B2.

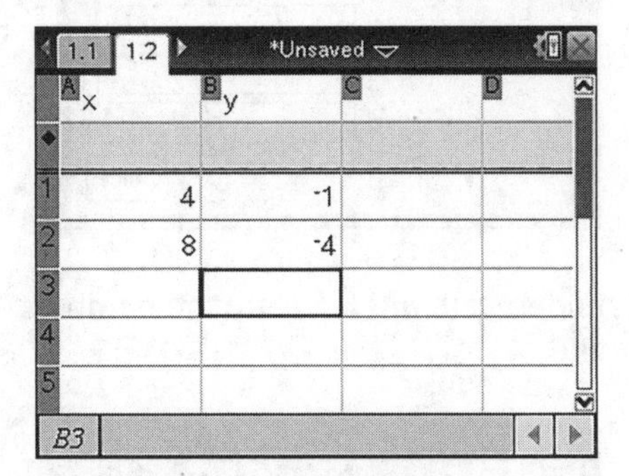

Press [menu], [4], and [1] to get to Stat Calculations.

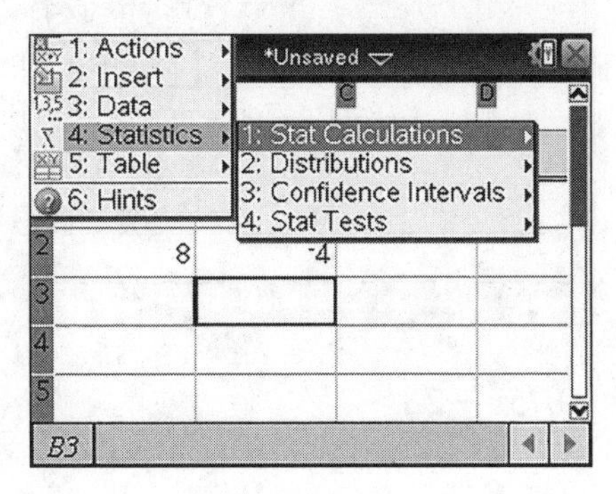

Press [3] for Linear Regression (mx + b).

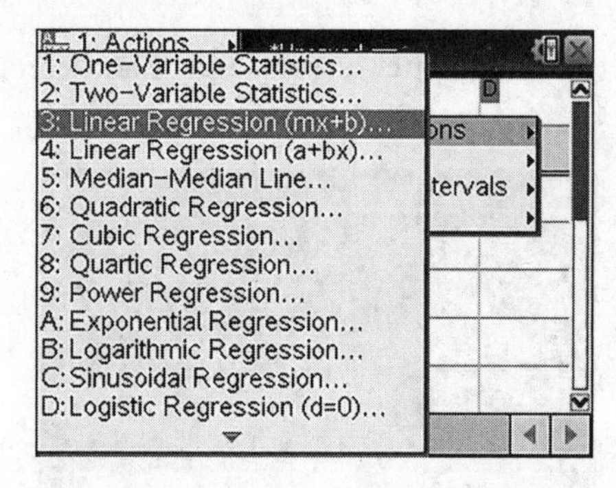

In the X List box, type x. In the Y List box, type y.

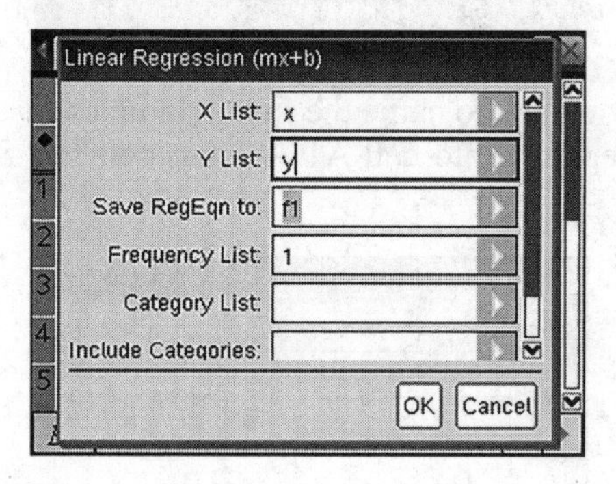

On the spreadsheet, underneath $m \cdot x + b$ are the numbers –0.75 and 2. These are the values of m and b.

	B y		C	D
◆				=LinRegM:
1	4	⁻1	Title	Linear Re..
2	8	⁻4	RegEqn	m*x+b
3			m	⁻0.75
4			b	2.
5			r²	1.

D1 ="Linear Regression (mx+b)"

The equation is $y = -0.75x + 2$.

Check Your Understanding of Section 5.5

A. Multiple-Choice

For each question, find the equation of the line that passes through the two given points.

1. (0, –7) and (5, 8)

(1) $y = 3x + 7$ (3) $y = \frac{1}{3}x + 7$

(2) $y = 3x - 7$ (4) $y = \frac{1}{3}x - 7$

2. (0, 5) and (–12, –1)

(1) $y = 2x + 5$ (3) $y = \frac{1}{2}x + 5$

(2) $y = 2x - 5$ (4) $y = \frac{1}{2}x - 5$

3. (0, 4) and (9, 10)

(1) $y = \frac{3}{2}x + 4$ (3) $y = \frac{2}{3}x + 4$

(2) $y = \frac{3}{2}x - 4$ (4) $y = \frac{2}{3}x - 4$

4. (2, 1) and (6, 9)

(1) $y = 2x - 3$ (3) $y = \frac{1}{2}x - 3$

(2) $y = 2x + 3$ (4) $y = \frac{1}{2}x + 3$

5. (3, 1) and (7, –3)

(1) $y = x + 4$ (3) $y = -x - 4$

(2) $y = x - 4$ (4) $y = -x + 4$

6. (4, –2) and (12, 4)

(1) $y = \frac{4}{3}x - 5$ (3) $y = \frac{3}{4}x - 5$

(2) $y = \frac{4}{3}x + 5$ (4) $y = \frac{3}{4}x + 5$

7. (5, 1) and (15, –3)

(1) $y = -\frac{2}{5}x - 3$

(2) $y = -\frac{2}{5}x + 3$

(3) $y = \frac{2}{5}x - 3$

(4) $y = \frac{2}{5}x + 3$

8. (–4, 1) and (4, 3)

(1) $y = 4x - 2$

(2) $y = 4x + 2$

(3) $y = \frac{1}{4}x - 2$

(4) $y = \frac{1}{4}x + 2$

9. (3, 7) and (6, 7)

(1) $y = -7$

(2) $y = 7$

(3) $x = -7$

(4) $x = 7$

10. (3, 5) and (3, 8)

(1) $x = -3$

(2) $x = 3$

(3) $y = -3$

(4) $y = 3$

B. *Show how you arrived at your answers.*

1. A line passes through the points (3, 4) and (9, 8). The line also passes through the point (5, a) for what value of a, rounded to the nearest tenth?

2. A line passes through the points (4, 7) and (8, 6). The line also passes through the point (a, 3.5) for what value of a?

3. Find the x-intercept and y-intercept of the line that passes through (5, 6) and (10, 2).

4. The equation $y = \frac{1}{3}x + 5$ passes through points (6, 7) and (12, a). What is the value of a?

5. The five points (5, 4), (6, 5), (7, 5), (9, 4), and (10, 5) are plotted on a coordinate plane. A line is drawn through the points (5, 4) and (10, 5). This line does not pass through either of the other three points. What is the sum of the three vertical line segments?

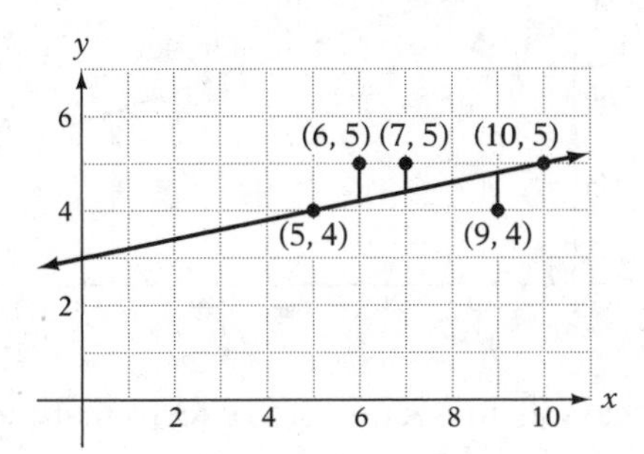

5.6 WORD PROBLEMS INVOLVING FINDING THE EQUATION OF A LINE

Usually in a scientific experiment, data are collected and then analyzed. If the graph of the collected data seems to make a line, two points on the line are enough to get an approximate equation describing the line. That equation can then be used to make predictions about the results of other experiments.

A Real-World Problem Involving Two Data Points

A party begins at 12:00 PM. At 12:21 PM, there were 56 pieces of cake remaining. At 12:36 PM, there were 46 pieces of cake remaining. Find an equation that relates the number of minutes past 12:00 PM, called T, and the number of pieces of cake left, called C. Use your equation to predict how many pieces of cake will remain after 51 minutes. Use your equation to predict how many minutes it will take before there are 10 pieces of cake remaining.

A graph of the scenario might look like this:

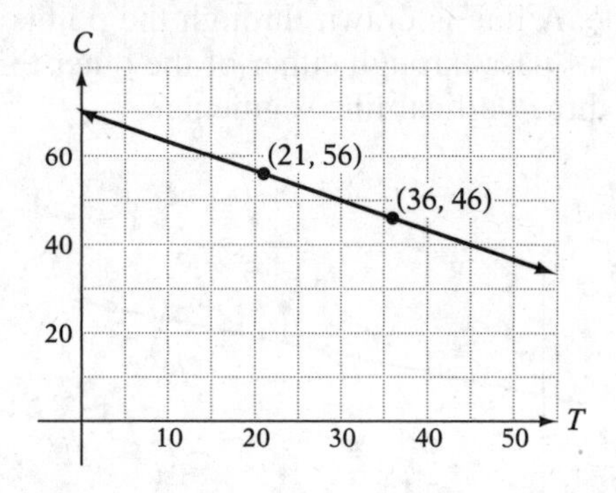

It uses the same process as in Section 5.5, except that the equation will not have x and y but will instead be $C = mT + b$.

The slope of the line is

$$m = \frac{46-56}{36-21} = \frac{-10}{15} = -\frac{2}{3}$$

For b, use one of the given points for T and C and the m just calculated.

$$\begin{aligned} C &= mT + b \\ 56 &= -\frac{2}{3}\cdot 21 + b \\ 56 &= -14 + b \\ +14 &= +14 \\ 70 &= b \end{aligned}$$

So the equation is $C = -\frac{2}{3}T + 70$.

For the second part of the question, substitute 51 for T and solve for C.

$C = -\frac{2}{3}\cdot 51 + 70 = -34 + 70 = 36$ so there will be 36 pieces of cake left after 51 minutes.

For the third part of the question, substitute 10 for C and solve for T with algebra.

$$10 = -\frac{2}{3}T + 70$$
$$-70 = -70$$
$$-60 = -\frac{2}{3}T$$
$$\left(-\frac{3}{2}\right)\cdot(-60) = \left(-\frac{3}{2}\right)\left(-\frac{2}{3}\right)T$$
$$90 = T$$

So it will take 90 minutes until just 10 pieces of cake remain.

Check Your Understanding of Section 5.6

A. Multiple-Choice

1. A balloon is held so that it is 6 feet above the ground. The balloon is released, and 10 seconds later it is 36 feet above the ground. Which equation can be used to relate the time since the balloon was released (T) to the height the balloon is above the ground (H)?
(1) $T = 3H - 6$
(2) $T = 3H + 6$
(3) $H = 3T - 6$
(4) $H = 3T + 6$

2. Lily has \$500 in the bank when the year begins. Each week she takes out the same amount of money. After 8 weeks she has \$436 left in the bank. Which equation can be used to model this scenario where W is the number of weeks that has passed since the beginning of the year and M is the amount of money remaining in the bank?
(1) $W = -8M + 500$
(2) $W = 8M - 500$
(3) $M = -8W + 500$
(4) $M = 8W - 500$

3. Marcus puts the same amount of money into the bank each week. Three weeks after the year begins, he has \$575 in the bank. Ten weeks after the year begins, he has \$750. Which equation can be used to model the amount of money (M) he has in the bank after (W) weeks?
(1) $M = 25W - 500$
(2) $M = 25W + 500$
(3) $W = 25M - 500$
(4) $W = 25M + 500$

4. A taxi costs a fixed amount to get into and then each mile is an additional fee. If a 4 mile ride costs \$8 and a 12 mile ride costs \$18, which equation models the cost (C) of a taxi ride compared to the number of miles (M) traveled?

(1) $C = 1.25M + 3$ (3) $M = 1.25C + 3$
(2) $C = 1.25M - 3$ (4) $M = 1.25C - 3$

5. Claire exercises each day by running around the track. She increases the amount she runs each day by the same amount. Fifteen days after she begins training, she is running six miles. Thirty-five days after she begins training, she is running ten miles. What equation relates the number of days she has been training (D) to the number of miles she runs that day (M)?

(1) $D = \left(\frac{1}{5}\right)M - 3$ (3) $M = \left(\frac{1}{5}\right)D - 3$

(2) $D = \left(\frac{1}{5}\right)M + 3$ (4) $M = \left(\frac{1}{5}\right)D + 3$

6. Miriam goes on a diet to lose weight. She loses the same amount of weight each week. If after four weeks she weighs 242 pounds and after seven weeks she weights 236 pounds, what equation relates her weight in pounds (P) to the number of weeks since she has been on the diet (W)?

(1) $W = -2P - 250$ (3) $P = -2W - 250$
(2) $W = -2P + 250$ (4) $P = -2W + 250$

7. When food is put into a special freezer, the temperature of the food decreases the same number of degrees each minute. If after six minutes the food is 67 degrees and after twenty minutes the food is 60 degrees, which equation relates the temperature of the food (T) to the number of minutes (M) the food has been in the freezer?

(1) $M = -\left(\frac{1}{2}\right)T + 70$ (3) $T = -\left(\frac{1}{2}\right)M + 70$

(2) $M = -\left(\frac{1}{2}\right)T - 70$ (4) $T = -\left(\frac{1}{2}\right)M - 70$

8. Jocelyn jumps out of an airplane with a parachute. She falls the same number of feet each second. Ten seconds after jumping she is 3,000 feet in the air. Fifteen seconds after jumping she is 2,000 feet in the air. Which equation relates her height in the air (H) to the number of seconds since she jumped (S)?

(1) $H = -200S + 5{,}000$ (3) $S = -200H + 5{,}000$
(2) $H = -200S - 5{,}000$ (4) $S = -200H - 5{,}000$

9. Isaiah and Jaxson are climbing up a mountain. Four hours after starting, they are 4,600 feet high. Nine hours after starting, they are 5,350 feet high. Which equation relates their height (A) to the number of hours they have been climbing (T)?

(1) $T = 150A - 4{,}000$ (3) $A = 150T - 4{,}000$
(2) $T = 150A + 4{,}000$ (4) $A = 150T + 4{,}000$

10. Waylon is on an elevator in the Empire State Building that is going down. After ten seconds, he is at the 104th floor. After thirty seconds he is at the 92nd floor. Which equation relates the number of seconds (S) he has been in the elevator to the floor (F) that he is at?

(1) $F = -\frac{3}{5}S - 110$ (3) $S = -\frac{3}{5}F - 110$

(2) $F = -\frac{3}{5}S + 110$ (4) $S = -\frac{3}{5}F + 110$

B. *Show how you arrived at your answers.*

1. A tree is planted in the ground. The tree grows the same amount each year. After two years the tree is 14 feet tall. After five years the tree is 26 feet tall. (a) Write an equation that relates the height of the tree (H) to the number of years since it was planted (Y). (b) Use your equation to determine how tall the tree will be after ten years?

2. Kendrick exercises by doing pull-ups. Each week he increases the number of pull-ups he can do by the same amount. After four weeks of training, he can do 9 pull-ups. After ten weeks of training, he can do 18 pull-ups. (a) Write an equation that relates the number of pull-ups he can do (P) to the number of weeks he has been training (W). (b) How many pull-ups could he do when he started his training? (c) After how many weeks will he be able to do 24 pull-ups?

3. Ava uploads to the Internet a video that goes viral. Each day the video is seen by a certain number more than the day before. On the sixth day, the video is watched 1,900 times. On the tenth day, the video is watched 3,100 times. (a) What equation relates the number of views (V) to the number of days (D) since the video was posted? (b) How many times will the video be watched on the 14th day? (c) How many times will the video be watched on the 31st day?

4. The population of Mathlandia is 300 million in 2007 and 380 million in 2015. The population increases by the same amount each year. (a) What equation can be used to relate the number of years since 2000 (T) to the population in millions (P)? (b) When will the population reach 730 million?

5. At 8:03 AM there are 188 empty seats on the train. At 8:10 AM there are 160 empty seats. The number of empty seats decreases by the same amount each minute. (a) What equation can be used to relate the number of minutes after 8:00 AM (M) to the number of empty seats on the train (E)? (b) How many seats will be empty at 8:15? (c) At what time will there be 120 empty seats? (d) At what time will there be no empty seats left?

Chapter Six GRAPHING SOLUTION SETS FOR QUADRATIC EQUATIONS

6.1 GRAPHING SOLUTION SETS TO QUADRATIC EQUATIONS

A two-variable quadratic equation is one where at least one of the variables has an exponent of 2. An example is the equation $y = x^2 - 6x + 8$. When the solution set to a two-variable quadratic equation is graphed, it does not become a line, but instead a *parabola*.

Graphing Solution Sets to Quadratic Equations by Making a Table

The equation $y = x^2$ has an infinite number of ordered pairs that satisfy it. Since $9 = 3^2$, (3, 9) is part of the solution set. Because it is also true that $9 = (-3)^2$, (–3, 9) is also in the solution set. With linear equations, only two points were needed since the graph was a line. With quadratic equations, five points are generally needed.

For the table, you can use the x values from –2 to +2.

x	y
–2	$(-2)^2 = 4$
–1	$(-1)^2 = 1$
0	$0^2 = 0$
1	$1^2 = 1$
2	$2^2 = 4$

The five ordered pairs, then, are (–2, 4), (–1, 1), (0, 0), (1, 1), and (2, 4).

Graph the five points on the coordinate plane.

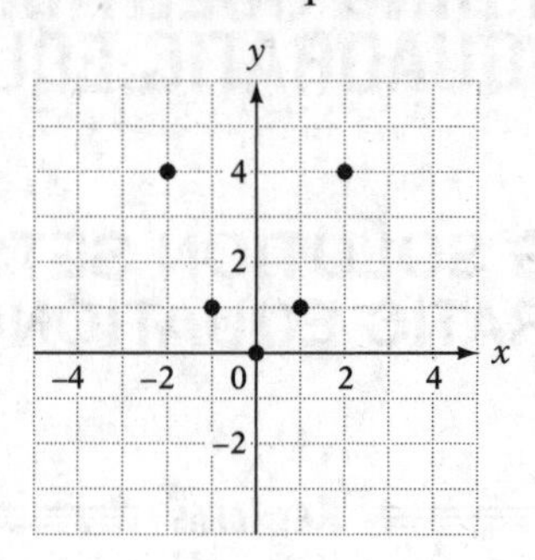

Connect the points with a U-shaped curve called a parabola. Make arrows on the ends to indicate that the curve continues forever in both directions.

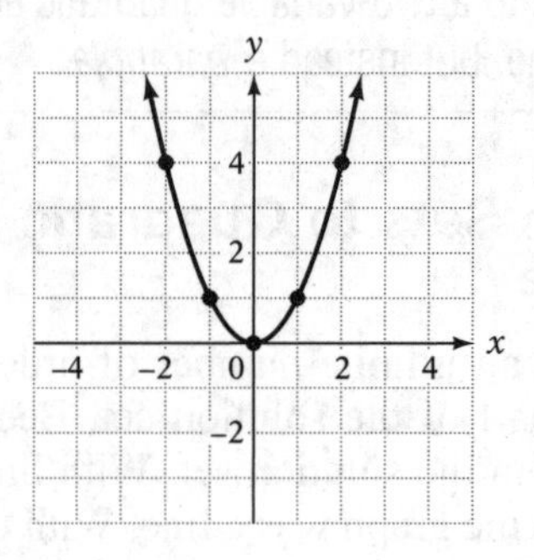

Example 1

Make a table of values and use them to graph the solution set of the equation $y = x^2 - 2x - 3$.

Solution: Using the five x values –2, –1, 0, 1, 2, the table becomes

x	y
–2	$(-2)^2 - 2(-2) - 3 = 4 + 4 - 3 = 5$
–1	$(-1)^2 - 2(-1) - 3 = 1 + 2 - 3 = 0$
0	$0^2 - 2(0) - 3 = 0 - 0 - 3 = -3$
1	$1^2 - 2(1) - 3 = 1 - 2 - 3 = -4$
2	$2^2 - 2(2) - 3 = 4 - 4 - 3 = -3$

Graphing the points (–2, 5), (–1, 0), (0, –3), (1, –4), and (2, –3) and connecting them with a parabola results in this graph.

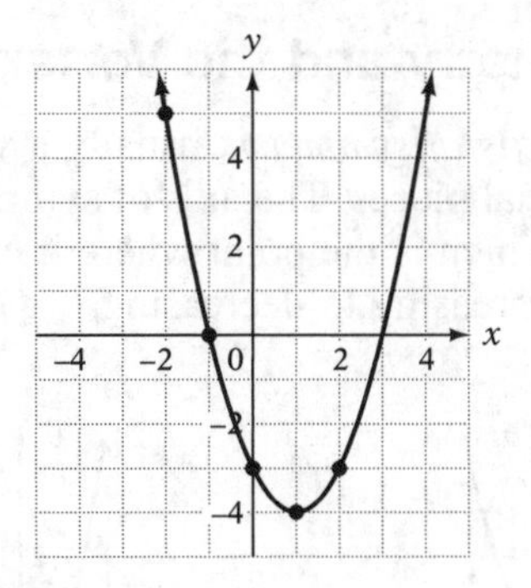

Graphing Solution Sets to Square Root Equations by Making a Table

The solution set to a square root equation is a sideways half-parabola. When making a table, it is easiest if x values are chosen so that what is under the square root sign is a perfect square.

To graph the solution set of $y = \sqrt{x} + 2$, the table could be

x	y
0	$\sqrt{0} + 2 = 0 + 2 = 2$
1	$\sqrt{1} + 2 = 1 + 2 = 3$
4	$\sqrt{4} + 2 = 2 + 2 = 4$
9	$\sqrt{9} + 2 = 3 + 2 = 5$

Since we can't (yet) take the square root of a negative number, and since the square root of something like 5 is not an integer, and would be more difficult to graph, the numbers 0, 1, 4, and 9 were good choices.

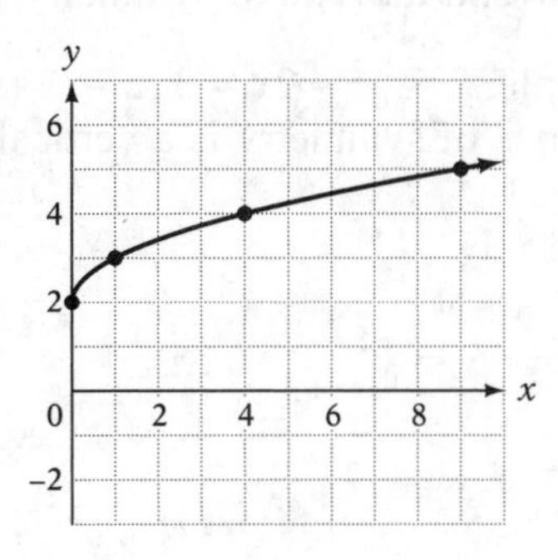

The Axis of Symmetry and the Vertex of a Parabola

Every parabola has an *axis of symmetry*, usually a vertical line that divides the parabola into two equal pieces. The axis of symmetry passes through the *vertex* of the parabola, which is the point where it changes from decreasing to increasing, or from increasing to decreasing.

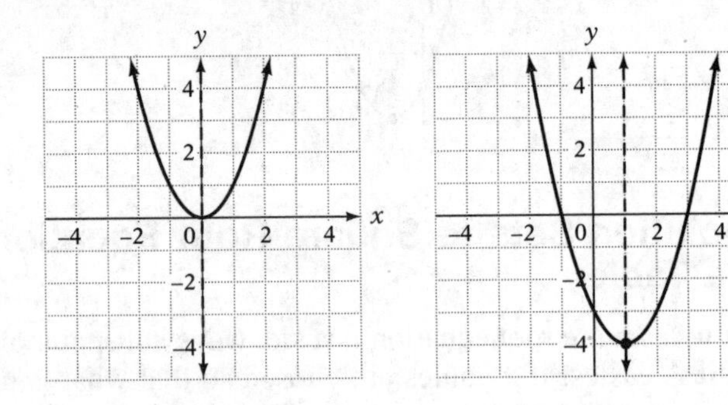

In the parabola for $y = x^2$, the axis of symmetry is vertical line $x = 0$ and the vertex is (0, 0). In the parabola for $y = x^2 - 2x - 3$, the axis of symmetry is the vertical line $x = 1$ and the vertex is (1, –4).

Using the Axis of Symmetry to Pick *x* Values for a Table

Rather than always use the five x values –2, –1, 0, 1, and 2, it is better to use as the middle number the x–intercept of the axis of symmetry. To determine the x–intercept of the axis of symmetry of a quadratic equation in the form $y = ax^2 + bx + c$, use the formula

$$x = -\frac{b}{2a}$$

For the first example, $y = x^2$, it is like $y = 1x^2 + 0x + 0$, so $a = 1$ and $b = 0$. So $-\frac{b}{2a} = -\frac{0}{2} = 0$ and the axis of symmetry was $x = 0$.

For the second example, $y = x^2 - 2x - 3$, $a = 1$ (since x^2 is the same as $1x^2$) and $b = -2$ so the axis of symmetry is a vertical line with the equation $x = -\frac{-2}{2(1)} = \frac{2}{2} = 1$.

Example 2

Create a table with five ordered pairs to graph the solution set to $y = x^2 + 6x + 8$.

Solution: The x-coordinate of the middle of the five points can be calculated with the formula $x = -\frac{b}{2a}$. Since $a = 1$ and $b = 6$, this is $\frac{-6}{2 \cdot 1} = -\frac{6}{2} = -3$.

The chart is

x	y
–5	$(-5)^2 + 6(-5) + 8 = 25 - 30 + 8 = 3$
–4	$(-4)^2 + 6(-4) + 8 = 16 - 24 + 8 = 0$
–3	$(-3)^2 + 6(-3) + 8 = 9 - 18 + 8 = -1$
–2	$(-2)^2 + 6(-2) + 8 = 4 - 12 + 8 = 0$
–1	$(-1)^2 + 6(-1) + 8 = 1 - 6 + 8 = 3$

Graph the five ordered pairs and use them to create a sketch of the parabola.

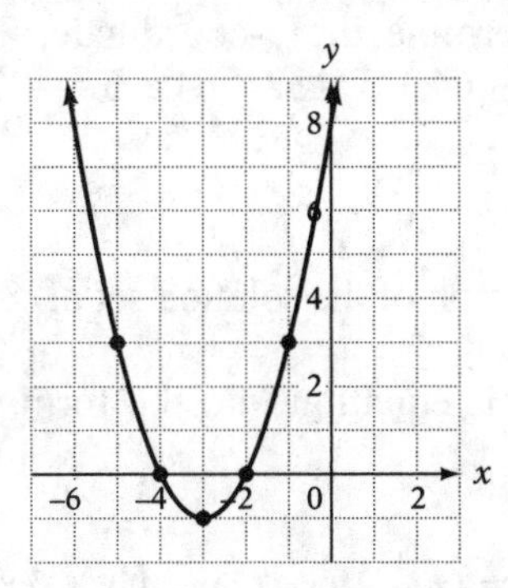

Determining the Coordinates of the Vertex of the Parabola

Because the vertex is on the axis of symmetry, the x-coordinate of the parabola that is the graph of the solution set of $y = ax^2 + bx + c$ is $-\frac{b}{2a}$. To determine the y-coordinate, substitute for x the value you get for $-\frac{b}{2a}$ into the equation $y = ax^2 + bx + c$.

If the equation is $y = x^2 + 6x + 8$, the x-coordinate of the vertex is $\frac{-6}{2 \cdot 1} = \frac{-6}{2} = -3$. The y-coordinate is $y = (-3)^2 + 6(-3) + 8 = 9 - 18 + 8 = -1$. So the vertex is $(-3, -1)$ as can be seen on the graph from the last example.

MATH FACTS

The x-coordinate of the vertex of the parabola defined by the equation $y = ax^2 + bx + c$ is $x = -\frac{b}{2a}$. To get the y-coordinate of the vertex, substitute the x-coordinate into the equation $y = ax^2 + bx + c$ and solve for y.

Example 3

What is the vertex of the graph of the solution set of $y = 2x^2 - 16x + 5$?

Solution: The x-coordinate is $-\frac{b}{2a} = -\frac{-16}{2 \cdot 2} = \frac{16}{4} = 4$. Substitute this 4 into the equation to determine the y-coordinate. The y-coordinate is $y = 2 \cdot 4^2 - 16 \cdot 4 + 5 = 2 \cdot 16 - 64 + 5 = 32 - 64 + 5 = -27$. The answer is $(4, -27)$.

Example 4

What is the vertex of the graph of the solution set of $y = (x - 2)^2 - 4(x - 2) + 7$?

Solution: First, get this equation into the form $y = ax^2 + bx + c$ before using the $x = -\frac{b}{2a}$ formula.

$$y = (x - 2)(x - 2) - 4(x - 2) + 7$$
$$y = x^2 - 4x + 4 - 4x + 8 + 7$$
$$y = x^2 - 8x + 19$$

The x-coordinate of the vertex is at

$$y = -\frac{-8}{2 \cdot 1} = \frac{8}{2} = 4$$

The y-coordinate of the vertex is

$$y = 4^2 - 8 \cdot 4 + 19$$
$$y = 16 - 32 + 19 = 3$$

So, the vertex is $(4, 3)$.

Graphing Parabolas with the Vertex and Intercepts Method

For the graph of a line, any two points will help produce an accurate graph. When a parabola has two x-intercepts, four points are very useful: the vertex, the y-intercept, and the two x-intercepts.

For the equation $y = x^2 + 6x + 8$ has a vertex, as calculated in the previous section, of (–3, –1). Since the y-intercept has an x-coordinate of 0, the y-intercept can be determined by substituting 0 in for x.

$$y = 0^2 + 6 \cdot 0 + 8 = 8 \text{ so the } y\text{-intercept is } (0, 8).$$

The x-intercepts each have a y-coordinate of 0. To calculate them, substitute 0 for y and solve.

$$0 = x^2 + 6x + 8$$

This is a quadratic equation covered in Chapter 3. This one can be solved by the factoring method.

$$0 = (x + 4)(x + 2)$$

$$\begin{array}{rcr} x + 4 = 0 & \text{or} & x + 2 = 0 \\ -4 = -4 & & -2 = -2 \\ x = -4 & \text{or} & x = -2 \end{array}$$

So the two x-intercepts are (–4, 0) and (–2, 0).

The four points, then, are (–3, –1), (0, 8), (–4, 0), and (–2, 0). Plot them and draw the parabola.

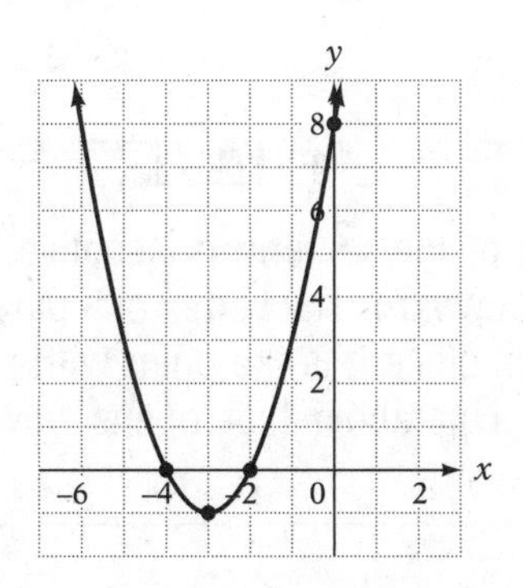

Example 5

Find the vertex, the y-intercept, and the two x-intercepts of the solution set of the equation $y = -x^2 + 4x + 12$.

Solution: The x-coordinate of the vertex is $x = \frac{-4}{2(-1)} = \frac{-4}{-2} = 2$.

The y-coordinate of the vertex is $y = -2^2 + 4 \cdot 2 + 12 = -4 + 8 + 12 = 16$.

The y-intercept is $y = -0^2 + 4 \cdot 0 + 12 = 12$.

The x-intercepts are the solutions to the equation

$$0 = -x^2 + 4x + 12$$
$$0 = -1(x^2 - 4x - 12)$$
$$0 = -1(x - 6)(x + 2)$$

$$\begin{array}{lcl} x - 6 = 0 & \text{or} & x + 2 = 0 \\ +6 = +6 & & -2 = -2 \\ x = 6 & \text{or} & x = -2 \end{array}$$

The vertex is (2, 16), the y-intercept is (0, 12), and the x-intercepts are (6, 0) and (–2, 0). Graph all four points to sketch the parabola.

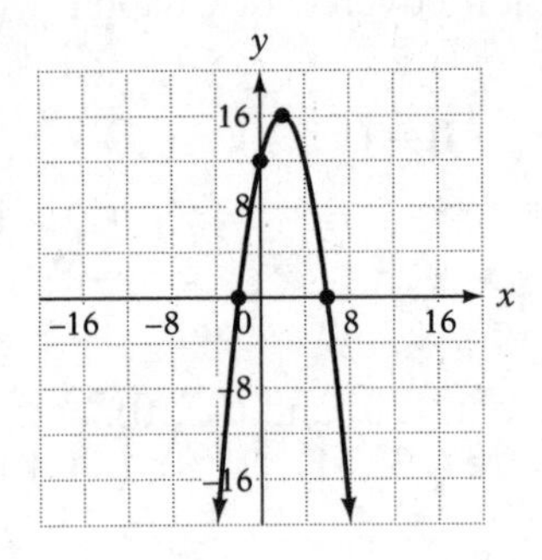

MATH FACTS

When the coefficient of the x^2 term is negative, the parabola opens downward instead of upward. The vertex of a parabola like this is also known as a *maximum* since it is the point with the highest y-coordinate. The vertex of a parabola that opens upward is known as a *minimum.*

Graphing Solution Sets to Quadratic Equations on the Graphing Calculator

In Section 5.4 the process for graphing linear equations on the TI-84 and TI-Nspire calculators was explained. For quadratic equations, it is the same process. Below are directions for graphing $y = x^2 + 6x + 8$ on both calculators.

For the TI–84:

Press [y=] key and enter $x^2 + 6x + 8$ into the first line.

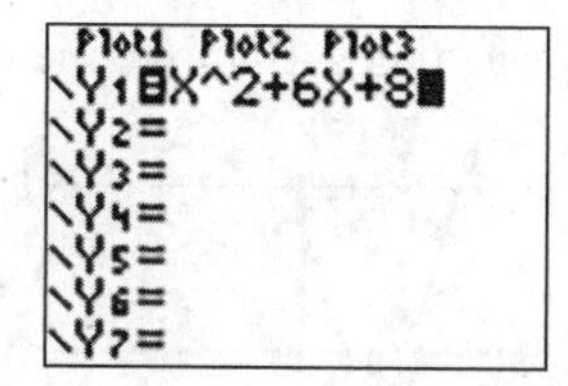

Push [ZOOM] and then [6] for a graph that has a minimum value of –10 and a maximum value of +10 for both x and y.

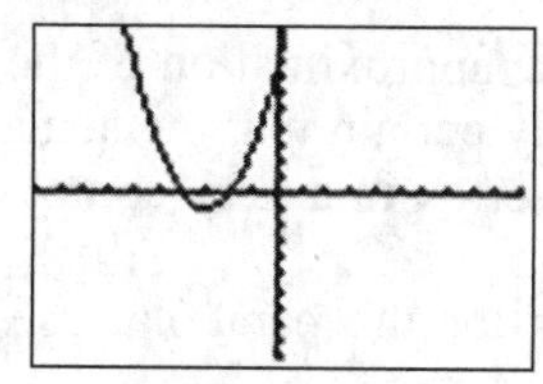

To obtain the coordinates of the vertex, press [2ND] and then [TRACE] for the CALCULATE menu. Then press [3] for minimum (or [4] for maximum if the vertex is the highest point on the curve.)

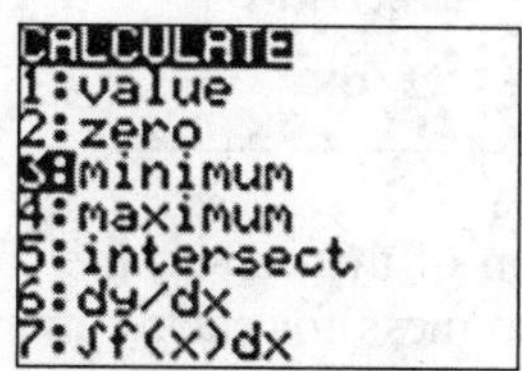

When the calculator asks for "Left Bound?" move the cursor to the left of the vertex and press [ENTER].

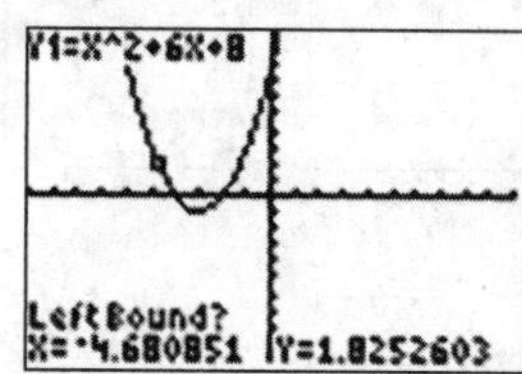

When the calculator asks for "Right Bound?" move the cursor to the right of the vertex and press [ENTER].

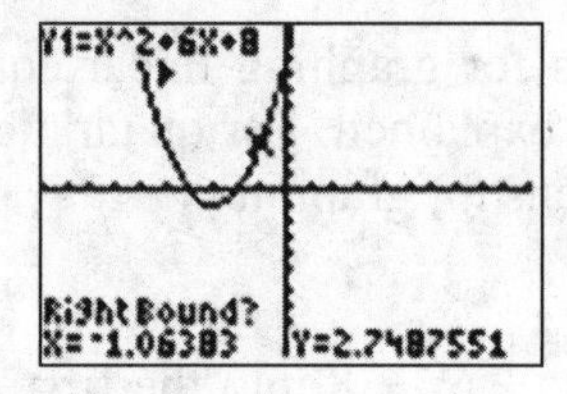

When the calculator asks for a "Guess?" move the cursor near the vertex and press [ENTER].

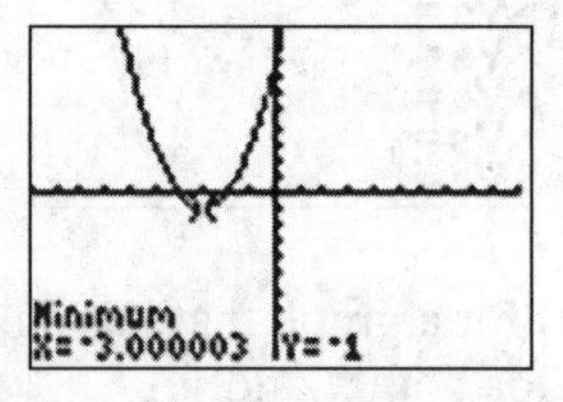

The calculator gives a close approximation of the vertex. Since this is very close to (–3, –1), it is nearly certain that the actual answer is (–3, –1). (The calculator figures out the vertex in a very complicated way that could lead to rounding errors.)

To find the *x*-intercepts for the parabola, use the zeros feature on the CALCULATE menu.

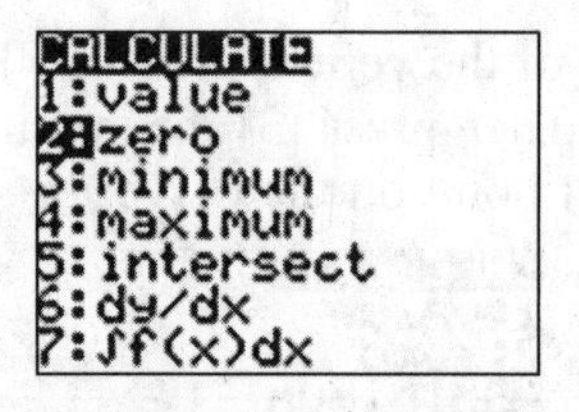

Just like finding the location of the vertex, the calculator will ask for a left bound, a right bound, and a guess for each *x*-intercept.

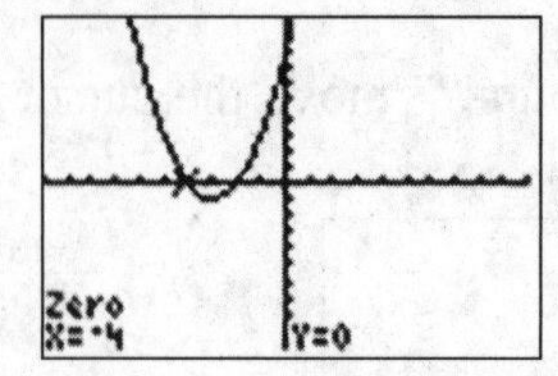

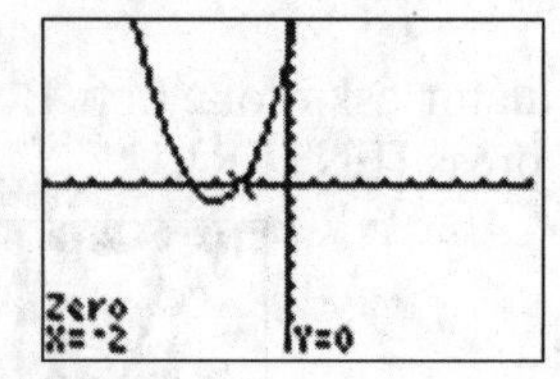

To make a copy of the graph onto your graph paper, press [2ND] and then [GRAPH] to see a table of values.

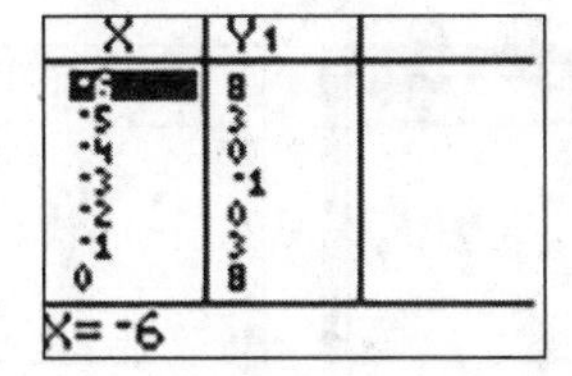

For the TI-Nspire:

From the home screen, press [B] for graphing on the Scratchpad. Enter $x^2 + 6x + 8$ on the entry line, and press [enter].

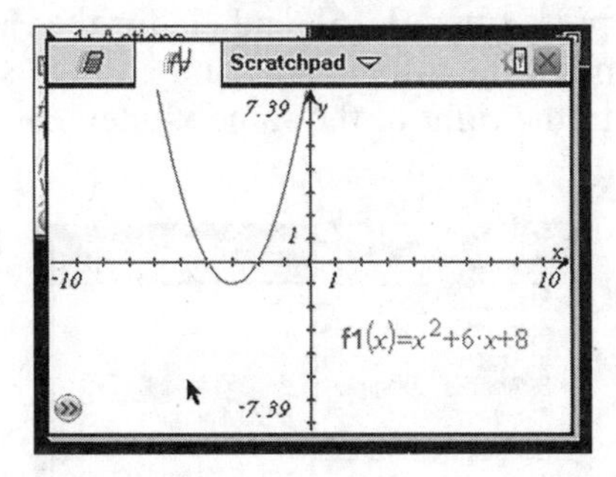

To find the coordinates of the vertex press [menu], [6], and [2] for Minimum (or [3] if the parabola opens downward and the vertex is the maximum point of the parabola).

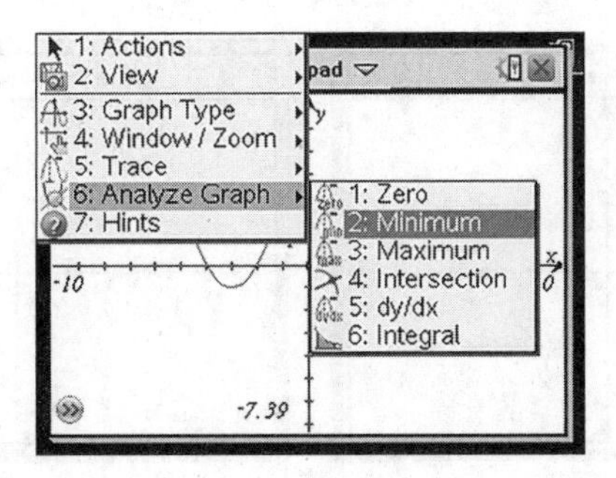

Move the hand to the left of the vertex and click to set the lower bound; then move the hand to the right of the vertex and click to set the upper bound.

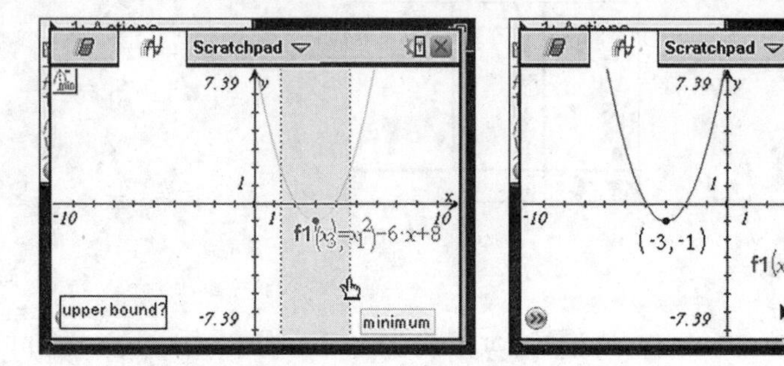

For the *x*-intercepts, press [menu], [6], and [1] for the Zero finder. Move the hand to the left of one of the *x*-intercepts and click to set the lower bound; then move the hand to the right of the same *x*-intercept for the upper bound and click.

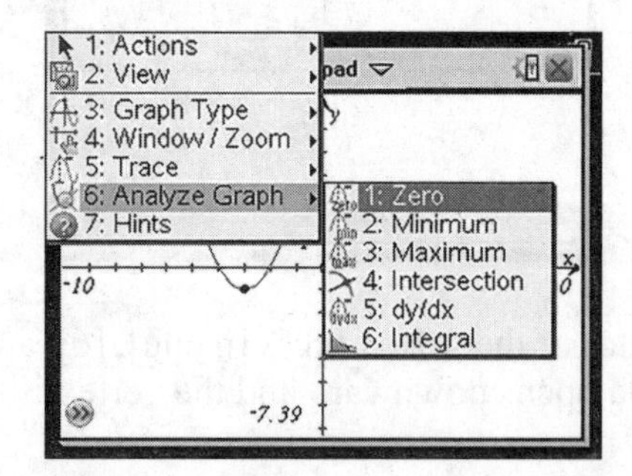

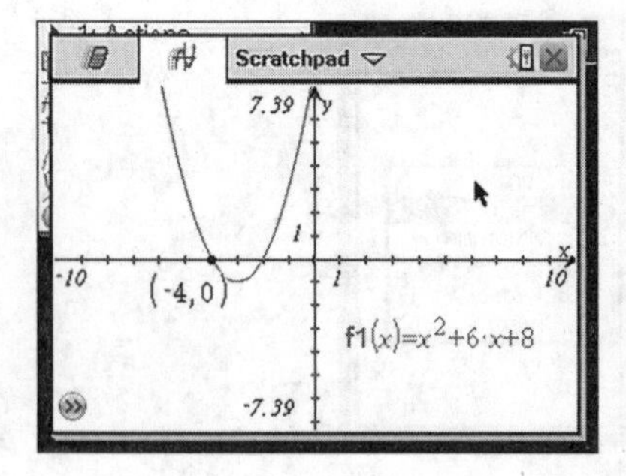

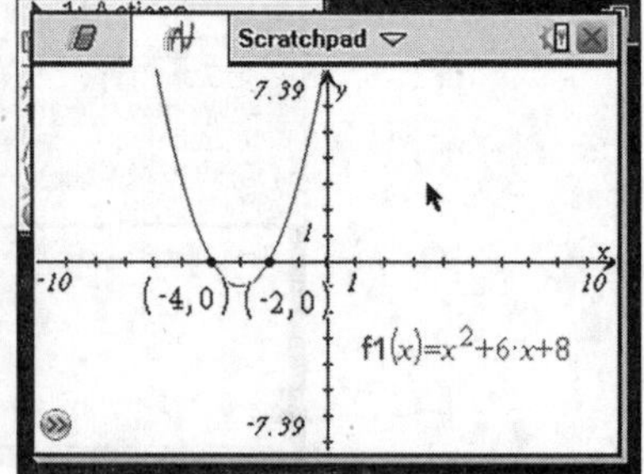

To see a table of values, press and hold the [ctrl] button and then press [t].

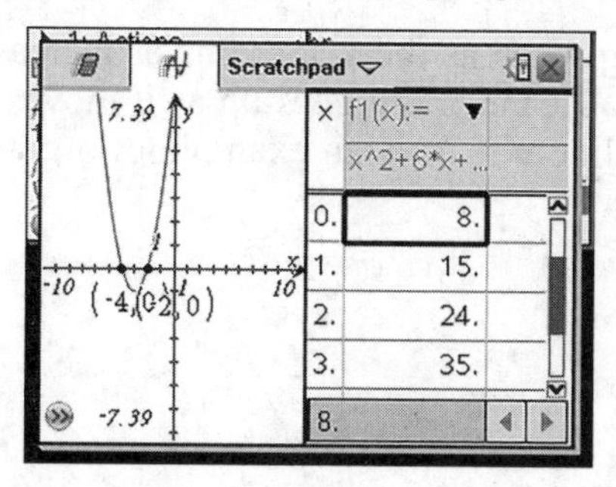

Graphing Solution Sets to Square Root Equations on the Graphing Calculator

To get the $\sqrt{}$ symbol on either TI calculator, press [2ND] key and then the [x^2] key. This symbol can be entered into an equation for graphing.

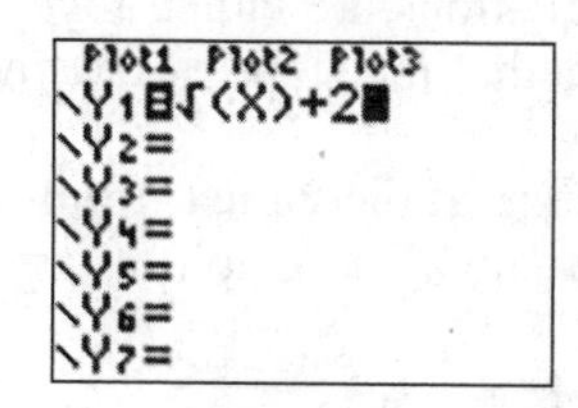

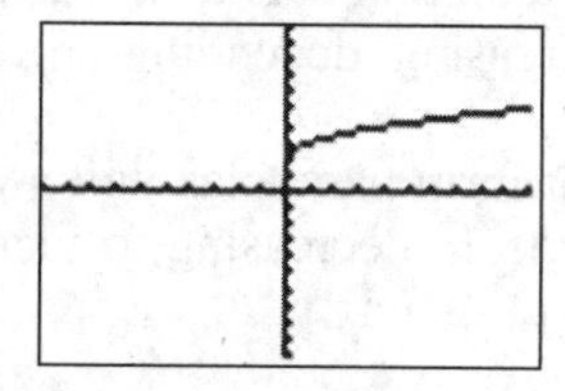

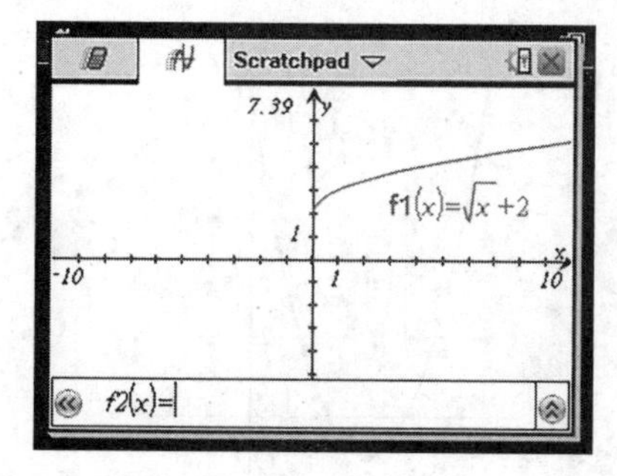

Intervals When a Graph Is Increasing or Decreasing

A graph is *increasing* when as the x-coordinates increase, the y-coordinates increase too. Informally, the graph goes up as it moves to the right.

The graph of the line $y = 2x$ is an example of a graph that is increasing everywhere.

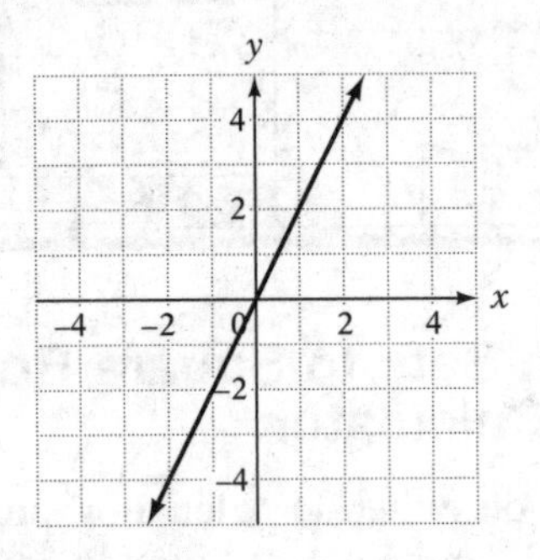

Graphs of the solution sets of linear equations are either always increasing or always decreasing, depending on whether the slope of the line is positive or negative.

Graphs that form parabolas will switch, at the vertex, from increasing to decreasing or from decreasing to increasing. The graph of $y = x^2 + 6x + 8$ has a vertex at (–3, –1).

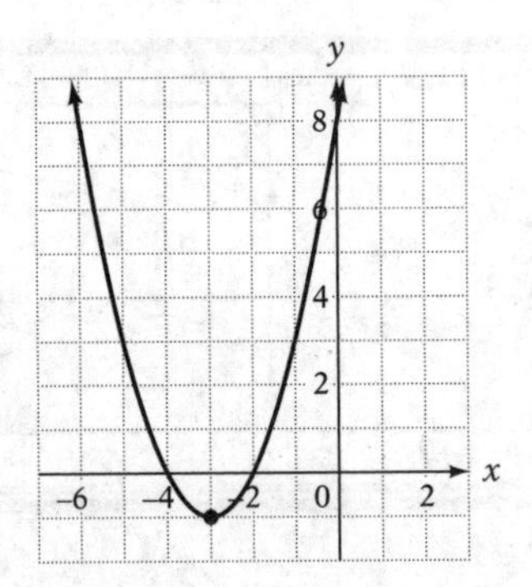

This curve is decreasing when $x < -3$ and is increasing when $x > -3$.

Check Your Understanding of Section 6.1

A. Multiple-Choice

1. Which is a point on the graph of the solution set of $y = x^2 + 5x - 2$?
   (1) (3, 19) (3) (3, 21)
   (2) (3, 20) (4) (3, 22)

2. The parabola defined by the equation $y = x^2 - 8x + 12$ has a y-intercept at
   (1) (0, –12) (3) (12, 0)
   (2) (0, 12) (4) (–12, 0)

3. Which is an x-intercept of the parabola defined by the equation $y = x^2 - 2x - 15$?
   (1) (0, –15) (3) (–5, 0)
   (2) (0, 15) (4) (5, 0)

4. What are the coordinates of the vertex of the parabola defined by the equation $y = x^2 - 4x - 1$?
   (1) (–2, 5) (3) (–2, –5)
   (2) (2, 5) (4) (2, –5)

5. What is the equation of the axis of symmetry of the parabola defined by the equation $y = x^2 - 6x - 2$?
   (1) $y = 3$ (3) $x = 3$
   (2) $y = -3$ (4) $x = -3$

6. $x = -4$ is the x-coordinate of the vertex for the parabola defined by which equation?
   (1) $y = x^2 + 8x + 3$
   (2) $y = x^2 - 8x + 3$
   (3) $y = x^2 + 4x + 3$
   (4) $y = x^2 - 4x + 3$

7. What could be the equation that determines this parabola?

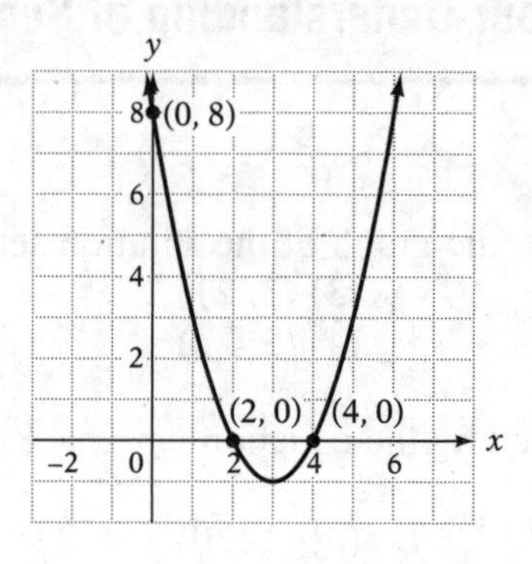

(1) $y = x^2 - 6x - 8$ (3) $y = x^2 + 6x - 8$
(2) $y = x^2 - 6x + 8$ (4) $y = x^2 + 6x + 8$

8. The axis of symmetry of the parabola defined by the equation $y = 3x^2 + 42x + 8$ is
(1) $x = -7$ (3) $x = -14$
(2) $x = 7$ (4) $x = 14$

9. Which is the graph of $y = -x^2 + 2x + 3$?

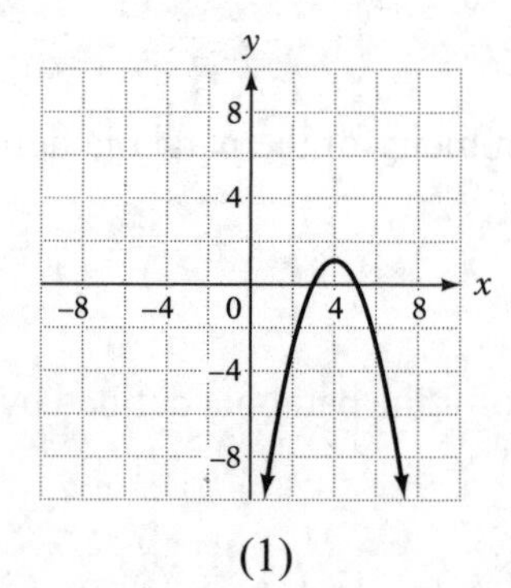

(1)

(3)

(2)

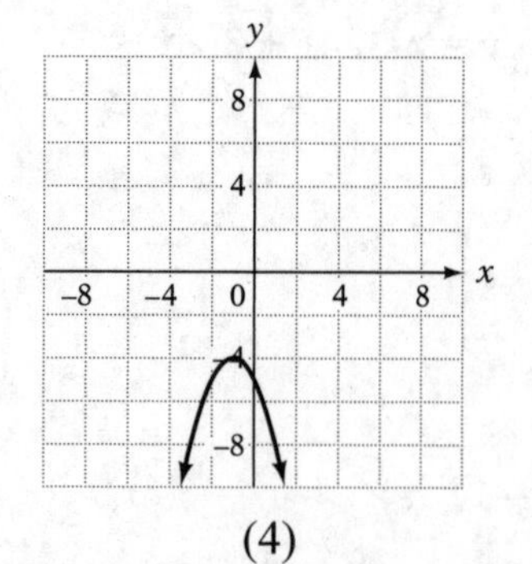

(4)

10. Based on this graph, what are the two solutions to the equation $x^2 - 2x - 1 = 2$?

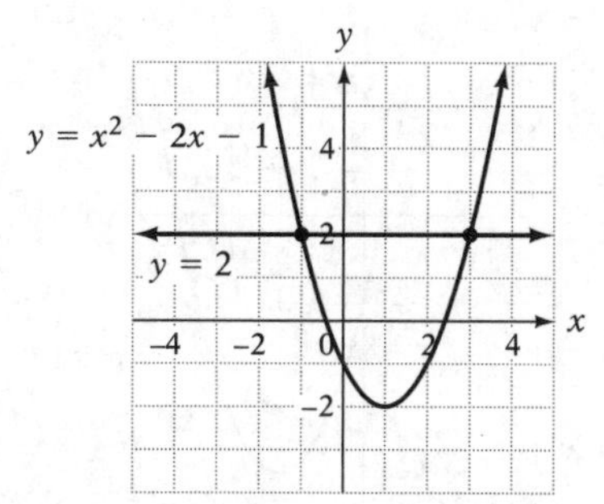

(1) $x = -3$ and $x = 1$
(2) $x = 3$ and $x = 1$
(3) $x = 3$ and $x = -1$
(4) $x = -3$ and $x = -1$

B. *Show how you arrived at your answers.*

1. The graph of the parabola defined by the equation $y = x^2 + bx + c$ has an axis of symmetry at $x = -3$. Find possible values for b and c.

2. The graph of the parabola defined by the equation $y = x^2 + bx + c$ has x-intercepts at (1, 0) and (–4, 0). What are possible values for b and c?

3. A portion of a parabola is graphed below. It will pass through the three points (1, 5), (6, 0), and vertex (4, –4). What are two other points on this parabola?

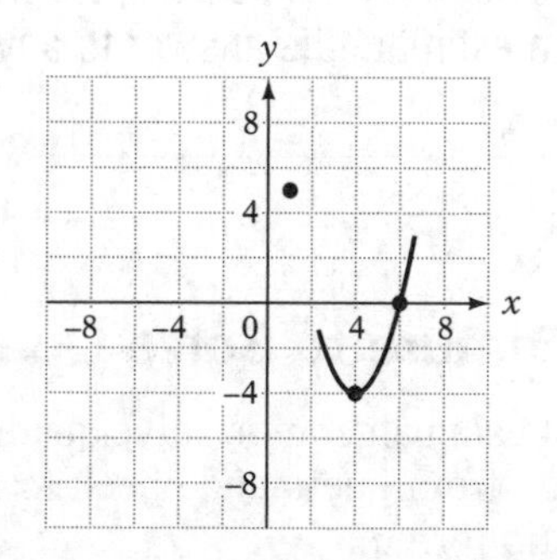

4. What are the coordinates of the vertex and the x-intercept(s) of the parabola defined by the equation $y = x^2 - 6x + 9$?

5. Below is the graph of $y = x^2 - 4x - 2$. What is the equation of the axis of symmetry of the graph of $y = (x - 1)^2 - 4(x - 1) - 2$?

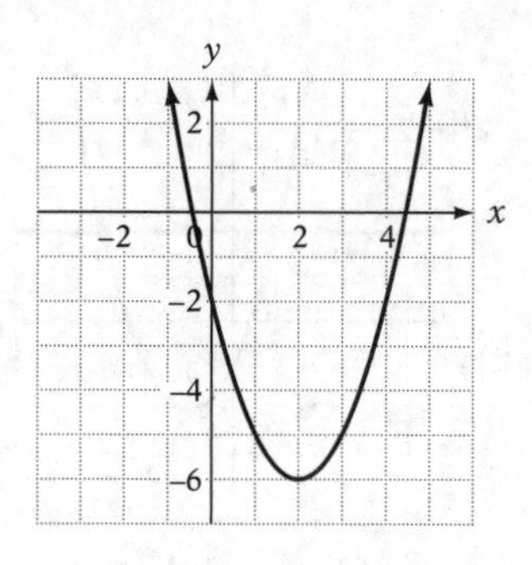

6.2 USING THE GRAPHING CALCULATOR TO SOLVE QUADRATIC EQUATIONS

If a quadratic equation does not require a solution that uses algebra, either because it is a multiple-choice question that does not require showing work, or because it is a free response question that does not say "only an algebraic solution will be accepted," it is possible for the graphing calculator to estimate the answer to any quadratic equation very quickly.

Using the Zeros Feature to Solve Quadratic Equations

There are several ways of using algebra to solve quadratic equations, outlined in Chapter 3. These range from guess and check to factoring to completing the square to the quadratic formula.

If there is an equation already in the form $ax^2 + bx + c = 0$, the solutions can be obtained by graphing the parabola and using the zeros feature of the graphing calculator to find the x-intercepts.

Example 1

Find the two solutions to the equation $2x^2 + 5x - 3 = 0$ on the graphing calculator.

Solution:

For the TI-84:

Plot the graph of $y = 2x^2 + 5x - 3$.

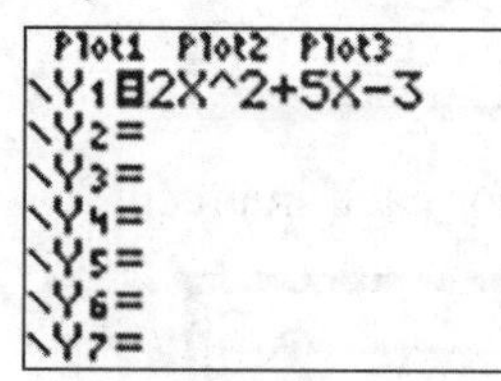

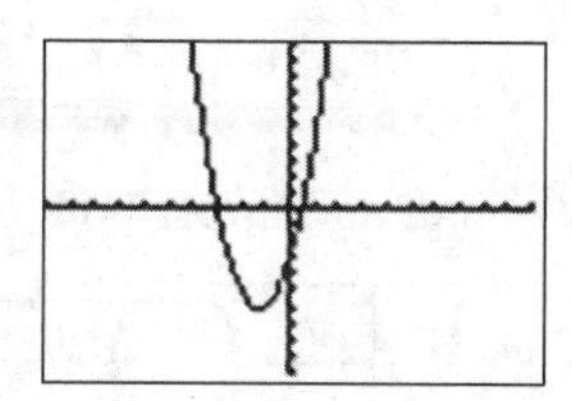

Press [2ND] and [TRACE] to get to the CALCULATE menu. Press [2] for zero. To find the first intercept, move the cursor to the left of it and press [ENTER] to set a lower bound; then move the cursor to the right of the intercept and press [ENTER] to set an upper bound. Move near the intercept and press [ENTER] again. Do this for the other intercept, too.

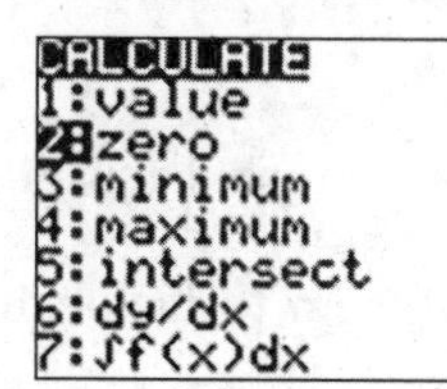

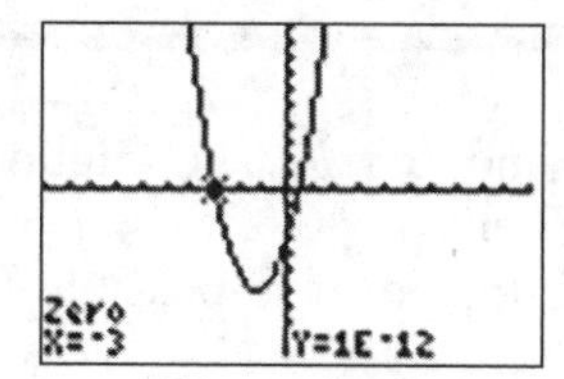

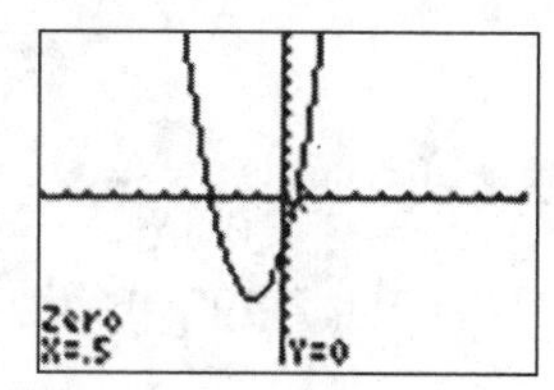

The first zero is at $x = -3$. The $y = 1E-12$ is a fancy way of saying $1 \cdot 10^{-12}$, which is very close to zero. The second zero is at $x = 0.5$. The x-coordinates of the x-intercepts are also the solutions to the equation $2x^2 + 5x - 3$ so the solutions to this quadratic equation are $x = -3$ or $x = 0.5$.

For the TI-Nspire:

Select [B] from the home screen to get to the graphing Scratchpad. Enter $2x^2 + 5x - 3$ on the entry line after $f1(x)$= and press [enter].

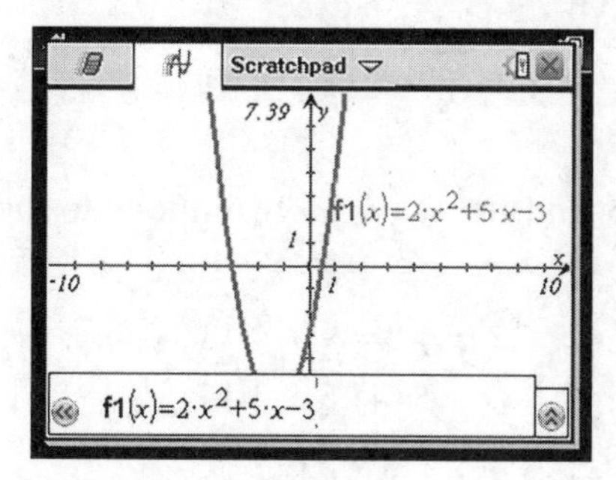

Press [menu], [6], and [1] to select the Zero finding feature.

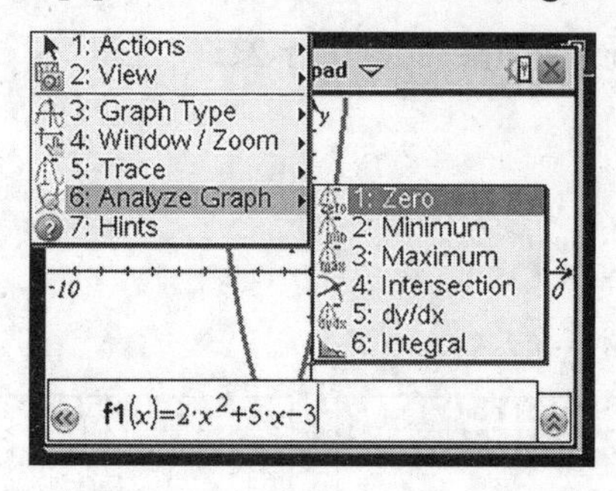

Select a lower bound and upper bound for each intercept.

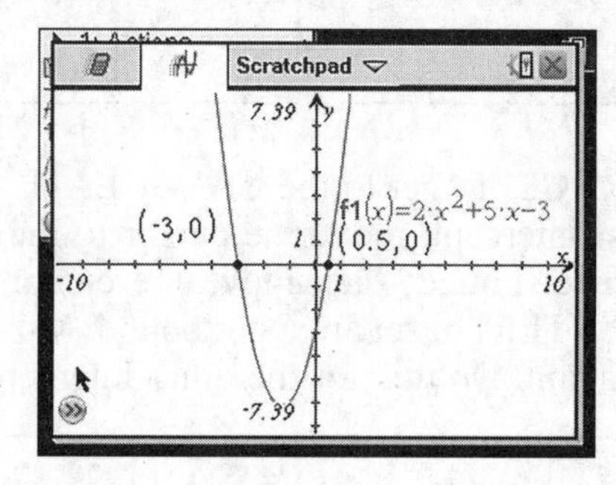

The x-intercepts are (–3, 0) and (0.5, 0). The solutions to the original equation $2x^2 + 5x - 3 = 0$ are $x = -3$ and $x = 0.5$.

After the zeros are determined, it is possible to use them to factor the original equation. A quadratic equation with zeros of –3 and $\frac{1}{2}$ can have factors of $(x + 3)$ and $\left(x - \frac{1}{2}\right)$.

But since $(x+3)\left(x - \frac{1}{2}\right)$ will not get the $2x^2$ as the first term, the factored quadratic must have been $2(x + 3)\left(x - \frac{1}{2}\right)$. Distribute the 2 through the $\left(x - \frac{1}{2}\right)$ term to become $(2x - 1)$.

So $2x^2 + 5x - 3 = (2x - 1)(x + 3)$.

Example 2

Use the calculator to find the two *exact* solutions to the quadratic equation $x^2 - 4x - 7 = 0$.

(1) $3 \pm \sqrt{11}$ (3) $4 \pm \sqrt{11}$

(2) $2 \pm \sqrt{11}$ (4) $5 \pm \sqrt{11}$

Solution: Using the Zero feature of the calculator the two x-intercepts of the parabola $y = x^2 - 4x - 7$ are approximately (5.31, 0) and (–1.32, 0). Check each answer choice to see which one is equivalent to these two answers.

Choice 1: $3 + \sqrt{11} = 6.32$, which is not either of the solutions.

Choice 2: $2 + \sqrt{11} = 5.31$, which is one of the answers. $2 - \sqrt{11} = -1.32$, which is the other answer for choice 2.

Solving Quadratic Equations in Different Forms with the Graphing Calculator

The method just described for solving quadratic equations of the form $ax^2 + bx + c = 0$ only works when the equation is in exactly that form. The graphing calculator can also quickly solve equations of the form $ax^2 + bx + c = d$.

Example 3

What are the two solutions to the equation $-x^2 - 4x + 6 = 1$?

Solution: One method for this equation is to subtract 1 from both sides of the equation and use the process for equations of the form $ax^2 + bx + c = 0$. An easier approach is to graph the two sides of the equation separately.

For the TI–84:

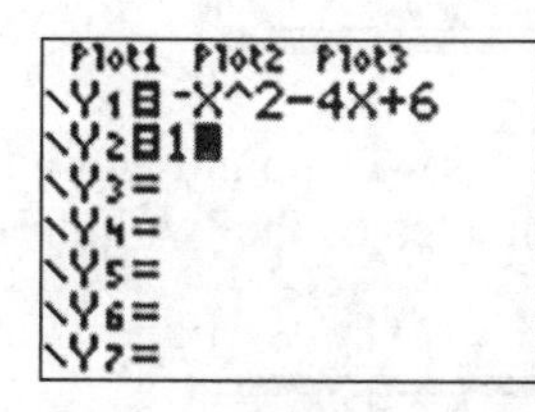

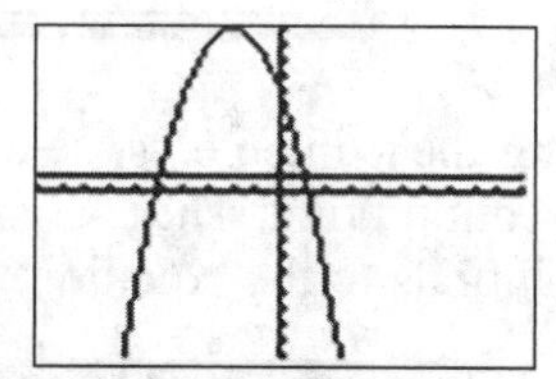

To find the intersection, press [2ND], [TRACE], and [5] for the INTERSECT feature. Press enter twice to set which curves to intersect and then move the cursor near the intersection and press [ENTER]. Do this for both intersection points.

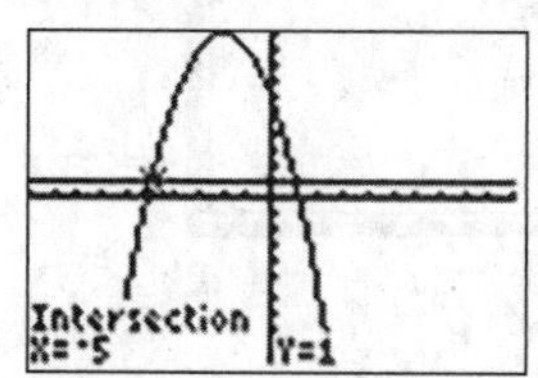

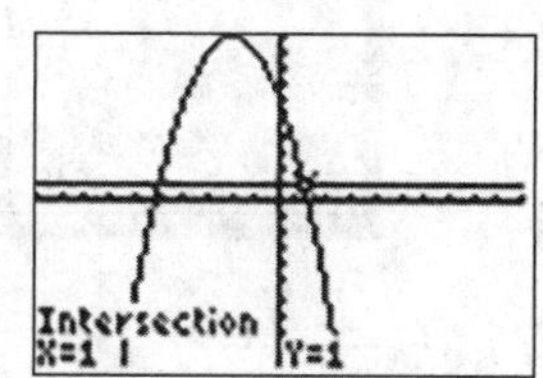

For the TI-Nspire:

Enter the equation $-x^2 - 4x + 6$ after $f1(x)$= and 1 after $f2(x)$=

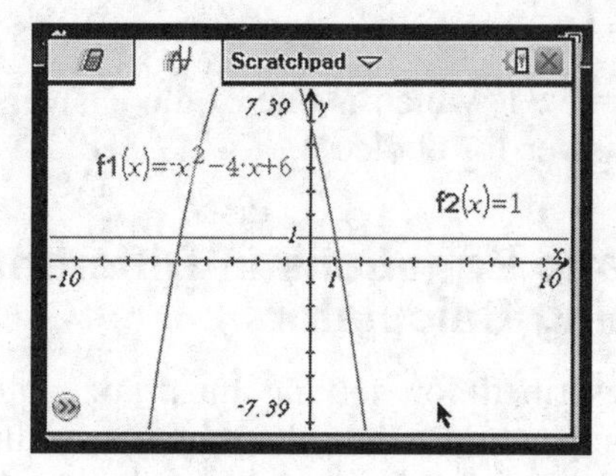

Find the intersection of the parabola with the horizontal line by pressing [menu], [6], and [4].

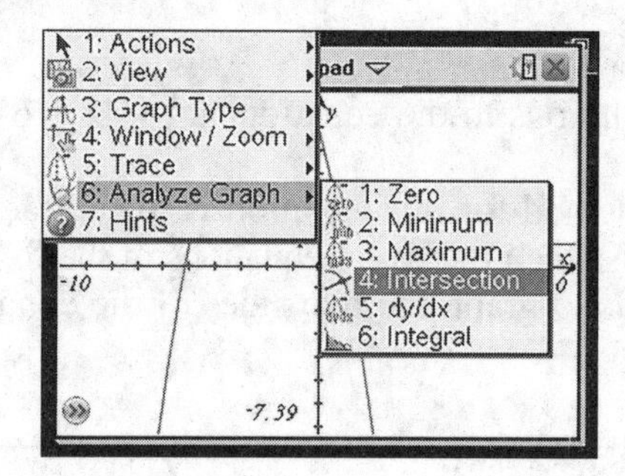

Just like finding the x-intercepts, select a lower bound and an upper bound for each intersection point. The x-coordinates of the two intersection points are the two solutions to the equation.

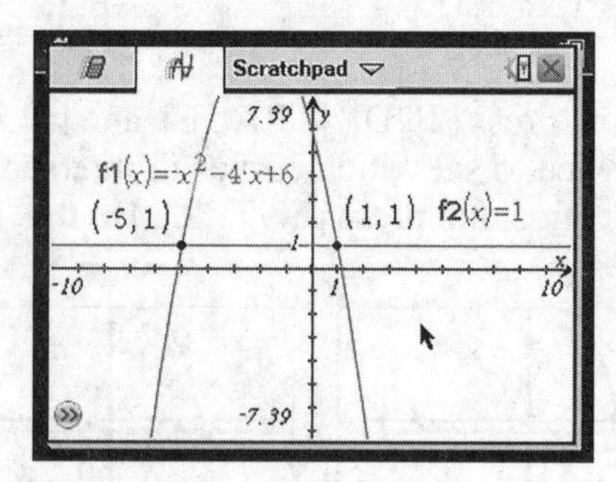

Check Your Understanding of Section 6.2

A. Multiple-Choice

1. What are the solutions to $x^2 + 3x - 4 = 0$?
(1) $x = 1, x = 4$
(2) $x = 1, x = -4$
(3) $x = -1, x = 4$
(4) $x = -1, x = -4$

2. What are the solutions to $x^2 - 3x - 10 = 0$?
(1) $x = 2, x = 5$
(2) $x = 2, x = -5$
(3) $x = -2, x = 5$
(4) $x = -2, x = -5$

3. Solve for all values of x, rounded to the nearest hundredth, $x^2 - 4x + 1 = 0$.
(1) 3.65, 0.24
(2) 3.81, 0.34
(3) 3.73, 0.27
(4) 3.96, 0.39

4. Solve for all values of x, rounded to the nearest hundredth, $x^2 + 10x + 23 = 0$.
(1) –3.41, –6.58
(2) –3.59, –6.41
(3) –3.31, –6.64
(4) –3.62, –6.18

5. Solve for all values of x: $x^2 + 6x + 2 = 0$.
(1) $-2 + \sqrt{7}, -2 - \sqrt{7}$
(2) $-3 + \sqrt{7}, -3 - \sqrt{7}$
(3) $-3 + \sqrt{11}, -3 - \sqrt{11}$
(4) $-2 + \sqrt{11}, -2 - \sqrt{11}$

6. Which graph could be used to find the solutions to the equation $x^2 - 4x = 5$?

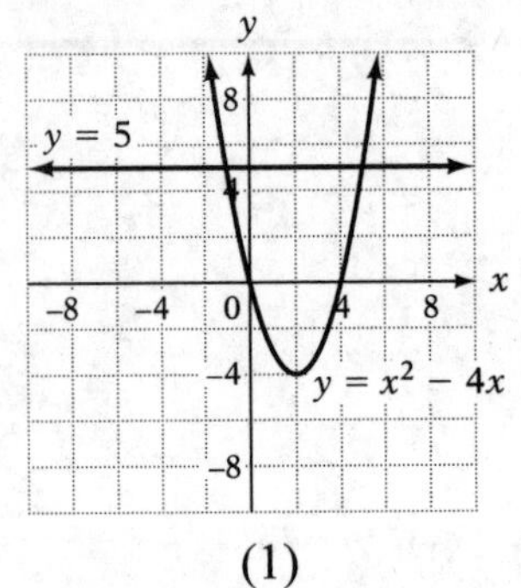

(1)

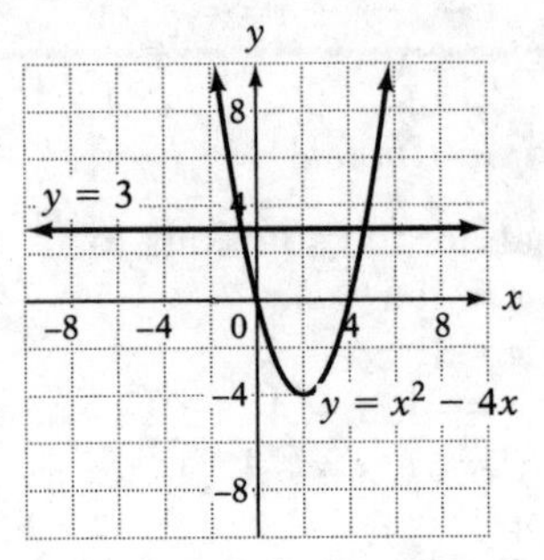

(3)

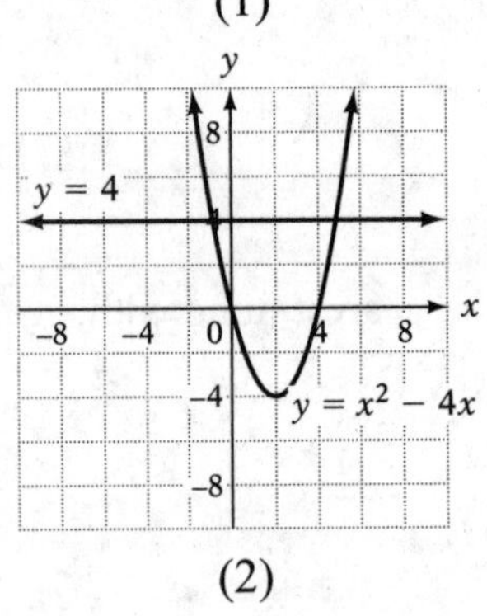

(2)

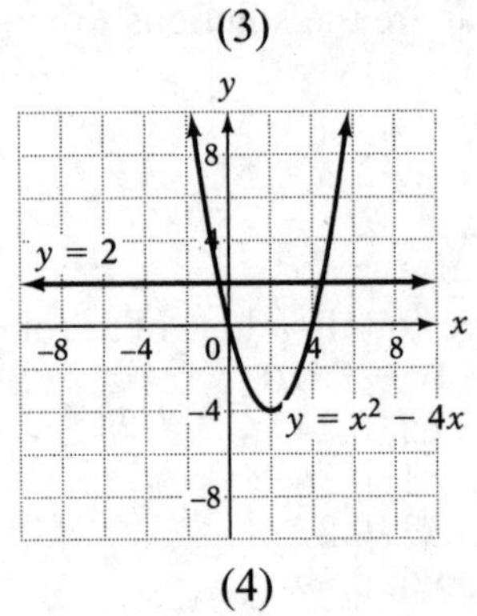

(4)

7. The x-intercepts of the parabola defined by which equation are the solutions to the equation $x^2 + 5x = 15$?

(1) $y = x^2 + 5x + 15$
(2) $y = x^2 + 5x$
(3) $y = x^2 + 5x - 15$
(4) $y = x^2 - 5x - 15$

8. The x-coordinates of the intersection of the line and the parabola are the solutions to which equation?

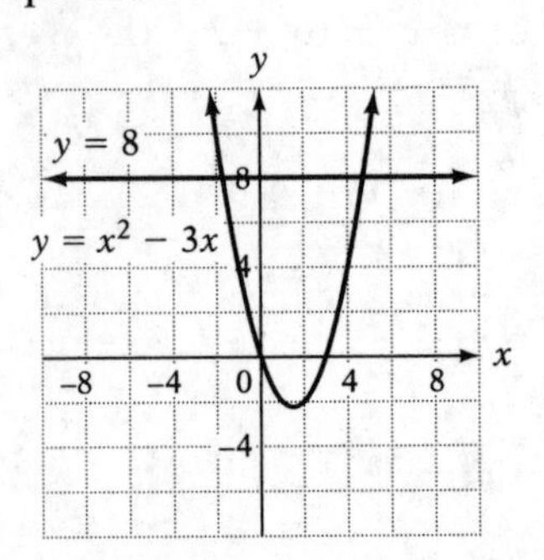

(1) $x^2 + 3x = 8$
(2) $x^2 + 8x = 3$
(3) $x^2 - 3x = 8$
(4) $x^2 - 8x = 3$

9. Which of these equations does not have the same solutions as the others?
(1) $x^2 - 8x - 3 = 0$
(2) $x^2 - 8x + 1 = 2$
(3) $x^2 - 8x = 3$
(4) $x^2 - 8x + 1 = 4$

10. This is a portion of the graph of the solution set of $y = x^2 - 14x + 47$. What are the approximate solutions to the equation $x^2 - 14x + 47 = 0$?

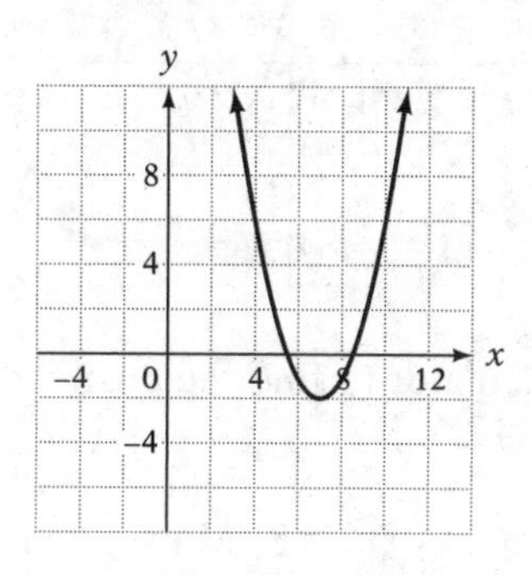

(1) 4 and 10
(2) 4.5 and 9.5
(3) 5 and 9
(4) 5.5 and 8.5

B. Show how you arrived at your answers.

1. Below is the graph of the equation $y = x^2 - 10x + 22$. Use it to estimate the solutions to the equation $x^2 - 10x + 22 = 0$.

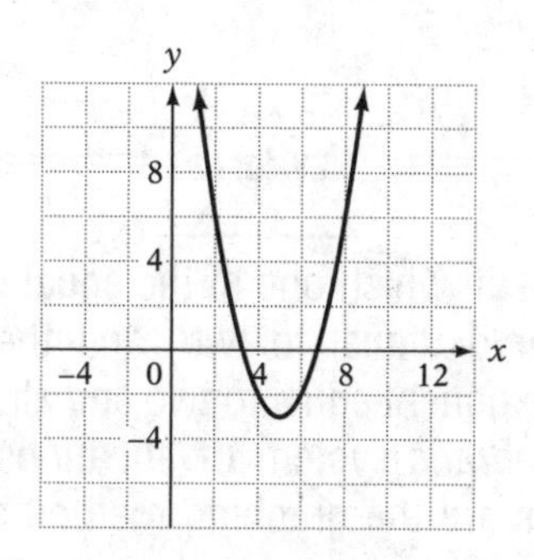

2. Use the graphing calculator to find the two solutions to the equation $x^2 = x + 1$.

3. The x-coordinates of the two intersection points of the line and the parabola are the solutions to what quadratic equation?

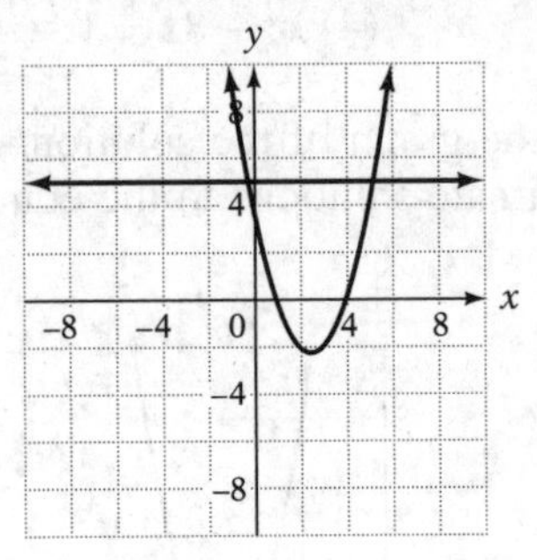

4. Use the graphing calculator to find the positive solution to the equation $0 = -16T^2 + 96T + 112$.

5. To solve the equation $x^2 - 5x = 3$, Mary graphs $y = x^2 - 5x$ and $y = 3$. Sofia solves it by graphing $y = x^2 - 5x - 3$. Christopher solves it by graphing $y = x^2$ and $y = 5x + 3$. Who is solving the question correctly?

6.3 SOLVING LINEAR-QUADRATIC SYSTEMS OF EQUATIONS

A system of equations in which one of the equations is linear (having no exponents greater or equal to two) and the other is quadratic (having at least one exponent equal to two and all others equal to one) is called a *linear-quadratic* system of equations. Two methods for solving such systems are the graphing method and the substitution method.

If the system of equations is

$$y = 2x^2$$
$$y = 2x + 4$$

All the ordered pairs that satisfy both equations at the same time are in the solution set for this system of equations. Unlike a system where both equations are linear and there is at most one ordered pair in the solution set, there are often two ordered pairs in a solution set for a system like this.

Solving a Linear-Quadratic System of Equations with a Table

If you were to make a table of values from $x = -3$ to $x = +3$, it would look like this:

x	$2x^2$	$2x + 4$
-3	$2(-3)^2 = 2(9) = 18$	$2(-3) + 4 = -6 + 4 = -2$
-2	$2(-2)^2 = 2(4) = 8$	$2(-2) + 4 = -4 + 4 = 0$
-1	$2(-1)^2 = 2(1) = 2$	$2(-1) + 4 = -2 + 4 = 2$
0	$2(0)^2 = 2(0) = 0$	$2(0) + 4 = 0 + 4 = 4$
1	$2(1)^2 = 2(1) = 2$	$2(1) = 4 = 2 + 4 = 6$
2	$2(2)^2 = 2(4) = 8$	$2(2) + 4 = 4 + 4 = 8$
3	$2(3)^2 = 2(9) = 18$	$2(3) + 4 = 6 + 4 = 10$

Just from this table, it can be seen that for $x = -1$, both $2x^2$ and $2x + 4$ equals 2. And for $x = 2$, both $2x^2$ and $2x + 4$ equals 8. So, two solutions are $(-1, 2)$ and $(2, 8)$.

Solving a Linear-Quadratic System of Equations with Algebra

If the equations are in the form $y = x^2 + bx + c$ and $y = mx + b$, set the expressions equal to each other and solve for x with algebra.

For the system

$$y = 2x^2$$
$$y = 2x + 4$$

form the equation $2x^2 = 2x + 4$.

Then solve the quadratic equation by first subtracting $2x$ and 4 from both sides so it is in the form where one of the sides of the equation is zero.

$$2x^2 = 2x + 4$$
$$-2x = -2x$$
$$2x^2 - 2x = 4$$
$$-4 = -4$$
$$2x^2 - 2x - 4 = 0$$
$$2(x^2 - x - 2) = 0$$
$$2(x - 2)(x + 1) = 0$$
$$x - 2 = 0 \text{ or } x + 1 = 0$$
$$x = 2 \text{ or } x = -1$$

To get the y values of the solutions, substitute the x values into either equation.

For $x = 2$, $y = 2(2)^2 = 8$ so $(2, 8)$ is one solution.

For $x = -1$, $y = 2(-1)^2 = 2 \cdot 1 = 2$ so $(-1, 2)$ is the other solution. These are the same solutions obtained from the table method used above.

Example 1

Use an algebraic method to find the two solutions to the system of equations:

$$y = x^2 - 5x + 3$$
$$y = 2x - 3$$

Solution: Since both expressions are equal to y, they are equal to each other.

$$x^2 - 5x + 3 = 2x - 3$$

To get this quadratic equation into the form with zero on the right-hand side of the equation, subtract $2x$ from both sides and add 3 to both sides.

$$\begin{aligned} x^2 - 5x + 3 &= 2x - 3 \\ -2x &= -2x \\ x^2 - 7x + 3 &= -3 \\ +3 &= +3 \\ x^2 - 7x + 6 &= 0 \\ (x - 6)(x - 1) &= 0 \\ x &= 6 \text{ or } x = 1 \end{aligned}$$

For $x = 6$, $y = 2(6) - 3 = 12 - 3 = 9$
For $x = 1$, $y = 2(1) - 3 = 2 - 3 = -1$

The solutions are $(6, 9)$ and $(1, -1)$.

Solving a Linear-Quadratic System of Equations with a Graph

Using the table at the top of the previous page, the graph can be produced below. The two points where the line and the parabola intersect are $(-1, 2)$ and $(2, 8)$.

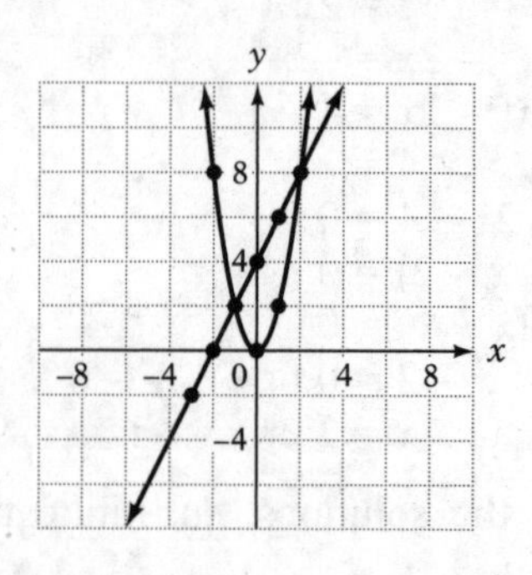

Solving a Linear-Quadratic System of Equations with the Graphing Calculator

Just as with a system of two linear equations, by graphing the two equations on the graphing calculator and using the INTERSECT feature, the calculator can estimate the points of intersection that correspond to the solution set to the system of equations.

For the TI-84:

Press [y=] and enter both equations.

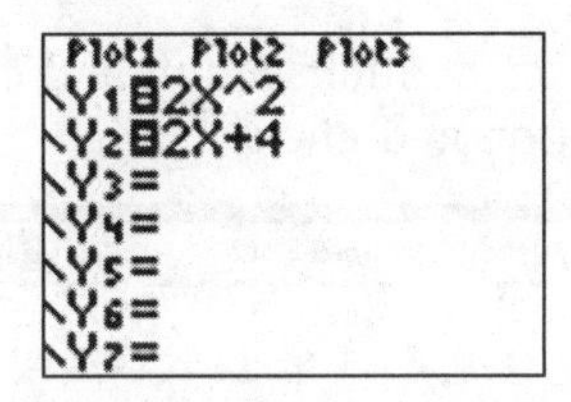

Press [GRAPH] then [2ND], [TRACE], and [5]. Select the two curves and move the cursor near one of the intersections and press [ENTER]. Then do the same for the other intersection.

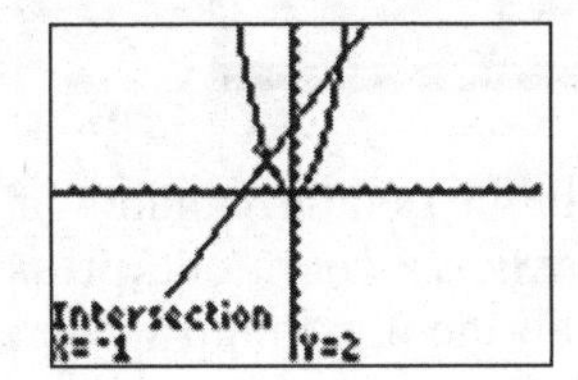

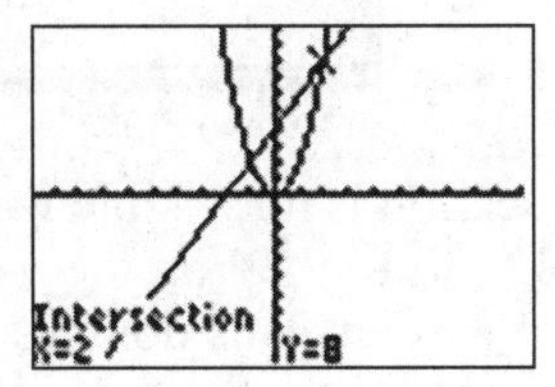

For the TI-Nspire:

From the home screen, press [B] for the Scratchpad Graph. Enter $2x^2$ after $f1(x)$= and $2x + 4$ after $f2(x)$=

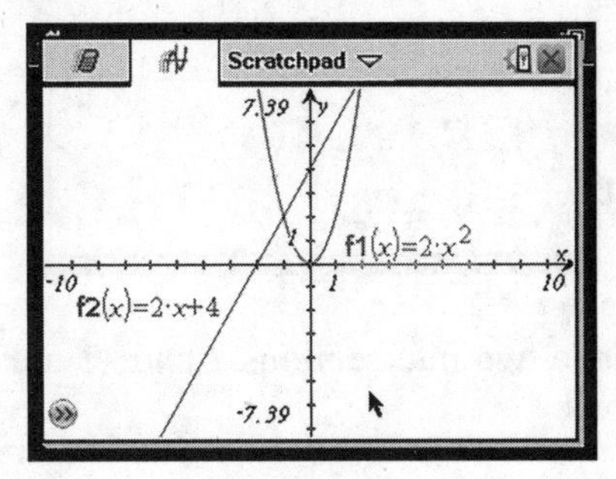

Press [menu], [6], and [4] for Intersection.

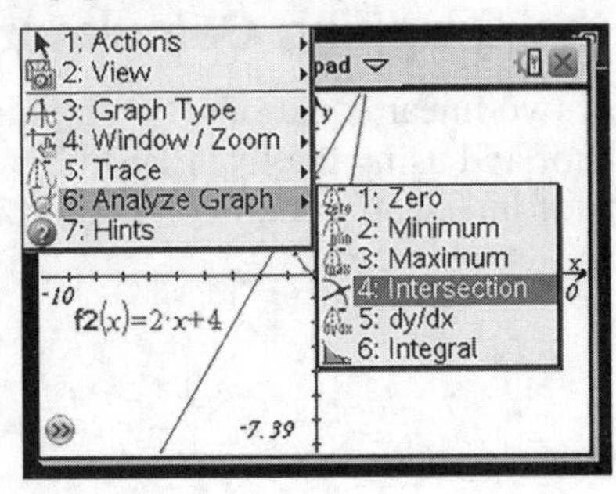

Move the hand to the left of the intersection and click; then move the hand to the right of the intersection and click again.

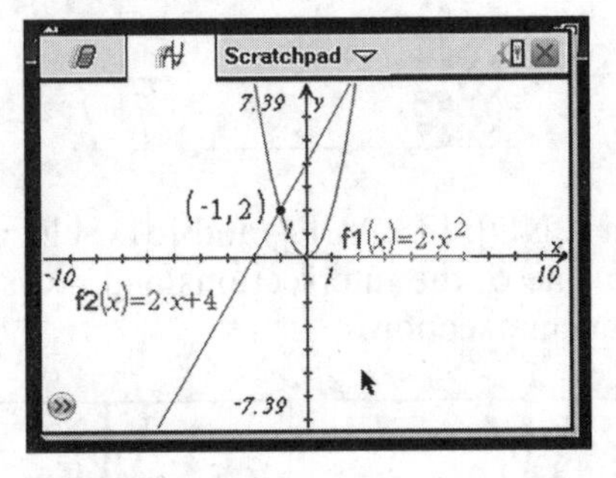

The other intersection is not visible with the default window of –10 to 10 on the x-axis and –7.39 to 7.39 on the y-axis. To zoom out, press [menu], [4], and [4] for Zoom—Out. This doubles all the maximum values. Now follow the same steps you did for the first intersection point.

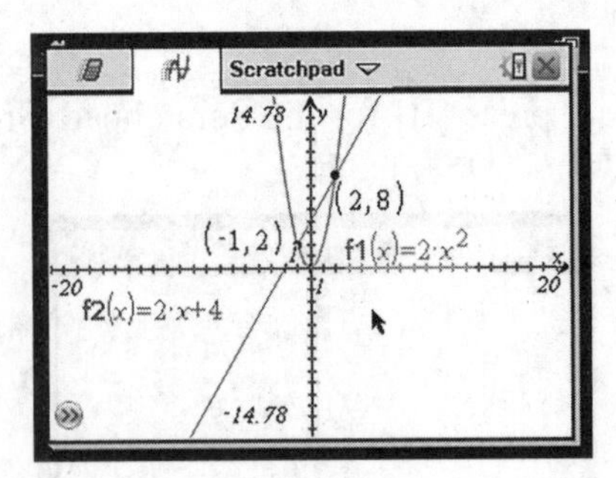

The x-coordinates of the two intersections 2 and –1 are the two solutions to the system of equations.

$$y = 2x^2$$
$$y = 2x + 4$$

Example 2

Use the graphing calculator to find the solutions to the system of equations.

$$y = x^2$$
$$y = 2x + 3$$

Solution: There are two intersections (–1, 1) and (3, 9). So $x = -1, y = 1$ is one solution and $x = 3, y = 9$ is the other solution.

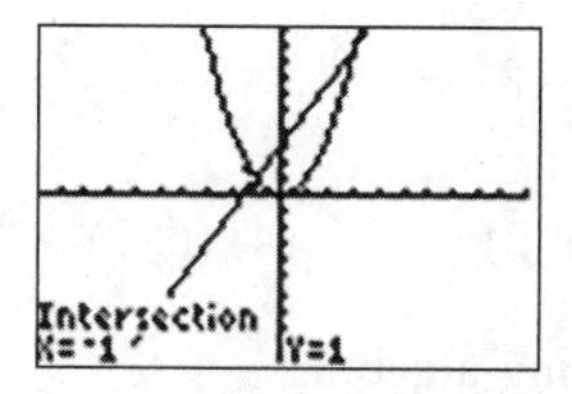

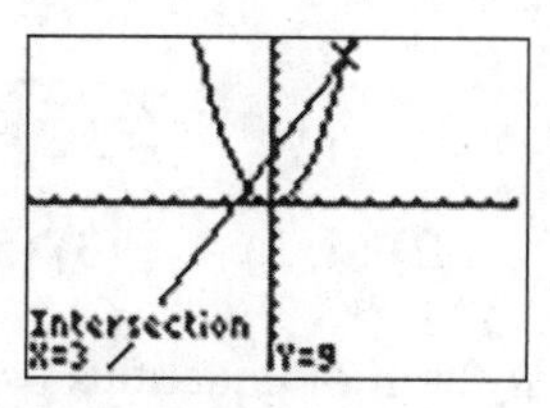

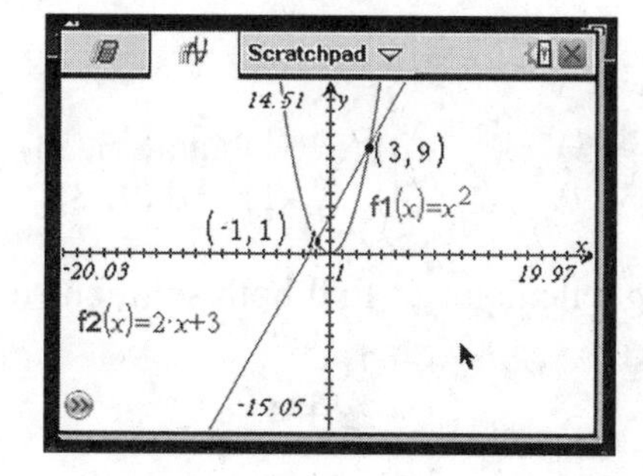

Check Your Understanding of Section 6.3

A. Multiple-Choice

1. Use the graphing calculator to find which ordered pair is a solution to the system.

$$y = x^2$$
$$y = x + 2$$

(1) (1, 4) (2) (2, 4) (3) (3, 4) (4) (4, 4)

2. Use the graphing calculator to find which ordered pair is a solution to the system.

$$y = 2x^2$$
$$y = -x + 3$$

(1) (1, 1) (2) (1, 2) (3) (1, 3) (4) (1, 4)

3. Use the graphing calculator to find which ordered pair is a solution to the system.

$$y = x^2 - 8x + 17$$
$$y = -x + 7$$

(1) (3, 2) (2) (4, 2) (3) (5, 2) (4) (6, 2)

4. Solve this system of equations using algebra.

$$y = x^2$$
$$y = 2x + 3$$

(1) (–1, 1) and (9, 3) (3) (–1, 1) and (–3, 9)
(2) (–1, 1) and (3, 9) (4) (1, –1) and (9, 3)

5. Use the graphing calculator to find both solutions to the system.

$$y = 2x^2$$
$$y = 2x + 4$$

(1) (–1, –2) and (–2, 8) (3) (1, 2) and (2, 8)
(2) (1, –2) and (2, –8) (4) (–1, 2) and (2, 8)

6. Use the graphing calculator to find both solutions to the system.

$$y = 3x^2$$
$$y = 3x$$

(1) (0, 0) and (–1, 3) (3) (0, 0) and (–1, –3)
(2) (0, 0) and (1, –3) (4) (0, 0) and (1, 3)

7. Solve this system of equations algebraically.

$$y = x^2 - 2x + 2$$
$$y = -x + 8$$

(1) (3, 5) and (2, 10)
(2) (3, 5) and (–2, 10)
(3) (–3, 5) and (–2, 10)
(4) (3, –5) and (2, –10)

8. Solve this system using any method you want.

$$y = x^2 + 2x - 2$$
$$y = x + 4$$

(1) (2, 6) and (–3, 1)
(2) (–2, –6) and (3, –1)
(3) (2, –6) and (–3, –1)
(4) (–2, 6) and (–3, 1)

9. Which system of equations could this graph be used to solve?

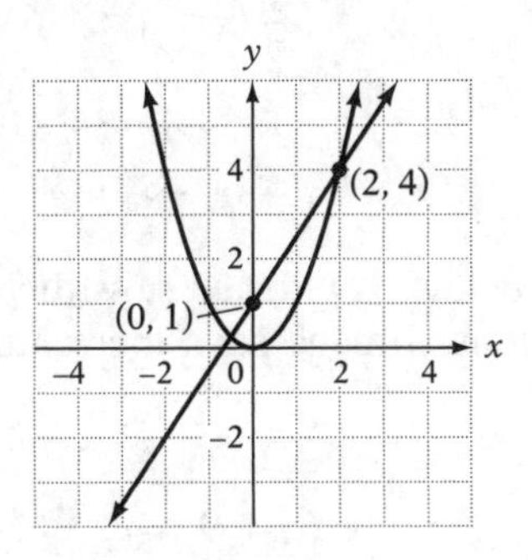

(1) $y = x^2$
$y = \frac{2}{3}x + 1$

(2) $y = x^2$
$y = \frac{3}{2}x - 1$

(3) $y = x^2$
$y = \frac{3}{2}x + 1$

(4) $y = x^2$
$y = \frac{2}{3}x - 1$

10. Graphically find all solutions to the system of equations.

$$y = x^2$$
$$y = 4x - 4$$

(1) (2, 4) only
(2) (2, 4) and (–2, 4)
(3) (2, 4) and (0, 0)
(4) (2, 4) and (2, –4)

B. *Show how you arrived at your answers.*

1. Find the two solutions to the system of equations using the graphing calculator.

$$y = \frac{1}{2}x^2$$
$$y = x + 4$$

2. Solve this system using an algebraic method. Any other method will not be accepted.

$$y = -x^2 + 4x + 5$$
$$y = x + 5$$

3. Use a graphing calculator to find the two solutions to the system of equations.

$$y = -\frac{1}{2}(x - 4)^2 + 6$$

$$y = -\frac{1}{3}x + 6$$

4. Based on the graph below, Kaydence says there is just one solution to the system of equations. Jimena says there are two solutions. Who is correct, and why?

$$y = 3x^2$$
$$y = 6x + 9$$

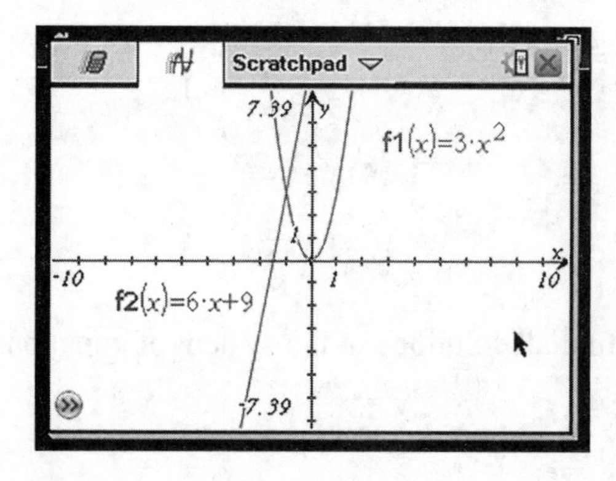

5. Find a value of b so that the system of equations has exactly (a) 2 solutions and (b) 0 solutions.

$$y = x^2$$
$$y = 2x + b$$

6.4 GRAPHING QUADRATIC EQUATIONS FOR REAL-WORLD APPLICATIONS

Many real-world scenarios can be modeled with quadratic equations. When a quadratic equation is given as a model for something, the techniques of solving quadratic equations either with algebra or with the graphing calculator can be used to solve the real-world problem.

The Height of a Projectile

When an object is thrown in the air, it rises to a maximum height and then drops back down. When x is the amount of time that has passed since the projectile was thrown and y is the height of the projectile at time x, the graph is a parabola. Its equation will be a quadratic equation with a negative coefficient for the x^2.

Example

A football is thrown in the air. Its height in the air after x seconds is determined by the equation $y = -16x^2 + 80x + 96$. Use the graphing calculator to determine (a) when the football will be 160 feet high and (b) when the football will land on the ground.

Solution:

For the TI–84:

Enter the equation into the graphing calculator.

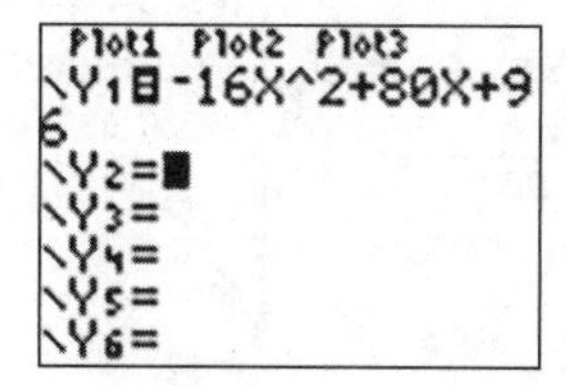

If you try to graph by pressing [ZOOM] and [6] for ZStandard, the graph doesn't look much like a parabola. This is because only a small part of the parabola fits in this viewing *window*.

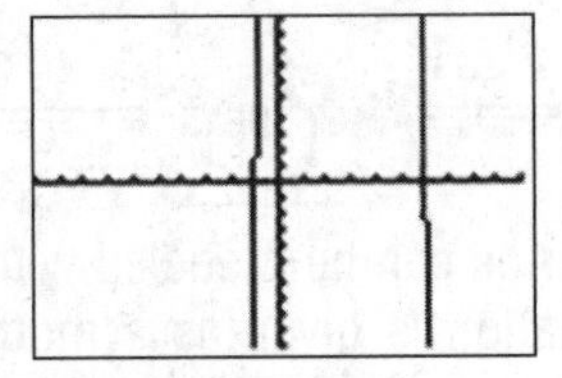

To fix this, the window needs to be adjusted. Press the [WINDOW] key.

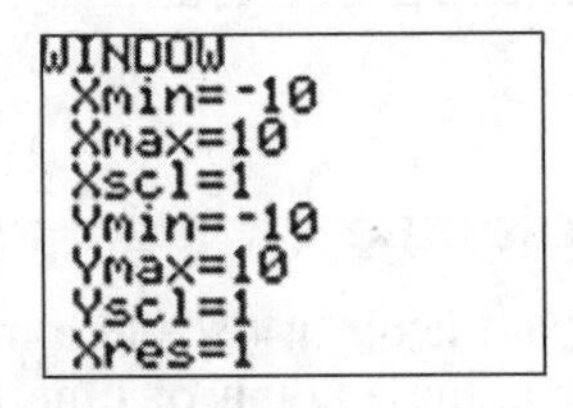

These six values need to be adjusted. Since we are only concerned with what happens after the ball is thrown, the Xmin can be set to 0. Since the ball does not go underground, set the Ymin to 0 also. The Xmax can be set to 6 since in the graph it can be seen that the ball is on the ground at time $x = 6$. Finally, the Ymax needs to be increased since we cannot see where the ball has its maximum height. Set the Ymax to 100. Xscl is the "scale" for how often it puts marks on the x-axis. Since there are only 6 units on the x-axis, an Xscl of 1 is appropriate. Since there will be 100 units on the y-axis, the marks need to be more spread out. A Yscl of 10 is appropriate.

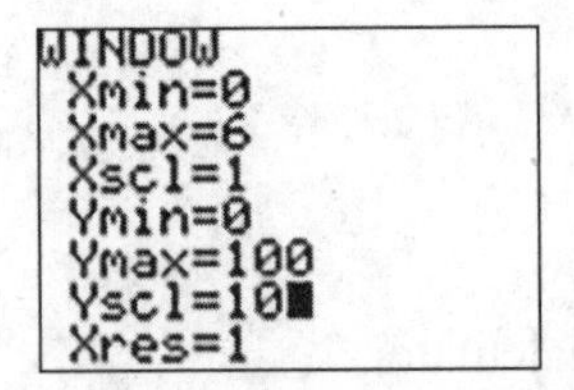

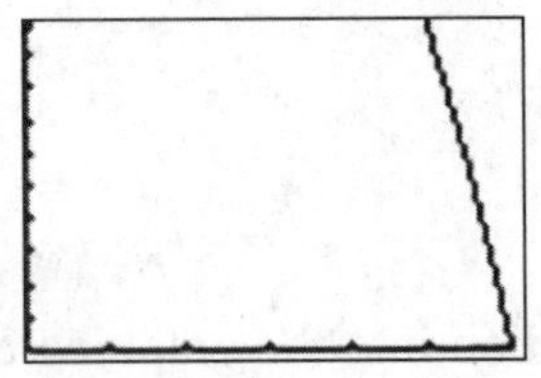

This looks better, but the Ymax needs to be increased even more. Try 200.

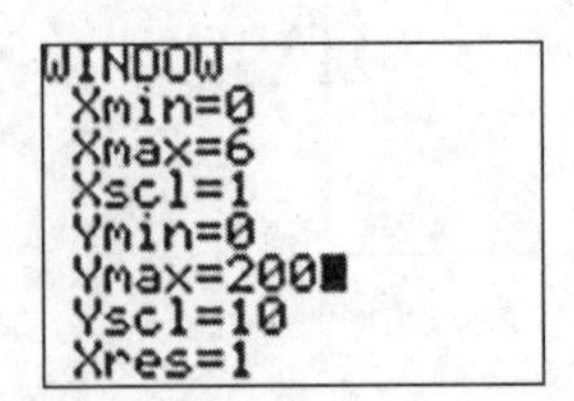

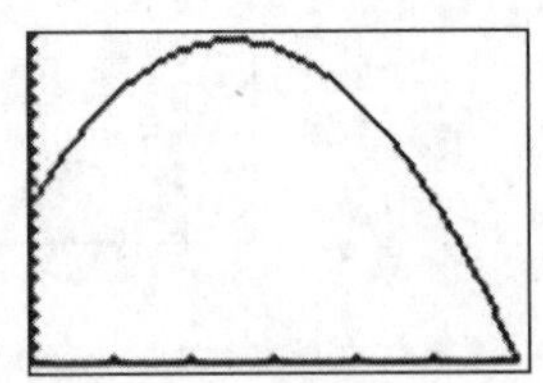

To find when the football will be 160 feet high, graph the line $y = 160$ on the same coordinate plane.

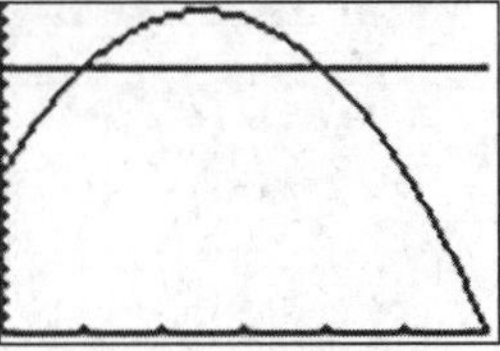

Use the INTERSECT feature to find the two times that the football is 160 feet high, once on the way up and once on the way down.

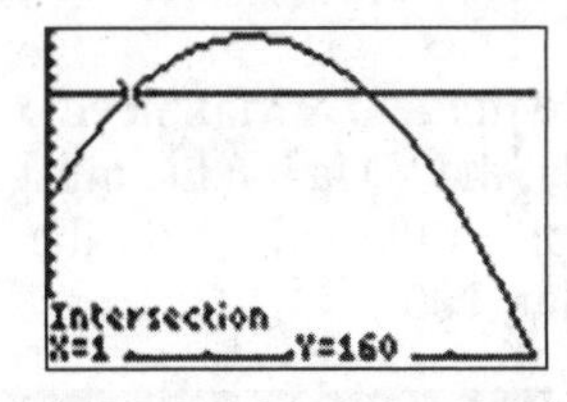

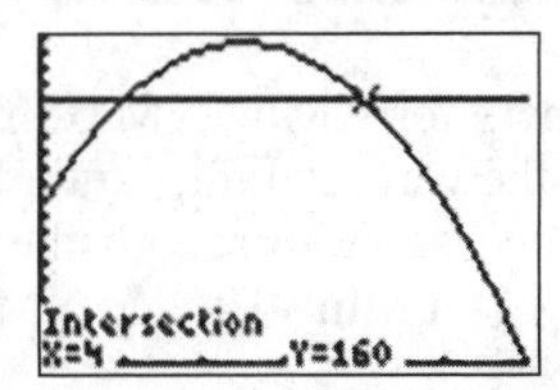

It is 160 feet high after 1 second and again after 4 seconds.

To find when the football is on the ground, use the ZERO feature.

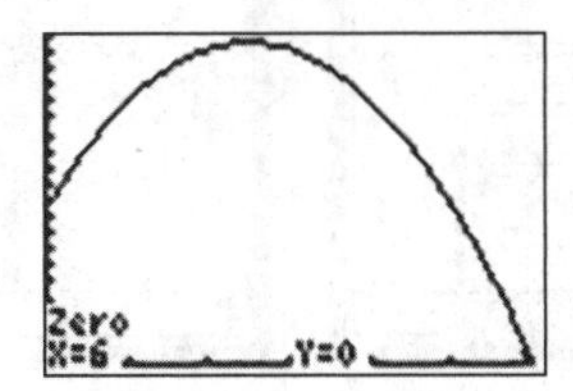

For the TI-Nspire:

Type $16x^2 + 80x + 96$ after the $f1(x) =$ in the entry line and press [enter].

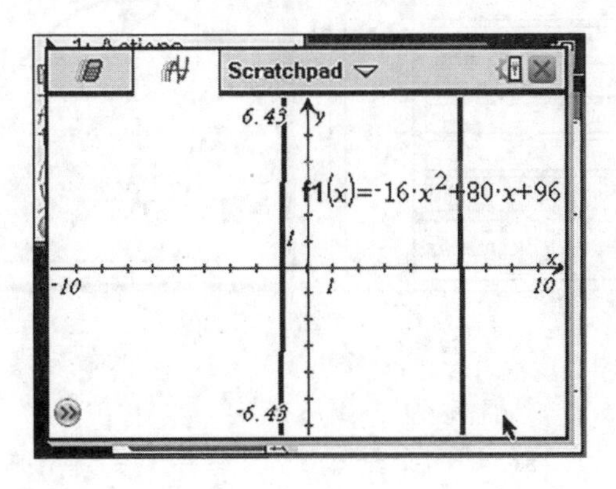

This is just a small portion of the parabola because the viewing window needs to be adjusted. Press [menu], [4], and [1] for Window Settings.

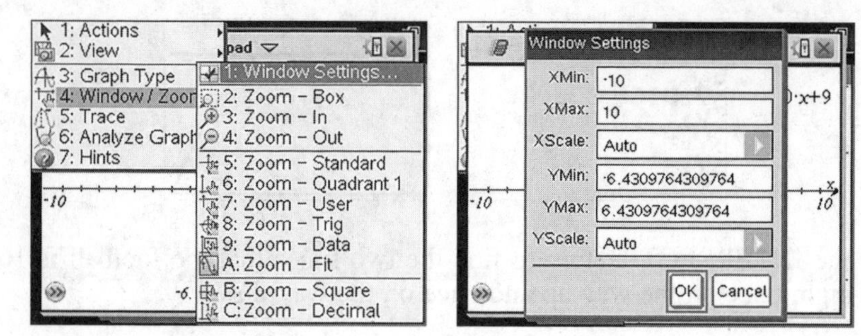

The four numbers for XMin, XMax, YMin, and YMax need to be adjusted. In order to see the vertex of the parabola, the YMax value must be increased. Also it is not necessary to see all the negative values on the x-axis. Make XMin 0, XMax 7, YMin –10, and YMax 100.

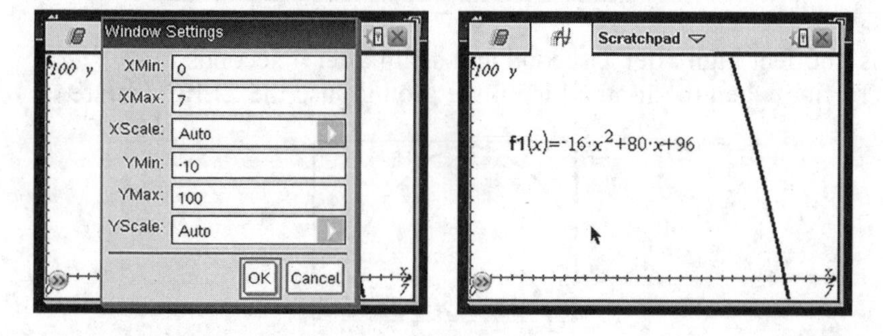

The YMax was not high enough. Adjust it to 200.

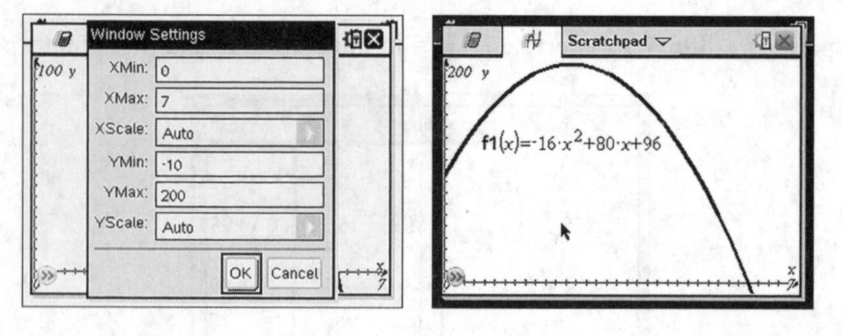

To find when the football is 160 feet high, graph $y = 160$ by typing 160 after $f2(x) =$.

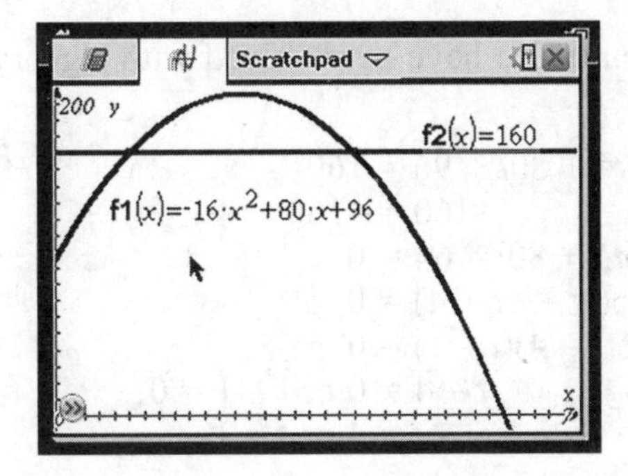

Now press [menu], [6], and [4] for Intersection and locate the two intersection points.

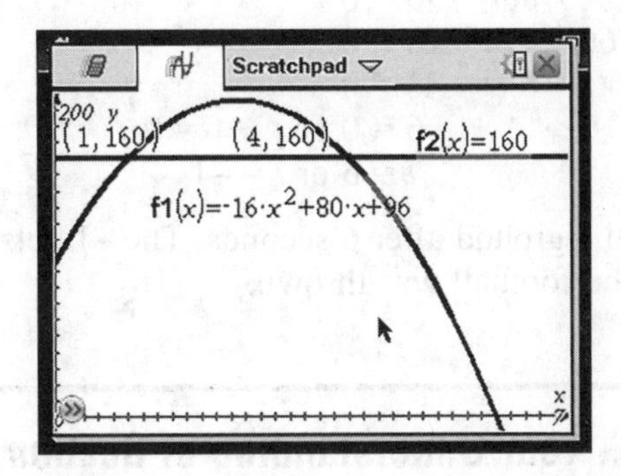

The x-coordinates of the intersection points are the two answers 1 and 4. The football is 160 feet high after 1 second and after 4 seconds.

To find when the football lands on the ground, press [menu], [6], and [1] for Zero and locate the x-intercept.

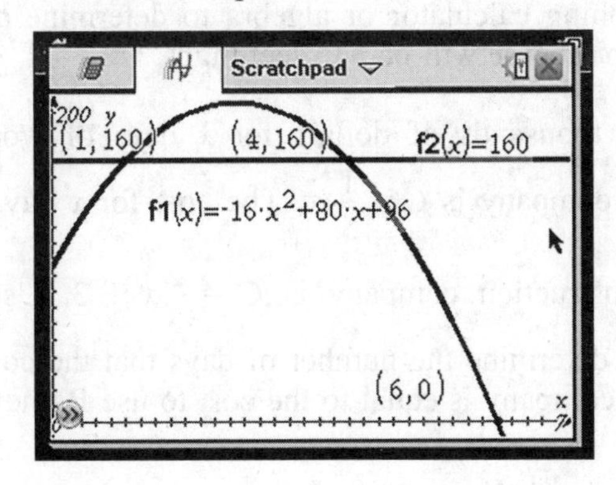

The x-intercept is (6, 0), which means that the height of the football after 6 seconds is 0 feet, which means that it has landed on the ground after 6 seconds.

This question could also have been solved with algebra.

For the first part,

$$\begin{aligned} -16t^2 + 80t + 96 &= 160 \\ -160 &= -160 \\ -16t^2 + 80t - 64 &= 0 \\ -16(t^2 - 5t + 4) &= 0 \\ -16(t - 4)(t - 1) &= 0 \\ t - 4 = 0 &\text{ or } t - 1 = 0 \\ t = 4 &\text{ or } t = 1 \end{aligned}$$

The football is 160 feet high after 1 second and after 4 seconds.

For the second part,

$$\begin{aligned} -16t^2 + 80t + 96 &= 0 \\ -16(t^2 - 5t - 6) &= 0 \\ -16(t - 6)(t + 1) &= 0 \\ t - 6 = 0 &\text{ or } t + 1 = 0 \\ t = 6 &\text{ or } t = -1 \end{aligned}$$

The football is on the ground after 6 seconds. The –1 gets rejected since –1 seconds is before the football was thrown.

Check Your Understanding of Section 6.4

***B.** Show how you arrived at your answers.*

1. A projectile's height is modeled by the equation $h = -16t^2 + 96t + 256$. Use the graphing calculator or algebra to determine after how many seconds the projectile will be 336 feet high.

2. The cost in thousands of dollars for x days of work from Fred's construction company is $C = \frac{1}{5}x^2$. The cost for x days of work from Barney's construction company is $C = \frac{2}{5}x + 3$. Use the graphing calculator to determine the number of days that the cost to use Fred's construction company is equal to the cost to use Barney's construction company.

3. A stunt man drops from the top of a 50 foot elevator shaft. His height above ground is determined by the equation $h = -16t^2 + 50$. At the moment the stunt man begins to drop, an elevator goes up so that its height above ground is determined by the equation $h = 34t$. After how many seconds will the stunt man land on top of the elevator?

4. A car falls off a bridge that is 100 feet high. The height of the car off the ground is $h = -16t^2 + 100$. At the moment the car falls, a superhero flies up from the ground. Her height off the ground is $h = 18t$. After how many seconds does the superhero catch the car, and how high in the air will the superhero be when she catches it?

5. In a new video game "Irate Iguanas," two iguanas are shot from slingshots at the same time. Both are traveling to the right at a speed of one foot per second. The height off the ground of the first iguana is determined by the equation $h = -t^2 + 6t + 27$. The height off the ground of the other iguana is $h = \frac{3}{2}t + 18$, where t is the amount of time in seconds. When will the two irate iguanas crash into each other?

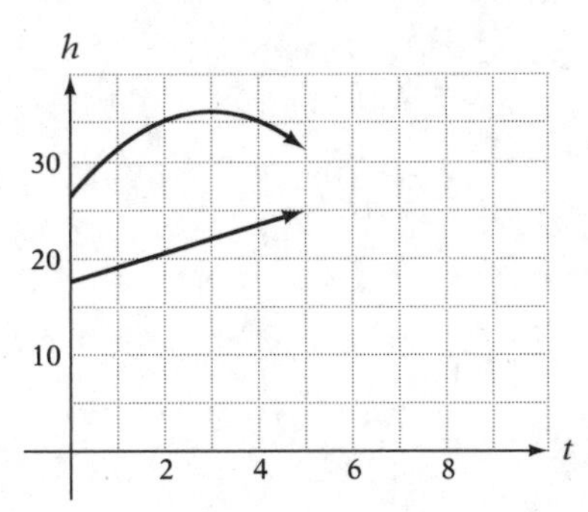

Chapter Seven LINEAR INEQUALITIES

7.1 ONE-VARIABLE LINEAR INEQUALITIES

A linear equation line $x + 2 = 10$ has just one solution, $x = 8$. This changes when the = is replaced with a > or a <. The inequality $x + 2 > 10$ does not have 8 as a solution, but it does have many other solutions including 9, 10, and 11. Solving one-variable linear inequalities is very similar to solving one-variable linear equations with one important exception, dealing with negative coefficients.

Solving an inequality like $x + 2 > 10$ is a lot like solving the equality $x + 2 = 10$. Just as the addition property of equality and subtraction property of equality allow you to add or subtract to both sides of an equation and keep it true, you can add to, or subtract from, both sides of an inequality and keep it true.

Use algebra to isolate the variable by subtracting 2 from both sides of the inequality.

$$\begin{aligned} x + 2 &> 10 \\ -2 &= -2 \\ x &> 8 \end{aligned}$$

The solution set includes all the numbers greater than 8. Integers like 9, 10, and 11 are part of the solution set as well as number like 8.1, 8.01, and 8.001. The number 8 is not part of the solution set since it is not true that $8 > 8$.

For an inequality that has a positive coefficient, like $2x > 10$, isolate the variable by dividing both sides of the inequality by the coefficient.

$$\frac{2x}{2} > \frac{10}{2}$$
$$x > 5$$

All numbers greater than 5 satisfy the inequality $2x > 10$.

When both sides of an inequality are divided by a negative number, the direction of the inequality sign must be reversed. For example, it is true that $6 > 4$ but if both sides are divided by -2, the inequality would say $-3 > -2$, which is not true.

MATH FACTS

When an inequality has a positive coefficient, you can divide both sides of the inequality by that coefficient without changing the direction of the inequality sign. When an inequality has a negative coefficient, divide both sides by it and reverse the direction of the inequality sign.

For the inequality $-2x > 10$, divide both sides by -2 and reverse the direction of the inequality sign.

$$\frac{-2x}{-2} > \frac{10}{-2}$$
$$x < -5$$

Two-step algebra inequalities are solved the same as two-step algebra equalities, still reversing the direction of the inequality sign when dividing both sides by a negative.

Example 1

What is the solution set to the inequality $-3x + 2 \geq 17$?

(1) $x > -5$ (2) $x \geq -5$ (3) $x < -5$ (4) $x \leq -5$

Solution:

$$-3x + 2 \geq 17$$
$$-2 = -2$$
$$\frac{-3x}{-3} \geq \frac{15}{-3}$$
$$x \leq -5$$

The answer is choice (4).

Example 2

What is the smallest integer that is a solution for x in the inequality $-5x - 2 < -22$?

(1) 3
(2) 4
(3) 5
(4) There is no smallest integer solution.

Solution: First, isolate the x to determine the entire solution set.

$$-5x-2<-22$$
$$+2=+2$$
$$\frac{-5x}{-5}<\frac{-20}{-5}$$
$$x>4$$

All numbers greater than 4 satisfy the original inequality. Of all the integers greater than 4, 5 is the smallest so the answer is 5, choice (3). If the last line had been $x \geq 4$, then the smallest value that would be a solution would be 4.

Example 3

If $3x + ax + 4 > 8$, and if $x = -2$, what is the greatest integer value for a that satisfies this inequality?

(1) –6 (2) –5 (3) –4 (4) –3

Solution: Substitute $x = -2$ into both x values in the inequality.

$$3(-2)+a(-2)+4>8$$
$$-6-2a+4>8$$
$$-2a-2>8$$
$$+2=+2$$
$$\frac{-2a}{-2}>\frac{10}{-2}$$
$$a<-5$$

Of all the integers less than –5, –6 is the greatest one. The answer is choice (1).

Check Your Understanding of Section 7.1

A. Multiple-Choice

1. What is the solution set for $2x > 10$?
(1) $x < 5$
(2) $x \leq 5$
(3) $x > 5$
(4) $x \geq 5$

2. What is the solution set for $3x < -18$?
(1) $x < -6$
(2) $x > -6$
(3) $x \geq -6$
(4) $x \leq -6$

3. What is the solution set for $-4x \geq 20$?
(1) $x \geq -5$
(2) $x > -5$
(3) $x \leq -5$
(4) $x < -5$

4. What is the solution set for $-5x + 12 \leq -8$?
(1) $x \leq -4$
(2) $x \leq 4$
(3) $x \geq 4$
(4) $x > 4$

5. What is the smallest integer that satisfies the inequality $-6x < -18$?
(1) 3
(2) 4
(3) 5
(4) 6

6. What is the smallest integer that satisfies the inequality $-6x \leq -18$?
(1) 2
(2) 3
(3) 4
(4) 5

7. What is the greatest integer that satisfies the inequality $-3x + 4 > -17$?
(1) 4
(2) 5
(3) 6
(4) 7

8. What is the greatest integer that satisfies the inequality $-5x - 2 \geq -37$?
(1) 6
(2) 7
(3) 8
(4) 9

9. Which number satisfies the inequality $-4x + 3 < 23$?
(1) -7
(2) -6
(3) -5
(4) -4

10. What is the smallest number that satisfies the inequality $-\frac{2}{3}x - 4 \leq 6$?

(1) –15
(2) –16
(3) –17
(4) –18

B. *Show how you arrived at your answers.*

1. Graph the solution set to the inequality $3x + 4 \geq 19$ on a number line.

2. Alejandra says that $-2x > 6$ has the solution set $x > -3$. Lorenzo says that it has the solution set $x < -3$. Which student is correct, and why?

3. Genevieve solves the inequality $-4x < 8$ the following way:

$$-4x < 8$$
$$+4x = +4x$$
$$0 < 4x + 8$$
$$-8 = -8$$
$$-\frac{8}{4} < \frac{4x}{4}$$
$$-2 < x$$
$$x > -2$$

Is this a valid way of finding the solution set? Explain why or why not.

4. What is the solution set of the inequality $5x - 2 \geq 2x + 10$?

5. Josephine has \$60 and wants to go to a restaurant. The bill, including a 20% tip, must be no more than \$60. The equation for what the price of the meal can be without the tip is $x + 0.20x \leq 60$. What is the solution set for this inequality?

7.2 GRAPHING TWO-VARIABLE LINEAR INEQUALITIES

The graph of a linear equation, like $x + y = 10$, is a line. When the = is replaced with a >, <, ≤, or a ≥, the graph becomes a *half-plane*. The process of creating the graph for the solution set to the inequality first requires graphing a line and then deciding which side of the line to shade in.

Graphing Inequalities Involving ≤ or ≥ Signs

The equation $x + y = 10$ had an infinite number of solutions including (2, 8), (3, 7), and (4, 6). When the = is replaced with the ≥ symbol, it becomes the two-variable inequality $x + y \geq 10$.

(2, 8), (3, 7), and (4, 6) are still solutions to this inequality, but also (2, 9), (3, 8), and (4, 7) since $2 + 9 = 11 \geq 10$, $3 + 8 = 11 \geq 10$, and $4 + 7 = 11 \geq 10$. The solution set for the inequality must contain the line and also these other points.

To graph the solution set to the inequality $x + y \geq 10$, first graph the line for the equation $x + y = 10$ using any of the methods from Chapter 5.

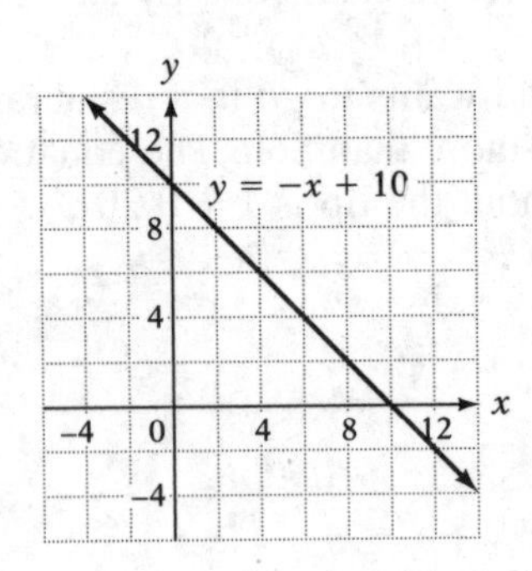

Every line divides the plane into two halves. Unless the line passes through (0, 0), one of those halves contains the point (0, 0) and the other one does not.

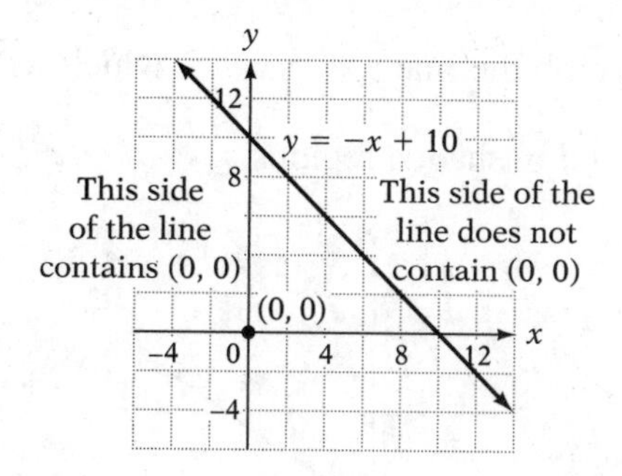

One of the sides of this line will be shaded. To determine which side to shade, pick a point that is not on the line, generally (0, 0) (as long as the line doesn't pass through (0, 0)) and substitute into the inequality to see if it makes the inequality true.

Testing (0, 0)

$$x + y \geq 10$$
$$0 + 0 \geq 10$$
$$0 \geq 10$$

Since 0 is not greater than or equal to 10, (0, 0) is not part of the solution set so the side of the line that does not contain (0, 0) gets shaded.

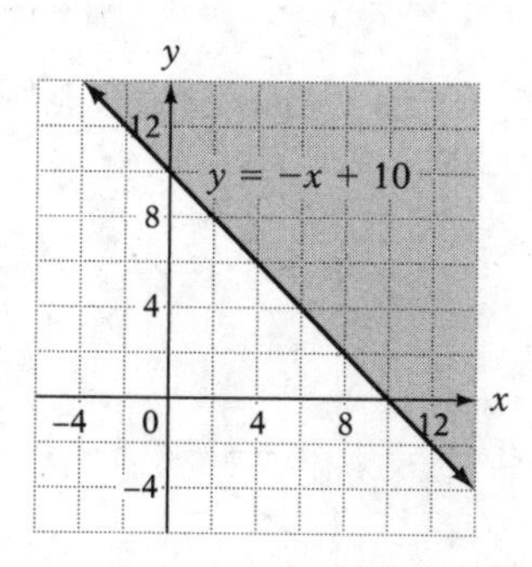

Example 1

Graph the solution set to the inequality $y \geq \frac{2}{3}x - 5$.

Solution: First, graph the line $y = \frac{2}{3}x - 5$, which will be the boundary between the shaded and unshaded regions.

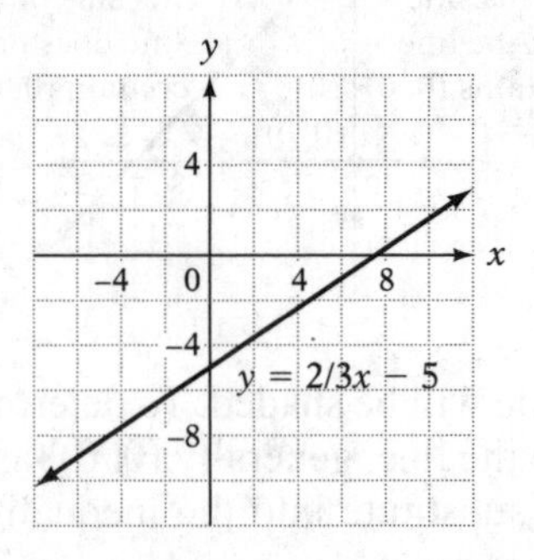

Since (0, 0) is not on the line, test to see if (0, 0) satisfies the inequality.

$$y \geq \frac{2}{3}x - 5$$

$$0 \geq \frac{2}{3}(0) - 5$$

$$0 \geq 0 - 5$$

$$0 \geq -5$$

Since 0 is greater than –5, (0, 0) is part of the solution set. Shade the side of the line that contains (0, 0).

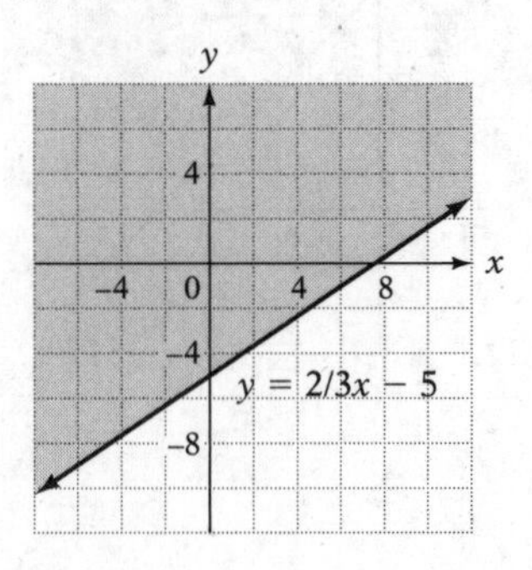

Graphing Inequalities Involving > or < Signs

When the two-variable inequality contains a > or < sign, the only difference is that when the line is first graphed it must be a dashed line. For the inequality $x + y > 10$, the ordered pair (2, 8) does not make it true since 2 + 8 is equal to 10, not greater than 10. None of the ordered pairs that satisfy the equation $x + y = 10$ also satisfies the equation $x + y > 10$. But the points (2, 9), (2, 8.1), and (2, 8.01) do make the inequality true so the rule for shading still applies.

To graph $x + y > 10$, first graph $x + y = 10$ (or $y = -x + 10$ if first changed to slope-intercept form) but instead of making the line solid, make it dashed.

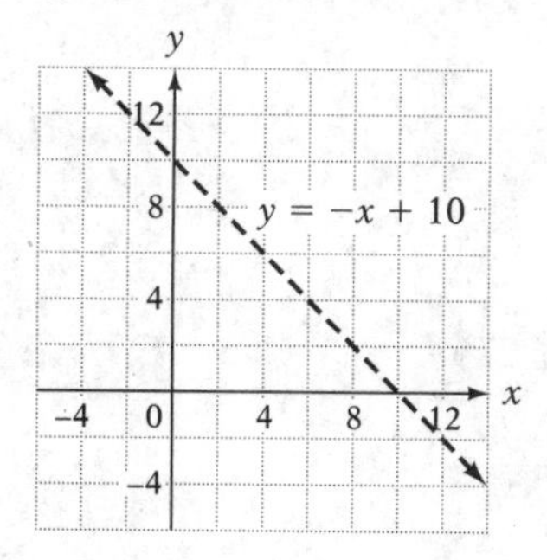

Check (0, 0) to determine whether to shade the side containing (0, 0) or the other side of the dashed line.

$$0 + 0 > 10$$
$$0 > 10$$

Since this is not true, shade the side of the dashed line that does not contain (0, 0).

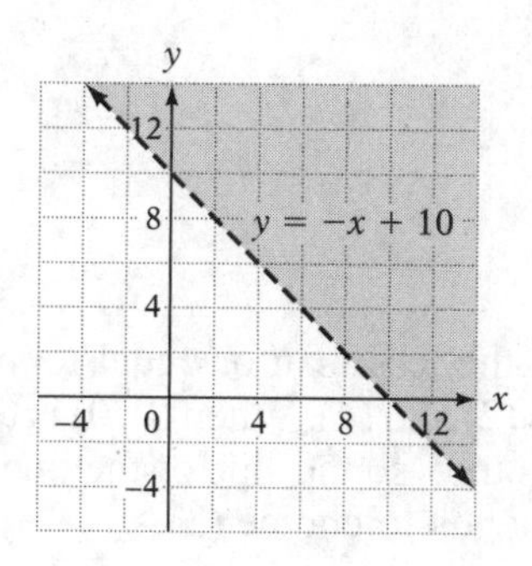

MATH FACTS

If the symbol in a two-variable inequality is a greater than sign (>) or a less than sign (<), the line separating the shaded half plane from the unshaded half plane will be dashed. If the symbol is a greater than or equal to sign (≥) or a less than or equal to sign (≤), the line will be solid.

Example 2

Which of the following could be the graph of $y \leq -2x + 5$?

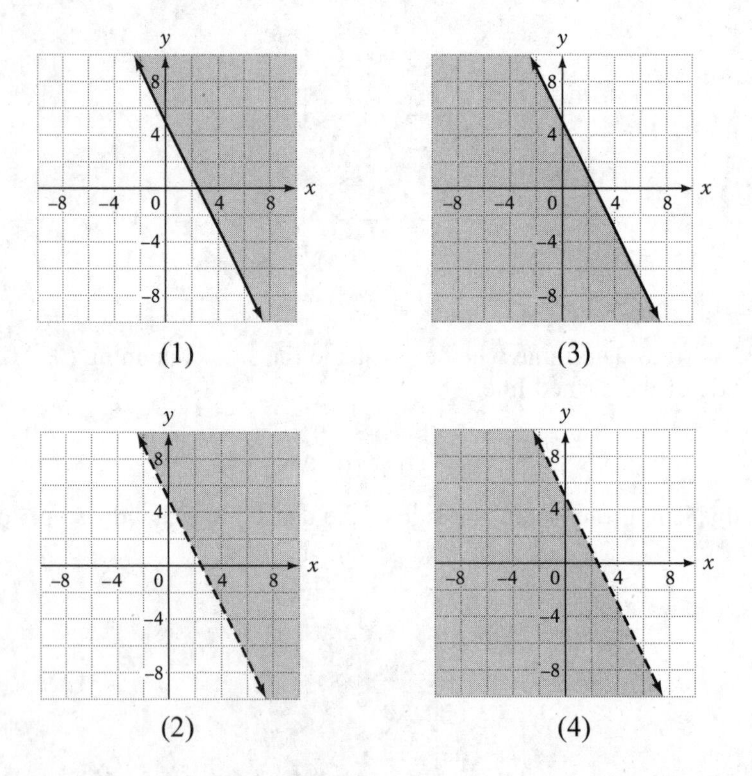

Solution: Because of the less than or equal to symbol, the line must be solid, eliminating choices (2) and (4). When (0, 0) is tested into the inequality, it becomes $0 \leq 5$, which is true, so the side of the line containing (0,0) should be shaded, which corresponds to choice (3).

Graphing Two-Variable Inequalities on the Graphing Calculator

Both the TI–84 and the TI–Nspire can graph inequalities. Before entering the inequality into the calculator, it must be in slope-intercept form. For the inequality $2x + 4y > 12$, isolate the y first.

$$2x + 4y > 12$$
$$-2x = -2x$$
$$\frac{4y}{4} > \frac{-2x+12}{4}$$
$$y > -\frac{2}{4}x + 3$$
$$y > -\frac{1}{2}x + 3$$

For the TI–84:

To graph inequalities on the TI–84, press the [APPS] button and then scroll down to an application called Inequalz and press [ENTER] twice. This opens the inequalities application and the Y= menu. When the cursor is moved to the equals sign, the options for the inequality signs appears at the bottom of the screen.

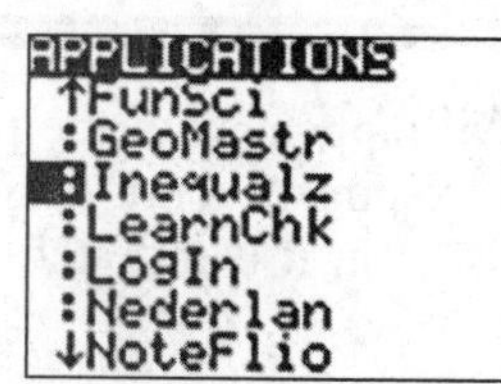

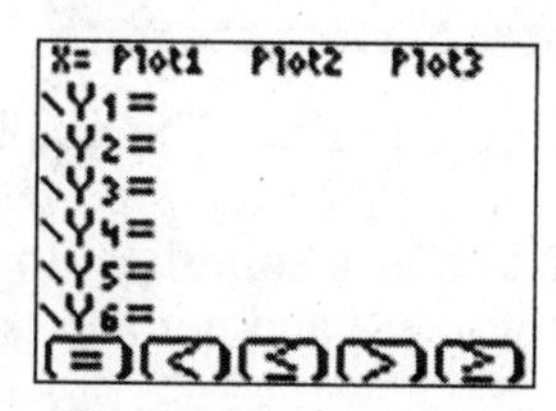

Move the cursor over the = sign after Y1 and press [ALPHA] and then [TRACE] to get the > symbol. Then enter $-\frac{1}{2}x + 3$ and press [GRAPH].

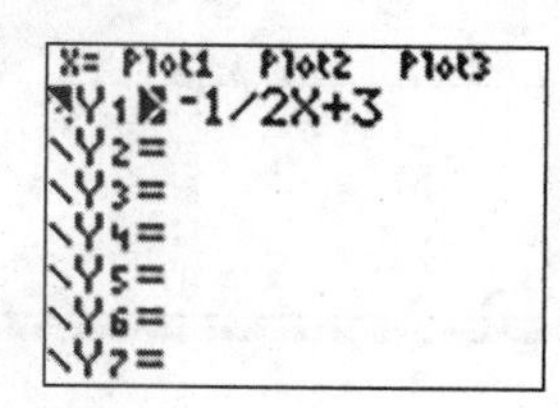

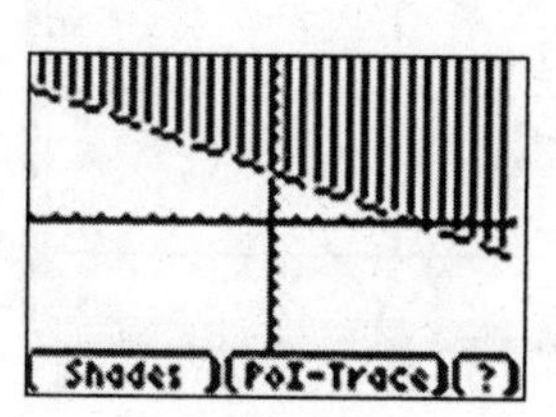

This is the graph of $y > -\left(\frac{1}{2}\right)x + 3$ with shading on the proper side of the line and also with the line being a dotted line since there is a > symbol and not a ≥ symbol.

To turn off the Inequalz application, press [APPS] scroll to it, press [ENTER], and then select choice [2] to quit.

Note: If the memory of the TI–84 is completely cleared before the test, the Inequalz application will be deleted also. If just the RAM is cleared, the application will remain on the calculator.

For the TI–Nspire:

Press [b] from the home screen to get to the Scratchpad graph. In the entry line where it says *f1*(*x*)=, push the [del] key to delete the = sign. A menu of the different inequality signs will appear.

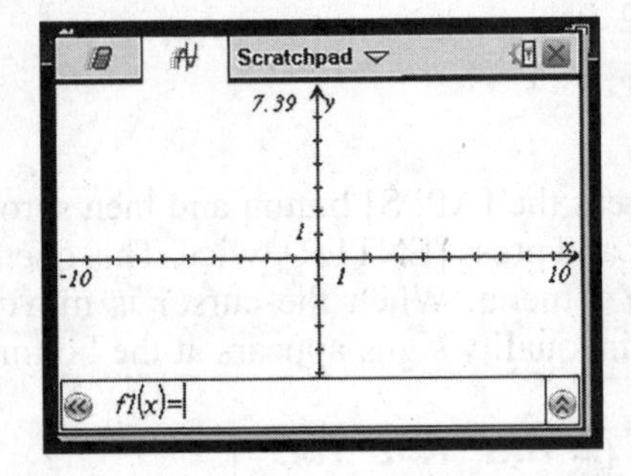

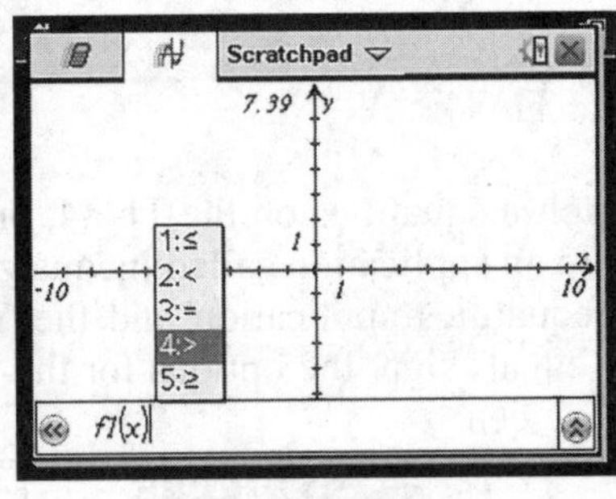

Select [4] and press [enter]. The graph will appear with the proper side of the line shaded in. Notice that on the TI–Nspire it is not easy to see if the line is a dotted line or a solid line. In this case, it is a dotted line since the inequality sign was a > and not a ≥.

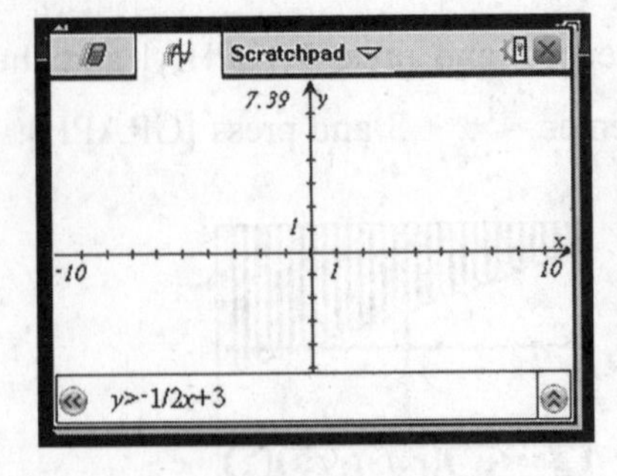

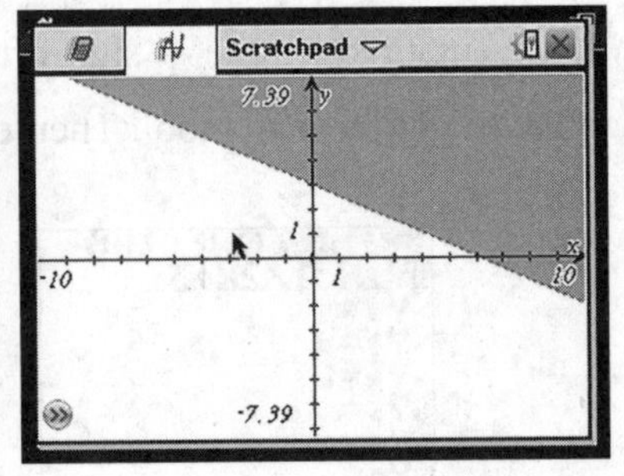

Check Your Understanding of Section 7.2

A. Multiple-Choice

1. Which is the graph of $y < x + 4$?

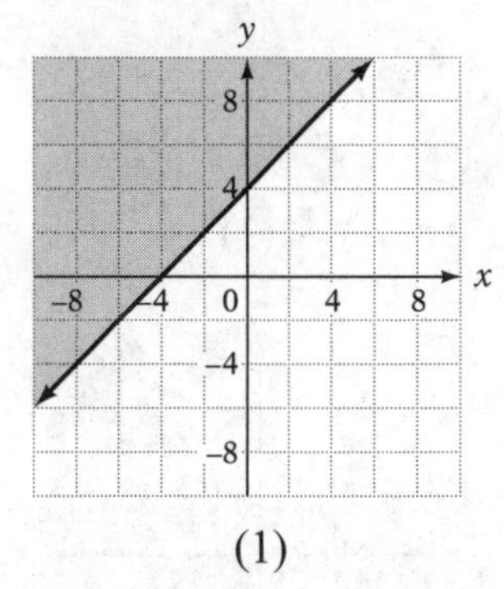

(1)

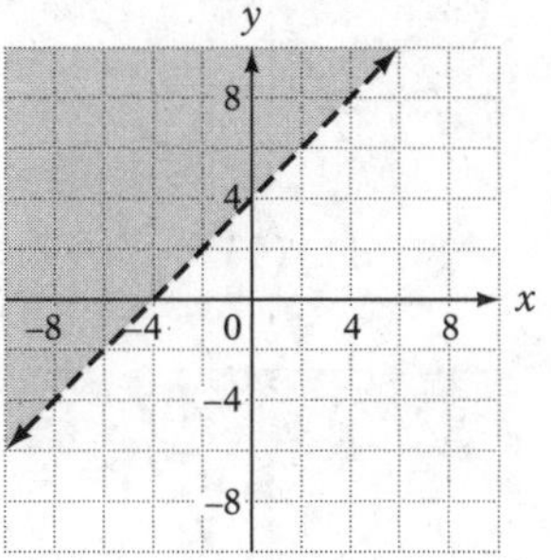

(3)

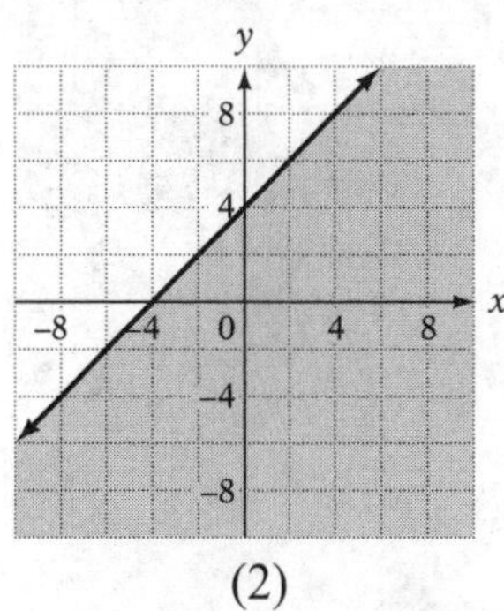

(2)

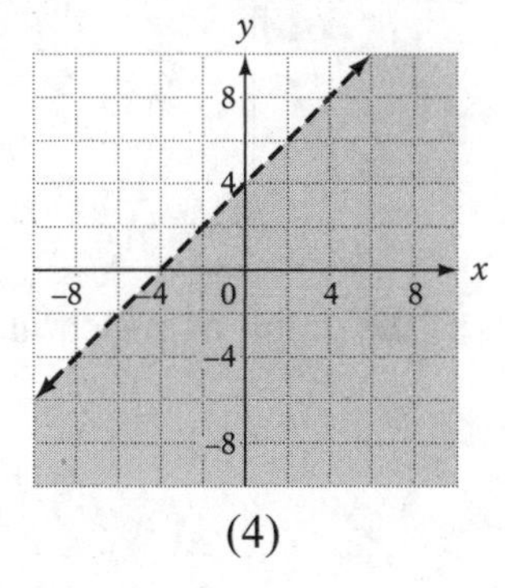

(4)

2. Which is the graph of $y \geq 2x - 1$?

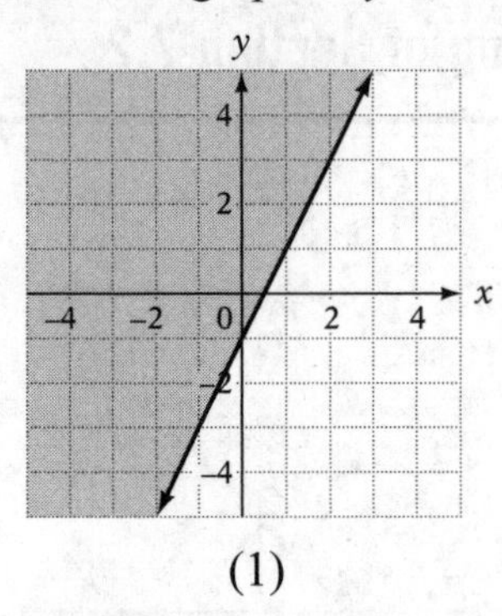

(1)

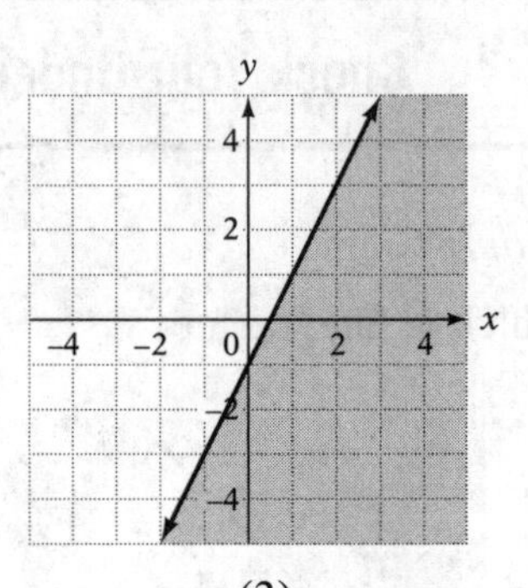

(3)

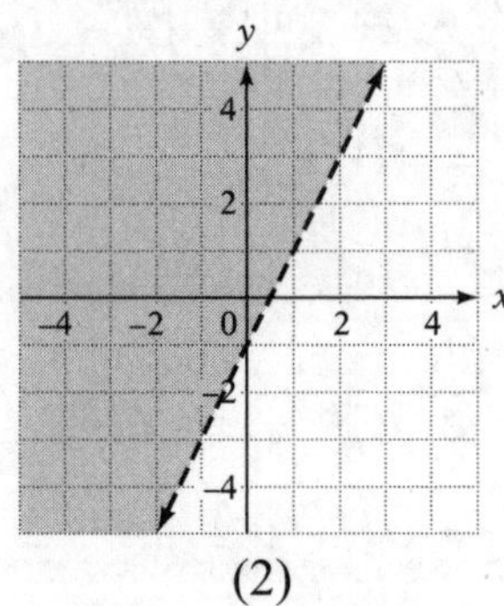

(2)

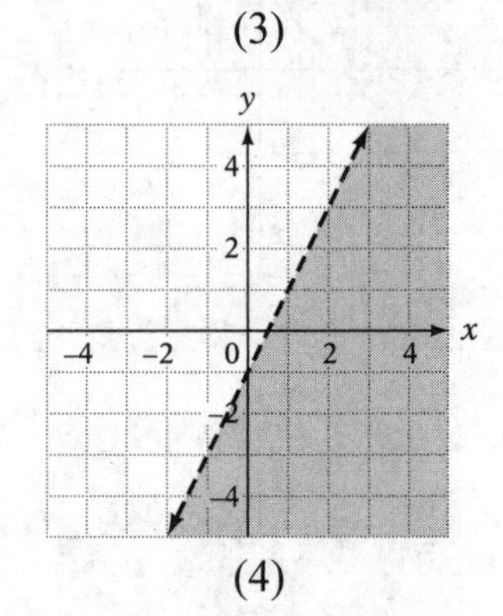

(4)

3. This is the graph of which inequality?

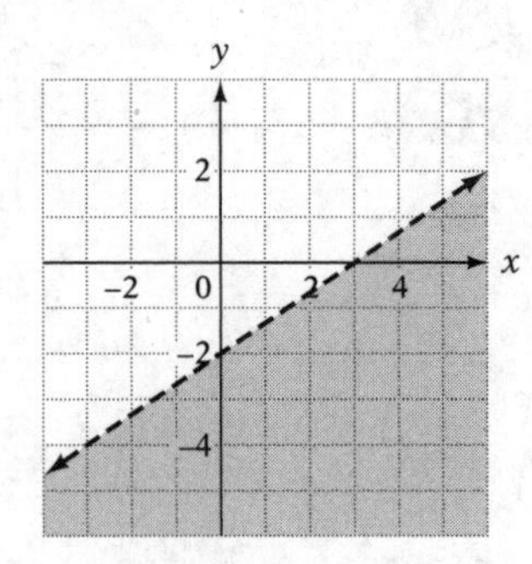

(1) $2x - 3y > 6$

(2) $2x - 3y < 6$

(3) $2x - 3y \geq 6$

(4) $2x - 3y \leq 6$

4. Which is the graph of the inequality $x + 2y > 10$?

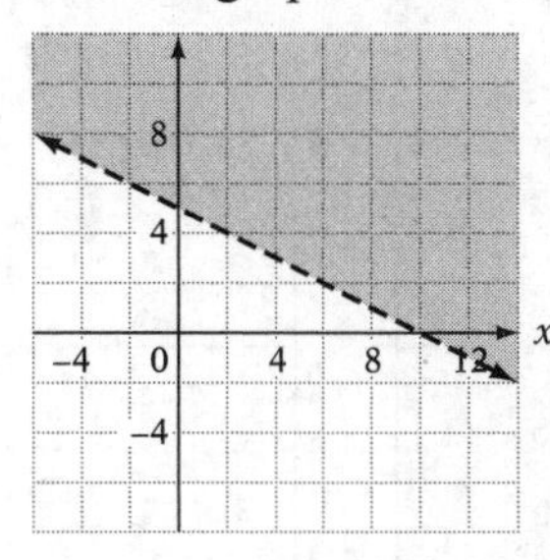

(1)

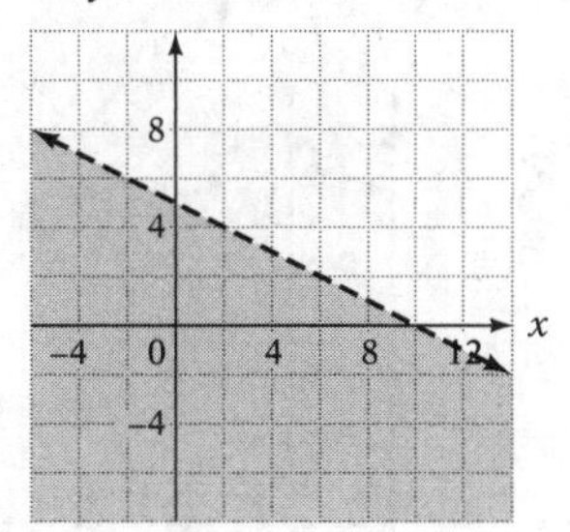

(3)

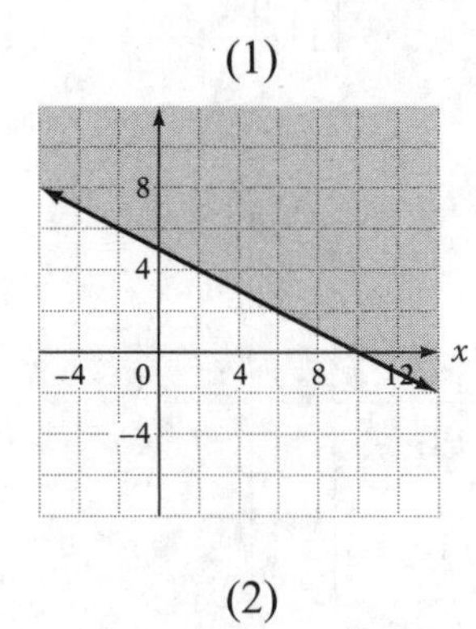

(2)

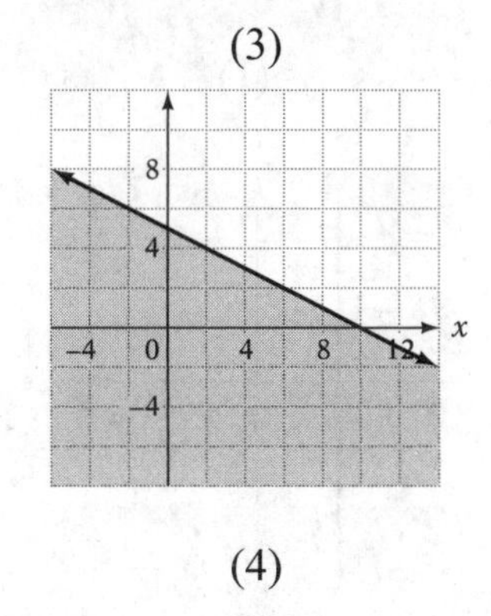

(4)

5. Which is the graph of $y > 4$?

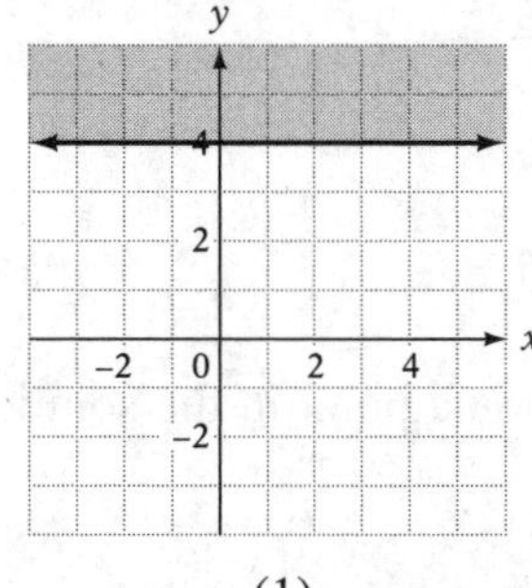

(1)

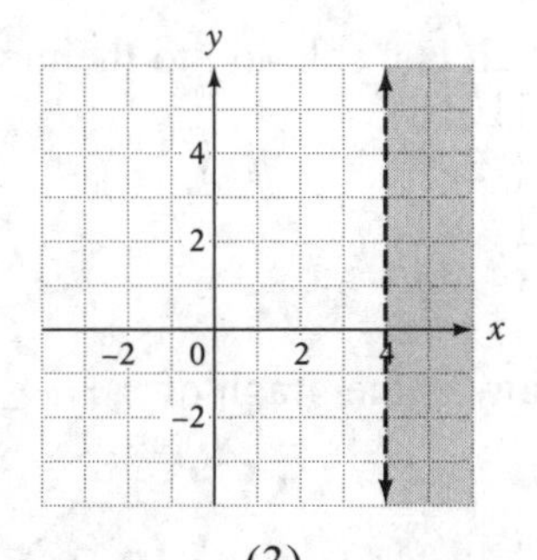

(3)

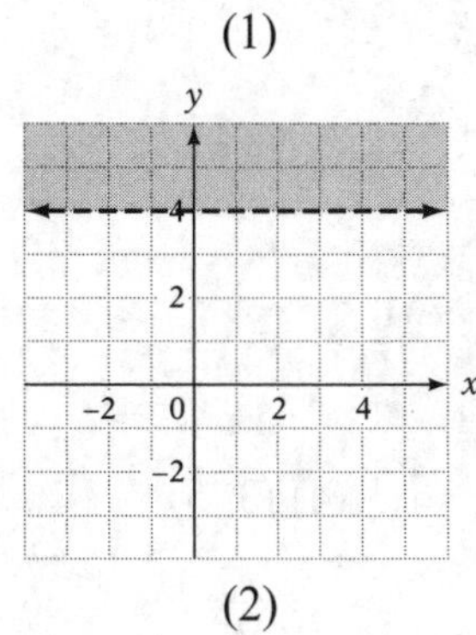

(2)

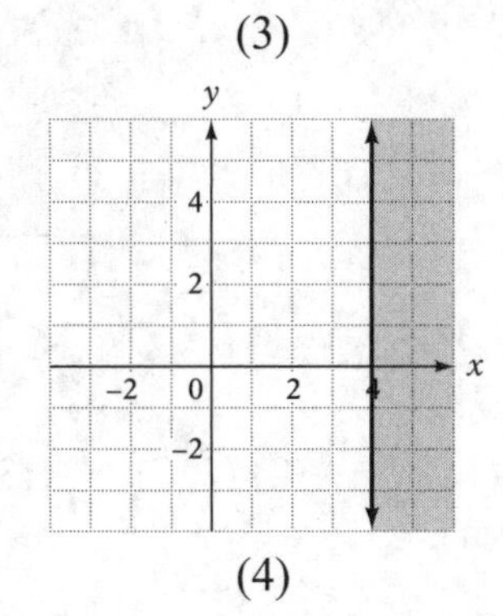

(4)

6. Which is the graph of $x \leq -2$?

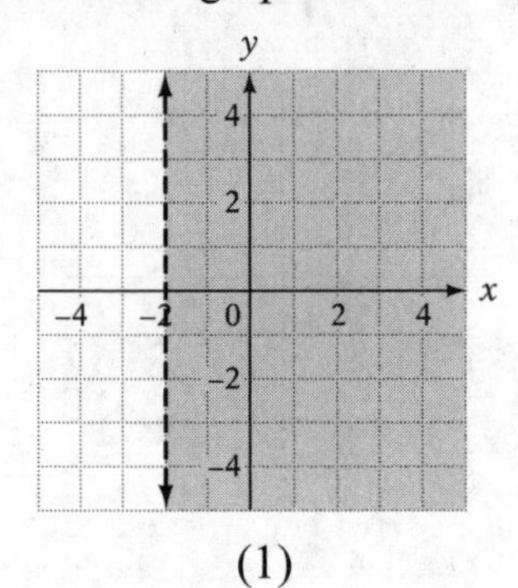

(1)

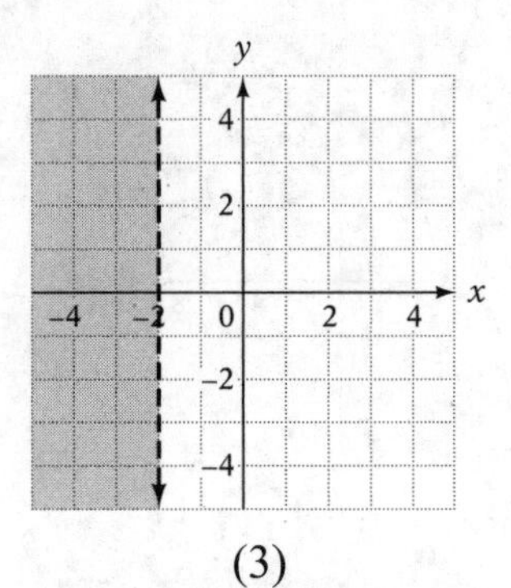

(3)

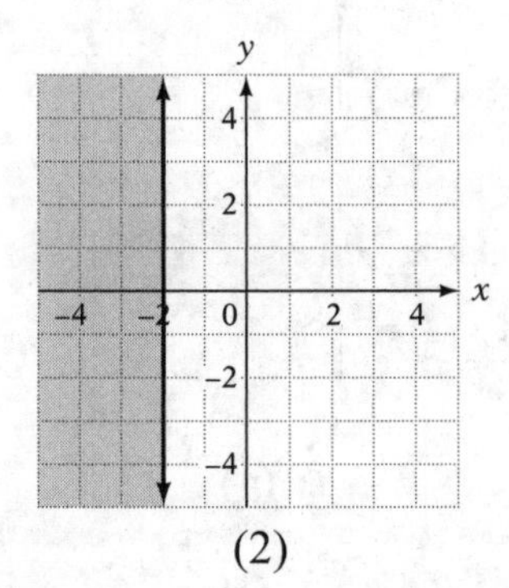

(2)

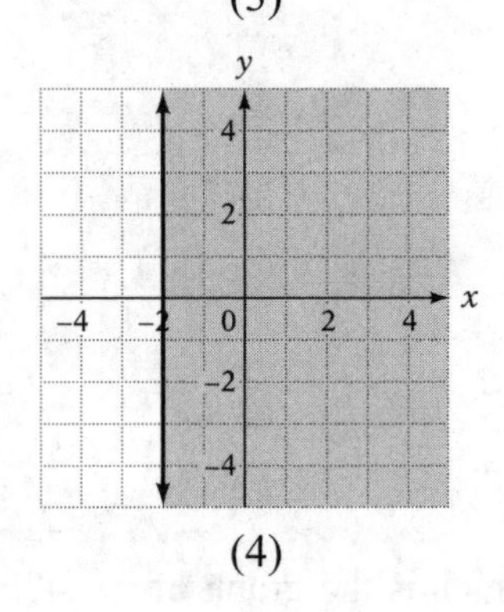

(4)

7. Which is a solution to the inequality $x - y > 3$?
(1) (10, 8)
(2) (12, 8)
(3) (7, 5)
(4) (4, 1)

8. Below is the graph of $x + y \leq 8$. Which is a point in the solution set?

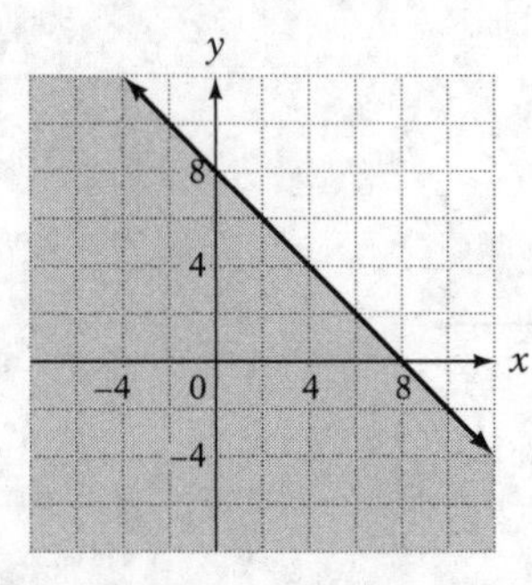

(1) (1, 8) (2) (2, 7) (3) (4, 5) (4) (3, 5)

9. All of these ordered pairs are part of the shaded region for the graph of $2x + y < 12$ except

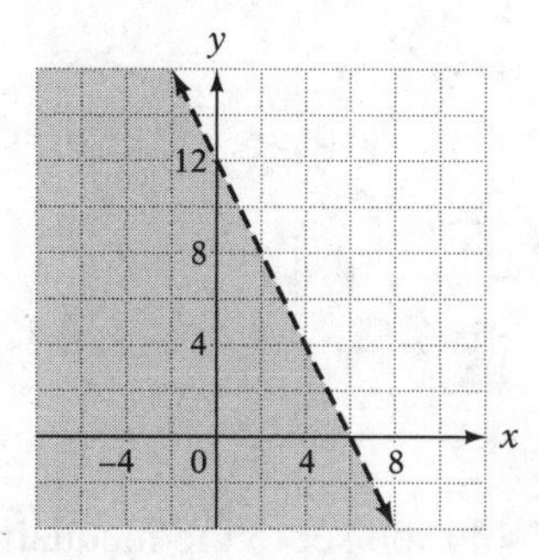

(1) (2, 7) (2) (3, 4) (3) (5, 3) (4) (5, 0)

10. Which point would not be a good point for checking which side of the line to shade for the inequality $y \leq -x + 4$?

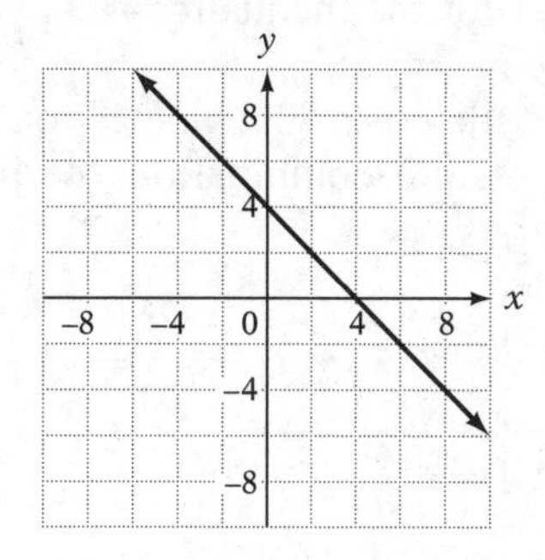

(1) (0, 5) (2) (0, 4) (3) (0, 3) (4) (0, 2)

***B.** Show how you arrived at your answers.*

1. A small elevator has a capacity of 4 people. On the graph below, make points indicating valid combinations of men and women who could be in the elevator.

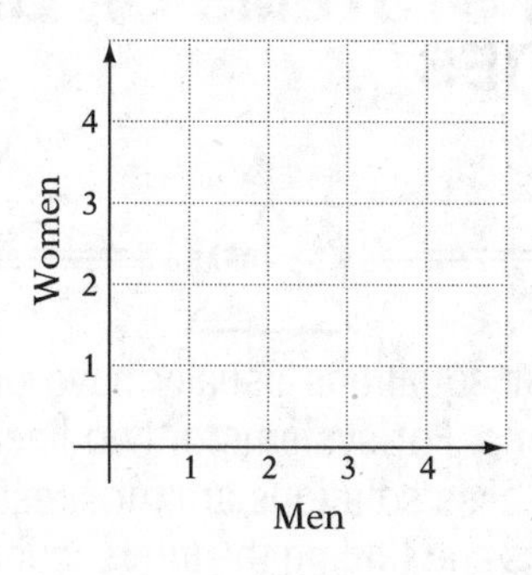

2. What is the inequality that has this solution set?

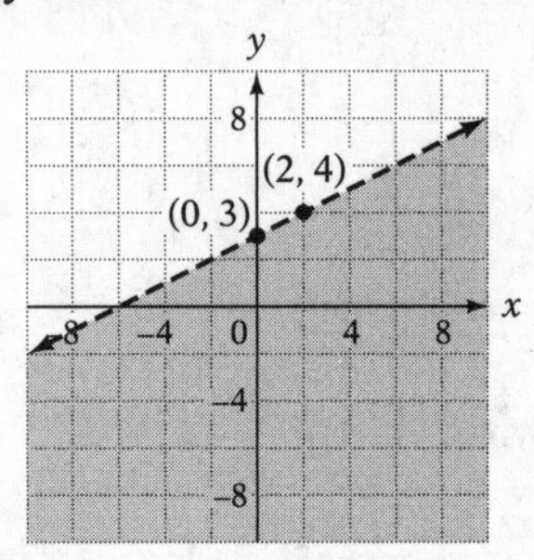

3. Ryan is graphing the solution set to the inequality $y \geq 3x$. After drawing the line $y = 3x$, he tests the point (0, 0) and finds that $0 \geq 3(0)$ is true. Does he now have enough information about which side of the line to shade in? Explain.

4. Graph the solution set for the inequality $-y < 2x - 4$.

5. Juice costs \$2 a pint. Milk costs \$3 a pint. If you have \$12, make a graph that shows all the possible combinations of juice and milk you can purchase.

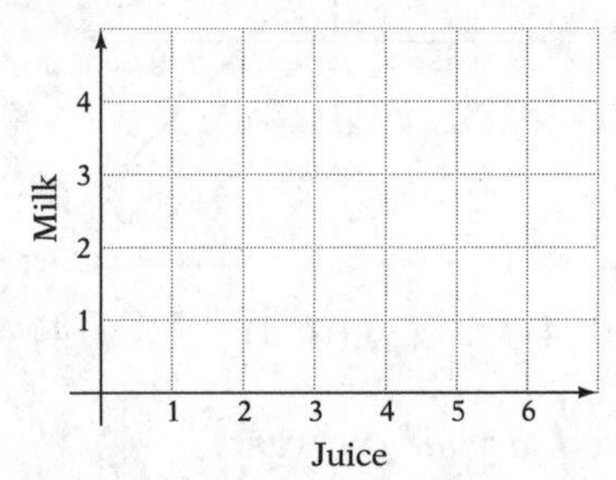

7.3 GRAPHING SYSTEMS OF LINEAR INEQUALITIES

Systems of two linear equations usually have one ordered pair that satisfies both equations. For systems of two linear inequalities, there are many solutions. These solutions are most easily solved by graphing the two inequalities and finding the intersections of the two shaded regions. The solution will look like a triangle.

Systems of Linear Inequalities

The system of equations

$$x + y = 10$$
$$x - y = 2$$

had the single solution (6,4). In Chapter 4 this was determined with algebra, and in Chapter 5 it was solved by graphing.

The system of inequalities

$$x + y < 10$$
$$x - y \geq 2$$

should be solved by graphing. For the two inequalities, use either two different colors for the shading or two different patterns.

First, graph the inequality $x + y < 10$. Pick a color to shade with if you have different colors, or use a pattern, like horizontal lines.

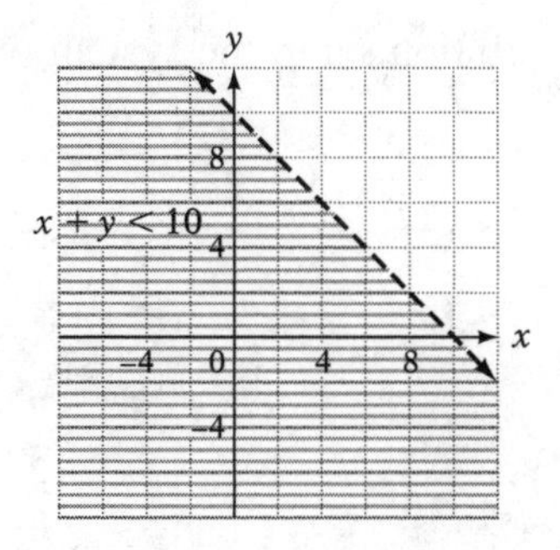

Then, on the same set of axes, graph $x - y \geq 2$, using either a different color for the shading or a different pattern, like vertical lines.

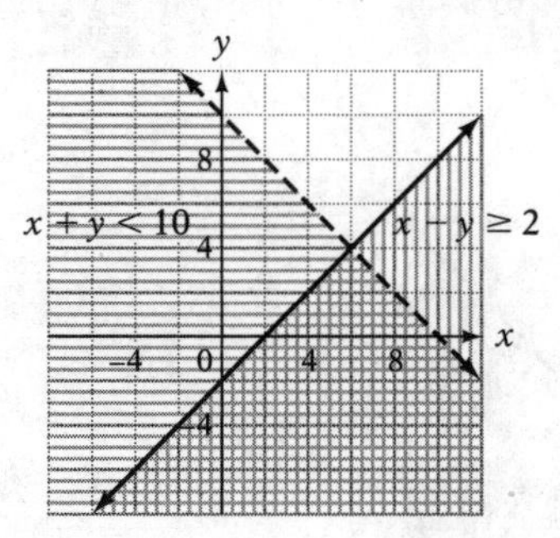

The two lines form an X that divides the plane into four regions. One of those regions will be shaded with both kinds of shading. This is the solution to the system of linear inequalities. Label it with an S.

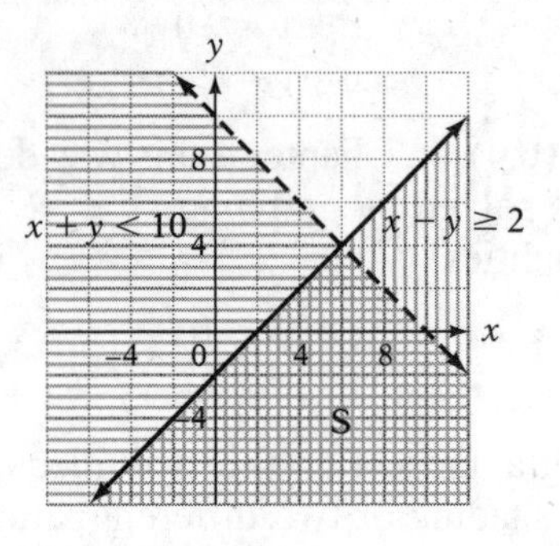

Example

Which graph shows the solution set to the system of inequalities?

$$y > -3x + 5$$
$$y \leq x + 2$$

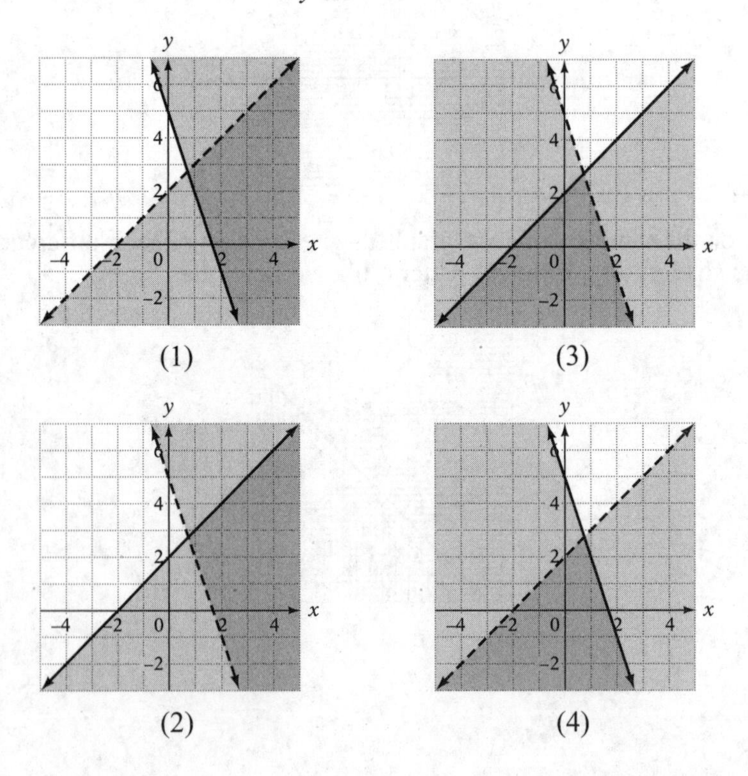

Solution: The line that has a negative slope must be dashed because the inequality $y > -3x + 5$ has a > symbol. This eliminates choices (1) and (4). Since (0, 0) is not a solution to $y > -3x + 5$, the side of the downward sloping line to be shaded with vertical lines will not contain the point (0, 0). This eliminates choice (1). The answer is choice (2).

Graphing Systems of Inequalities with the Graphing Calculator

To graph systems of inequalities, follow the same steps as you did for graphing single inequalities, but this time enter both inequalities.

Let's consider the system from Example 1:

$$y > -3x + 5$$
$$y \leq x + 2$$

For the TI–84:

With the Inequalz application open, enter both equations, and the calculator will shade one with horizontal lines and the other with vertical lines.

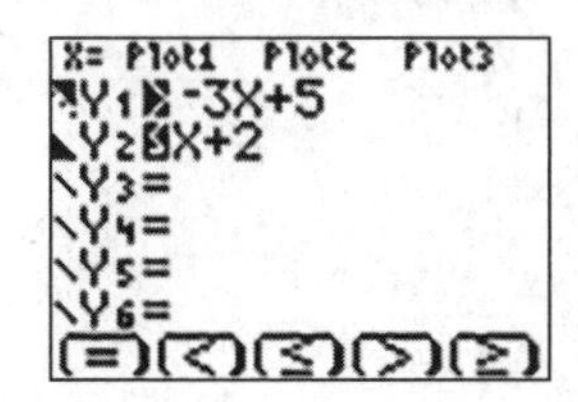

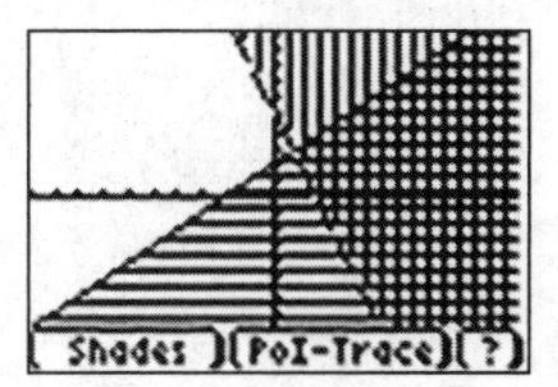

For the TI–Nspire:

Graph both inequalities, and the calculator will use different colors for the shadings and a third color for the overlapping region.

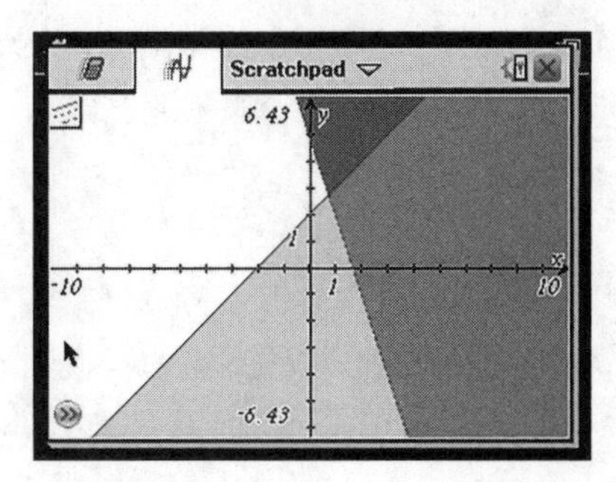

Check Your Understanding of Section 7.3

A. Multiple-Choice

1. (7, 1) is in the solution set for which system of inequalities?

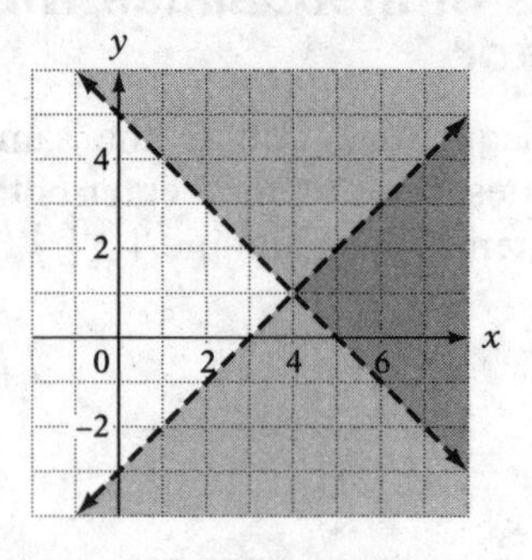

(1) $x + y > 5$
$x - y > 3$

(2) $x + y > 5$
$x - y < 3$

(3) $x + y < 5$
$x - y > 3$

(4) $x + y < 5$
$x - y < 3$

2. Which graph shows the solution to the system of inequalities?

$$y < 2x + 1$$
$$y > \left(\frac{1}{3}\right)x + 4$$

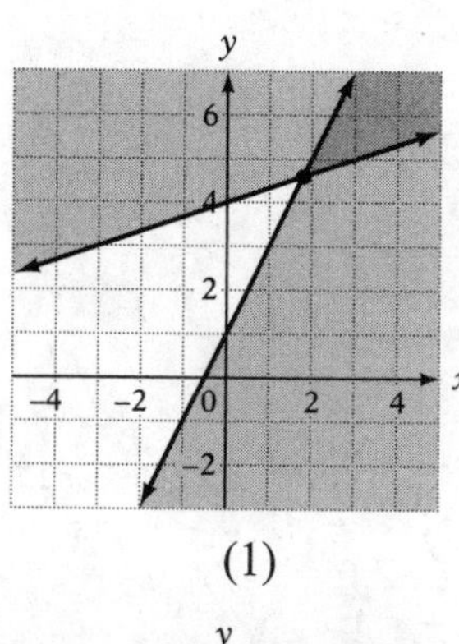

(1)

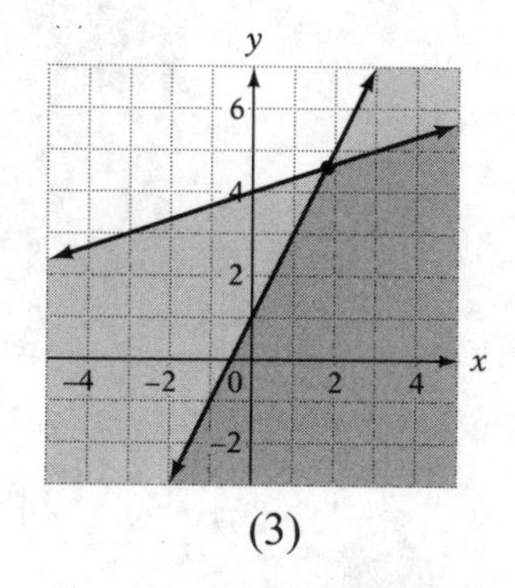

(3)

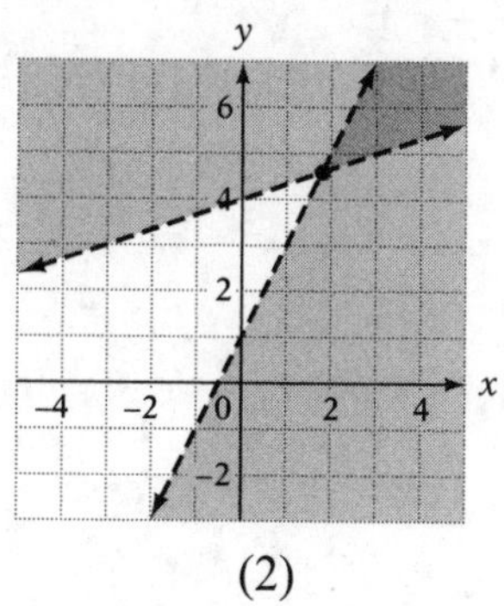

(2)

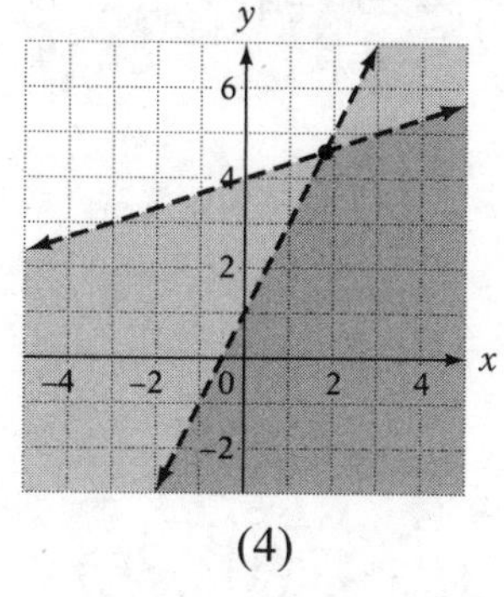

(4)

3. Which system of inequalities does the following graph show the solution for?

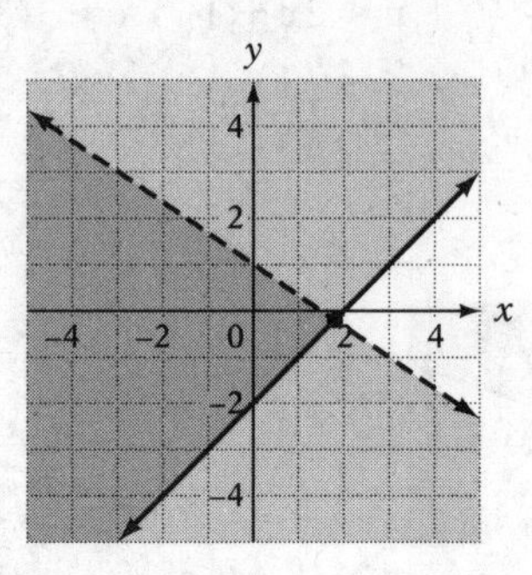

(1) $y \geq x - 2$

$y < -\frac{2}{3}x + 1$

(2) $y \leq x - 2$

$y > -\frac{2}{3}x + 1$

(3) $y < x - 2$

$y \geq -\frac{2}{3}x + 1$

(4) $y > x - 2$

$y \leq -\frac{2}{3}x + 1$

4. Which graph shows the solution to the system of inequalities?

$$y > \left(\frac{1}{2}\right)x + 4$$

$$y > 2x - 1$$

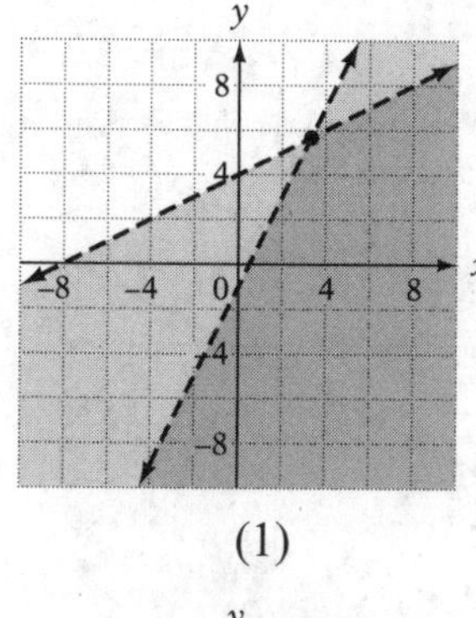

(1)

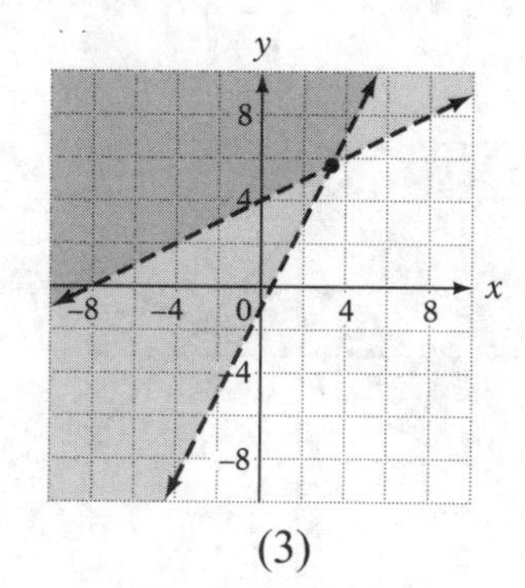

(3)

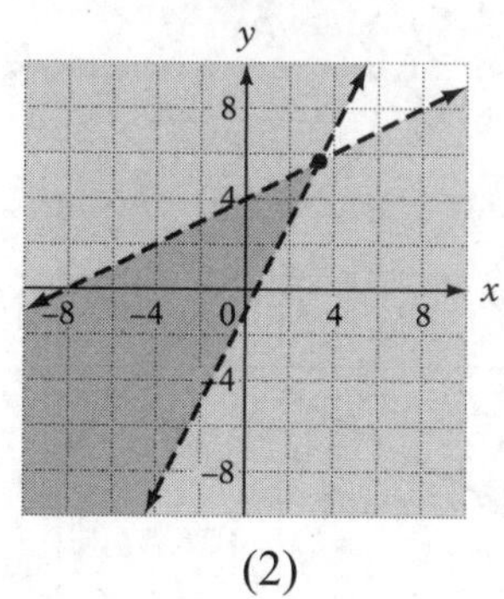

(2)

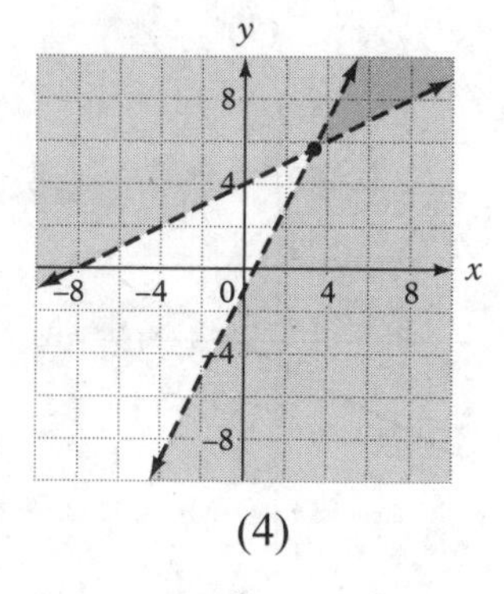

(4)

Chapter Seven **LINEAR INEQUALITIES**

5. Which graph has the solution set shaded in for the following system of inequalities?

$$y \leq -x + 6$$

$$y \geq \left(\frac{1}{2}\right)x - 1$$

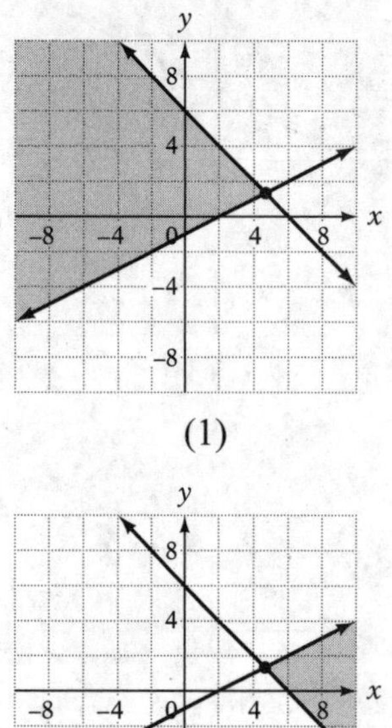

(1)

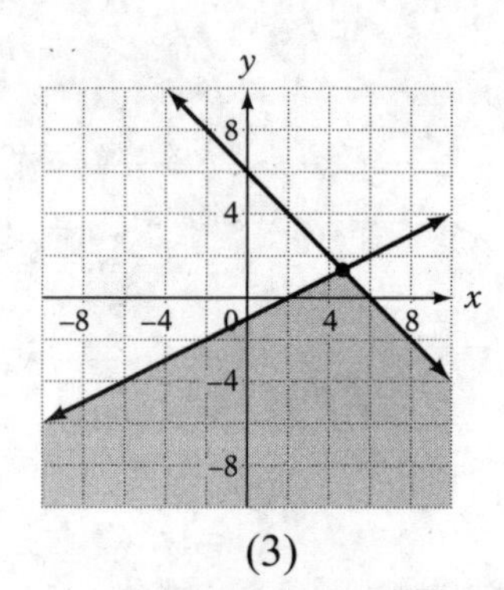

(3)

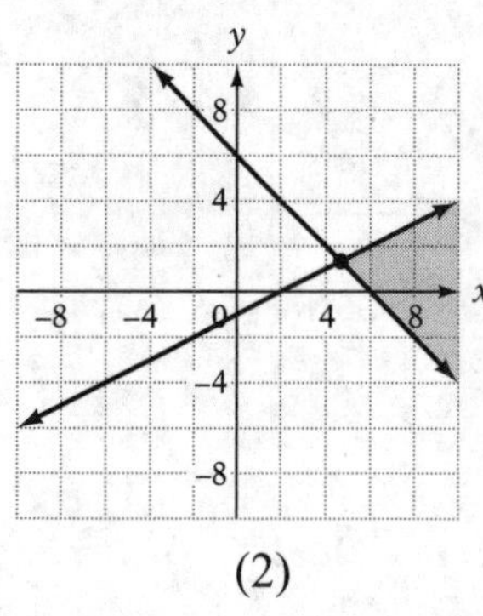

(2)

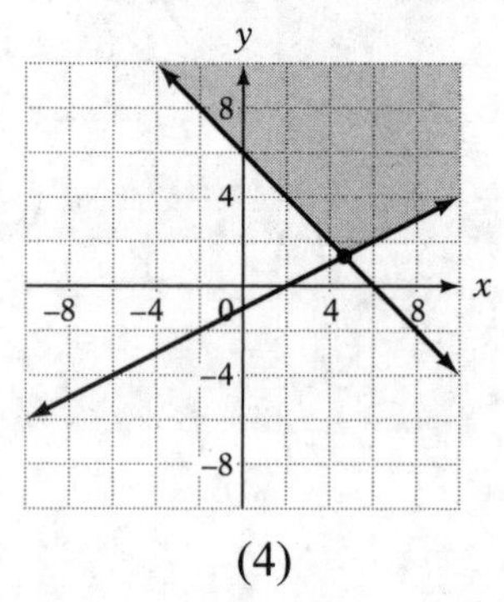

(4)

6. Which graph shows the solution set shaded in for the following system of inequalities?

$$y < x - 2$$

$$y > -\left(\frac{3}{4}\right)x + 3$$

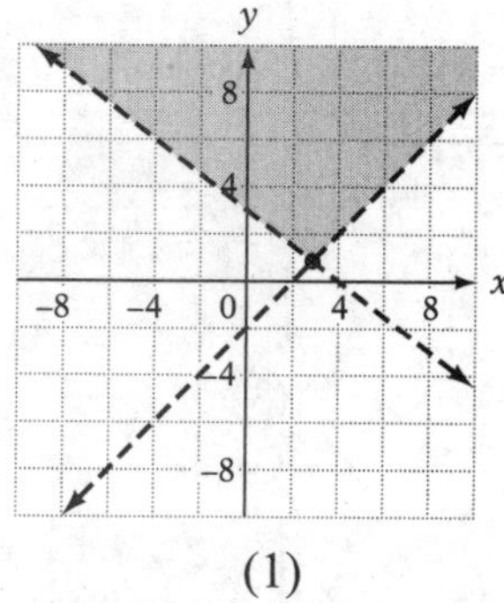

(1)

(3)

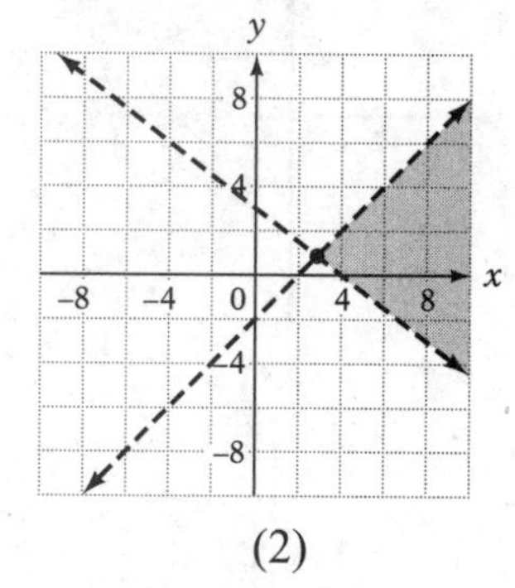

(2)

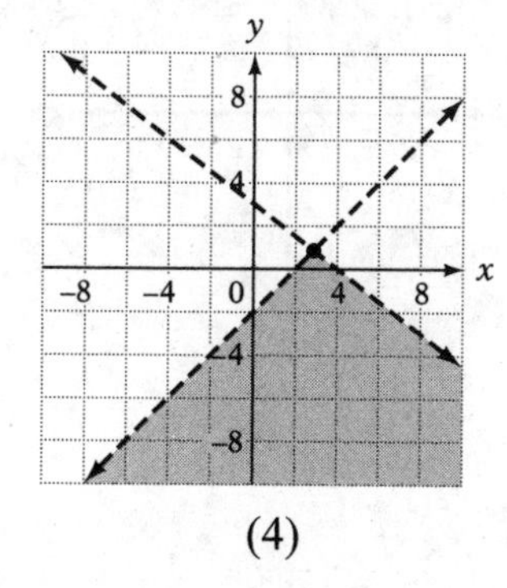

(4)

7. Which is the graph of this system of inequalities?

$$y \geq 0$$
$$x \leq 0$$

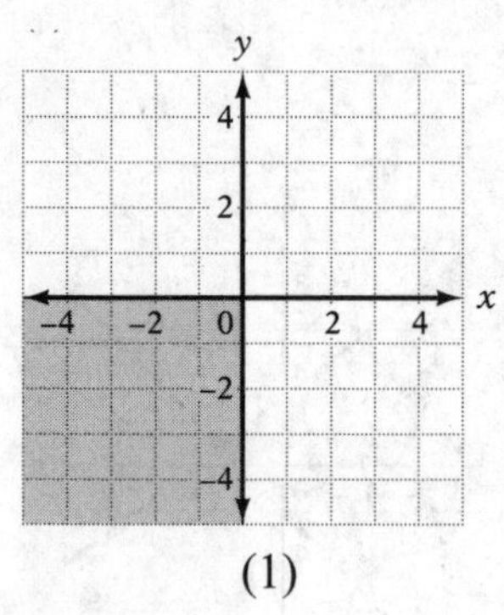

(1)

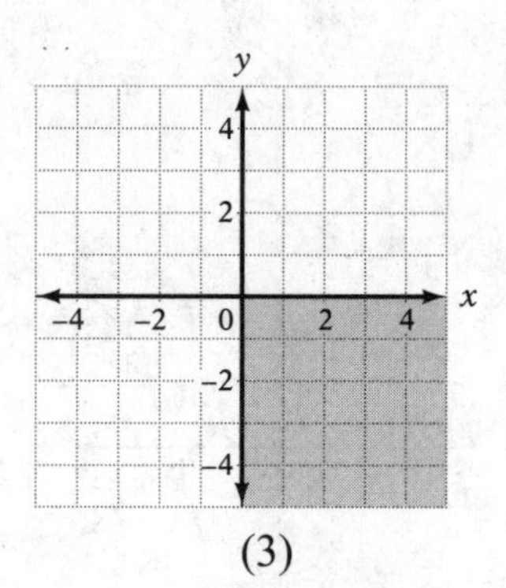

(3)

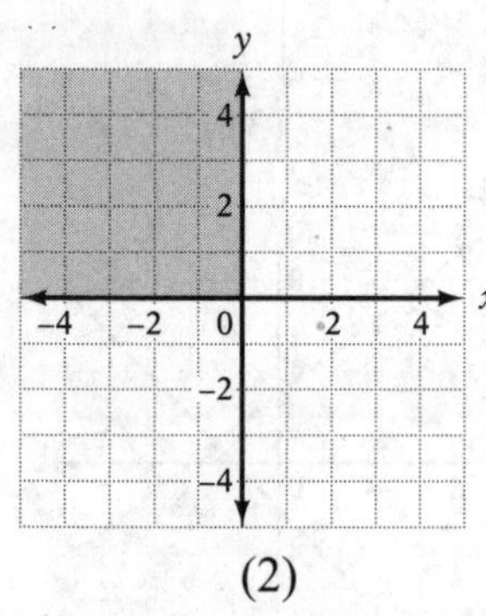

(2)

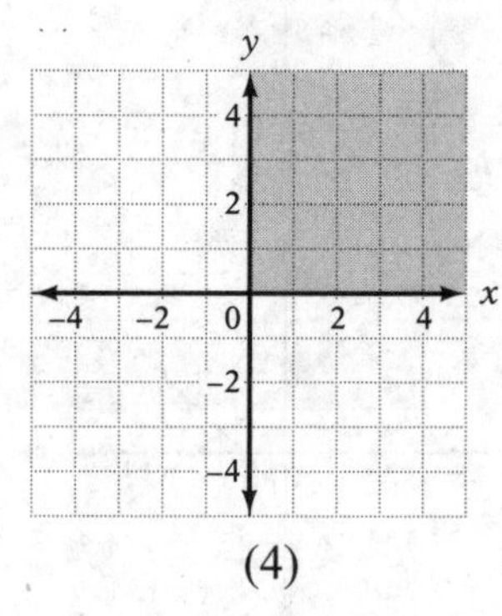

(4)

8. What system of inequalities does the graph below show the solution for?

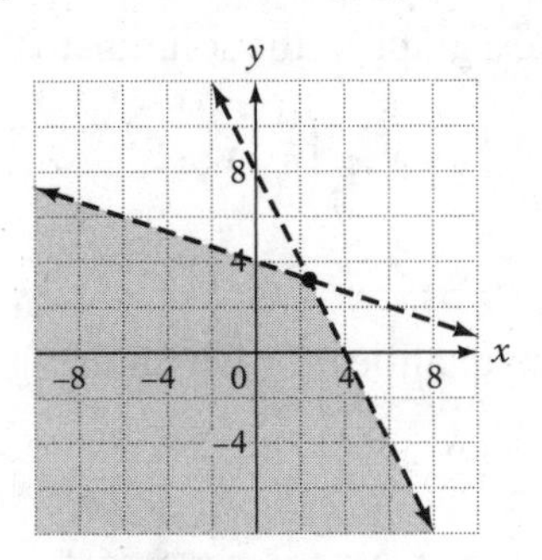

(1) $y > -\left(\frac{1}{3}\right)x + 4$
$y < -2x + 8$

(2) $y > -\left(\frac{1}{3}\right)x + 4$
$y > -2x + 8$

(3) $y < -\left(\frac{1}{3}\right)x + 4$
$y < -2x + 8$

(4) $y < -\left(\frac{1}{3}\right)x + 4$
$y > -2x + 8$

9. An elevator can hold at most 10 people and at most 1,200 pounds. Children weigh an average of 60 pounds each and adults weigh an average of 130 pounds each. If x is the number of children and y is the number of adults, which system of inequalities can be used to model all the allowable combinations of children and adults that can ride the elevator?

(1) $x + y \leq 10$
$130x + 60y \leq 1{,}200$

(2) $x + y \leq 10$
$60x + 130y \leq 1{,}200$

(3) $60x + 130y \leq 10$
$x + y \leq 1{,}200$

(4) $x - y \geq 10$
$130x - 60y \geq 1{,}200$

10. The graph below is the solution set to a system of inequalities where $y \leq 3x - 4$ is one of the equations. What is the other equation?

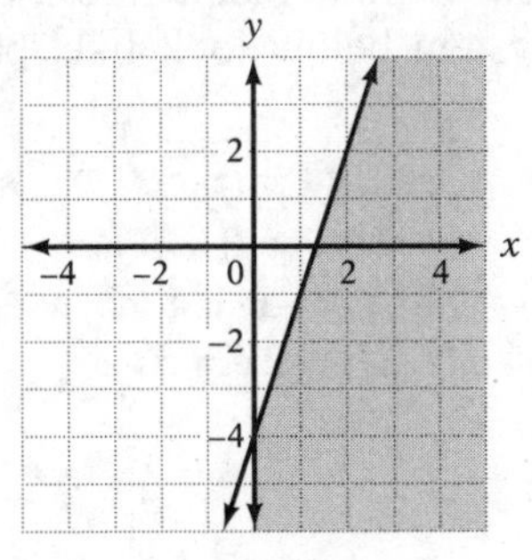

(1) $y \geq 0$
(2) $y > 0$
(3) $x < 0$
(4) $x \geq 0$

B. *Show how you arrived at your answers.*

1. On graph paper, make a graph of the solution to the system of inequalities.

$$y > \left(\frac{2}{3}\right)x - 5$$
$$y \geq -2x + 6$$

2. Find the system of inequalities for which this graph shows the solution.

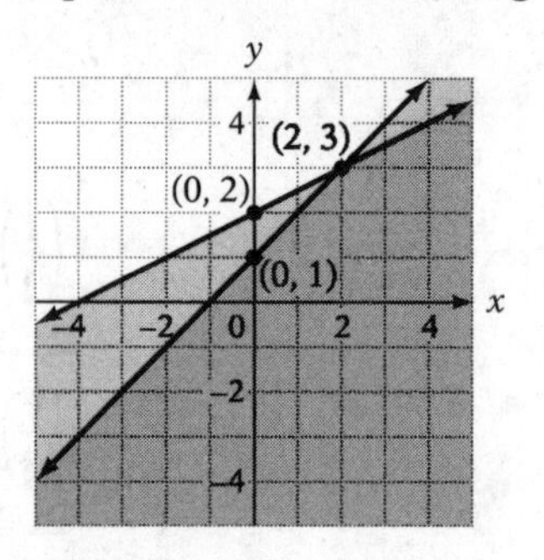

3. Johan goes to a used bookstore. He wants to buy at most 50 books and spend at most \$90. Paperback books cost \$0.75 and hard cover books cost \$2.50. Create a system of linear inequalities that could be used to determine all the different combinations of paperback and hard cover books he can purchase.

4. Movie tickets cost \$8 for children and \$12 for adults. A movie theater has 300 seats. The theater wants to collect at least \$3,200. What system of equations could be used to determine all the different combinations of children and adult tickets that would fit in the theater and produce the money the theater wants to collect?

5. In economics, the feasible region for production can sometimes be modeled with a system of four inequalities. What would the graph of the solution of the system below look like, based on a problem from economics?

$$x \geq 0$$
$$y \geq 0$$
$$y \leq -x + 4$$
$$y \leq -2x + 5$$

Chapter Eight EXPONENTIAL EQUATIONS

8.1 SOLVING EXPONENTIAL EQUATIONS

An *exponential equation* is one where the variable is an exponent. An example of a one-variable exponential equation is $2^x = 8$. Examples of two-variable exponential equations are $y = 3^x$ and $y = 2 \cdot 5^x$.

Finding Solutions to Exponential Equations

In a two-variable equation like $y = 3^x$, substitute values for x, and find the corresponding value for y to get a solution.

For example, if $x = 2$, then $y = 3^2 = 9$.

On the TI graphing calculators, raising a number to a power requires using the [^] button.

When the equation has a number multiplied by the exponential expression, be careful not to multiply the number by the base. For the equation $y = 2 \cdot 5^x$, if $x = 2$, then $y = 2 \cdot 5^2 = 2 \cdot 25 = 50$. It would NOT be correct to multiply the 2 and 5 together to get $y = 10^2 = 100$.

MATH FACTS

When a number other than 0 is raised to the 0 power, it becomes 1, not 0. So $5^0 = 1$, $10^0 = 1$, and $100^0 = 1$. When a number other than 0 is raised to a negative power, it becomes the reciprocal of that number raised to the positive version of that power. So $3^{-2} = \frac{1}{3^2} = \frac{1}{9}$ and not -9.

Solving for the Exponent in an Exponential Equation

In a one-variable exponential equation, where the exponent is unknown, isolate the exponential expression and then use guess and check.

Example

Solve for x in the equation $2 \cdot 4^x = 32$.

Solution: Isolate the exponential part by dividing both sides of the equation by 2.

$$\frac{2 \cdot 4^x}{2} = \frac{32}{2}$$

$$4^x = 16$$

$$x = 2$$

The last step was done by guess and check.

Check Your Understanding of Section 8.1

A. Multiple-Choice

1. If $x = 2$ and $y = 3^x$, solve for y.
(1) 8 (2) 9 (3) 10 (4) 11

2. If $x = 3$ and $y = 2 \cdot 3^x$, solve for y.
(1) 216 (2) 27 (3) 54 (4) 18

3. If $x = 0$ and $y = 4^x$, solve for y.
(1) $\frac{1}{4}$ (2) 0 (3) 1 (4) 4

4. If $x = 0$ and $y = 5 \cdot 2^x$, solve for y.
(1) 0 (2) 1 (3) 3 (4) 5

5. If $x = -2$ and $y = 5^x$, solve for y.
(1) -25 (2) 25 (3) $\frac{1}{25}$ (4) $-\frac{1}{25}$

6. If $y = 64$ and $y = 4^x$, solve for x.
(1) 16 (2) 8 (3) 5 (4) 3

7. If $y = 512$ and $y = 4 \cdot 2^x$, solve for x.
(1) 7 (2) 5 (3) 3 (4) 1

8. If $y = \frac{1}{125}$ and $y = 5^x$, solve for x.

(1) $\frac{1}{3}$ (2) -3 (3) 3 (4) $-\frac{1}{3}$

9. If $x = -3$ and $y = \left(\frac{1}{2}\right)^x$, solve for y.

(1) $-\frac{1}{8}$ (2) $\frac{1}{8}$ (3) -8 (4) 8

10. If $x = 100$ and $y = \left(1 + \frac{1}{x}\right)^x$, solve for y rounded to the nearest hundredth.

(1) 2.70 (2) 2.72 (3) 2.73 (4) 2.74

B. *Show how you arrived at your answers.*

1. Phoebe put \$500 in the bank. The amount of money she has after t years is determined by the equation $A = 500(1.05)^t$. After 4 years, how much money will Phoebe have in the bank?

2. The population of a town after t years can be approximated by the equation $P = 10{,}000(1.023)^t$. (a) According to the formula, what will the population of the town be after ten years? (b) In what year will the population become 14,065?

3. Zoe drinks a cup of coffee that has 100 mg of caffeine. The amount of caffeine in the bloodstream after t hours can be determined by the equation $C = 100 \cdot .5^{\frac{t}{5}}$. How much caffeine will be left in her bloodstream after 20 hours?

4. Food that is 110 degrees is put into a 30 degree freezer. The temperature of the food is related to the number of hours the food is in the freezer by the equation $T = 80 \cdot 0.7^h + 30$. Between which two hours will the food be 32 degrees?

5. Daphne says that 6^x is always greater than 5^x. Julia says that this is not true and that 5^x sometimes is greater than 6^x. Which student is correct? Explain.

8.2 GRAPHING SOLUTION SETS TO TWO-VARIABLE EXPONENTIAL EQUATIONS

The solution set to a two-variable exponential equation, like $y = 3 \cdot 2^x$, can be produced with a table or with a graphing calculator. The shape of the graph is not a line or a parabola, but a distinctive shape that looks a bit like a playground slide.

Graphing the Solution Set to an Exponential Equation with a Table

Five points are generally sufficient for graphing the solution set for an exponential equation. x values of 2, 1, 0, –1, and –2 should produce enough points for an accurate graph.

Example 1

Make a table of solutions to $y = 3 \cdot 2^x$, and use them to make a sketch of the graph of the solution set.

Solution: When making the chart, remember that anything raised to the zero power (except zero!) is equal to 1. So $2^0 = 1$. Also, anything raised to a negative power is the reciprocal of the same number raised to the positive version of that power. So $2^{-2} = \left(\frac{1}{2}\right)^2 = \frac{1}{4}$. You can use the [^] button on your calculator to do these calculations if you are not sure.

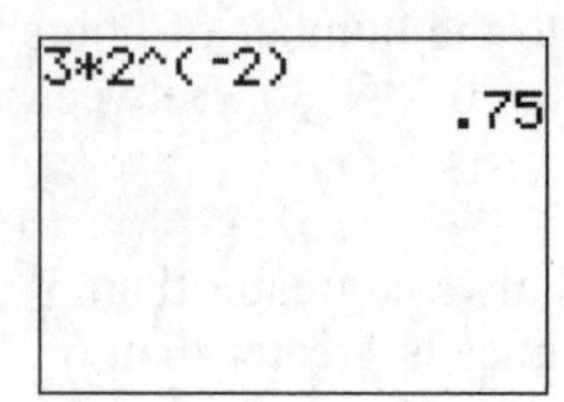

x	y
2	$3 \cdot 2^2 = 3 \cdot 4 = 12$
1	$3 \cdot 2^1 = 3 \cdot 2 = 6$
0	$3 \cdot 2^0 = 3 \cdot 1 = 3$
–1	$3 \cdot 2^{-1} = 3 \cdot \left(\frac{1}{2}\right) = \frac{3}{2} = 1.5$
–2	$3 \cdot 2^{-2} = 3 \cdot \left(\frac{1}{4}\right) = \frac{3}{4} = 0.75$

Graph the points that fit on the coordinate plane and join to make the curve.

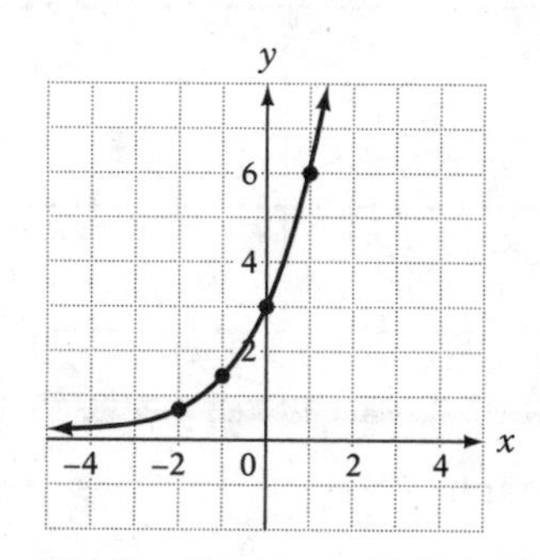

Making Tables and Graphs on the Graphing Calculator

For the TI–84:

Enter the equation into the Y= menu and then press [2ND] and [GRAPH]; the calculator will produce the chart.

```
Plot1 Plot2 Plot3
\Y1=3*2^X
\Y2=
\Y3=
\Y4=
\Y5=
\Y6=
\Y7=
```

X	Y1
-2	.75
-1	1.5
0	3
1	6
2	12
3	24
4	48

Y1=48

Press [ZOOM] and [6] to see the graph on a 20 by 20 coordinate plane.

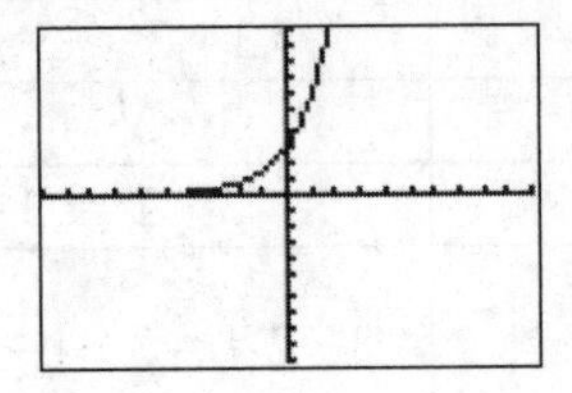

For the TI–Nspire:

From the home screen press [B], select the graphing Scratchpad and type $3 \cdot 2^x$ after the $f1(x)$= in the entry line.

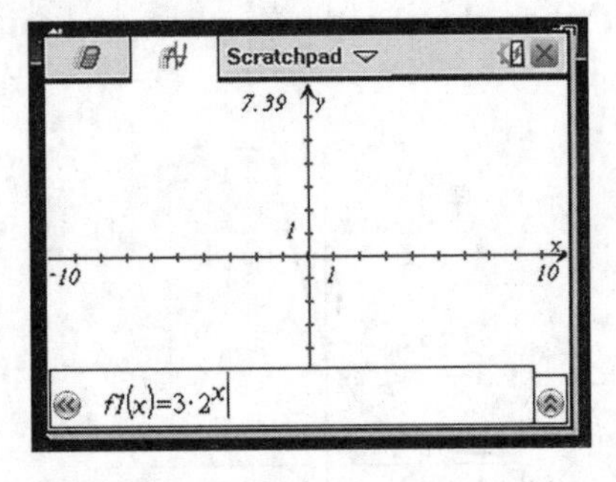

Press [enter] to see the graph.

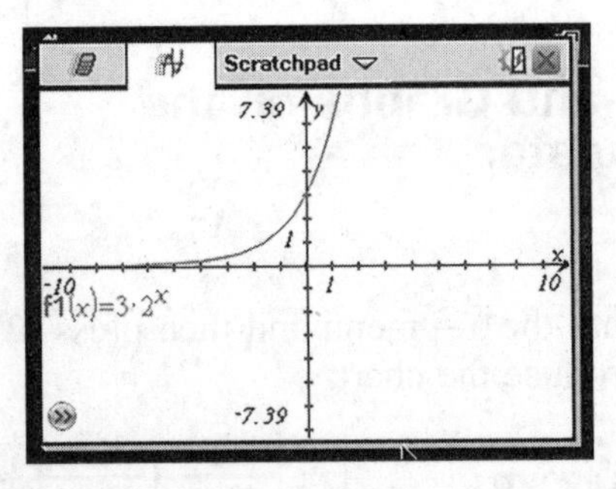

Press [menu], [2], and [5] to show the table.

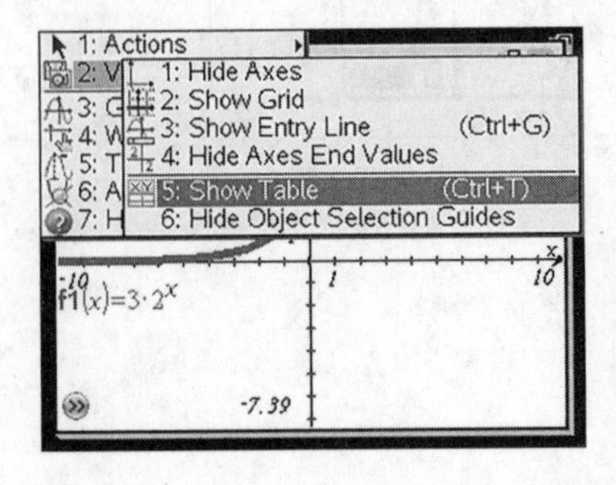

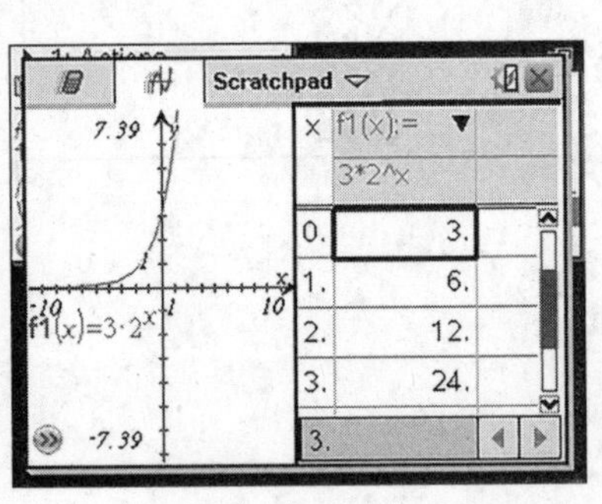

Example 2

Make a table of solutions to the equation $y = 4 \cdot 0.5^x$. Use the table to produce a graph of the solution set.

Solution:

x	y
2	$4 \cdot 0.5^2 = 1$
1	$4 \cdot 0.5^1 = 2$
0	$4 \cdot 0.5^0 = 4$
–1	$4 \cdot 0.5^{-1} = 8$
–2	$4 \cdot 0.5^{-2} = 16$

The graph looks like this:

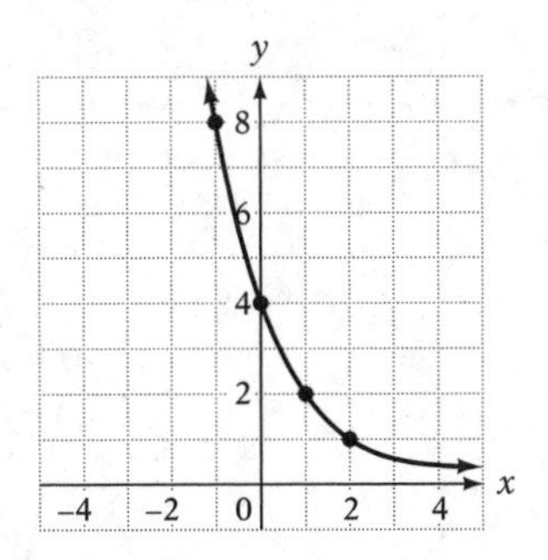

MATH FACTS

In an exponential equation, the thing being raised to the power is called the *base*. When the base of an exponential equation is between 0 and 1, the graph shows exponential *decay*. When the base is greater than 1, the graph shows exponential *growth*.

Check Your Understanding of Section 8.2

A. Multiple-Choice

1. Which ordered pair is in the solution set of $y = 3^x$?
(1) (0, 0) (2) (3, 1) (3) (9, 2) (4) (4, 81)

2. Which ordered pair is in the solution set of $y = 5 \cdot 2^x$?
(1) (0, 0) (2) (2, 25) (3) (3, 40) (4) (3, 100)

3. Which is the graph of $y = 2^x$?

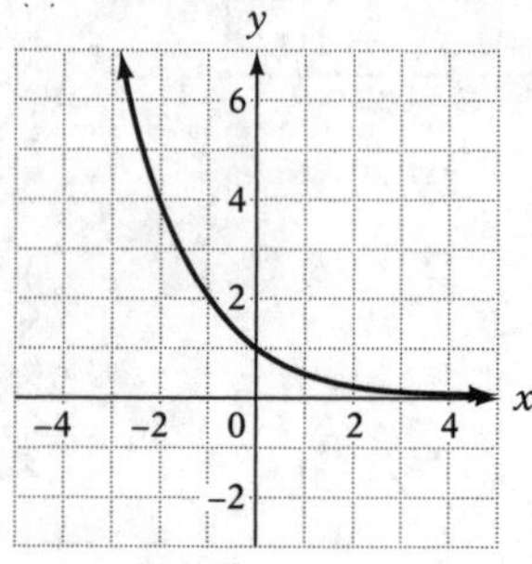

(1)

(3)

(2)

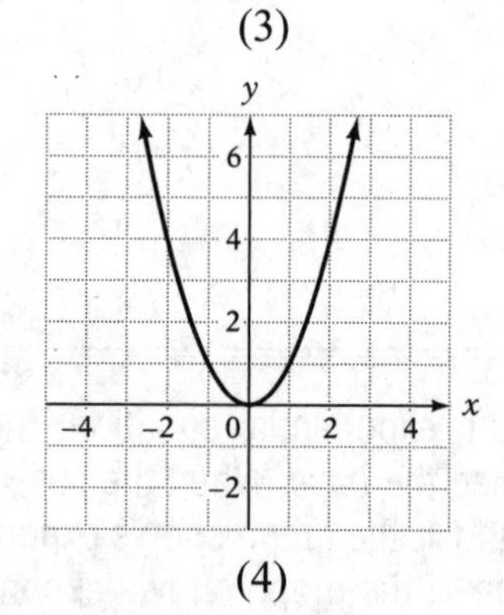

(4)

4. Which is the graph of $y = \left(\frac{1}{2}\right)^x$?

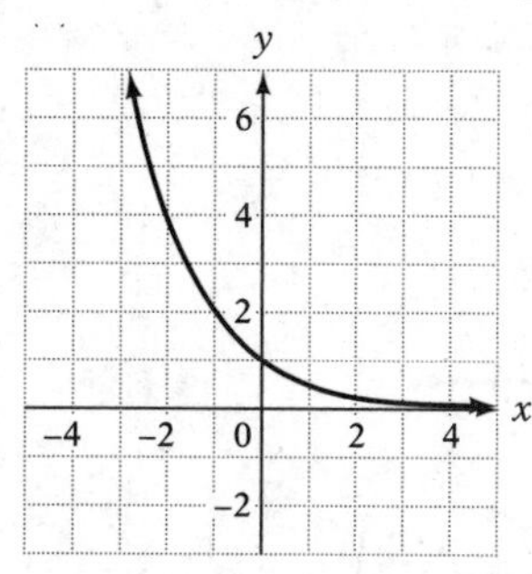

(1)

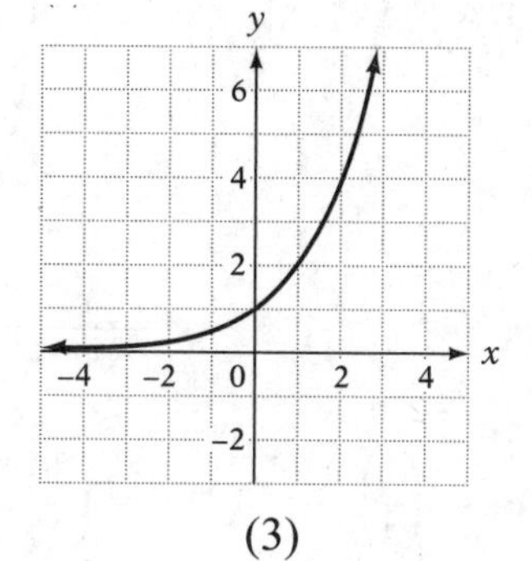

(3)

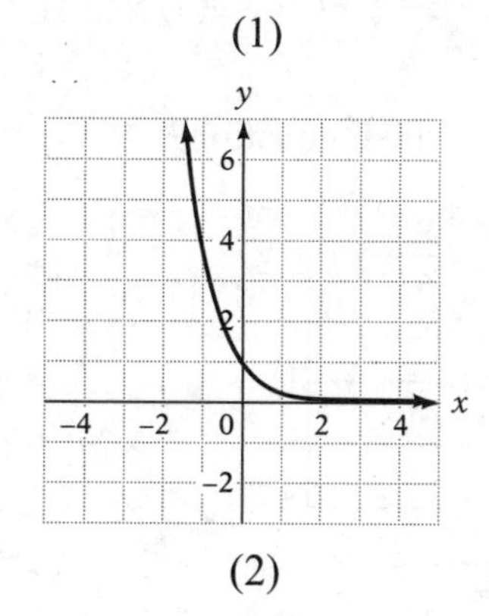

(2)

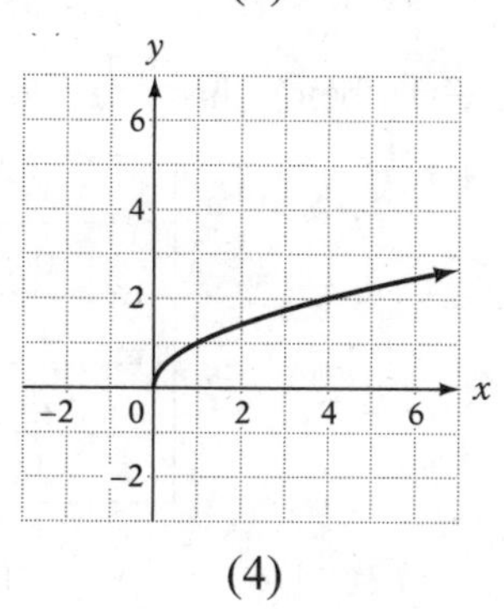

(4)

5. Below is the graph of which equation?

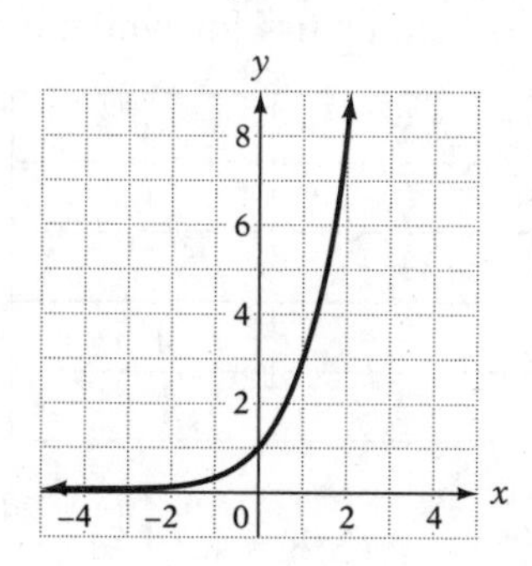

(1) $y = \left(\frac{1}{3}\right)^x$

(2) $y = 10^x$

(3) $y = 4^x$

(4) $y = 3^x$

6. Below is the graph of

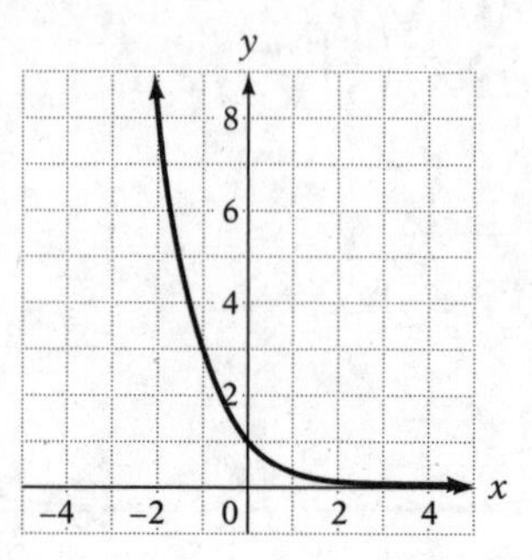

(1) $y = \left(\frac{1}{3}\right)^x$ (2) $y = \left(\frac{1}{2}\right)^x$ (3) $y = \left(\frac{1}{4}\right)^x$ (4) $y = 3^x$

7. The chart below has ordered pairs for which equation?

x	y
0	6
1	12
2	24
3	48

(1) $y = 12x^2 + 6$
(2) $y = 6x + 6$
(3) $y = 12^x$
(4) $y = 6 \cdot 2^x$

8. The chart below has ordered pairs for which equation?

x	y
0	6
1	3
2	1.5
3	.75

(1) $y = 6 \cdot 2^x$
(2) $y = 6 \cdot \left(\frac{1}{2}\right)^x$
(3) $y = 6 \cdot \left(\frac{1}{3}\right)^x$
(4) $y = -3x + 6$

9. In what interval is the graph of $y = 1.5^x$ increasing?
(1) Always
(2) Never
(3) When $x \geq 0$
(4) When $x \leq 0$

10. Below is the graph of $y = b^x$. What is true about the value of b?

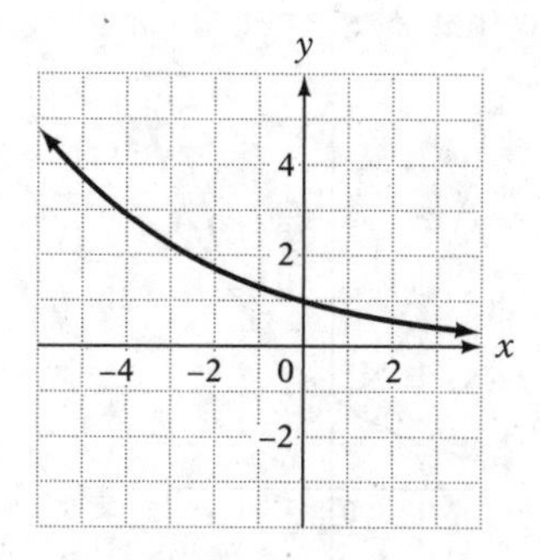

(1) b must be greater than 1.
(2) b must be less than 1 and greater than 0.
(3) b must be less than 0 and greater than −1.
(4) b must be less than −1.

***B.** Show how you arrived at your answers.*

1. The graph of $y = b^x$ passes through the point (5, 243). What must the value of b be?

2. After putting $200 into the bank, the amount of money Xavier has after t years is $P = 200 \cdot 1.15^t$. Make a graph showing how the money grows for 5 years. Choose appropriate units for the horizontal t-axis and the vertical P-axis.

3. Below is the graph of $y = 2^x$ and $y = 3^x$ on the same set of axes.

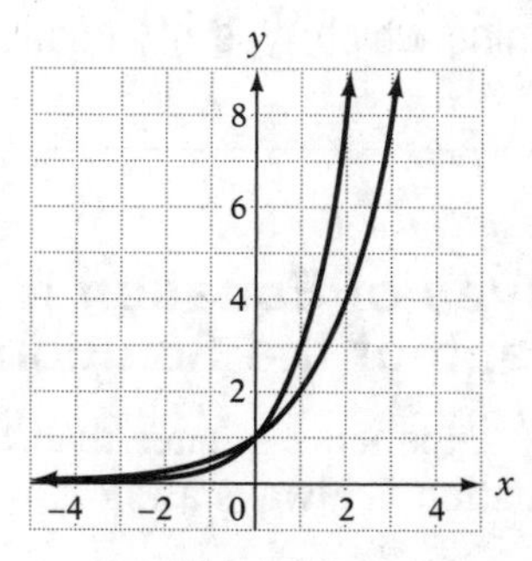

Is it true that $3^x \geq 2^x$ for all values of x? Explain.

4. What is the solution to this system of equations?

$$y = 2^x$$
$$y = 3^x$$

5. Below are the graphs of $y = x^2 + 1$ and $y = 2^x$. They both contain the point (0, 1). Is it true that $2^x < x^2 + 1$ for all values of $x > 0$? Explain.

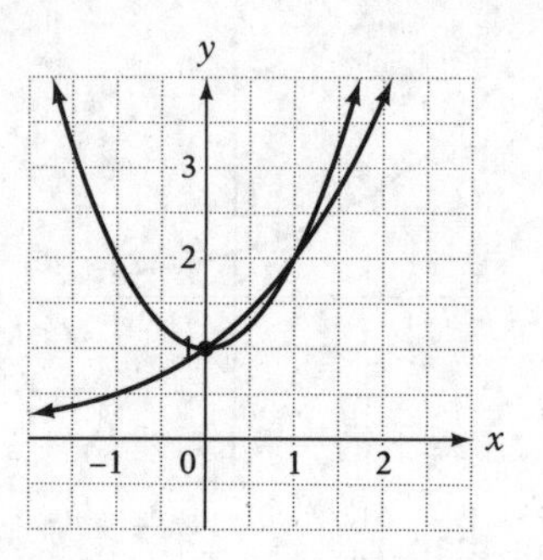

8.3 DISTINGUISHING BETWEEN LINEAR, QUADRATIC, AND EXPONENTIAL EQUATIONS

The graphs of the solution sets of linear, quadratic, and exponential equations are very distinctive so it is possible to tell what type of equation a graph was produced from by just looking at it. When the solution set is given in table form, it is possible to graph the data from the table or to determine which type of equation it came from with fast calculations.

Identifying What Type of Equation It Is By Looking at the Graph of the Solution Set

Linear equations have no exponents greater than or equal to one in them. The graph of a linear equation is always a line.

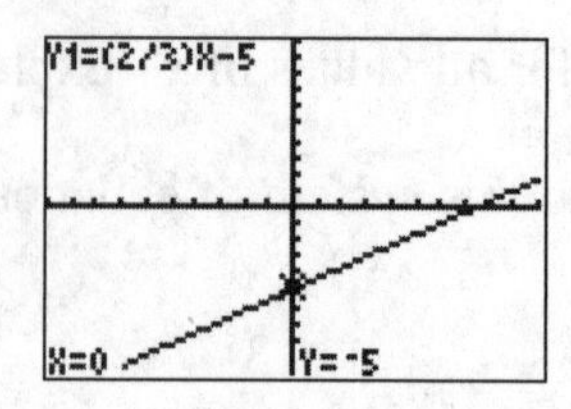

Quadratic equations have either an x^2 or a y^2 term. The graph of a quadratic equation is a parabola.

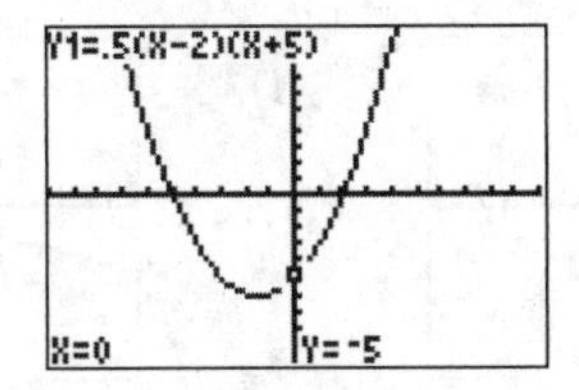

Exponential equations have an x as an exponent. The graph of an exponential equation looks like a playground slide.

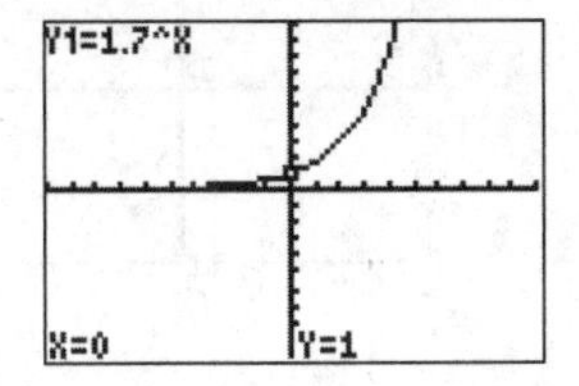

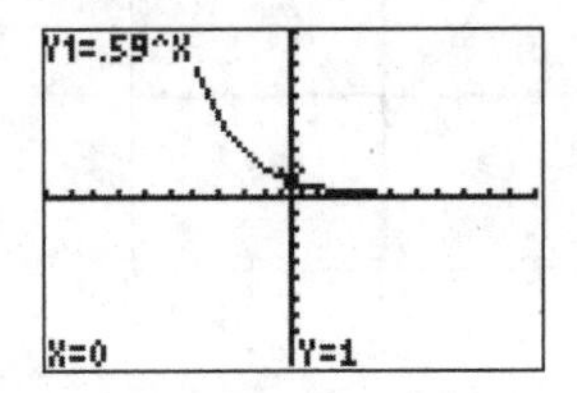

Example 1

Below is the graph of the solution set of which equation?

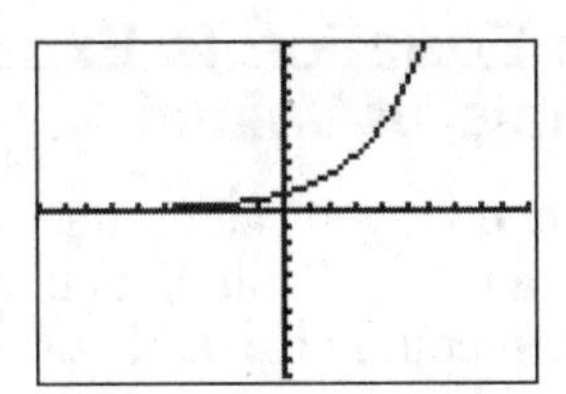

(1) $y = x^2 + 1$ (2) $y = 3x + 1$ (3) $y = 0.4^x$ (4) $y = 1.5^x$

Solution: Choice (4). Because of the shape of the graph, the equation must be an exponential equation so choices (1) and (2) can be eliminated. Since the graph is increasing for all values, the base must be greater than 1 so choice (3) can also be eliminated.

Example 2

Which could be the graph of the solution set of $y = 0.4^x$?

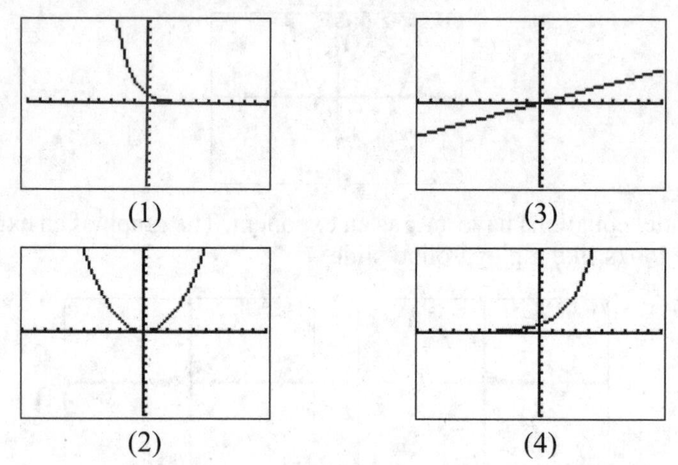

Solution: Choice (1). Choice (2) is the graph of a quadratic equation. Choice (3) is the graph of a linear equation. Choice (4) is the graph of an exponential equation where the base is greater than 1.

Identifying That an Equation Is Exponential By Looking at a Table of Values

When you are given an equation, you can produce a table of solutions to that equation. If you are given a table of solutions to an equation and are given choices of what the equation could be, a table of values for each choice can be made until one matches the question. Graphing the points could help eliminate some choices.

Example 3

Below is a table with solutions to a two-variable equation. Which choice could be the equation?

x	y
2	7
3	25
4	79
5	241

(1) $y = 2^x + 3$ (2) $y = 3^x - 2$ (3) $y = x^2 + 3$ (4) $y = 3x + 1$

The table of values for choice (1) matches the given chart for the first row, but not for the rest.

x	y
2	7
3	12
4	19
5	28

The table of values for choice (2) matches for all the values.

x	y
2	7
3	25
4	79
5	241

Using the Graphing Calculator to Graph Tables of Values

If equations are not given as choices, the quickest way to identify the type of equation a table was based on is by producing a graph of the points and seeing if it has the shape of a line, a parabola, or an exponential equation (like a slide).

Example 4

The following table has solutions to what kind of equation?

x	y
0	250.00
20	435.00
40	756.90
60	1,317.01
80	2,291.59

(1) Linear increasing
(2) Linear decreasing
(3) Exponential growth
(4) Exponential decay

Solution:

For the TI-84:

Press [STAT] key and then [1] to edit the lists. If data are already in the lists, move the cursor to the L1 in the top row and press [CLEAR]. Do the same for L2.

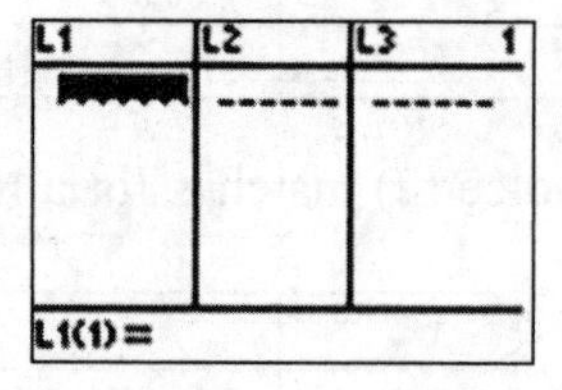

Enter the five solutions with the x in the L1 column and the y in the L2 column.

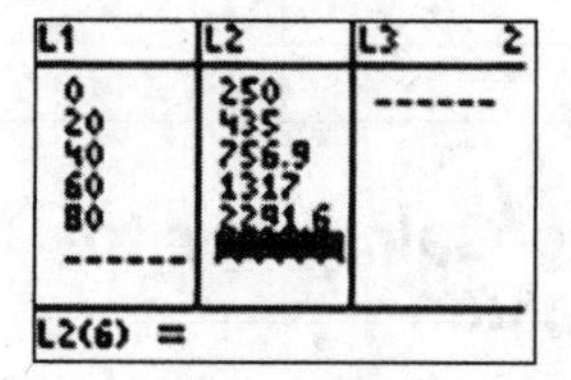

Press the [2ND] and then the [Y=] key to get to the STAT PLOT menu. Select option 1. Using the arrow keys, move to On and press [ENTER]. Now press [ZOOM] and then [9].

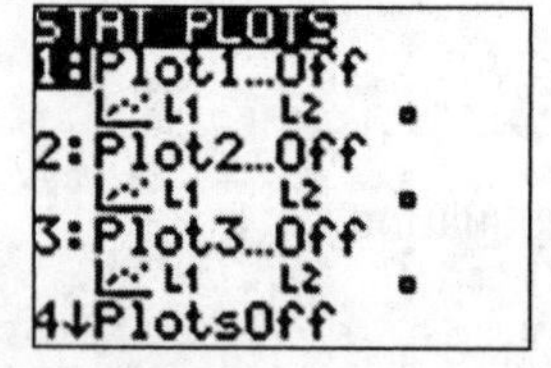

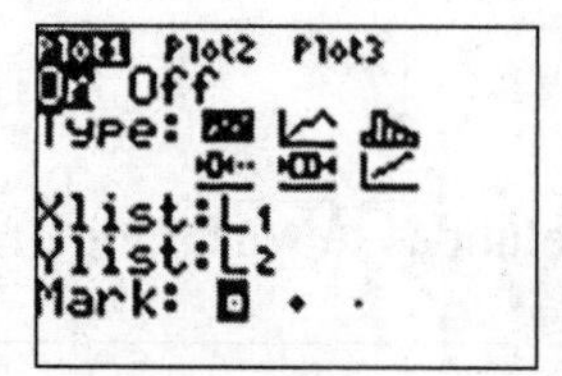

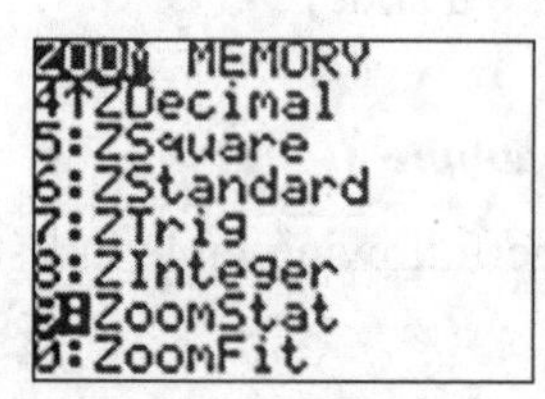

The calculator will graph the points from the lists and set the window so all the points can be seen.

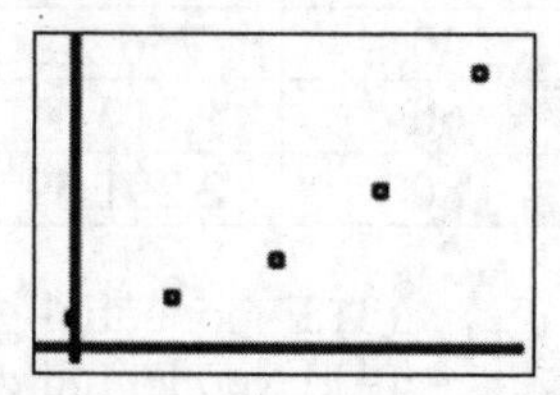

For the TI-Nspire:

From the home screen select the Add Lists & Spreadsheet icon,

the first column x, and the second column y.

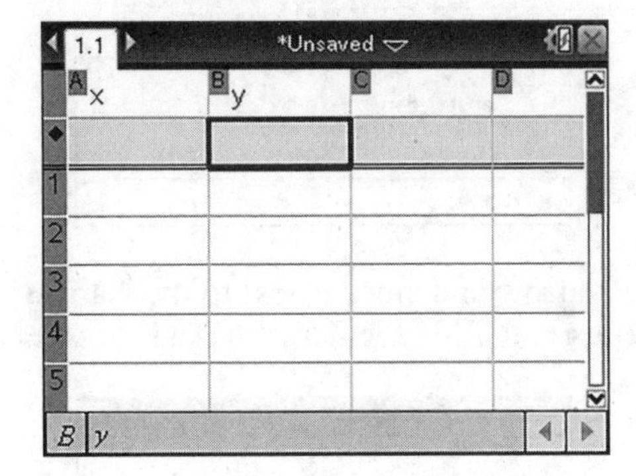

Enter the five x-values into the x column and the five y-values into the y column.

	A x	B y	C	D
1	0	250		
2	20	435		
3	40	756.9		
4	60	1317		
5	80	2291.6		

Go back to the home screen and select Add Graphs icon.

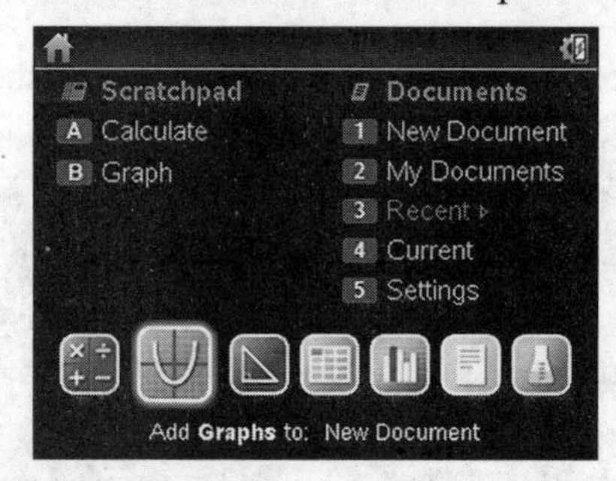

Press [menu], [3], and [4] for Scatter Plot.

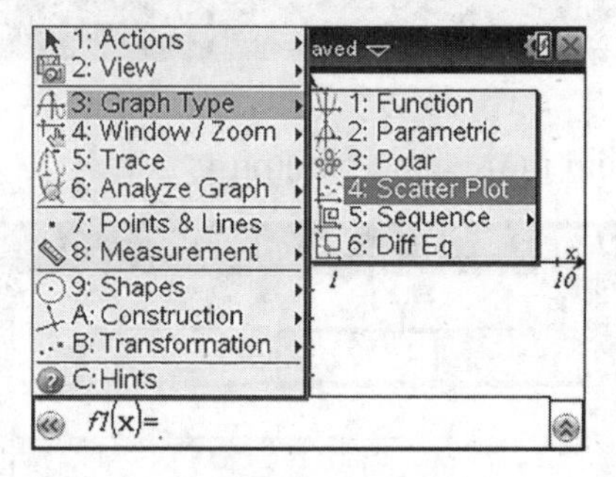

Enter *x* next to the *x* arrow and enter *y* next to the *y* arrow and press [enter]. Then press [menu], [4], and [9] for Zoom—Data.

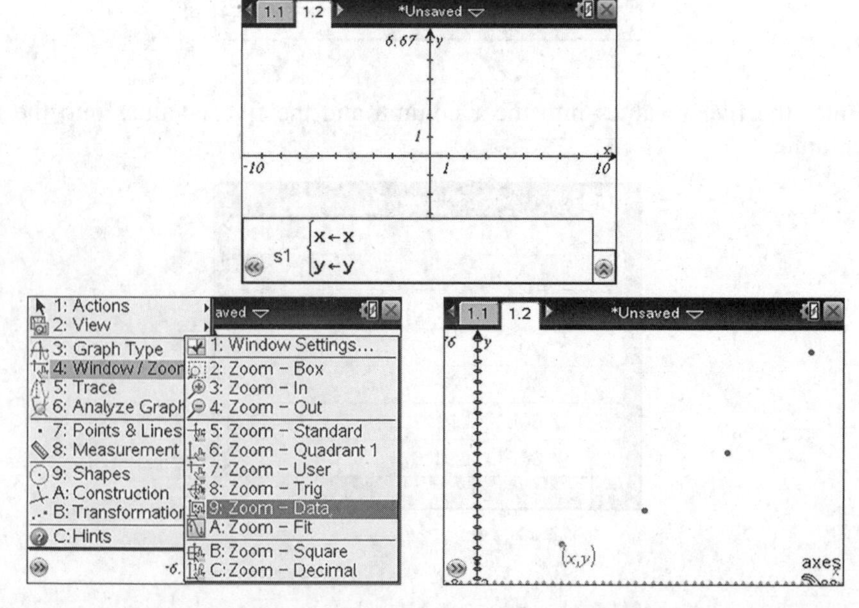

Of the four choices, this looks most like exponential growth. Choice (3).

Identifying the Type of Equation from a Table Without Graphing

If the x values in the chart are evenly spaced, there is a very fast way to determine if it is an exponential or a linear equation.

If the *difference* between consecutive y values is the same for each pair, the equation is a linear equation.

x	y
0	130.00
15	224.35
30	318.70
45	413.05
60	507.40

Since each x value is 15 more than the previous one, check to see the difference between successive y values:

$$224.35 - 130.00 = 94.35$$
$$318.70 - 224.35 = 94.35$$
$$413.05 - 318.70 = 94.35$$
$$507.40 - 413.05 = 94.35$$

Since all these differences are the same, this was a linear equation.

If the *quotient* (what you get by dividing one y value by the previous y value) of two consecutive y values is the same for each pair, the equation is an exponential equation.

The table from Example 4 was

x	y
0	250.00
20	435.00
40	756.90
60	1,317.01
80	2,291.59

Each x-value is 20 more than the previous x-value. The differences between consecutive y-values are not the same.

$$435.00 - 250.00 = 185$$
$$756.90 - 435.00 = 321.90$$

When the quotients are calculated, they become

$$\frac{435.00}{250.00} = 1.74 \qquad \frac{1317.01}{756.90} = 1.74$$

$$\frac{759.90}{435.00} = 1.74 \qquad \frac{2291.59}{1317.01} = 1.74$$

Since these quotients are the same, this is an exponential equation, and since the y values are increasing, it is an example of exponential growth.

Check Your Understanding of Section 8.3

A. Multiple-Choice

1. What type of equation has a graph like the one below?

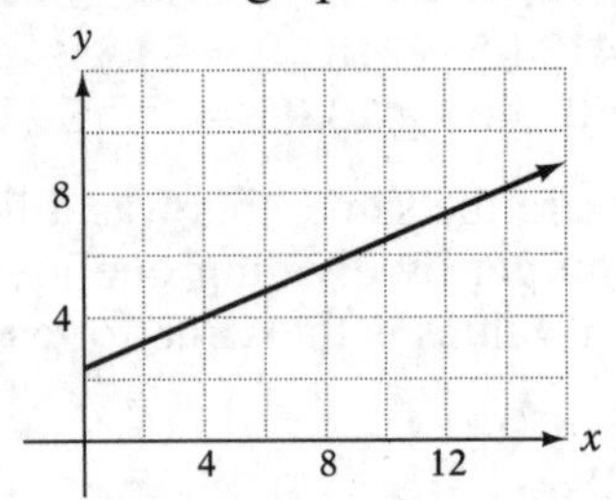

(1) Linear
(2) Quadratic
(3) Exponential
(4) None of the above

2. What type of equation has a graph like the one below?

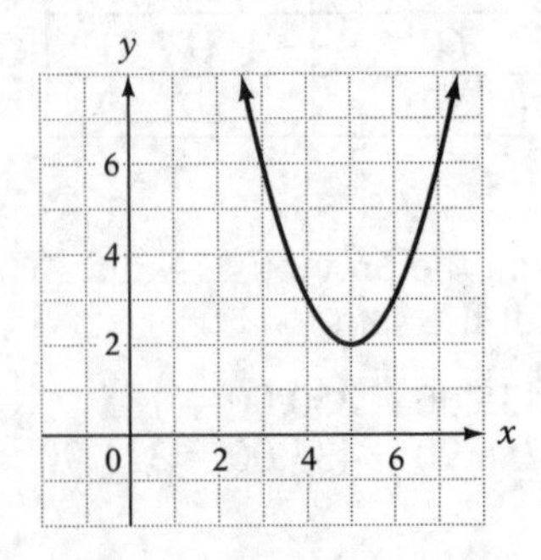

(1) Quadratic
(2) Linear
(3) Exponential
(4) None of the above

3. What type of equation has a graph like the one below?

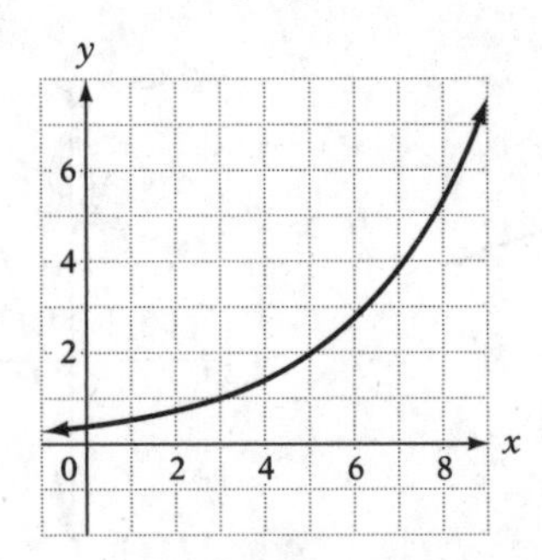

(1) Linear
(2) Exponential
(3) Quadratic
(4) None of the above

4. Which of the graphs below corresponds to a quadratic equation?

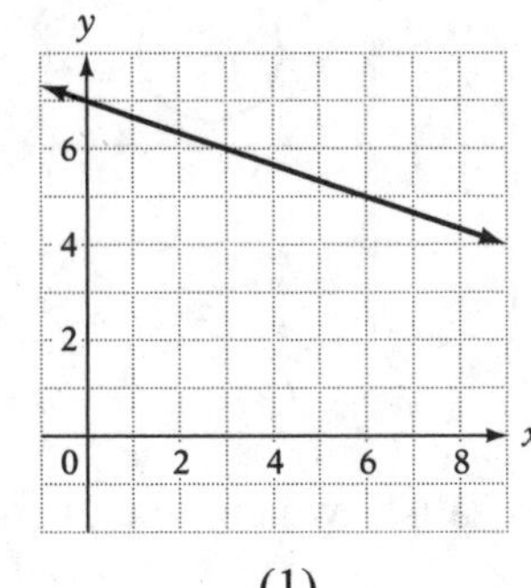

(1)

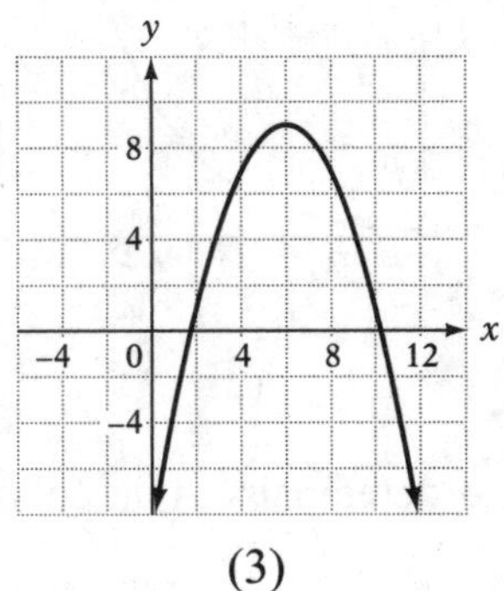

(3)

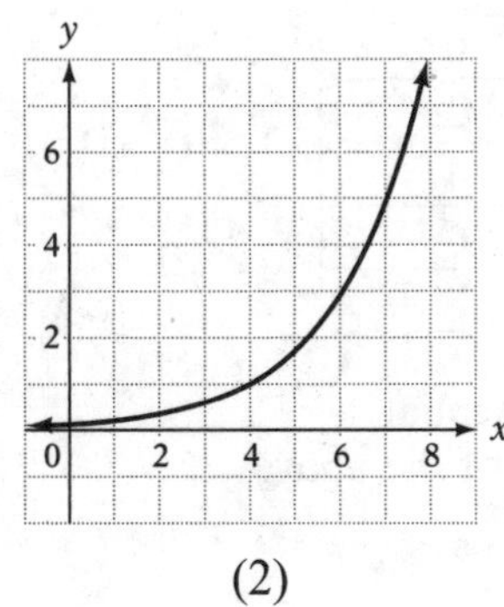

(2)

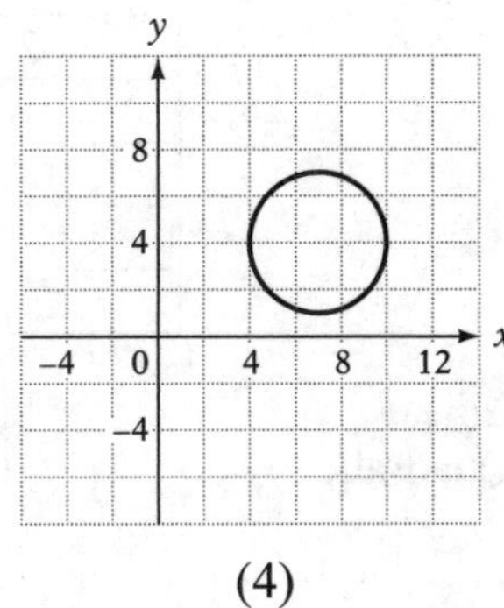

(4)

5. Which of the graphs below corresponds to an exponential equation?

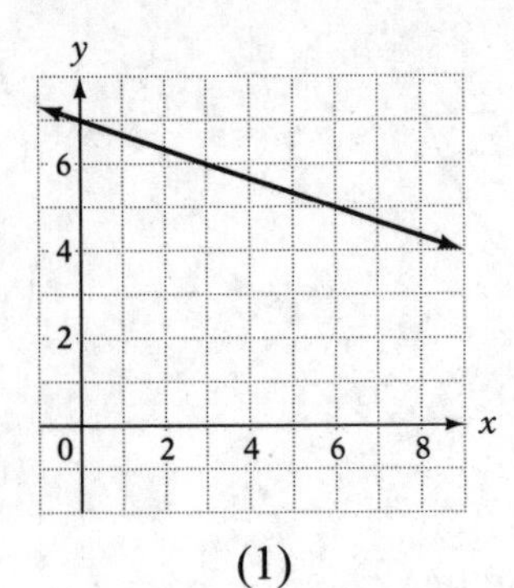

(1)

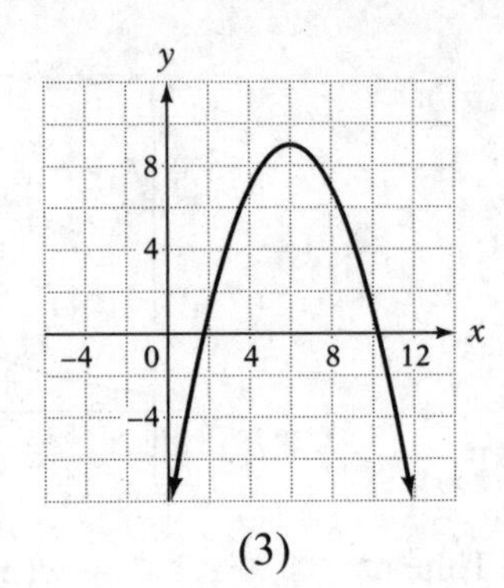

(3)

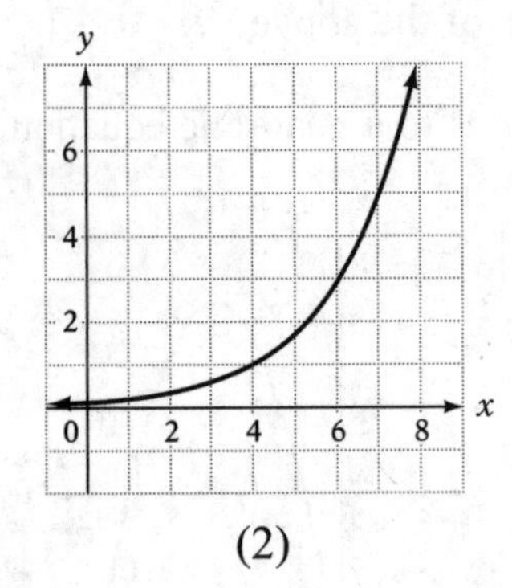

(2)

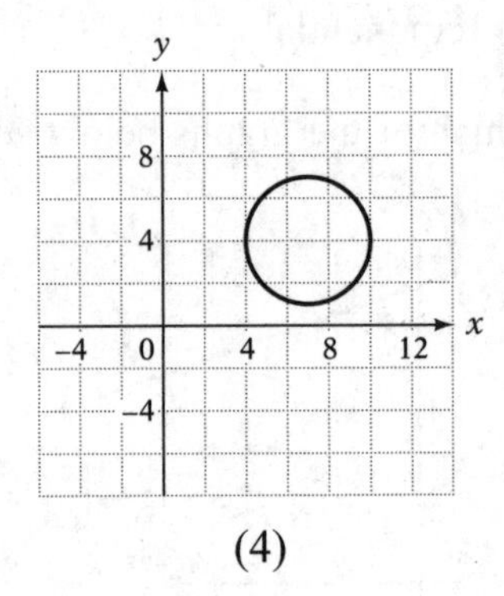

(4)

6. The ordered pairs in the following chart satisfy which type of equation?

x	y
–2	0
–1	2
0	4
1	6
2	8

(1) Linear
(2) Quadratic
(3) Exponential
(4) None of the above

7. Which of the graphs below corresponds to a linear equation?

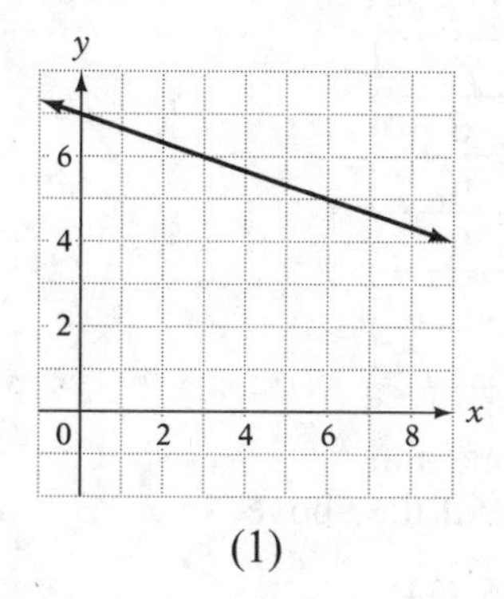

(1)

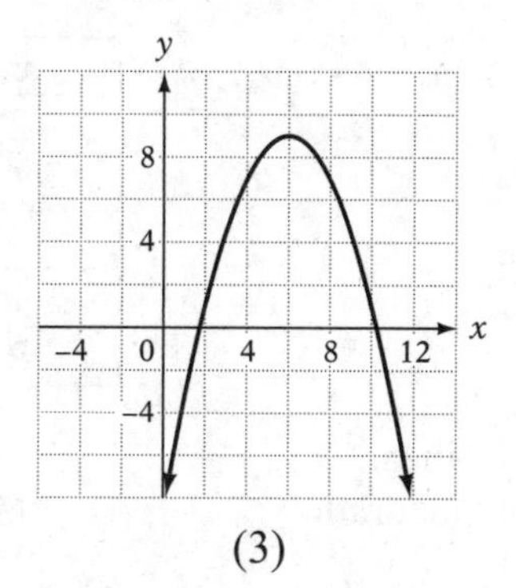

(3)

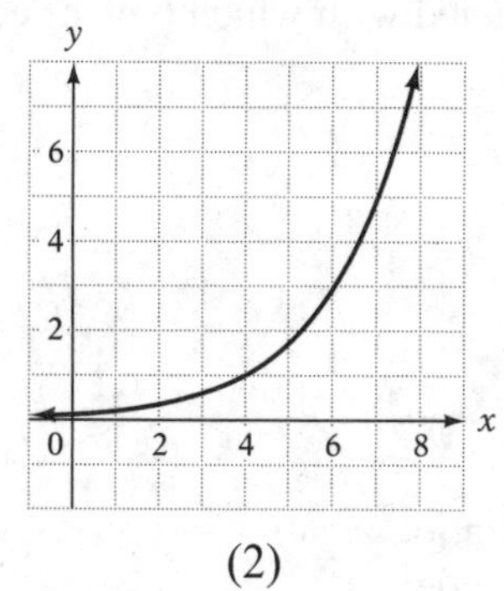

(2)

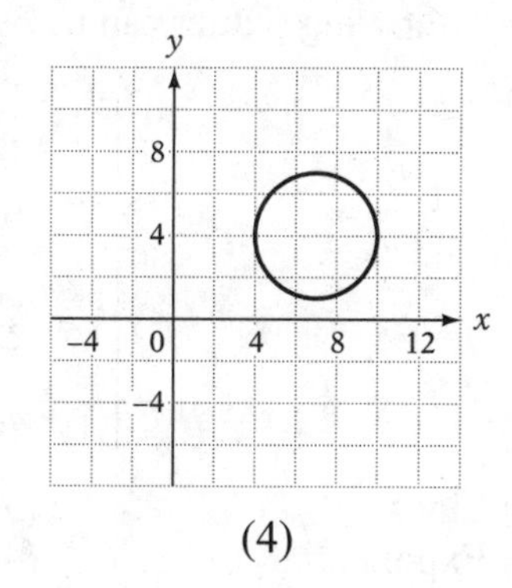

(4)

8. The ordered pairs in the following chart satisfy which type of equation?

x	y
–2	$\frac{2}{9}$
–1	$\frac{2}{3}$
0	2
1	6
2	18

(1) Linear
(2) Quadratic
(3) Exponential
(4) None of the above

9. The ordered pairs in the following chart satisfy which type of equation?

x	y
–2	–6
–1	6
0	0
1	–6
2	6

(1) Linear
(2) Quadratic
(3) Exponential
(4) None of the above

10. The following scatter can be best modeled with which type of equation?

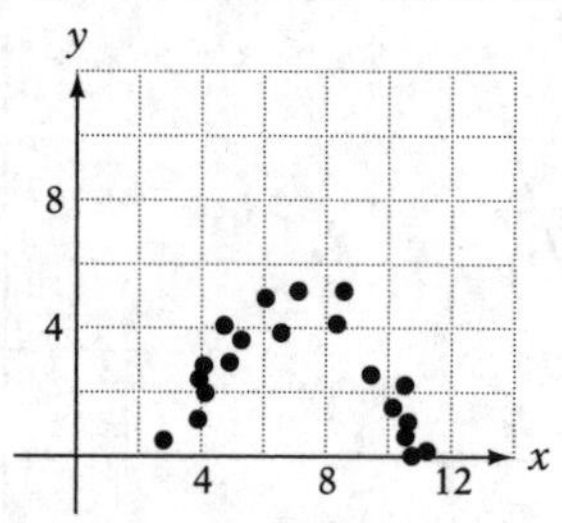

(1) Linear
(2) Exponential
(3) Quadratic
(4) Sinusoidal

B. *Show how you arrived at your answers.*

1. The population of a county for several different times is plotted on a graph where the x-axis represents years since the year 1990 and the y-axis represents the population in millions. The graph is below. What type of equation could be used to model this data? Explain.

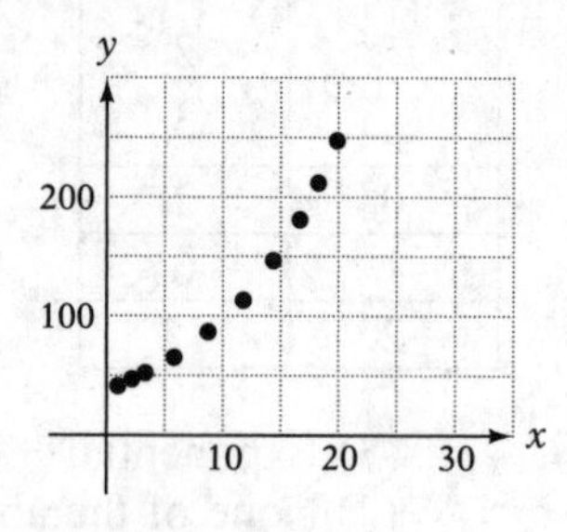

2. Some data points on a graph are plotted. Myah believes that this is a portion of a quadratic graph. Chloe believes that this is a portion of an exponential graph. Is there enough information to determine who is correct? If not, what more information would be needed?

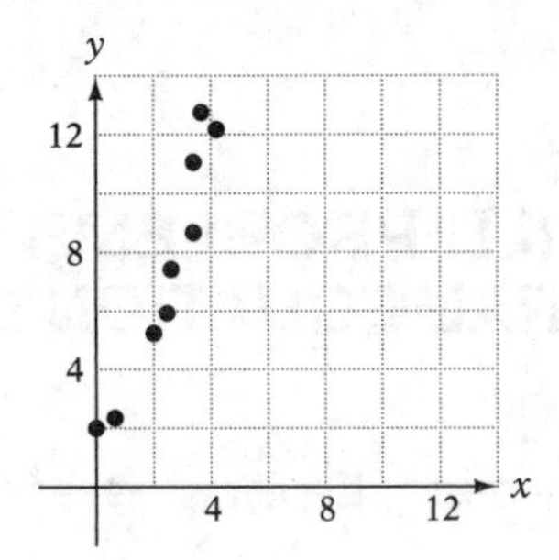

3. When the number of empty seats on a train is graphed as the y-coordinate and the time since 8:00 A.M. is graphed as the x-coordinate, it makes the graph below. Which type of equation could be used to model this data? What details do you know about the equation besides just the type of equation it is?

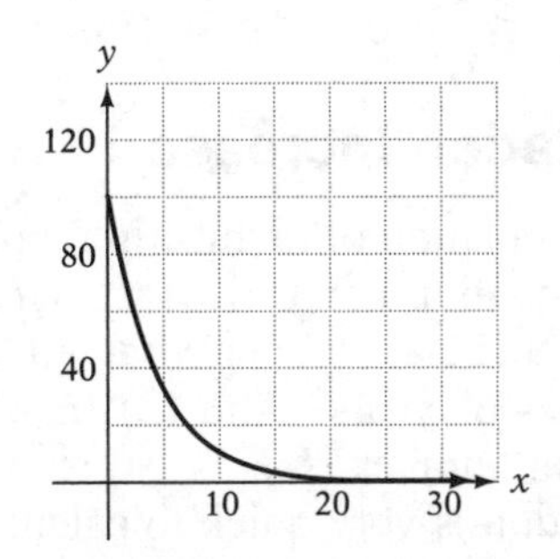

4. Below is a chart with ordered pairs that satisfy an equation. The equation is either linear, quadratic, or exponential. Determine which type of equation it is, and fill in the missing number.

x	y
1	50
2	85
3	?
4	245.65
5	417.605

5. A tennis ball is dropped from a height of 30 feet. After each bounce, the highest point of the next bounce is 80% as high as the bounce before. If the height of each bounce is graphed as the y-coordinate and the bounce number is the x-coordinate, will the graph be linear, quadratic, or exponential?

8.4 REAL-WORLD PROBLEMS INVOLVING EXPONENTIAL EQUATIONS

KEY IDEAS

Exponential equations can be used to model many real-life situations ranging from the way that a cup of hot chocolate cools down to the way that the world population increases. If you are given an exponential equation that is a model for something in the real world, that equation can be used to answer questions about it.

Modeling a Population Increase

In a certain forest, the population of rabbits, in thousands, x years from now can be modeled by the equation $y = 3 \cdot 1.618^x$. With this equation, there are two main questions that can be asked: (1) What will the population of rabbits be after a certain number of years? And (2) In what year will the rabbit population reach a certain number?

The first type of question is very quick to calculate. The second (for this course) requires guess and check or the graphing calculator.

If they ask what the population will be in 5 years, substitute $x = 5$ into the equation $y = 3 \cdot 1.618^x \approx 33.26$. Since this number is in thousands, this represents around 33,000 rabbits.

If they ask how many years it will take for the rabbit population to reach 150,000 rabbits, test numbers until you get close to 150.

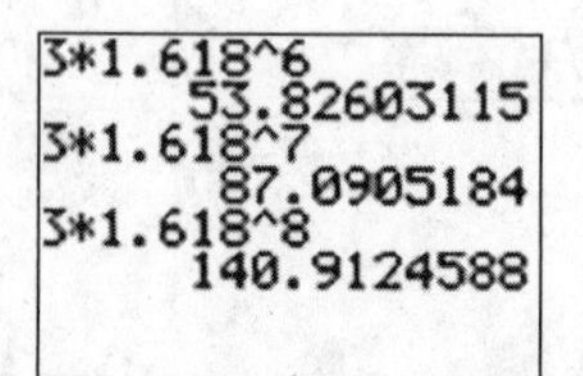

Scratchpad

$3\cdot(1.618)^6$ 53.826

$3\cdot(1.618)^7$ 87.0905

$3\cdot(1.618)^8$ 140.912

3/99

This can also be done by intersecting the graph of $y = 3 \cdot 1.618^x$ with the graph of the line $y = 150$.

For the TI-84:

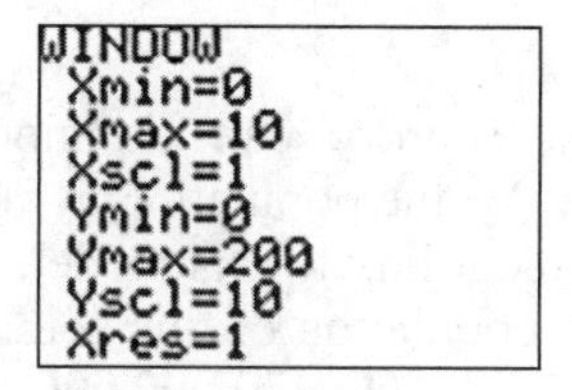

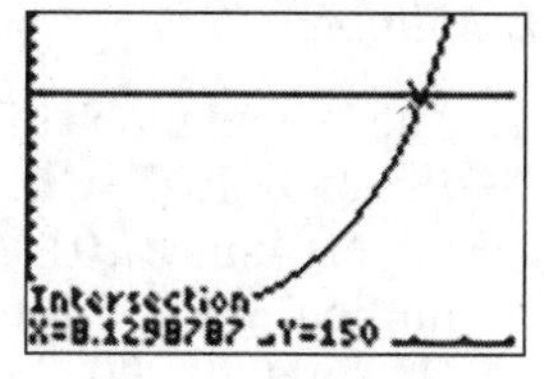

For the TI-Nspire:

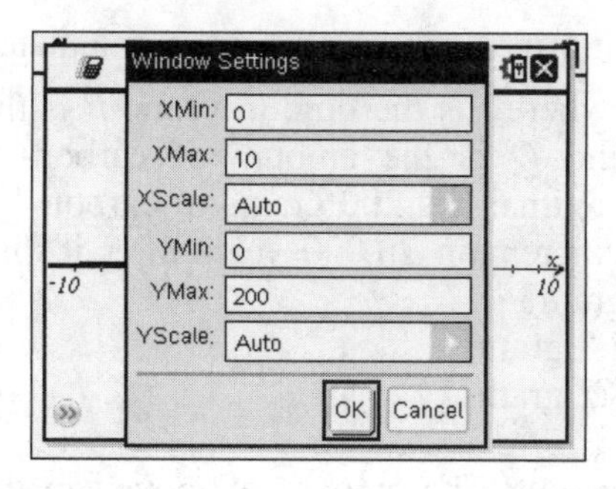

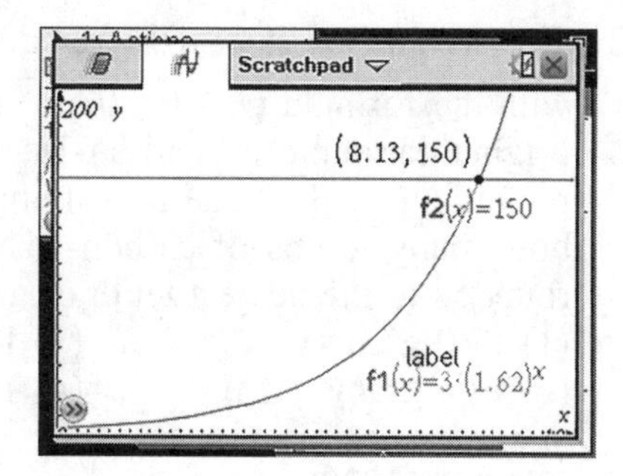

Example

The temperature of a slice of pizza that just came out of the oven can be modeled by the equation $y = 280 \cdot .73^x + 70$, where y is the temperature in degrees and x is the number of minutes since the pizza was removed from the oven. What will the temperature of the pizza be after 10 minutes?

Solution: Substitute 10 for x in the equation.

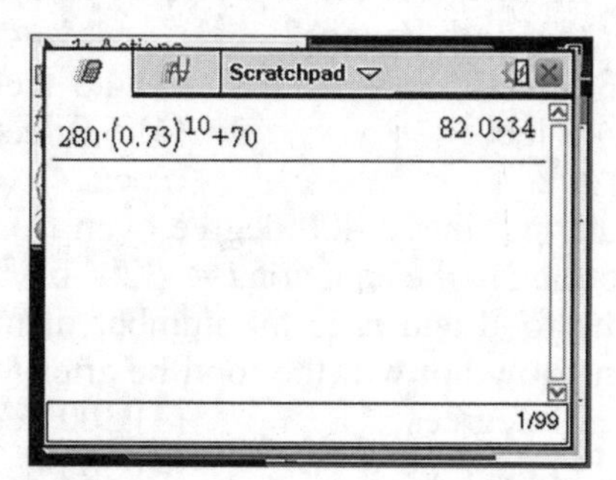

The answer is approximately 82 degrees.

Check Your Understanding of Section 8.4

A. Multiple-Choice

1. The population of a country can be modeled with the equation $P = 250 \cdot 1.07^t$, where P is the population in millions and t is the number of years since 2010. According to this model, rounded to the nearest ten million, what will the population of this country be in 2019?
(1) 450,000,000 (3) 470,000,000
(2) 460,000,000 (4) 480,000,000

2. The amount of carbon-14 remaining in a substance can be calculated with the formula $Q = P \cdot 0.5^{\frac{t}{5,730}}$, where t is the time in years, P is the original amount of carbon-14, and Q is the amount of carbon-14 remaining. If the bone in a living animal has 200 grams of carbon-14, how many grams of carbon-14 will remain 20,000 years after it dies rounded to the nearest tenth of a gram?
(1) 17.0 grams (3) 17.8 grams
(2) 17.4 grams (4) 18.2 grams

3. Kyle puts \$300 into a savings account. The amount of money in the account after t years can be calculated with the formula $A = P \cdot 1.03^t$, where P is the original amount of money and A is the amount of money after t years. How much money will Kyle have after 10 years?
(1) \$397.17 (3) \$401.17
(2) \$399.17 (4) \$403.17

4. A ball is dropped from a window 30 feet above the ground. The height of the nth bounce can be calculated with the formula $h = 30 \cdot .7^n$ where h is the height at the top of the nth bounce. How high will the ball go after the sixth bounce?
(1) 3.5 feet (3) 4.5 feet
(2) 4.0 feet (4) 5.0 feet

5. A cake put into a 400 degree oven from an 80 degree kitchen heats up according to the equation $t = -320 \cdot 0.8^m + 400$ where t is the temperature of the food and m is the number of minutes since it was put into the oven. How hot will the food be after 5 minutes in the oven?
(1) 290 degrees (3) 300 degrees
(2) 295 degrees (4) 305 degrees

6. A cup of tea that is 200 degrees is put into a room that is 80 degrees. The temperature of the tea can be calculated with the formula $t = 120 \cdot 0.9^m + 80$ where m is the number of minutes since the tea was put into the room. What will the temperature of the tea be after 10 minutes?
(1) 116 degrees
(2) 118 degrees
(3) 120 degrees
(4) 122 degrees

7. The grade Jerry gets on a test is related to the number of hours he studies by the equation $g = -60 \cdot 0.7^h + 100$ where h is the number of hours studied and g is the grade the student gets. According to this equation, what grade will Jerry get if he studies for 6 hours?
(1) 89
(2) 91
(3) 93
(4) 95

8. Jude weighs 250 pounds and goes on a diet. His weight is determined by the equation $p = 80 \cdot 0.95^t + 170$ where t is the number of weeks since he started the diet and p is his weight in pounds. How much will Jude weigh after 12 weeks?
(1) 203 pounds
(2) 208 pounds
(3) 213 pounds
(4) 218 pounds

9. A YouTube video of a cat wearing rain boots goes viral. The number of views d days after the video was posted can be calculated with the equation $V = 50 \cdot 1.7^d$. How many times will the video be viewed 20 days after it was posted, rounded to the nearest million?
(1) 2,000,000
(2) 3,000,000
(3) 4,000,000
(4) 5,000,000

10. The number of shaded triangles in the first picture is 3. The number of shaded triangles in the second picture is 9. If this pattern continues, the number of shaded triangles in the nth picture can be calculated with the formula $T = 3^n$. How many shaded triangles would be in the fifth picture?

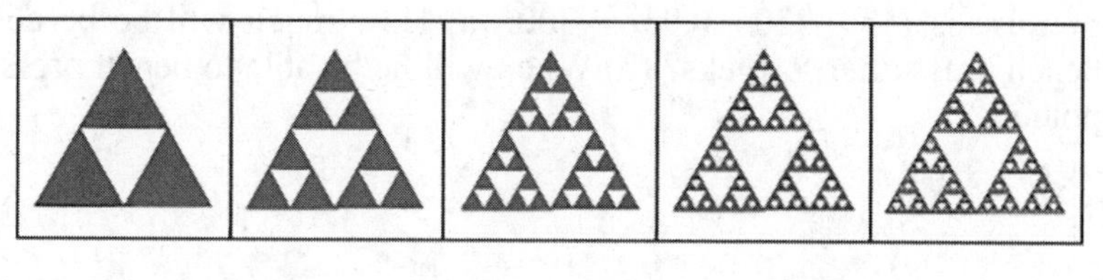

(1) 27 (2) 81 (3) 243 (4) 729

***B.** Show how you arrived at your answers.*

1. The population of a country can be determined by the equation $P = 300 \cdot 1.13^t$ where P is the population in millions and t is the number of years since 2010. (a) Use this formula to determine the population of the country in 2019. (b) Use this formula to determine when the population of the country will be 800 million.

2. Ice cream is removed from a freezer and put into an 80 degree room. The formula that can be used to determine the temperature of the ice cream m minutes after being removed from the freezer is $t = -48 \cdot 0.9^m + 80$. (a) Use this formula to determine what the temperature of the ice cream will be after 5 minutes. (b) After how many minutes will the ice cream be 72 degrees?

3. Lucas wants to put his \$100 into a bank. For the first bank, the formula for how much money he will have after t years is $A = P + t \cdot 0.4P$, where P is the amount of money originally deposited. For the second bank, the formula for how much money he will have after t years is $A = P(1.06)^t$. (a) If he is going to withdraw his money after 10 years, which bank is the better choice? (b) If he is going to withdraw his money after 54 years, which bank is the better choice?

4. The first picture has 4 line segments. The second picture has 16 line segments. If the equation for the number of line segments of the nth picture is $s = 4^n$, how many line segments will be in the fifth picture?

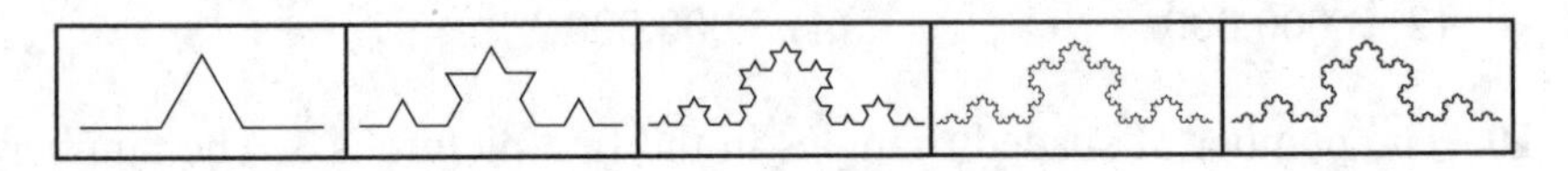

5. Zion has a goal to bench press 250 pounds. The equation that can be used to determine the amount he can bench press after t weeks of exercise is $P = -220 \cdot 0.93^t + 300$. (a) How much will he be able to bench press after 5 weeks? (b) When will he be able to bench press 250 pounds?

Chapter Nine CREATING AND INTERPRETING EQUATIONS FROM REAL-WORLD SCENARIOS

9.1 CREATING AND INTERPRETING LINEAR EQUATIONS

Many real-world scenarios can be modeled with two-variable linear equations. The scenarios generally have a *fixed* part, which is the same for every situation, and a *variable* part, which changes depending on the situation. These equations often resemble the slope-intercept form $y = mx + b$.

Identifying the Fixed and Variable Parts of a Real-World Scenario

At a carnival, it costs people \$10 to enter and an additional \$3 for each ride they go on. The total amount people spend at the carnival depends on how many rides they choose to go on. If they go on no rides, they pay \$10. If they go on one ride, they pay \$13. If they go on 20 rides, they pay \$70. The goal is to create an equation that relates the variable C, for total cost, to the variable R, for number of rides.

One way to create this equation is to first pick some number of rides and calculate how much it would cost to enter the carnival and go on that number of rides. It doesn't matter which number you pick for this method. If you picked 5 rides, you would figure out how much that costs.

For five rides, the person would spend \$10 to enter and then $\$3 \cdot 5 = \$15$ for rides, for a total of \$25. This could be written as $10 + 3 \cdot 5 = 25$. If the number of rides was 20, the calculation would be $10 + 3 \cdot 20 = 70$.

For the variable R rides, it would still cost \$10 to get into the carnival. That is the fixed part. But to go on R rides would cost $\$3R$. So the total cost is $10 + 3R$.

As a two–variable equation with C for cost and R for rides, the equation can be written several ways:

$$C = 10 + 3R \text{ or } C = 3R + 10 \text{ or } 3R + 10 = C \text{ or } 10 + 3R = C$$

The $C = 3R + 10$ is the most common form since it resembles $y = mx + b$ form.

Example 1

An empty ice cream cone costs \$3. Each scoop of ice cream on the cone costs \$1.50. Create an equation that relates the cost of the cone with the ice cream, C, with the number of scoops, S.

Solution: If there are no scoops, the cost is \$3. If there is one scoop, the cost is \$3 + \$1.50 · 1. If there are 5 scoops (careful not to let it tip over!), the cost is \$3 + \$1.50 · 5.

For S scoops, the cost is $C = 3 + 1.50S$ or, more commonly, $C = 1.50S + 3$.

Example 2

Briggs has \$150 in the bank. Each week he takes out \$25. Create an equation that relates the total amount he has in the bank, called B, and the number of weeks that have happened, called W.

Solution: After one week he has \$150 – \$25 = \$125 in the bank. After two weeks he has \$150 – \$25 · 2 = \$100 in the bank, and after three weeks he has \$150 – \$25 · 3 = \$75 in the bank. If W is substituted for the number of weeks in any of these equations, the equation becomes $150 - 25W = B$. This can be written as $B = -25W + 150$, though it does not have to.

Example 3

Lydia has \$21.00. After she buys one banana, she has \$20.25 left. After she buys a second banana, she has \$19.50 left. Create an equation that relates the amount of money she has left, M, to the number of bananas she has purchased, B. Use this equation to determine (a) how much money she will have left after buying 10 bananas and (b) how many bananas she can buy before she runs out of money.

Solution: For this question you are not told what the cost of each banana is, but you are given enough information to determine it yourself. The price of each banana is \$0.75 since $21.00 - 20.25 = 0.75$. After buying B bananas, she will have $21.00 - 0.75B$ dollars left so the equation is $M = 21 - 0.75B$, which may also be written as $M = -0.75B + 21$.

(a) By substituting 10 for B in the equation, the amount of money she has left after purchasing 10 bananas can be calculated.

$$M = 21 - 0.75 \cdot 10$$
$$M = 21 - 7.50$$
$$M = 13.50$$

After buying 10 bananas, she will have \$13.50 remaining.

(b) By substituting 0 for M in the equation, the number of bananas she can purchase can be calculated.

$$0 = 21 - 0.75B$$
$$-21 = -21$$
$$\frac{-21}{-.75} = \frac{-.75B}{-.75}$$
$$28 = B$$

Interpreting Linear Equations

Just as a real-life scenario can sometimes be modeled as a two-variable linear equation, it is also possible to reverse the process. Given a two-variable linear equation that resembles $y = mx + b$ form, it is possible to *interpret* what the constant and the coefficient represent.

For this topic, you will be given a scenario and a two-variable equation. For example: At a pizzeria, the cost of a medium pizza with T toppings is $C = 2.50T + 8.00$. The two numbers 2.50 and 8.00 represent different things about the price of the pizza. One way to approach a question like this is to substitute $T = 0$ into the equation. This would lead to $C = 2.50 \cdot 0 + 8.00 = 8.00$ so a zero-topping pizza would be \$8.00. Notice that 8.00 is the constant in the equation.

If you substitute $T = 1$ into the equation, it would become $C = 2.50 \cdot 1 + 8.00 = 2.50 + 8.00 = 10.50$. A pizza with one topping is \$10.50, which is \$2.50 more than a pizza with zero toppings. Notice that 2.50 is the coefficient of the T.

The 8.00 in the equation $C = 2.50T + 8.00$ must represent the cost of the pizza, not including the toppings, and the 2.50 must represent the cost of each topping.

Example 4

The cost of riding a taxi M miles can be modeled with the equation $C = 0.75M + 2.50$. What does the 0.75 in the equation represent?

(1) The cost of getting into the taxi
(2) The number of miles traveled
(3) The cost of the entire taxi ride
(4) The cost of traveling each mile

Solution: Choice (4). If $M = 0$, the cost is \$2.50, which is the price to get into the taxi. If $M = 1$, the cost is $0.75 + 2.50 = 3.25$, which is \$0.75 more than it cost to go zero miles. So 0.75 is the cost for each mile.

Example 5

A video store makes a profit for each game sold. The store begins each month by paying rent, however, and has no other expenses. The total profit after paying rent that a video store makes selling G games can be modeled by the equation $P = 5V - 250$. What does the 250 represent in this equation?

(1) The profit for each game
(2) The rent the store has to pay
(3) The number of games sold
(4) The total amount of profit after paying rent

Solution: Choice (2). Substitute 0 for V to get the amount of total profit after selling 0 games. The equation becomes $P = 5 \cdot 0 - 250 = -250$. So if they make no money from selling the games, they pay $250 so 250 must be the cost of the rent.

MATH FACTS

When an equation has the form $y = mx + b$, b generally represents the *fixed* part and the mx is the *variable* part. The variable part is the part that changes depending on the value of x.

Check Your Understanding of Section 9.1

A. Multiple-Choice

1. It costs $10 to go to the movies and $3 for each bag of popcorn. Which equation relates the total cost (C) to the number of bags of popcorn purchased (P)?

(1) $C = 3P + 10$
(2) $C = 10P + 3$
(3) $P = 3C + 10$
(4) $P = 10C + 3$

2. A salad costs $6 for the lettuce and $2 for each topping. Which equation relates the total cost (C) to the number of toppings purchased (T)?

(1) $T = 6C + 2$
(2) $T = 2C + 6$
(3) $C = 6T + 2$
(4) $C = 2T + 6$

3. A tablet computer costs $400 and $2 for each app. Which equation relates the total cost (C) to the number of apps purchased (A)?

(1) $A = 2 + 400C$
(2) $A = 400 + 2C$
(3) $C = 2 + 400A$
(4) $C = 400 + 2A$

4. An ice cream store collects \$5 for each ice cream. At the end of the day, the store must pay \$300 total for all the employees. Which equation relates the total amount of money remaining (M) after paying the employees to the number of ice creams sold (I)?

(1) $I = 5M - 300$
(2) $M = 5I - 300$
(3) $M = 5I + 300$
(4) $M = 300I - 5$

5. A cable TV plan costs \$80 a month plus \$10 extra for each premium channel. Which equation relates the monthly bill (B) to the number of premium channels ordered (C)?

(1) $B = 80C + 10$
(2) $B = 10C + 80$
(3) $C = 80B + 10$
(4) $C = 10B + 80$

6. Muhammad has saved \$40. Each week he earns \$15 delivering newspapers. Which equation relates the amount Muhammad has saved (S) to the number of weeks working (W)?

(1) $W = 15S + 40$
(2) $W = 40S + 15$
(3) $S = 15W + 40$
(4) $S = 40W + 15$

7. Lydia wants to buy a DVD player and some DVDs. The equation that relates the total cost for the DVD player and N DVDs is $P = 20N + 200$. What does the number 200 in the equation represent?

(1) The cost of the DVD player
(2) The cost of each DVD
(3) The cost of all N DVDs
(4) The total cost of the DVD player and all N DVDs

8. Amelia buys an empty sticker album and some sticker sheets. The equation that relates the total cost for the empty sticker album and N sticker sheets is $P = 0.75N + 3.00$. What does the number 0.75 in the equation represent?

(1) The cost of the empty sticker album
(2) The cost of each sticker sheet
(3) The cost of all N sticker sheets
(4) The total cost of the empty sticker album and all N sticker sheets

9. Riley has a cell phone that requires a certain amount of money to be paid for each gigabyte of data. If the equation that relates the cost of the phone together with N gigabytes of data is $C = 5N + 80$, what is the cost of one gigabyte of data?

(1) 80
(2) $5N$
(3) 5
(4) $5N + 80$

10. Colton is training to run a marathon. The first month he runs 1 mile a day. The second month he runs 3 miles a day. The third month he runs 5 miles a day. Which equation relates the number of months training (M) to the distance he runs each day (D) in the Mth month?

(1) $M = D + 2$
(2) $M = 2D - 1$
(3) $D = M + 2$
(4) $D = 2M - 1$

B. *Show how you arrived at your answers.*

1. A teacher has a starting salary of $40,000 and each year she gets a $5,000 raise. Create an equation that relates her salary (S) to the number of years (Y) she has been working.

2. For a taxi ride it costs a certain amount to get into the taxi and an additional amount for each mile driven. If the equation that relates the total cost (C) to travel M miles is $C = 2M + 4$, what do the numbers 2 and 4 represent?

3. A 5 foot tree was planted in the year 2000. This chart shows four sets of values where Y is the number of years since 2000 and H is the height of the tree in that year. Create an equation that relates H and Y.

Y	H
0	5
1	8
2	11
3	14

4. Owen, who weighs 260 pounds, goes on a diet where he loses the same amount of weight each month. Below is a chart where M represents the number of months Owen has been on the diet and W represents Owen's weight after M months on the diet. Create an equation that relates W and M.

M	W
0	260
1	258
2	256
3	254

5. Make up a real-world situation that can be represented with the equation $C = 6N + 20$.

9.2 CREATING AND INTERPRETING EXPONENTIAL EQUATIONS

Many real-life scenarios are more accurately modeled with a two-variable exponential equation rather than a linear equation. These include questions about population growth, money and interest, and cooling liquids. Many exponential equations can be written in the form $y = a \cdot (1 + r)^x$, where a is related to the starting value and r is related to how fast something is growing or decaying.

Percent Increase

Exponential equations are about something that is increasing (or decreasing) by the same percent each time period. Calculating how something changes when it is increased or decreased by a certain percent is the key to creating exponential equations.

Bank A offers 15% interest compounded annually. You put \$100 into the bank on January 1, 2015. One year later, January 1, 2016, you will get $\$100 \cdot 15 = \$15$ interest. Together with the original \$100, you have \$115 in the bank. Another year later, January 1, 2017, you will get $\$115 \cdot 0.15 = \$17.25$ interest. Together with the \$115 you already had, your total is now \$132.25.

It took two steps to calculate that you will have \$115 on January 1, 2016, multiplying \$100 by 0.15 to get \$15 and then adding the \$15 to the original \$100. There is a way to get this same answer in one step by multiplying the original \$100 by 1 + 0.15 which is 1.15.

$$\$100 \cdot 1.15 = \$115$$

To get the amount on January 1, 2017, you can multiply the \$115 by 1.15.

$\$115 \cdot 1.15 = \$132.25$.

To get the amount on January 1, 2018, you can multiply the \$132.25 by 1.15.

$$\$132.25 \cdot 1.15 = \$152.09$$

To get from the original \$100 to the amount 10 years later, you can calculate $\$100 \cdot 1.15 \cdot 1.15 \cdot 1.15 \cdot 1.15 \cdot 1.15 \cdot 1.15 \cdot 1.15 \cdot 1.15 \cdot 1.15 \cdot 1.15$, which can be done much more quickly as $\$100(1.15)^{10}$.

Example 1

Bank B offers 12% interest compounded annually. You deposit \$200 into the bank on January 1, 2015. Which expression could be used to calculate the amount of money in the bank 8 years later, January 1, 2023?

(1) $\$200(1.12)^8$
(2) $\$200 + 8(\$200 \cdot 0.12)$
(3) $\$200 + (1.12)^8$
(4) $\$200(8 \cdot \$200 \cdot 0.12)$

Solution: After one year, the amount will be $\$200 \cdot 1.12$. After two years, it will be $\$200 \cdot 1.12 \cdot 1.12 = 200(1.12)^2$, and after three years. it will be $\$200 \cdot (1.12)^3$. So after 8 years it will be $\$200 \cdot (1.12)^8$, which is choice (1).

Percent Decrease

If instead of getting interest of 15% at a bank you had to pay a fee of 15% of the amount you have, each year your savings would decrease. If you started with \$200 on January 1, 2015, then on January 1, 2016, you would have to pay $\$200 \cdot 0.15 = \$30$. You would then have $\$200 - \$30 = \$170$ left. This required two calculations, multiplying by 0.15 and then subtracting \$30 from \$200. Another way to calculate this is to multiply \$200 by $(1 - 0.15)$ which is $\$200(0.85) = \$170$. If you take 15% off, it leaves you with 85% left.

After 2 years, you will have $\$200(0.85 \cdot 0.85) = \$200(0.85)^2 = \$144.50$. In general, you will have $\$200(0.85)^T$ after T years.

Example 2

Mila weighs 170 pounds. She goes on a diet, which causes her to lose in a month 2% of whatever her weight was at the beginning of that month. Which expression can be used to determine her weight at the end of 10 months?

(1) $170(0.98)^{12}$
(2) $170(0.98)^{10}$
(3) $170(0.2)^{10}$
(4) $170(1.02)^{10}$

Solution: For 2% off, subtract $1 - 0.02$ to get 0.98, which is what needs to be multiplied each month. After 10 months, the expression will be $170(0.98)^{10}$, choice (2).

Two-variable Percent Increase Equations

If in a real-world scenario a starting value is given and the percent increase, or growth rate, is also given, a two-variable equation can be created to model the scenario.

The population in the United States in 2014 is approximately 317 million. Each year the population increases by approximately 1%. Write an equation that relates the population, P, to the number of years that have passed since 2014, T.

To increase something by 1%, multiply it by $1 + 0.01 = 1.01$. So after one year, the population will be 317(1.01). After two years, the population will be $317(1.01 \cdot 1.01) = 317(1.01)^2$.

The equation is $P = 317(1.01)^T$.

MATH FACTS

The exponential growth equation is $A = P(1 + r)^T$ where P is the starting value, r is the growth rate, and T is the amount of time periods that have happened.

Example 3

On the first day of school, there are 14 students who know a rumor. Each day the number of students who know the rumor increases by 25%. Create an equation that relates the number of students who know the rumor, N, to the number of days that have passed since the first day of school, T. Approximately how many people will know the rumor after 10 days?

Solution: Since the growth rate is 25%, each new day the number of people who know the rumor is $1 + 0.25 = 1.25$ times the number of people who knew it the day before.

After 3 days, the number of people who know the rumor is $14(1.25 \cdot 1.25 \cdot 1.25) = 14(1.25)^3$. In general, the equation that relates N and T is $N = 14(1.25)^T$.

Substituting 10 for T, it becomes

$$N = 14(1.25)^{10} \approx 130 \text{ people}$$

Two-Variable Percent Decrease Problems

The difference between a percent increase and a percent decrease equation is that instead of multiplying by 1 plus the growth rate each time, you multiply by 1 minus the growth rate.

When a tennis ball is dropped, the height of each bounce is 20% less than the previous bounce. If it is initially dropped from a height of 7 feet, what equation relates the height of the bounce, H, to the number of bounces, B?

To get the number inside the parentheses, calculate 1 minus the growth rate. For this example, this is $1 - 0.20 = 0.80$.

The height of the first bounce is $7(0.8) = 5.6$. The height of the second bounce is $7(0.8 \cdot 0.8) = 7(0.8)^2 = 4.48$. In general, the equation is $H = 7(0.8)^B$.

MATH FACTS

The exponential decay equation can be written as $A = P(1 - r)^T$ where P is the starting value, r is the decay rate, and T is the number of time periods that have happened.

Interpreting Exponential Equations

When a real-world scenario can be modeled with an exponential equation of the form $y = a \cdot (1 + r)^x$, the a and the r are related to the starting value and the growth rate. If the scenario has exponential growth, the r will be positive. If the scenario has exponential decay, r will be between -1 and 0 so that $1 + r$ will be between 0 and 1.

Interpreting Exponential Growth Equations

The equation $A = 500(1.07)^T$ can be used to model many things. If the problem says that A is the amount of money a customer will have in the bank after T years, it is possible to interpret what the numbers 500 and 1.07 represent.

In an equation like $y = a \cdot (1 + r)^x$, the a is the starting amount so the 500 must represent the starting amount of money. One less than what gets raised to the power, which is the growth rate. In this case, the growth rate is .07 or 7%.

Example 4

The equation $P = 25(1.31)^T$ models the number of rabbits in a pet store after T months. How many rabbits were there originally, and what is the growth rate of the rabbits?

Solution: First rewrite the equation as $P = 25(1 + 0.31)^T$. The number in front of the parentheses is the starting amount so there were 25 rabbits originally. The number added to the 1 is the growth rate so in this case the growth rate is 0.31 or 31% each month.

Interpreting Exponential Decay Equations

When an exponential equation represents an exponential decay scenario, the expression being raised to the power will be less than one. Two examples of expressions that represent exponential decay are $T = 80(0.75)^M$ and $H = 100(1 - 0.4)^N$. The number in front of the parentheses will represent the starting value. The number that is being subtracted from 1 is the growth rate. When the growth rate is between 0 and 1 it is also known as the decay rate or rate of decay.

Example 5

When you take a pizza out of an oven and put it into a freezer for M minutes, the temperature of the pizza, T, can be determined by the equation $T = 170(0.92)^M$. What was the temperature of the pizza when it came out of the oven? What is the rate of cooling?

Solution: If you substitute 0 for M, T evaluates to 170, so 170 is the starting value. 0.08 is the rate of decay since it is equivalent to $1 - 0.92$.

Check Your Understanding of Section 9.2

A. *Multiple-Choice*

1. There were 900 birds in a forest. Each year the bird population increases by 12%. Which equation relates the bird population (P) to the number of years that have passed (t)?
   (1) $P = 900(1.12)^t$ (3) $P = 900(0.88)^t$
   (2) $P = 900(0.12)^t$ (4) $t = 900(1.12)^P$

2. Clara deposits $300 into a bank. The bank offers 5% interest compounded annually. Which equation relates the amount of money in the bank (A) to the number of years that have passed (t)?
   (1) $A = 300(1.05)^t$ (3) $A = 300(0.95)^t$
   (2) $A = 300(1.5)^t$ (4) $t = 300(1.05)^A$

3. A bouncing ball is dropped from 20 feet high. After each bounce, the height of the next bounce is 65% as high as the last bounce. Which equation relates the height of the bounce (H) to the number of bounces that have happened (N)?
   (1) $H = 20(0.35)^N$ (3) $H = 20(0.65)^N$
   (2) $H = 20(1.65)^N$ (4) $N = 20(0.65)^H$

4. Food that is 80 degrees is put into a freezer. Each minute the temperature of the food decreases by 18%. Which equation relates the temperature of the food (T) to the number of minutes since the food was put into the freezer (M)?
(1) $T = 80(1 + 0.18)^M$
(2) $T = 80(1 - 0.18)^M$
(3) $T = 80(0.18)^M$
(4) $M = 80(1 + 0.18)^T$

5. The population (P) of a town after t years can be modeled with the equation $P = 20{,}000(1.07)^t$. What does the 20,000 represent?
(1) The growth rate
(2) The percent increase each year
(3) The population after t years
(4) The starting population of the town

6. Trinity deposits some money into a bank that offers interest compounded annually. The amount of money in the bank (A) after t years can be modeled with the equation $A = 700(1.03)^t$. What does 1.03 represent?
(1) The percent increase each year
(2) One plus the growth rate
(3) The amount of money in the bank after t years
(4) The starting amount of money

7. After Allie takes some medicine, each hour the number of milligrams of medicine (M) remaining in her body after t minutes can be modeled with the equation $M = 200(0.73)^t$. Which number represents the decay rate?
(1) 0.73
(2) 200
(3) 0.27
(4) 146

8. A ball is dropped from the window of a building. The height of the bounce (H) is related to the number of bounces (N) by the equation $H = 50(0.4)^N$. Which number represents the height the ball was originally dropped from?
(1) 50
(2) 0.4
(3) 0.6
(4) 20

9. Mason puts money into a bank that offers interest compounded annually. The formula relating the amount of money in the bank (A) to the number of years it has been in the bank (t) is $A = 800(1.2)^t$. What is the interest rate the bank offers?
(1) 1.2%
(2) 2%
(3) 20%
(4) 120%

10. An exponential equation $y = a(b)^x$ models exponential decay. What is known about the value of b?
(1) It is greater than 2.
(2) It is between –1 and 0.
(3) It is between 1 and 2.
(4) It is between 0 and 1.

B. *Show how you arrived at your answers.*

1. The population of a city is 300,000. If the population increases by 4% each year, create an equation that relates the population of the city (P) to the number of years that have passed (t).

2. Blake is trying to quit drinking soda. At the beginning of the year he drinks 64 ounces a day. Each week the amount of soda he drinks each day is 75% of the amount he drank each day the week before. Create an equation that relates the amount of soda he drinks each day (S) to the number of weeks (t) that have passed since he started trying to quit.

3. When Aria started school in kindergarten she had 40 minutes of homework each night. Each year the number of minutes of homework she had was equal to 1.15 times the number of minutes of homework she had the year before. Create an equation that relates the number of minutes of homework she has each night (H) to the number of years that have passed since kindergarten (t).

4. A company's annual profits can be modeled by the equation $P = 200{,}000(1.36)^t$, where P is the amount of profit and t is the number of years they have been in business. What do the numbers 200,000 and 0.36 represent?

5. Bailey is training to run a marathon. The number of minutes it takes to run the marathon after t weeks of training is $M = 300(0.95)^t$. What do the numbers 300 and 0.05 represent?

Chapter Ten FUNCTIONS

10.1 DESCRIBING FUNCTIONS

A *function* is like a machine that converts numbers into other numbers. Functions are often named with the letters f, g, and h. If function f converts the number 2 into the number 7, we write $f(2) = 7$. Functions can be described with an equation, a graph, or a list of ordered pairs.

What Is a Function?

A function can be thought of as a machine with an "in" slot into which you can put numbers on a card and another "out" slot from which numbers on cards come out. If the function's name is f and you put the number 2 on a card and feed it into in slot on the box, another card with the number 7 might come out of the out slot. If 7 comes out when 2 is fed into the function machine, we say $f(2) = 7$. Inside the parentheses tells what number is being fed into the function machine and does not have anything to do with multiplication.

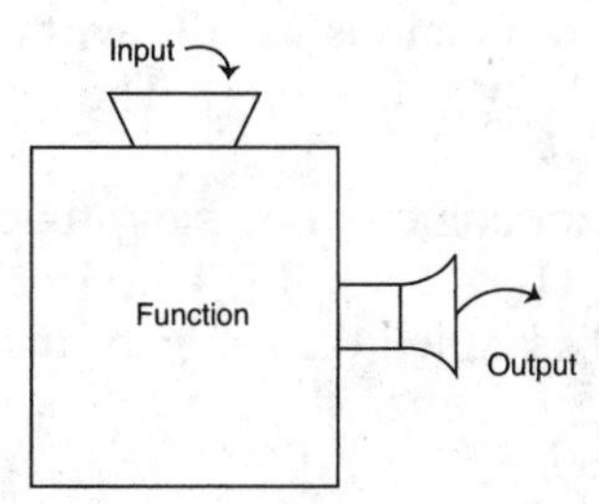

Representing Functions as a List of Ordered Pairs

The only thing we know about the function called f described above is that $f(2) = 7$. We do not know $f(3)$, $f(4)$, or anything else. One way to know more about the function is if a set is given like this:

$$F = \{(1, 4), (2, 7), (3, 10), (4, 13), (5, 16)\}$$

The ordered pair (1, 4) in this list indicates that if you put 1 into the in slot of the function machine, the number 4 will come out of the out slot. Also, if you put the number 2 into the in slot, the number 7 will come out of the out slot.

Since there are five ordered pairs in the set, there are five things we now know about this function:

$$f(1) = 4, f(2) = 7, f(3) = 10, f(4) = 13, \text{ and } f(5) = 16$$

Example 1

If the function g is defined as $g = \{(1, 7), (2, 9), (3, 11), (4, 13), (5, 15), (6, 17), (7, 19)\}$, what is the value of $g(7)$?

(1) 1 (2) 9 (3) 15 (4) 19

Solution: Since the ordered pair that has a 7 as the first coordinate is (7, 19), this means that $g(7) = 19$. Choice (4).

If the function g is defined as in the last example as $g = \{(1, 7), (2, 9), (3, 11), (4, 13), (5, 15), (6, 17), (7,19)\}$, and you are asked for the value of $g(8)$, there is no answer since there is no ordered pair in the set that has 8 as its first coordinate. When this happens we say that the function is *undefined* at 8.

Domain and Range of a Function Described by a Set of Ordered Pairs

The function g described in Example 1 is only defined for the values 1, 2, 3, 4, 5, 6, and 7. If any other number is put into the in slot of this function machine, it will be undefined. These seven numbers are said to be in the *domain* of function g. The domain is usually written as a set of values. The domain of function g is $\{1, 2, 3, 4, 5, 6, 7\}$. The function is only defined for numbers in its domain.

The only values that can come out of the g function's out slot described in Example 1 are 7, 9, 11, 13, 15, 17, and 19. The set of these values $\{7, 9, 11, 13, 15, 17, 19\}$ is called the *range* of the function.

Example 2

What is the domain of the function $h = \{(2, 5), (4, 17), (7, 50)\}$?

(1) $\{5, 17, 50\}$ (3) $\{5, 4, 7\}$
(2) $\{2, 17, 50\}$ (4) $\{2, 4, 7\}$

Solution: Choice (4). The set of all the first coordinates is the domain. $h(2) = 5$, $h(4) = 17$, and $h(7) = 50$. For any other input value, the function is undefined; for example, $h(3)$ is undefined.

Example 3

What is the range of the function $f = \{(2, 5), (3, 7), (6, 5)\}$?

Solution: There are only two different y values so the set for the range will just be $\{5, 7\}$. In a set, you do not list "repeats" so the answer is *not* $\{5, 5, 7\}$.

Domain and Range in Real-World Applications

In some real-world situations, certain numbers don't make sense. For example, you can't buy half of a car or eat negative 5 pieces of pizza. When a function is related to a real-world situation, there may be limits on the domain and range.

If a movie theater sells tickets for $10 and there is a function $f(x) = 10x$ for calculating the amount of money collected for x tickets, the domain is not going to be all real-numbers. Since someone cannot buy half a ticket or negative 10 tickets, the domain is just going to be the numbers $\{0, 1, 2, 3, 4, 5, \ldots\}$.

When a Set of Ordered Pairs Cannot Describe a Function

If in a function f, $f(2) = 7$, this means that any time the number 2 is put into the function machine, the number 7 comes out. There will never be a time that 2 is put in and the number 8 comes out. If it did, we would want to get our function machine repaired!

When a function is described by a list of ordered pairs, no two ordered pairs will have the same first coordinate. If (2, 7) is in the set, there will not be a (2, 8) or a (2, 9) or anything else with a 2 as the first coordinate.

Function	Not a Function
$f = \{(2, 7), (3, 9), (4, 13), (5, 13)\}$	$f = \{(2, 7), (2, 9), (4, 13), (5, 17)\}$
It is OK to have matching y-values	↑ ↑

If a function were described as $f = \{(2, 7), (2, 9), (4, 13), (5, 17)\}$ and they asked the value of $f(2)$, there would be two possible answers, 7 or 9. For this reason, a function cannot have different output values for the same input value.

Example 4

Which of the following cannot be the description of a function?

(1) {(1, 4), (3, 6), (5, 8), (7, 10)}
(2) {(1, 4), (3, 4), (5, 4), (7, 4)}
(3) {(1, 6), (3, 10), (5, 4), (7, 8)}
(4) {(1, 4), (3, 6), (5, 8), (5, 10)}

Solution: Since choice (4) has two ordered pairs that both have 5 as the first coordinate, it cannot be the description of a function.

Check Your Understanding of Section 10.1

A. Multiple-Choice

1. If a function f is defined as $f = \{(1, 2), (2, 3), (3, 1), (4, 4)\}$, what is $f(2)$?
(1) 1 (2) 2 (3) 3 (4) 4

2. If a function g is defined as $g = \{(1, 4), (3, 2), (4, 3), (5, 1)\}$, what is $g(2)$?
(1) 1 (2) 2 (3) 3 (4) Undefined

3. If $f(4) = 7$, which could *not* be the definition of the function?
(1) $f = \{(1, 3), (4, 7), (5, 6), (7, 4)\}$
(2) $f = \{(2, 5), (4, 7), (5, 1), (6, 5)\}$
(3) $f = \{(1, 3), (4, 8), (5, 6), (7, 4)\}$
(4) $f = \{(1,3), (4, 7), (5, 6), (4, 7)\}$

4. If $f = \{(1, 4), (2, 8), (3, 7), (4, 1)\}$ and $g = \{(1, 5), (2, 9), (3, 6), (4, 2)\}$, which of the following is true?
(1) $f(1) > g(1)$
(2) $f(2) > g(2)$
(3) $f(3) > g(3)$
(4) $f(4) > g(4)$

5. Which of the following *cannot* be the definition of a function?
(1) $f = \{(1, 5), (2, 7), (2, 8), (4, 9)\}$
(2) $f = \{(1, 2), (2, 2), (3, 2), (4, 2)\}$
(3) $f = \{(0, 0), (1, 1), (-1, 1), (2, 4), (-2, 4)\}$
(4) $f = \{(6, 1)\}$

6. What is the domain of the function defined as $f = \{(1, 4), (3, 7), (4, 8), (5, 8)\}$?
(1) {4, 7, 8}
(2) {1, 3, 4, 5, 7, 8}
(3) {1, 3, 4, 5}
(4) {4}

7. What is the range of the function defined as $f = \{(1, 4), (3, 7), (4, 8), (5, 8)\}$?
(1) $\{1, 3, 4, 5, 7, 8\}$
(2) $\{1, 3, 4, 5\}$
(3) $\{4\}$
(4) $\{4, 7, 8\}$

8. If $f = \{(1, 4), (2, 3), (3, 2), (4, 1)\}$ and $g = \{(1, 3), (2, 4), (3, 1), (4, 2)\}$ what is $f(g(1))$?
(1) 1 (2) 2 (3) 3 (4) 4

9. A function g takes as an input a number representing the number of gallons of gasoline purchased and outputs the price of that many gallons of gasoline. What is the domain of this function?
(1) All numbers greater than or equal to zero
(2) All numbers
(3) All nonnegative integers
(4) All numbers greater than zero

10. A function t takes as an input the day of the year in New York and outputs the average temperature for that day. What is a reasonable range for this function?
(1) Numbers between –80 and 200
(2) Numbers between –10 and 100
(3) Numbers between 60 and 80
(4) Numbers between 0 and 60

B. *Show how you arrived at your answers.*

1. Explain why this is or is not a definition of a function.
$f = \{(3, 5), (4, 5), (5, 5), (6, 5)\}$

2. If $f = \{(1, 4), (2, 1), (3, 2), (4, 3)\}$, calculate (a) $f(1) + f(2)$; (b) $f(1 + 2)$; (c) $f(f(1))$.

3. If $f = \{(1, 3), (4, 9), (5, 2), (6, 8)\}$ and $f(a) = 8$, what are all possible values for a?

4. Is $f = \{(3, 4), (3, 5), (4, 7), (5, 1)\}$ a function? Explain why or why not?

5. William says the range of a function always has either the same number of numbers as the domain or more numbers than the domain. Mia says the range of a function can either have the same number of numbers as the domain or fewer numbers than the domain. Alice says that the range of a function can have more numbers, fewer numbers, or the same number of numbers as the domain. Which of the three students is correct and why?

10.2 FUNCTION GRAPHS

A list of ordered pairs is not always the most efficient way to describe the input and output values of a function. A graph of these values is especially useful when there are an infinite number of numbers in the domain of the function. The function graph can be used to evaluate the function at different values and can also be used to determine the domain and range of the function.

Graphing a Function

If a function is represented as a set of ordered pairs, those ordered pairs can be graphed to form the graph of a function.

If function f is described as $f = \{(1, 1), (2, 3), (3, 5), (4, 7), (5, 9)\}$, the graph of the function will contain just five points corresponding to the five ordered pairs in the set. The graph of these five points looks like this.

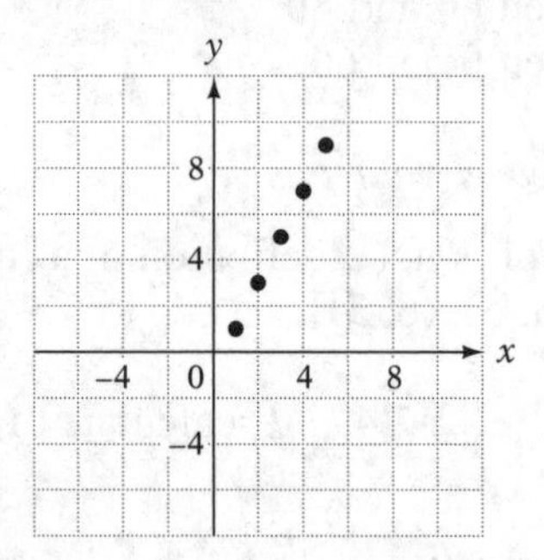

This graph contains the same information as the set of ordered pairs. Since there is a point at (1, 1), this means that $f(1) = 1$. Since there is a point at (2, 3), this means that $f(2) = 3$.

Sometimes you are given the graph of a function and not the set of ordered pairs. From this graph, it is possible to answer questions about the function. For example, on the next page is the graph of a different function called f.

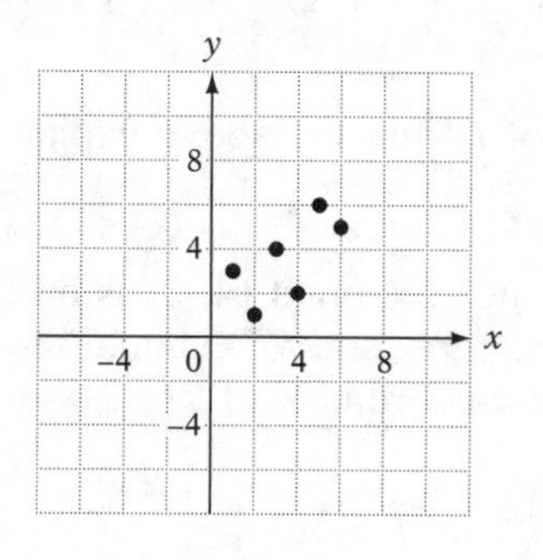

To find the value of $f(2)$, look for a point on the graph that has an x-coordinate of 2.

Step 1: Locate 2 on the x-axis.

Step 2: Look up and down the vertical line through the 2 on the x-axis to see if there is a point that has an x-coordinate of 2.

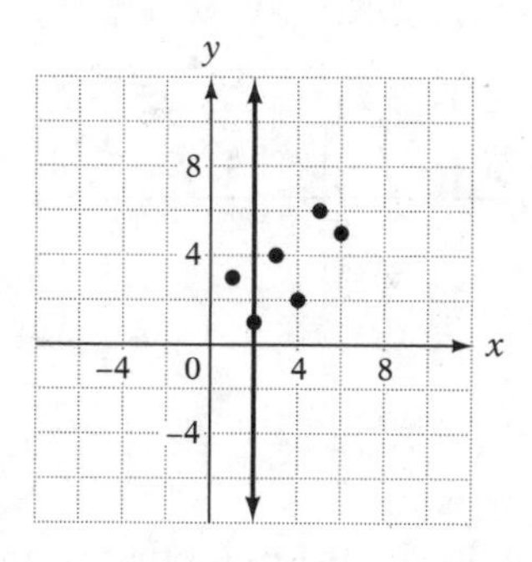

Step 3: If there is no point that has this x-coordinate, the function is undefined at that 2.

Step 4: If there is a point that has this x-coordinate, the y-coordinate of that point is the value of $f(2)$.

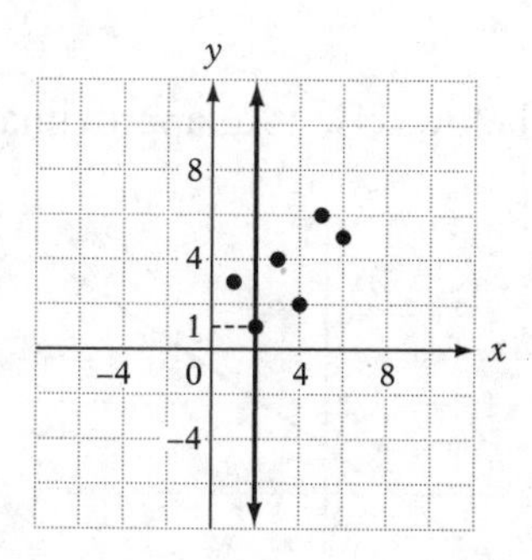

Since the point on the graph that has an x-coordinate of 2 has a y-coordinate of 1, $f(2) = 1$.

Example 1

Using the same graph of f from the above explanation, find the values of $f(4)$, $f(5)$, and $f(7)$.

Solution: $f(4) = 2$ since the point (4, 2) is on the graph. $f(5) = 6$ since the point (5, 6) is on the graph, and $f(7)$ is undefined since there is no point on the graph that has an x-coordinate of 7.

Example 2

Below is the graph of a function called g. Which point could be used to find the value of $g(4)$?

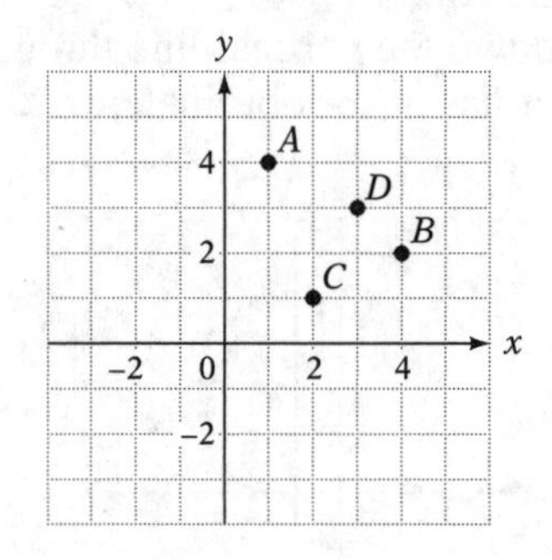

(1) A (2) B (3) C (4) D

Solution: To find $g(4)$, locate the point that is on the vertical line through (4, 0). This line would pass through point B. The value of $g(4) = 2$, though they are not asking for the value of $g(4)$, just the point that can be used to find $g(4)$, which is choice (2).

Example 3

The graph of h is given below. Use it to approximate the value of $h(4.5)$.

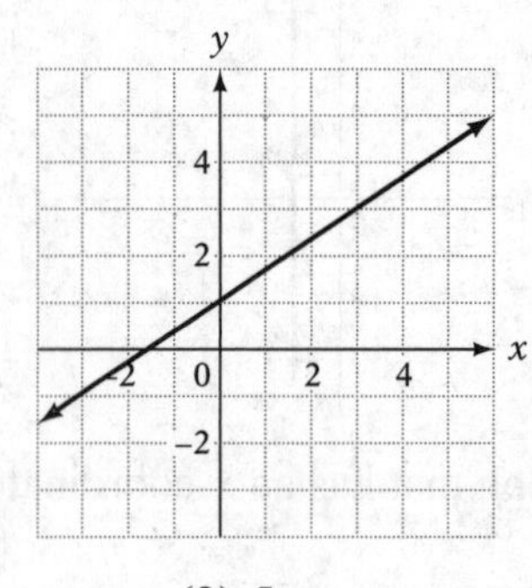

(1) 3 (2) 4 (3) 5 (4) 6

Solution: From the point (4.5, 0) on the x-axis, draw a vertical line up toward the line representing the function. When you reach the point on the line, draw a horizontal line to the left until you reach the y-axis. It seems to be near the point (4.5, 4), so there is only one choice that is feasible, choice (2).

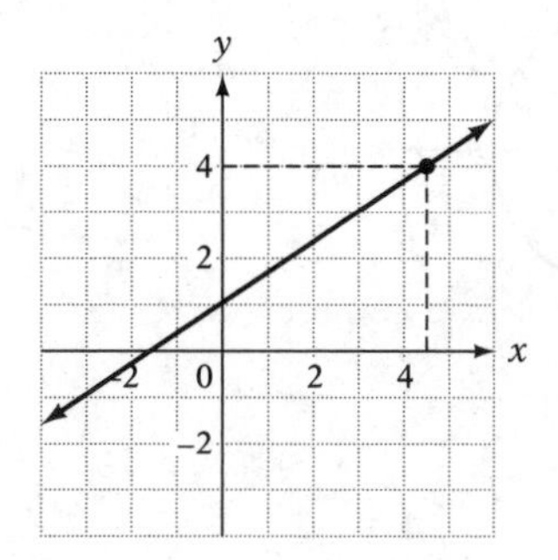

Determining if a Graph Represents a Function

The set $f = \{(1, 3), (1, 5), (2, 6)\}$ does not represent a function because two of the ordered pairs have the same first coordinate. $f(1)$ can only equal one thing, but from this set, it seems to be both 3 and 5. When these points from this nonfunction are graphed, the two points with the same x value will lie on the same vertical line.

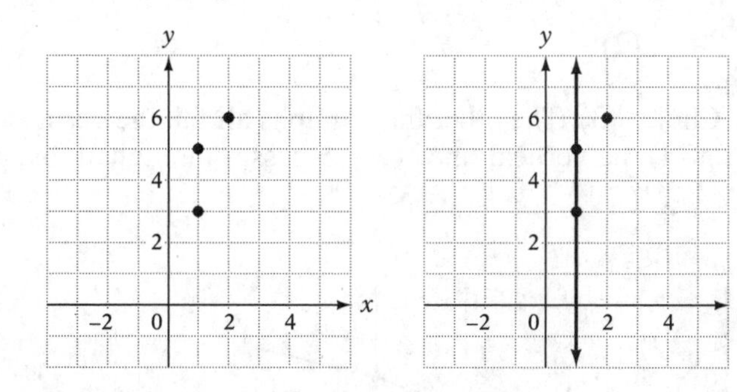

If the graph of a set of ordered pairs has at least two points that lie on the same vertical line, the graph fails the *vertical line test* and the set cannot represent a function.

Example 4

Which of the following can be the graph of a function?

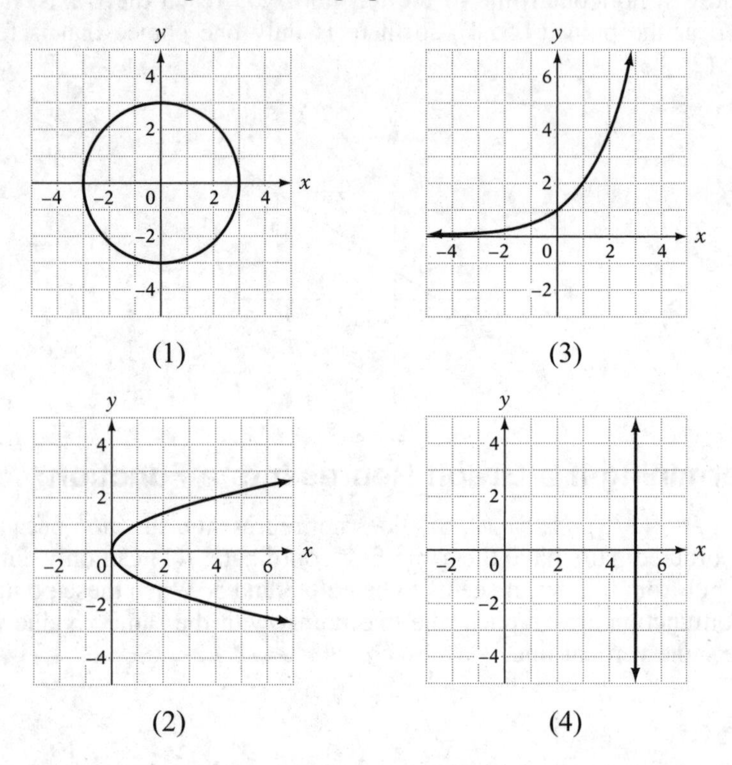

Solution: Choice (3). The other three graphs all fail the vertical line test. In the graph for (4) the vertical line at $x = 5$ passes through all the points on the graph.

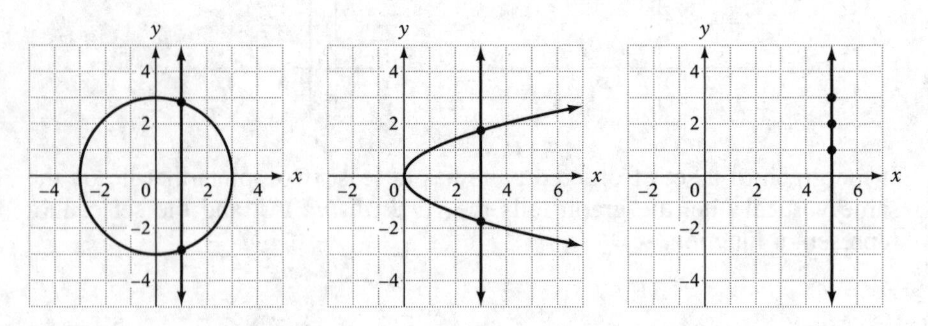

Finding the Domain and Range from the Graph of a Function

If the graph of a function is a bunch of points, the domain is the set of all the x-coordinates and the range is the set of all the y-coordinates of the points.

In the graph below, the domain is $\{1, 2, 4, 5\}$, and the range is $\{3, 4\}$.

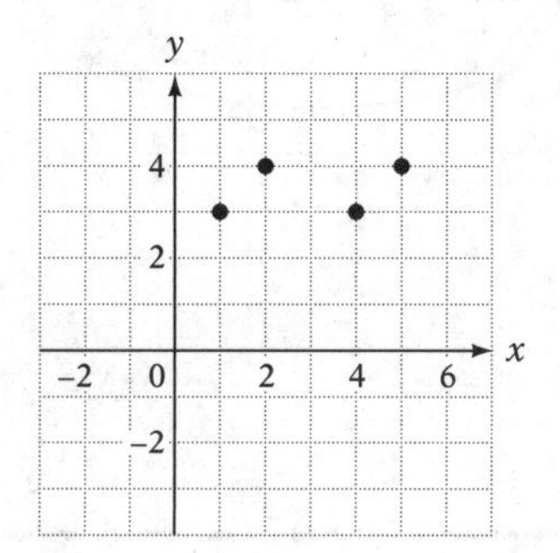

If the graph of the function is some kind of curve, the domain and range cannot be described by a list since there are an infinite number of values in each set.

In the graph of the function below, there are an infinite number of points on the line segment.

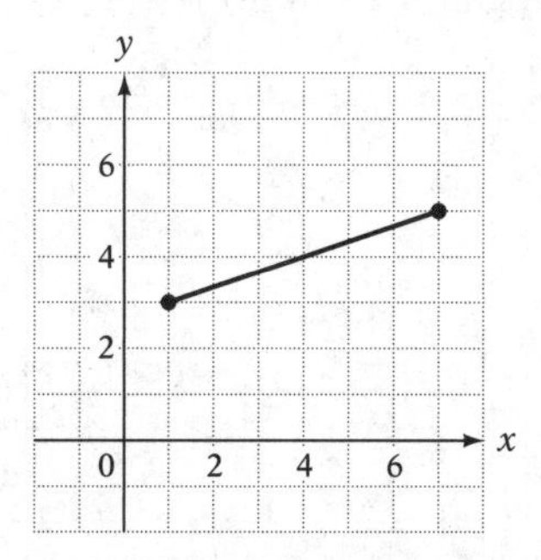

For every number between 1 and 7, including 1 and 7, there is a point on the line segment that has that as an x-coordinate. The domain can be described as $1 \leq x \leq 7$. Likewise, the range can be described as $3 \leq y \leq 5$.

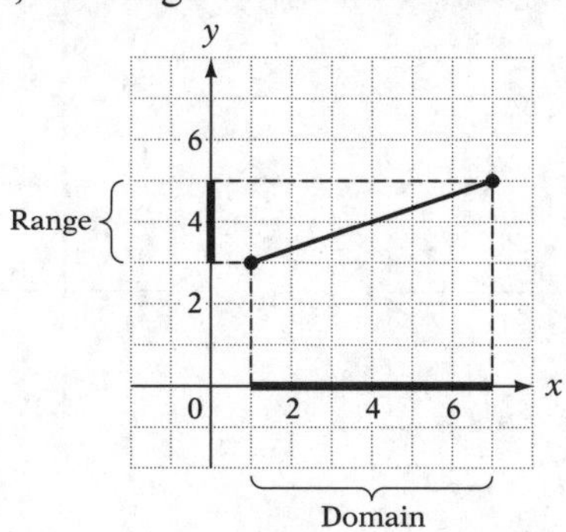

It is possible for the domain or the range to contain every real number. In this case, write "All real numbers." For example, if the graph of the function is given below, the domain is all real numbers, and the range is $y \geq 2$.

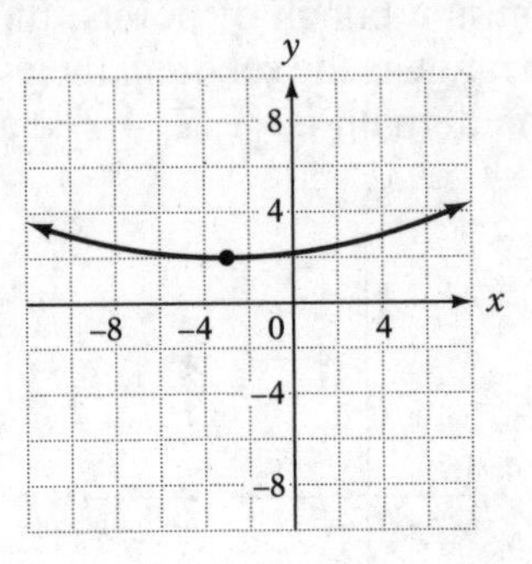

Check Your Understanding of Section 10.2

A. Multiple-Choice

1. Which is the graph of the function $f = \{(1, 2), (2, 5), (3, 1), (4, 1)\}$?

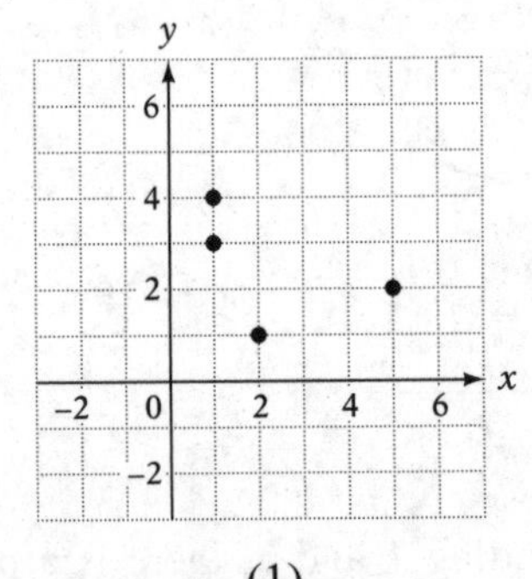

(1)

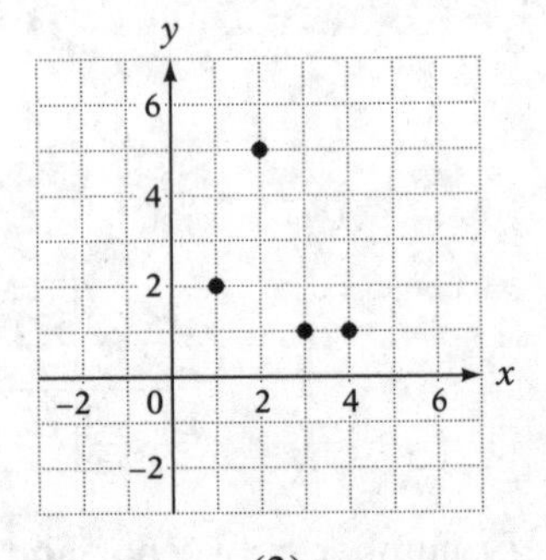

(3)

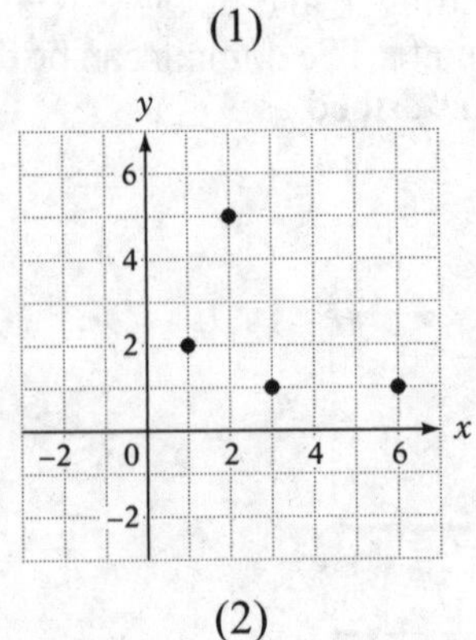

(2)

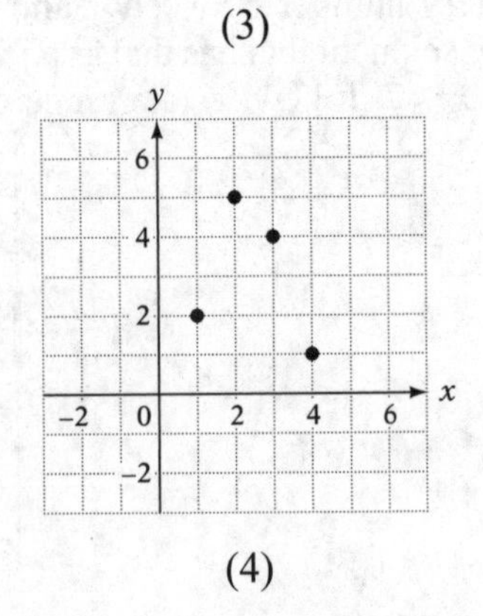

(4)

2. Below is the graph for which function?

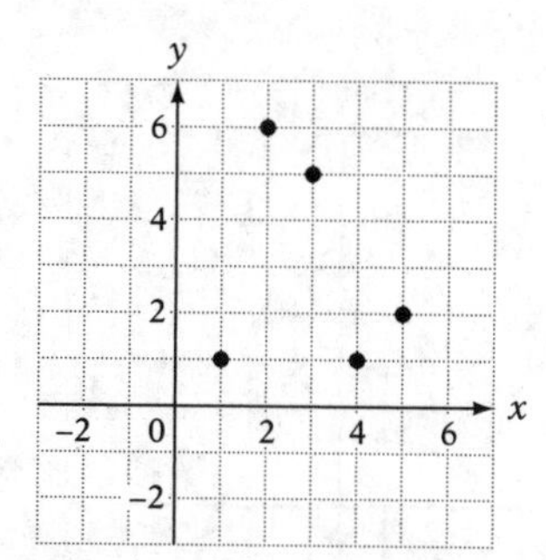

(1) $f = \{(1, 1), (2, 6), (3, 5), (4, 1), (5, 2)\}$
(2) $f = \{(1, 1), (6, 2), (5, 3), (6, 4), (2, 5)\}$
(3) $f = \{(1, 1), (2, 2), (3, 3), (4, 4), (5, 5)\}$
(4) $f = \{(1, 1), (2, 4), (3, 6), (4, 2), (5, 3)\}$

3. Below is the graph of $y = f(x)$. What is the value of $f(3)$?

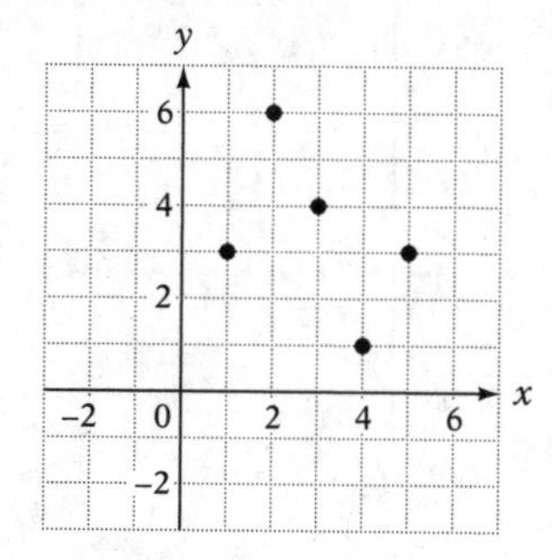

(1) 1 (2) 2 (3) 3 (4) 4

4. A portion of the graph of $y = f(x)$ is shown below. What is the value of $f(100)$?

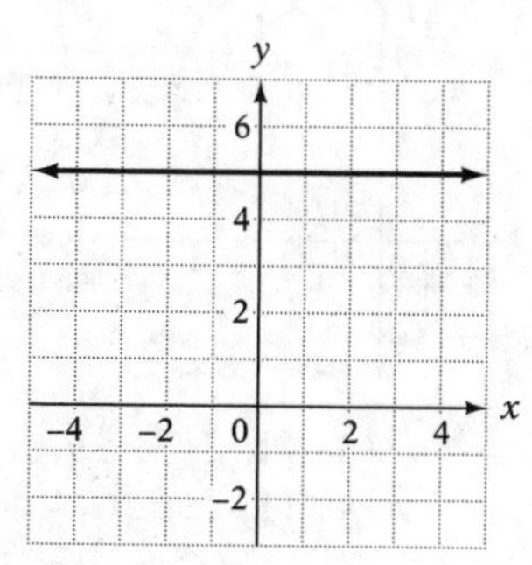

(1) 5 (2) 6 (3) 7 (4) 100

5. Below is the graph of $y = f(x)$. Which point could be used to determine the value of $f(4)$?

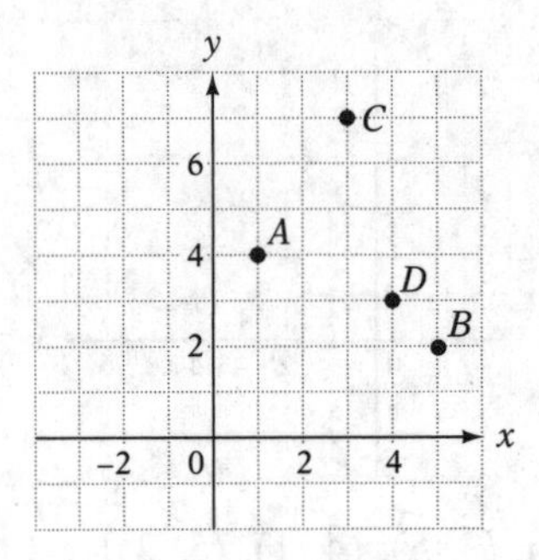

(1) A (2) B (3) C (4) D

6. Below is a graph of $y = g(x)$. What is the approximate value of $g(5)$?

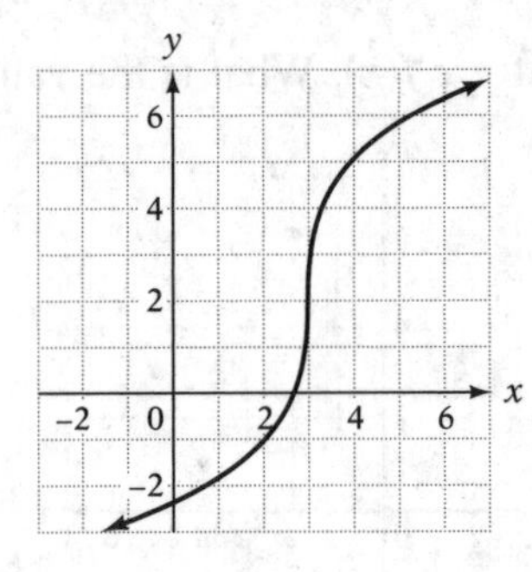

(1) 5 (2) 6 (3) 7 (4) 8

7. Below is the graph of $y = f(x)$. What is the domain of f?

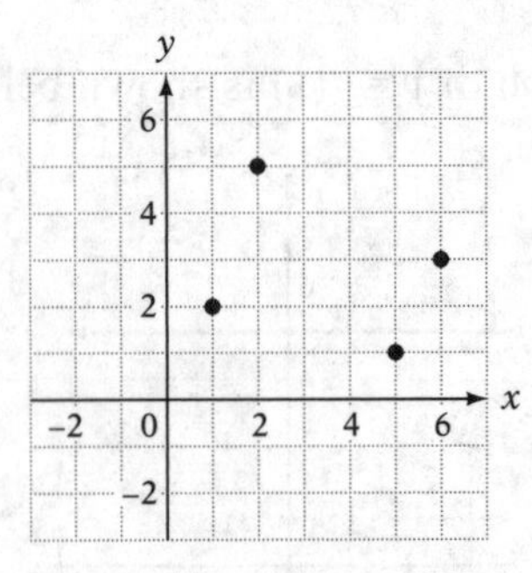

(1) {1, 2, 3, 4}
(2) {2, 3, 5, 6}
(3) {1, 2, 3, 5}
(4) {1, 2, 5, 6}

8. Below is the graph of $y = f(x)$. What is the range of f?

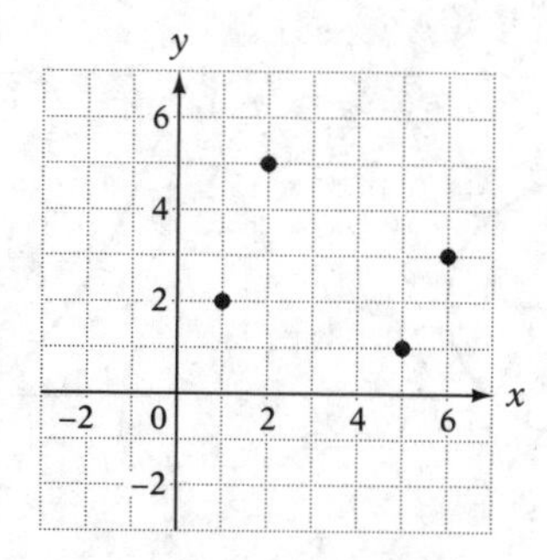

(1) $\{1, 2, 3, 5\}$
(2) $\{1, 2, 3, 6\}$
(3) $\{2, 3, 5, 6\}$
(4) $\{1, 2, 5, 6\}$

9. Below is the graph of $y = f(x)$. What is the domain and range of f?

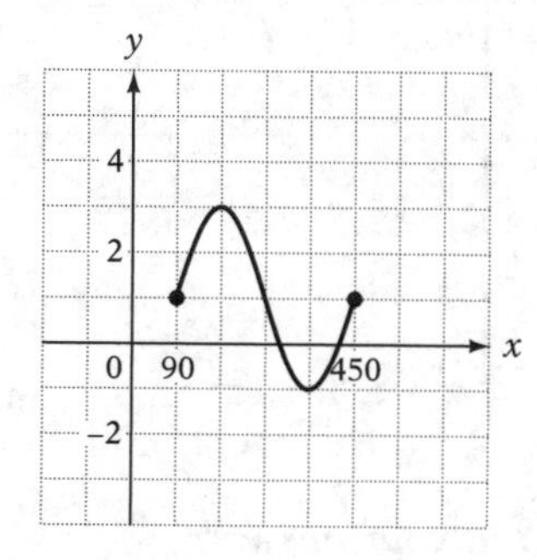

(1) Domain $90 < x < 450$, Range $-1 < y < 3$
(2) Domain $90 \leq x \leq 450$, Range, $-1 \leq y \leq 3$
(3) Domain $-1 < x < 3$, Range $90 < y < 450$
(4) Domain $-1 \leq x \leq 3$, Range $90 \leq y \leq 450$

10. Which is the graph of a function?

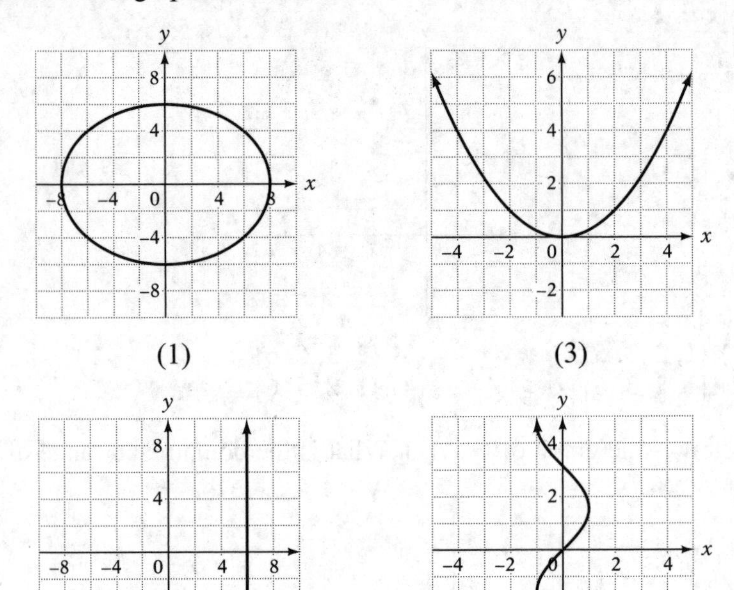

B. *Show how you arrived at your answers.*

1. Alisha says that this is not the graph of a function because there are two points that have the same y-coordinate, like (3, 9) and (–3, 9). Tyson says that it is the graph of a function. Who is right and why?

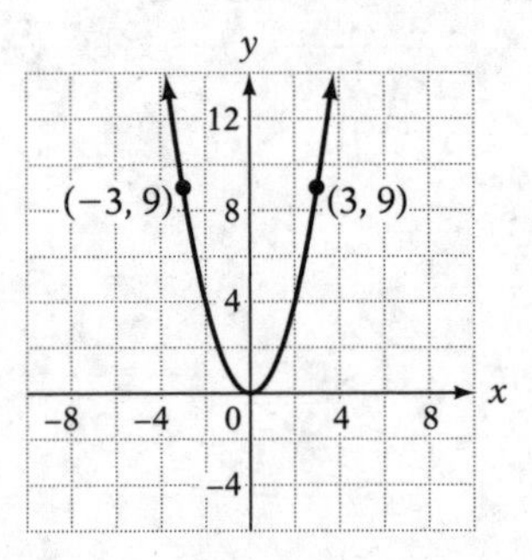

2. What is the domain and range of the function whose graph is below?

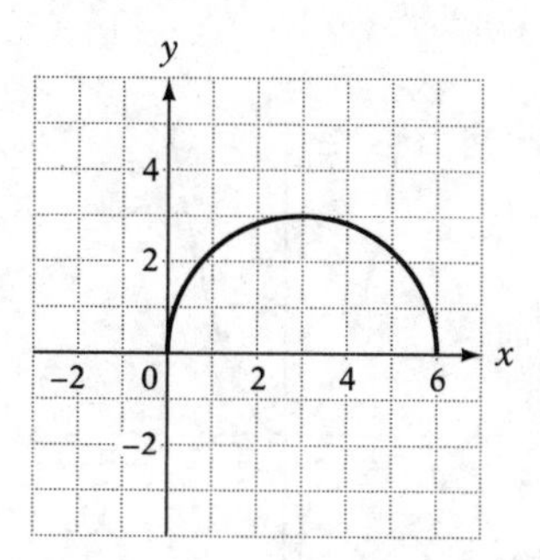

3. Using the two graphs below, determine the value of $f(g(3))$.

$y = f(x)$ $\qquad\qquad$ $y = g(x)$

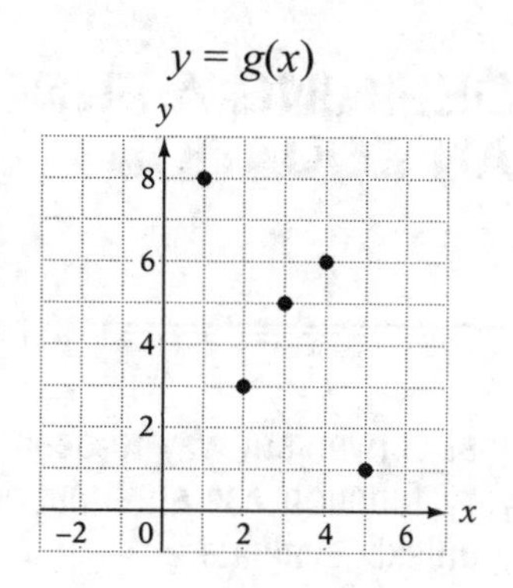

4. Below is the graph of $y = f(x)$. What is the value of $f(f(f(f(f(f(f(f(2))))))))$?

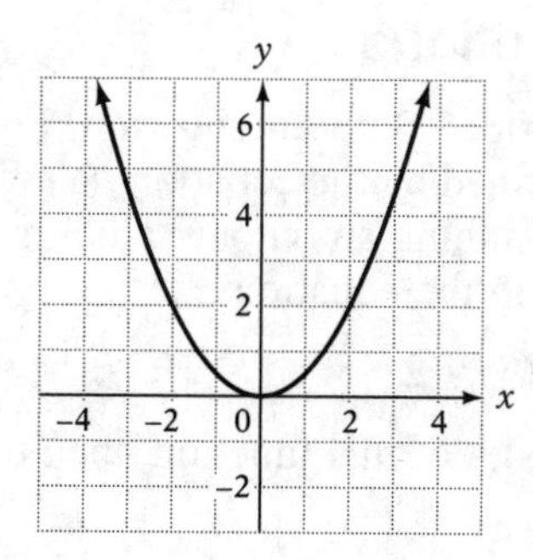

5. In the graph of the function below, list all values that satisfy the equation $f(a) = 4$.

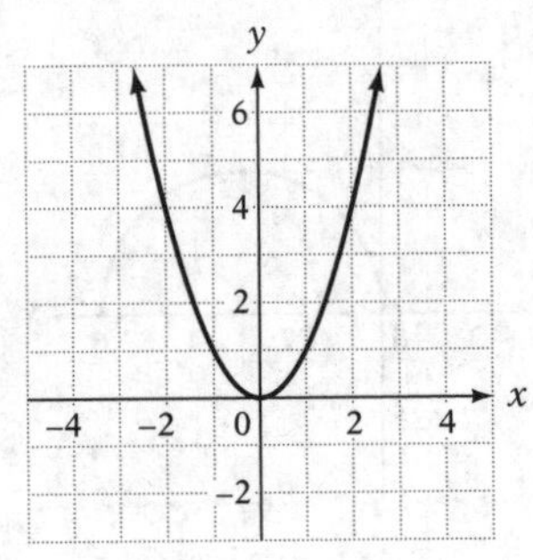

10.3 DEFINING A FUNCTION WITH AN EQUATION

The most convenient way to describe a function is with an equation. Not every function has an equation to describe it, but those that do can be quickly evaluated.

Equations for Functions

Often a function is described with an equation. For example, it might say that for a question f is defined by the equation $f(x) = 2x - 1$. When a function is defined this way, determining an output value for something like $f(5)$ just requires substituting 5 into the equation.

$$f(5) = 2 \cdot 5 - 1 = 10 - 1 = 9$$

If 5 is put into the input slot of this function, the number 9 will come out of the output slot.

Example 1

If the function g is defined as $g(x) = x^2 - 3x$, what is the value of $g(5)$?

Solution: 10 because $g(5) = 5^2 - 3 \cdot 5 = 25 - 15 = 10$.

Example 2

If the function h is defined as $h(x) = x^2 + 2x - 5$, write an expression for $h(x - 2)$.

Solution: Just as $h(3) = 3^2 + 2 \cdot 3 - 5 = 9 + 6 - 5 = 10$, and $h(a) = a^2 + 2a - 5$, when $(x - 2)$ is substituted in for the x in the function, it becomes $h(x - 2) = (x - 2)^2 + 2(x - 2) - 5$. This can be further simplified to $h(x) = x^2 - 2x - 5$.

Defining Functions Based on Other Functions

When a function's description includes another function, it requires two steps to evaluate it. An example of this is $g(x) = f(x + 2) - 1$. In order to determine the value of $g(5)$, substitute 5 in for x in all places. The equation would then become $g(5) = f(5 + 2) - 1 = f(7) - 1$. Unless f is defined somehow, the expression can't be simplified any further.

If it is known, for this example, that $f(x) = x^2 - 10$, then $g(5) = f(7) - 1 = 7^2 - 10 - 1 = 49 - 10 - 1 = 38$ would be the answer.

Example 3

If $f(x) = 2x - 5$ and $g(x) = f(x - 3)$, what is the value of $g(8)$?

Solution: $g(8) = f(8 - 3) = f(5)$ and $f(5) = 2 \cdot 5 - 5 = 10 - 5 = 5$ so $g(8) = 5$.

Graphing a Function from Its Equation

The function $f(x) = 2x - 5$ has the same graph as the solution set of the equation $y = 2x - 5$. This is the graph of $y = f(x)$. All the rules for graphing linear, quadratic, and exponential equations described already in this book apply to graphing linear, quadratic, and exponential functions. Functions can be graphed on the graphing calculator by pressing [Y=] and entering the equation for the function.

Example 4

Make a graph of $y = f(x)$ if $f(x)$ is defined as $f(x) = x^2 - 8x + 12$.

Solution: This is the same process as graphing the two-variable equation $y = x^2 - 8x + 12$. It can be done by hand or with the graphing calculator.

For the TI-84:

Use the y = menu.

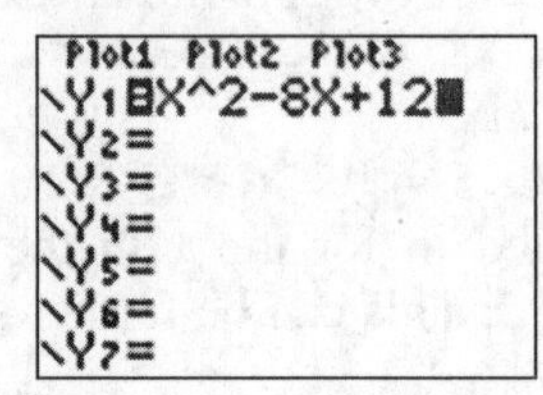

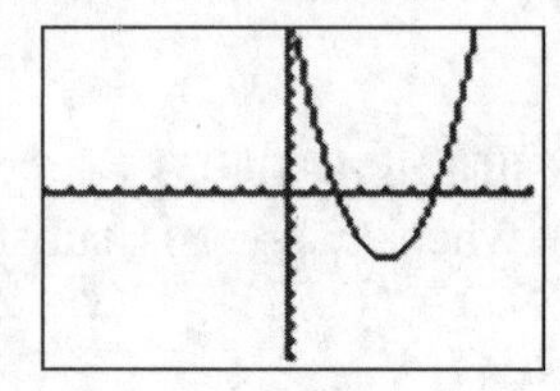

For the TI-Nspire:

Type the equation after $f1(x)$ = in the entry bar of the graphing Scratchpad.

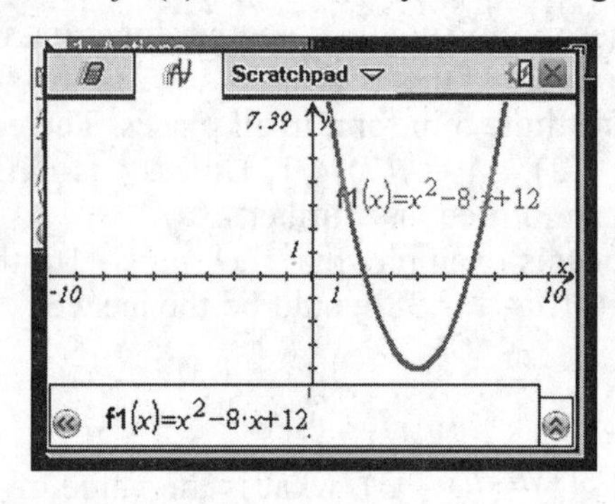

Using the Graphing Calculator to Determine Function Values

If a function is defined by an equation, there are three ways to use a graphing calculator to determine output values of the function for different input values. If the function is defined as $f(x) = x^2 - 8x + 12$, then to determine $f(3)$:

For the TI-84:

Method #1: Graph the equation and press [2ND] and then [TRACE] to get to the CALC menu. On the CALC menu, select [1] for value. The graph will appear and there will be an "X=" in the bottom left corner of the screen. Type 3, and it will calculate $f(3)$ and also show the corresponding point on the graph. In this example, $f(3) = -3$.

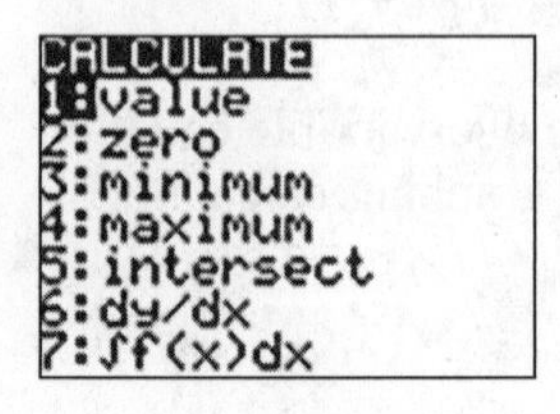

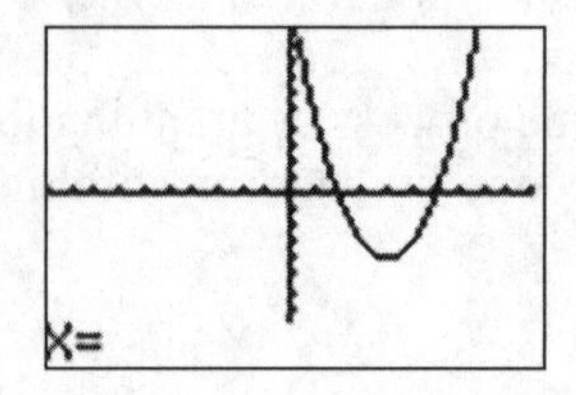

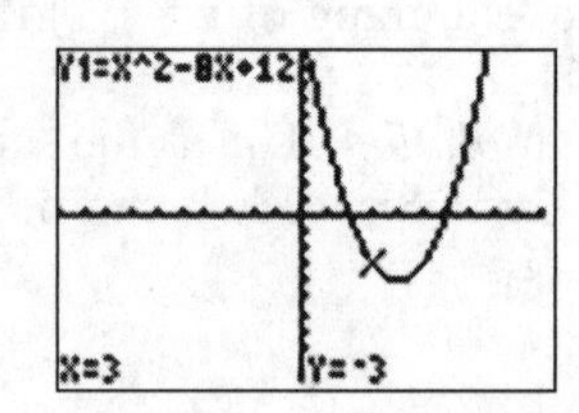

Method #2: After entering the equation for the function into the Y= menu, press [2ND] and then [GRAPH] to get to the TABLE menu. Use the up or down arrows to more the cursor to the x value you want to evaluate the function at.

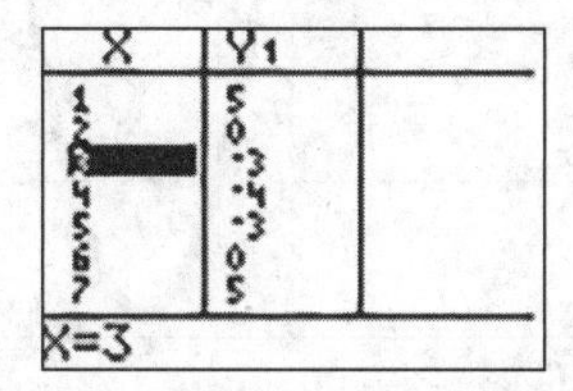

Method #3: After entering the equation for the function into the Y= menu, press [2ND] and then [MODE] to quit. Then press [VARS], select Y-VARS by pressing the right arrow, and then select [1] for Function and [1] again for Y_1. A "Y_1" will appear on the screen. To evaluate the function at a value, press [(], [3], [)] after the Y_1 and press [ENTER].

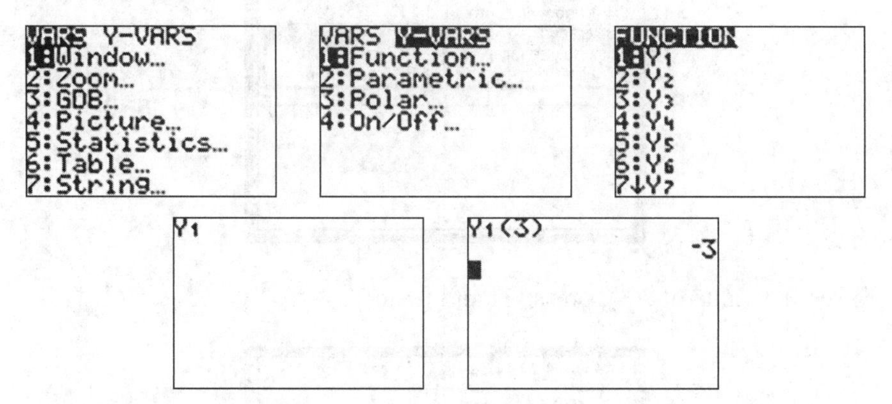

For the TI-Nspire:

Method #1: From the home screen press [B] for the Graph Scratchpad.

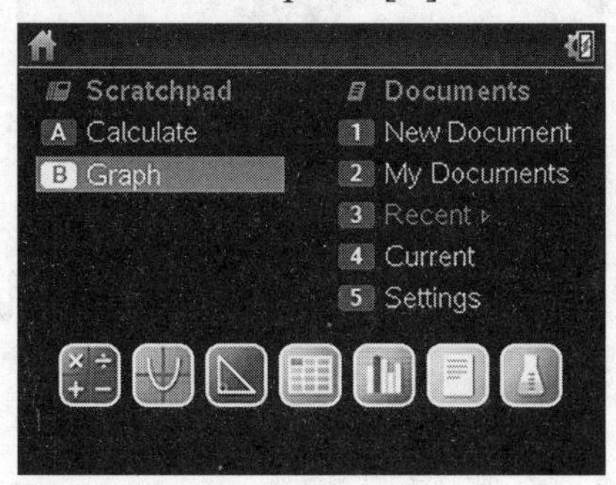

Type $x^2 - 8x + 12$ after the $f1(x)$= and press [enter].

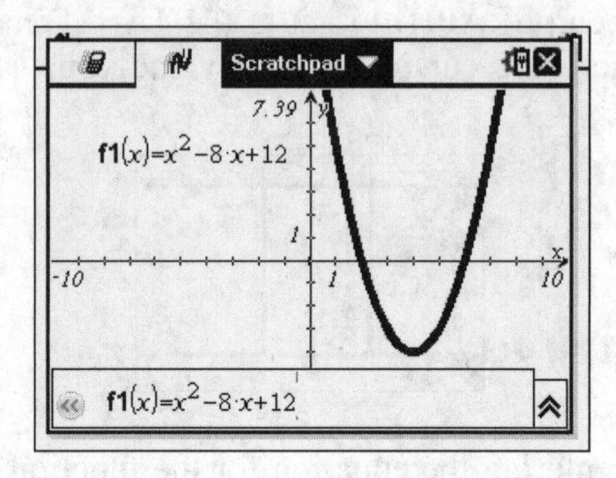

Press [menu], [5], and [1] for Graph Trace.

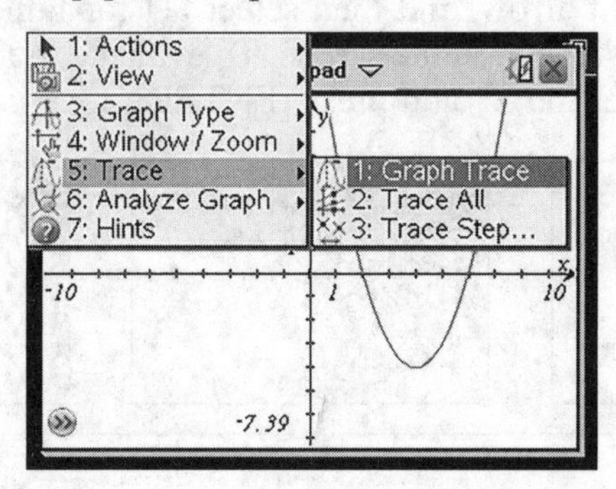

To see the value of $f(3)$, press [3] and [enter].

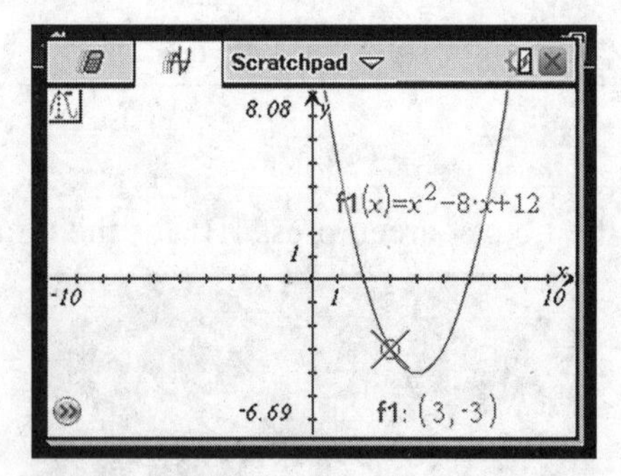

Method #2: After graphing $f(x) = x^2 - 8x + 12$, hold down the [ctrl] and press [t]. This will display a table of values. To see what $f(3)$ is, scroll down to the 3 in the x column.

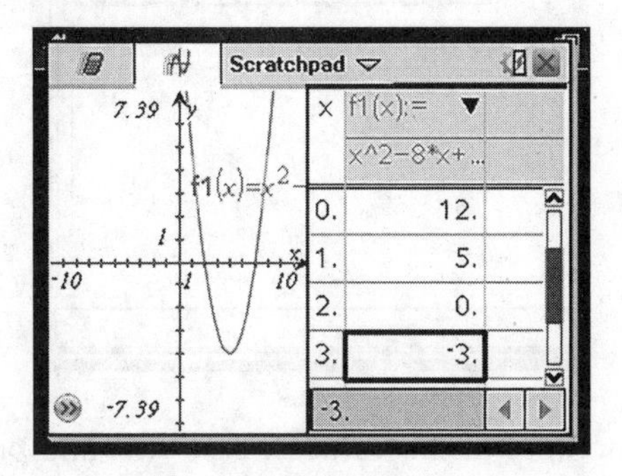

Method #3: From the home screen press [A] for the Calculator Scratchpad.

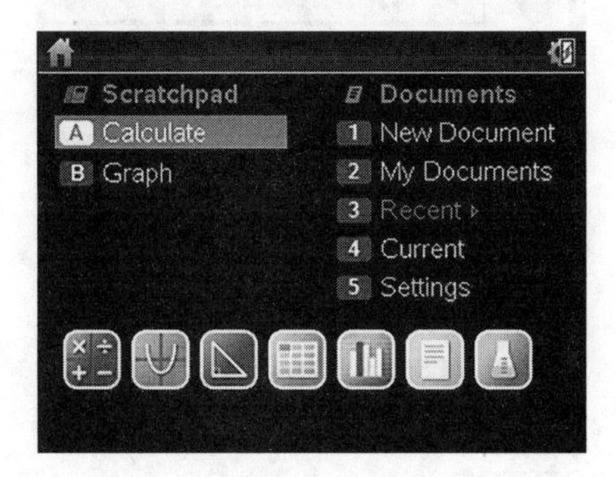

Press [menu], [1], and [1] for Define.

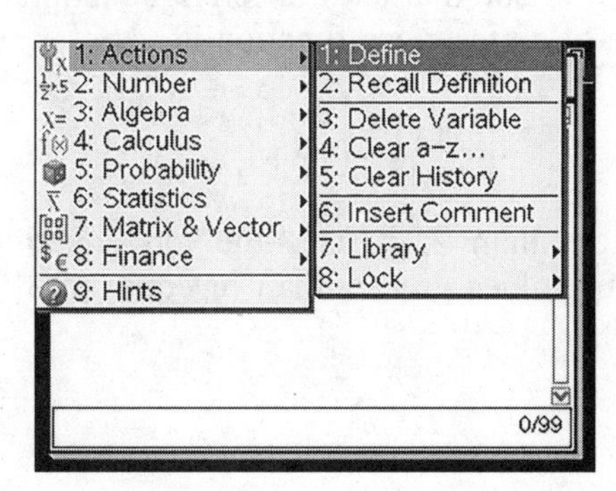

After the word "Define," type $f(x) = x^2 - 8x + 12$ and press [enter].

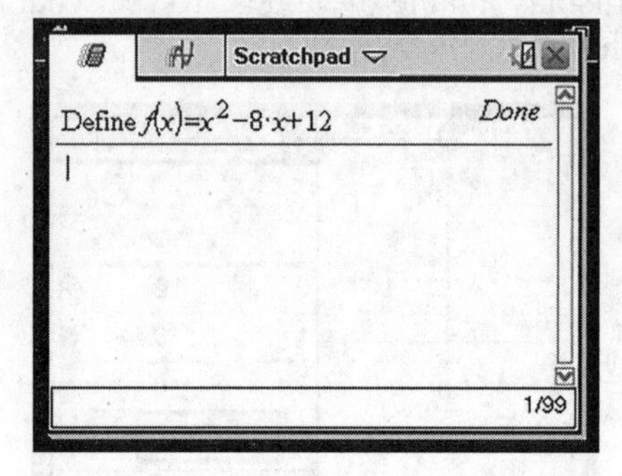

To evaluate the function f at the value 3, type $f(3)$ and press [enter].

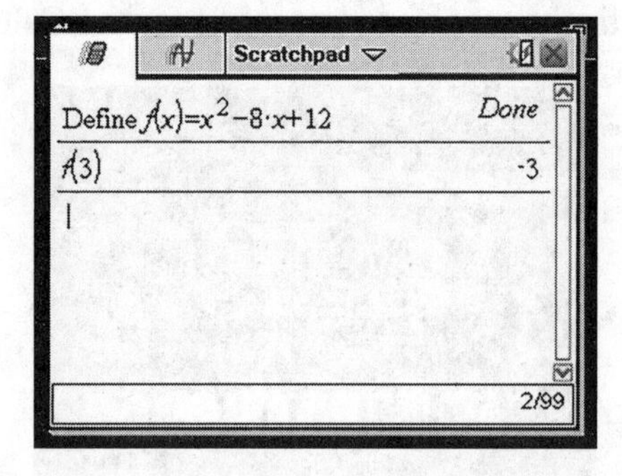

Piecewise Functions

A *piecewise function* is one that uses different equations for different input values. An example of a piecewise function is.

$$f(x) = \begin{cases} x-1, x<3 \\ 2x+1, x \geq 3 \end{cases}$$

For input values less than 3, it uses the top equation. $f(2)$ will equal $2 - 1 = 1$. For input values greater than or equal to 3, it uses the bottom equation. $f(4) = 2 \cdot 4 + 1 = 9$.

A portion of the chart for this piecewise function looks like this.

x	$f(x)$
-2	$-2 - 1 = -3$
-1	$-1 - 1 = -2$
0	$0 - 1 = -1$
1	$1 - 1 = 0$
2	$2 - 1 = 1$
3	$2 \cdot 3 + 1 = 7$
4	$2 \cdot 4 + 1 = 9$

Plotting these seven values looks like this:

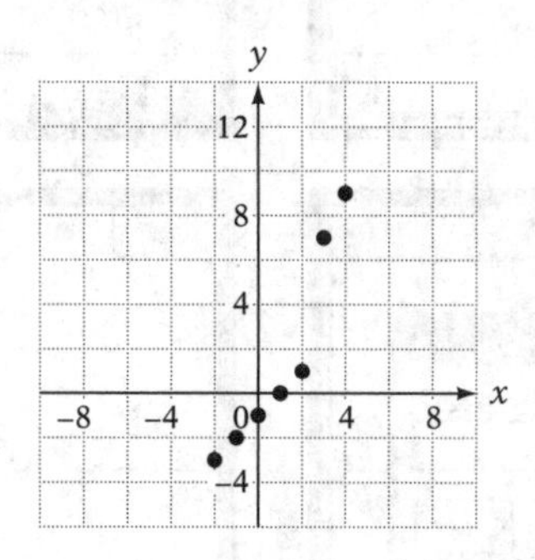

To fill in the rest of the graph, plot an open circle at the point (3, 2), which is what you get when you substitute 3 into the top equation. Even though this is not a point on the graph, it serves as the boundary between the two different pieces of the graph.

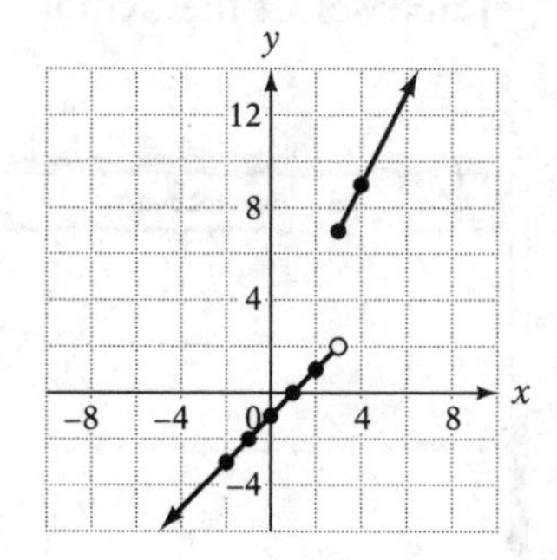

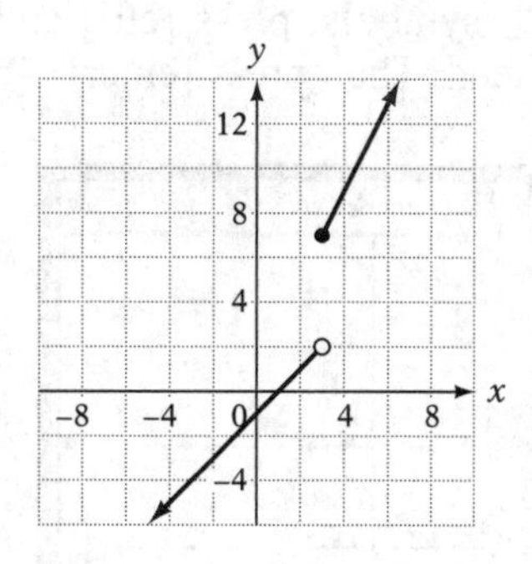

Graphing Piecewise Functions on the Graphing Calculator

With a TI-Nspire, piecewise functions can be quickly graphed. From the Graph Scratchpad in the entry line press the [template] key next to the [9] and select the piecewise template on the top row, third from the right. It will prompt you to enter the number of pieces in the function.

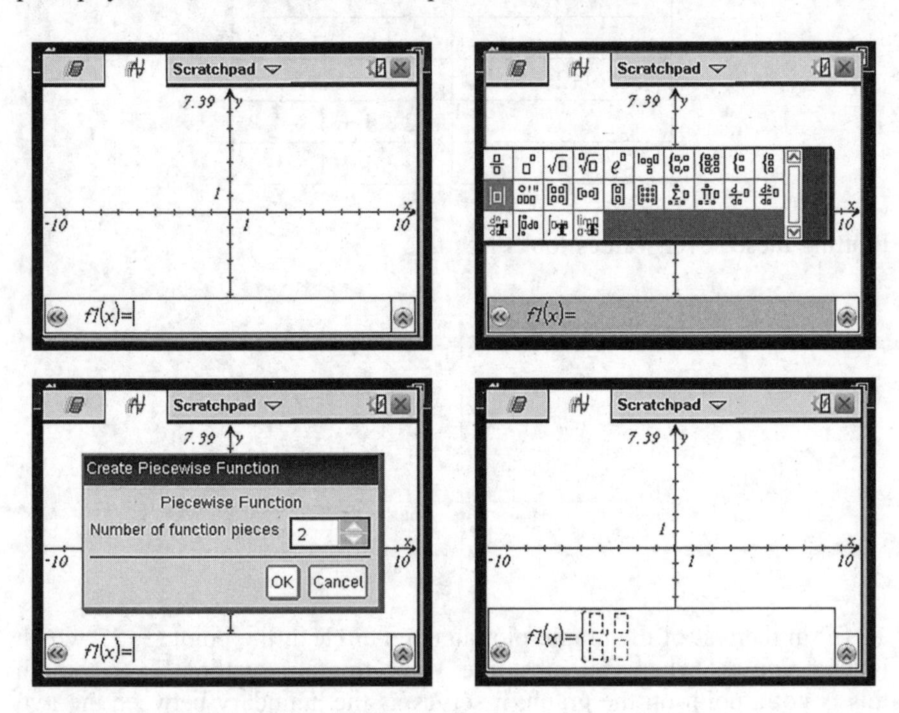

Enter the equations $x - 1$, $x < 3$, $2x + 1$, and $x \geq 3$ into the template. To get the < and ≥ symbols, press [ctrl] and [=] and select the symbol you want from the menu. Then press [enter].

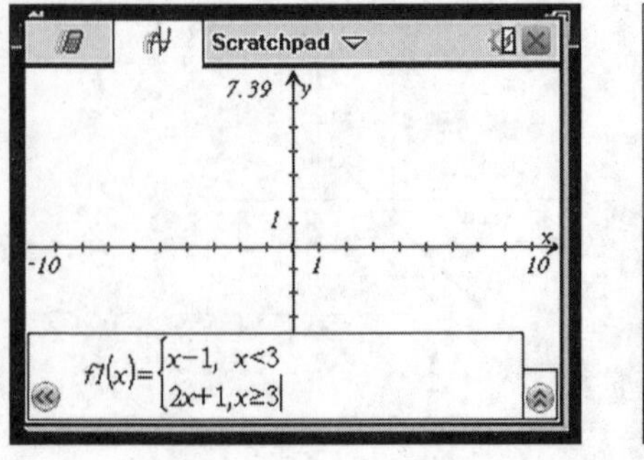

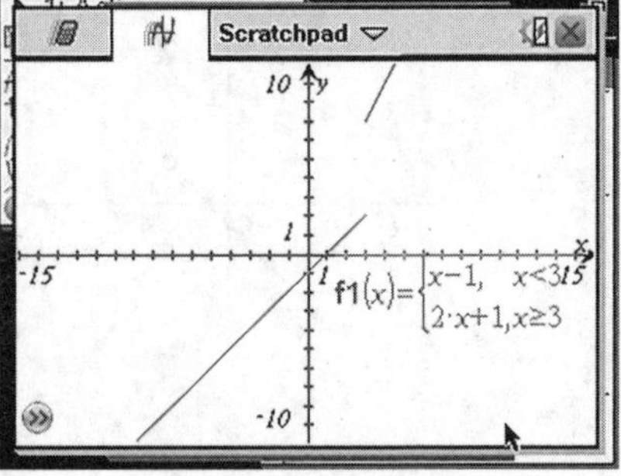

Notice that it is not clear which endpoint is an open dot and which is a closed dot on the graphing calculator.

Determining Intervals for When a Function Is Positive or Negative

Unlike numbers that are always either positive, negative, or 0, functions can be positive for some input values, negative for others, and 0 for others. The function $f(x) = 2x - 6$ can be negative, $f(0) = -6$; positive, $f(5) = 4$; or 0, $f(3) = 0$.

The quickest way to determine when a function is positive, negative, or 0 is to graph it and see where the graph is below the x-axis, above the x-axis, or crossing the x-axis.

Example 5

In what interval is the function $f(x) = 3x + 6$ negative?

Solution: The graph of $y = 3x + 6$ has an x-intercept of $(-2, 0)$.

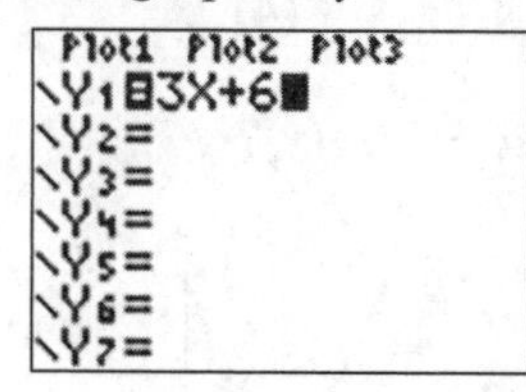

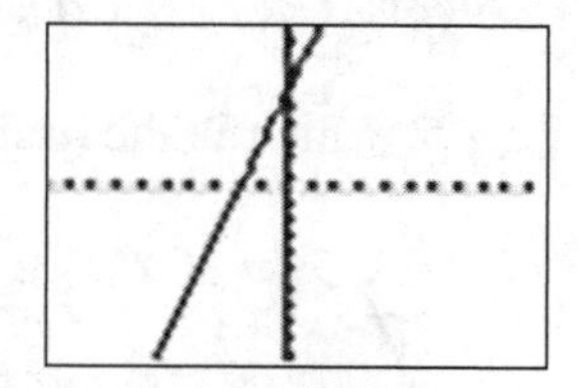

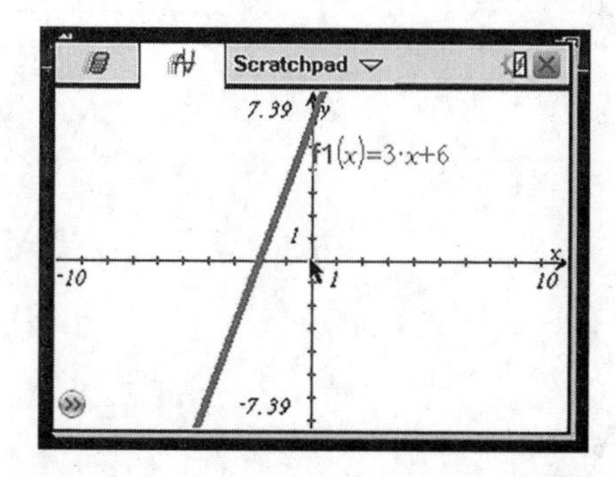

For every x value less than -2, $f(x)$ will be negative. For every x value greater than -2, $f(x)$ will be positive. The solution is that $f(x)$ will be negative when $x < -2$.

Check Your Understanding of Section 10.3

A. Multiple-Choice

1. If $f(x) = 3x + 7$, what is $f(2)$?
(1) 10 (2) 11 (3) 12 (4) 13

2. If $g(x) = -x^2 + 7x + 1$, what is $g(2)$?
(1) 11 (2) 19 (3) 27 (4) 35

3. If $f(x) = 2x + 5$, what is $f(a)$?
(1) 7 (2) $2a - 5$ (3) $2a + 5$ (4) $a + 5$

4. If $f(x) = 5x - 2$, what is $f(x - 1)$?
(1) $5x - 2$ (2) $5x - 3$ (3) $5x + 1$ (4) $5x - 7$

5. If $f(x) = x^2 + 5$ and $g(x) = f(x - 2)$, what is $g(6)$?
(1) 31 (2) 21 (3) 11 (4) 1

6. If $g(x) = 2x - 5$, which is the graph of $y = g(x)$?

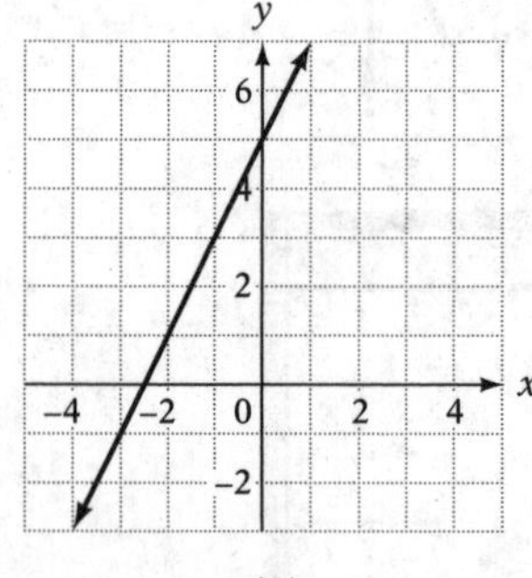

(1)

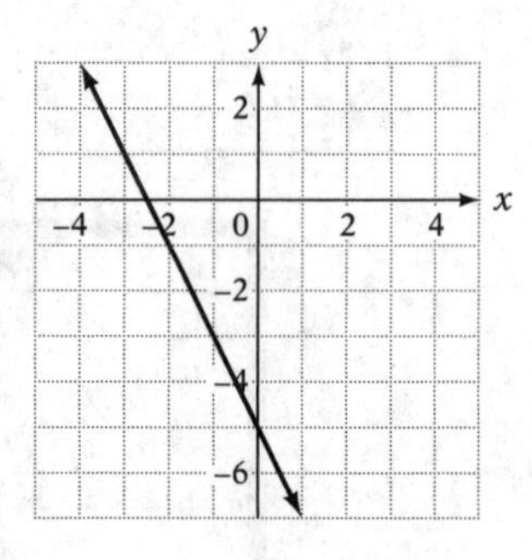

(3)

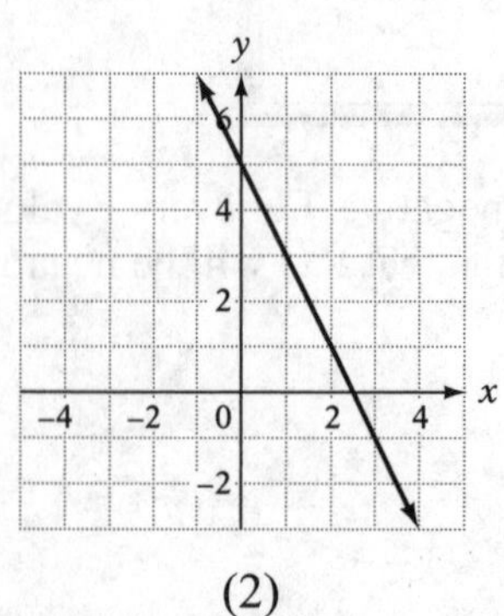

(2)

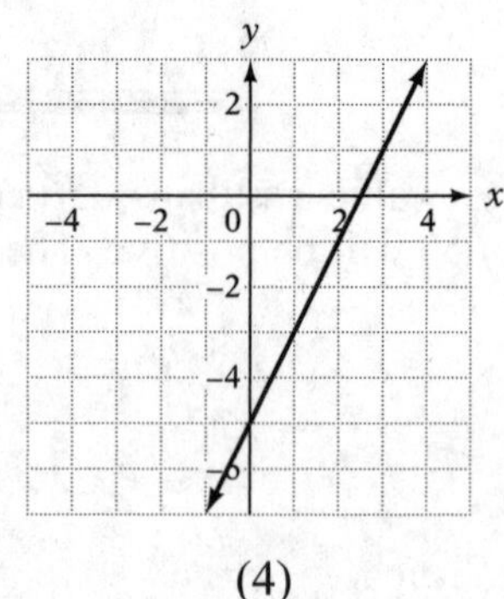

(4)

7. If $f(x) = 2x + 5$, for what value of a does $f(a) = 11$?
(1) 1 (2) 3 (3) 5 (4) 7

8. For what values of x is the graph of the function $h(x) = 2^x$ increasing?
(1) For $x > 0$
(2) For $x < 0$
(3) All values
(4) No values

9. What are the zeros of the function $f(x) = x^2 + 10x + 16$?
(1) –2 and –8
(2) 2 and 8
(3) 2 and –8
(4) –2 and 8

10. If $f(x) = x^2 - 6x + 5$, for what values of x is the function $f(x) < 0$?
(1) For values between 1 and 5
(2) For values greater than 5
(3) For values less than 1
(4) For all values

B. *Show how you arrived at your answers.*

1. If $f(x) = x^2 + 8x + 7$, what are (a) $f(3)$ and (b) $f(x + 1)$?

2. If $f(x) = 2x - 3$ and $g(x) = x^2 + 2$, what are (a) $f(g(3))$ and (b) $g(f(3))$?

3. $f(x) = x + 2$, $g(x) = 2x$, and $h(x) = x^2$. Manuel says that the expression $2(x + 2)^2$ can be written as $h(g(f(x)))$. Elizabeth says that this is not correct. Which student is right and why?

4. If $f(x) = x^2$ and $g(x) = f(x - 3)$, what does the graph of $g(x)$ look like?

5. If $g(x) = 2x + 1$ and $g(f(3)) = 15$, what is the value of $f(3)$?

10.4 FUNCTION TRANSFORMATIONS

The graph of the function $f(x)$ looks very similar to the graphs of $f(x+k)$, $f(x-k)$, $f(x)+k$, and $f(x)-k$. The only difference is that the graphs of the other four are *translated*, in other words shifted, left, right, up, or down. The graphs of the other four can be produced quickly if you know which direction to shift and what the original graph looks like.

Vertical Translations

The graph of $f(x) + k$ will be the same as the graph of $f(x)$ only shifted up by k units. If the graph of $y = f(x)$ is shown below, the graph of $y = f(x) + 3$ will have all points shifted 3 units up.

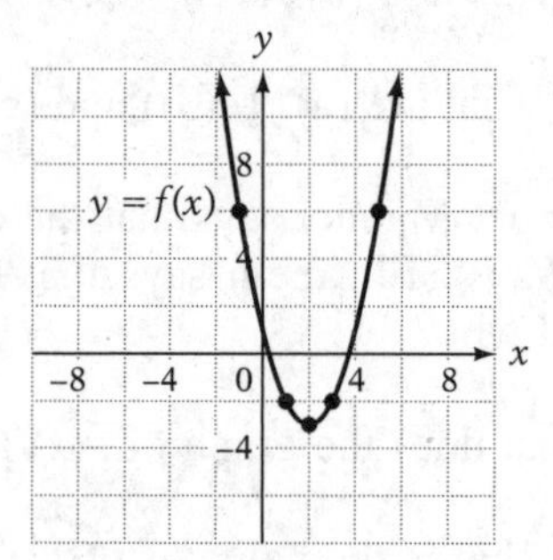

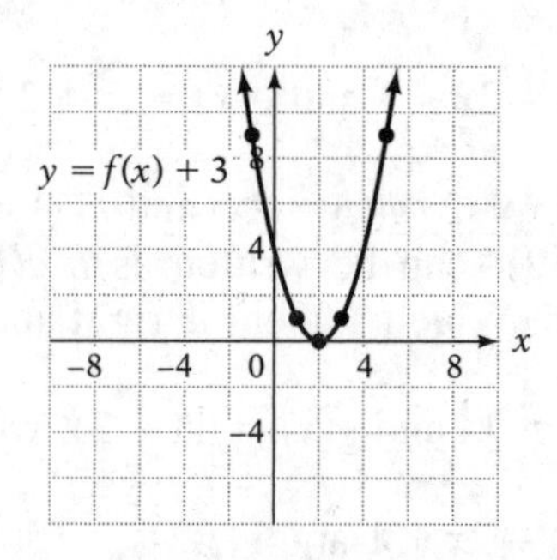

The graph of $f(x) - k$ will be the same as the graph of $f(x)$ only shifted down by k units. If the graph of $y = f(x)$ is shown below, the graph of $y = f(x) - 3$ will have all points shifted 3 units down.

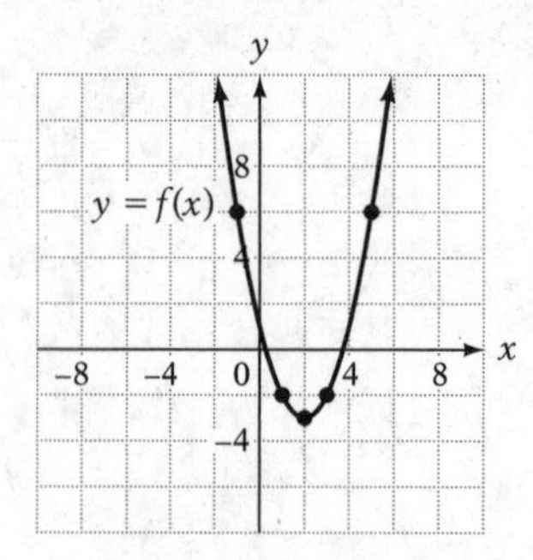

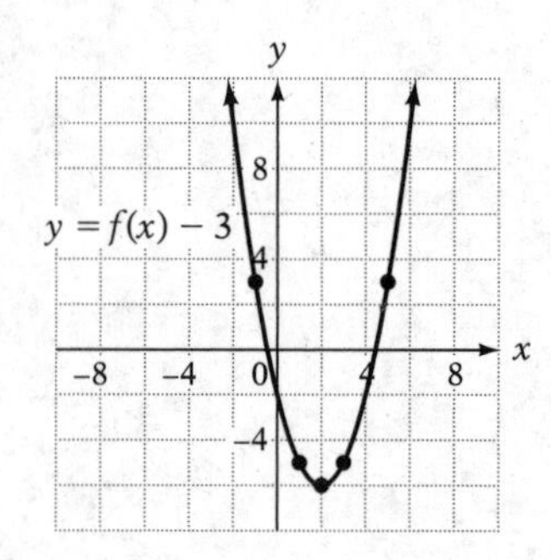

Horizontal Translations

The graph of the function $f(x + 3)$ is the same as the graph of $f(x)$ when translated 3 units to the left. The graph of the function $f(x - 3)$ is the same as the graph of $f(x)$ when translated 3 units to the right. In general, the graph of the function $f(x + k)$ is the same as the graph of $f(x)$ when translated to the left by k units. The graph of the function $f(x - k)$ is the same as the graph of $f(x)$ when translated to the right by k units.

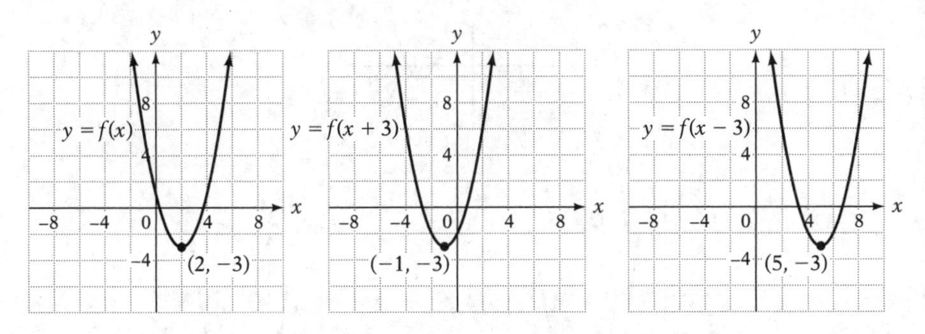

Example 1

The graph of $y = f(x)$ is on the left and the graph of $y = g(x)$ is on the right. How could $g(x)$ be defined?

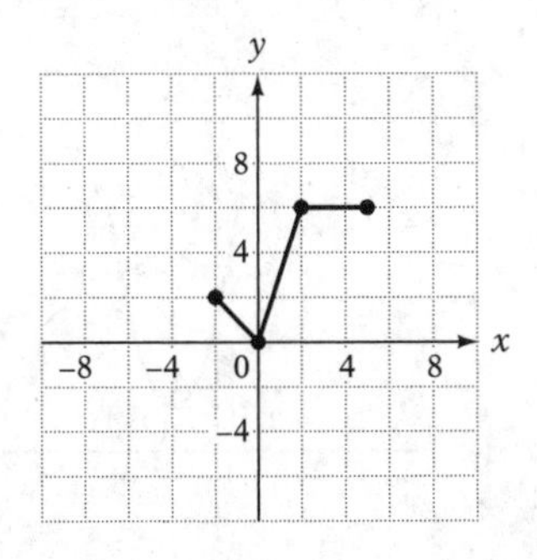

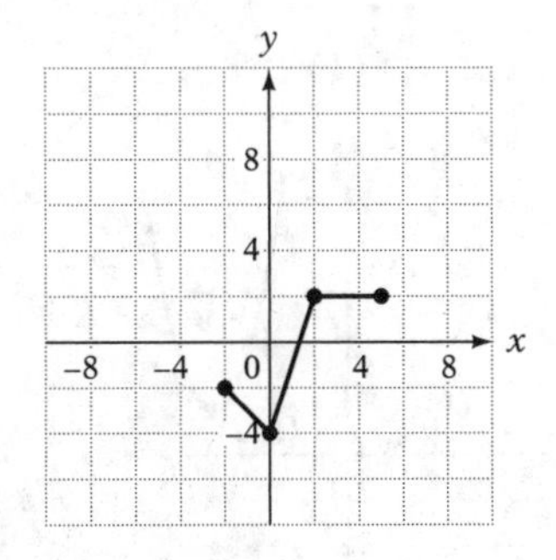

(1) $g(x) = f(x) + 4$
(2) $g(x) = f(x) - 4$
(3) $g(x) = f(x + 4)$
(4) $g(x) = f(x - 4)$

Solution: Since the graph on the right is like the graph on the left shifted down by 4 units, the solution is choice (2).

Example 2

If this is the graph of $y = f(x)$, which could be the graph of $y = g(x)$ if $g(x) = f(x + 2) + 3$?

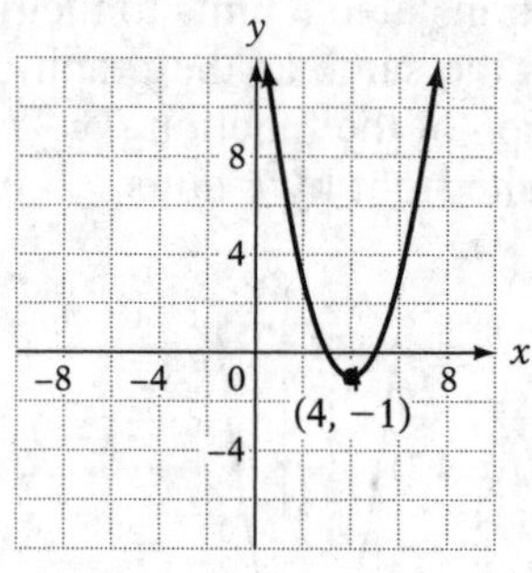

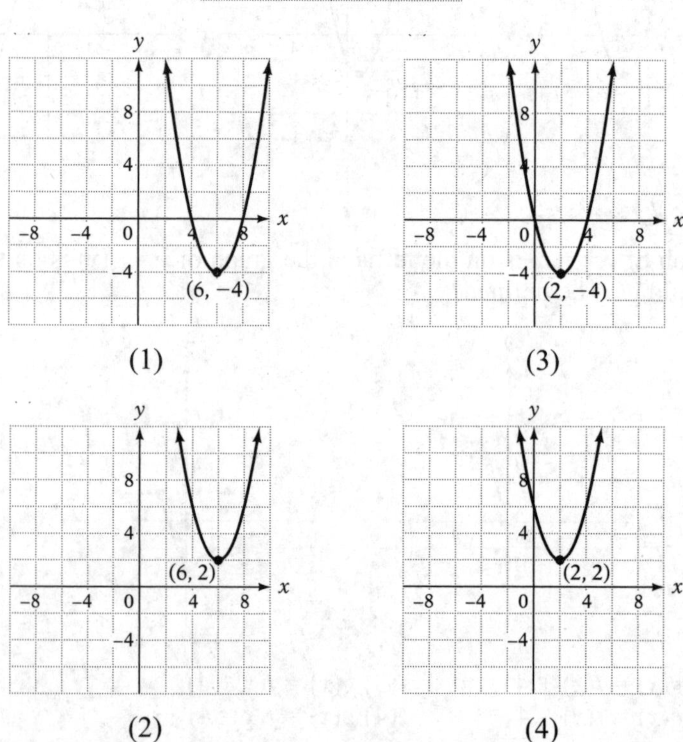

Solution: This has two transformations the +2 in the parentheses means the transformed graph will be shifted 2 units to the left. The +3 outside the parentheses means the transformed graph will be shifted 3 units up. The vertex of the original graph is (4, –1). The vertex of the transformed graph will be 2 to the left and 3 above (4, –1), which is at (2, 2). Choice (4).

Example 3

The graph of $y = f(x)$ is a parabola with a vertex at (5, 7). If $g(x) = f(x - 4)$, what are the coordinates of the vertex of the parabola that is the graph of $y = g(x)$?

(1) (1, 7) (2) (9, 7) (3) (5, 11) (4) (5, 3)

Solution: Every point on the graph of $y = f(x - 4)$ is four units to the right of the corresponding point on $y = f(x)$. The vertex of the original function is at (5, 7) so the vertex of the translated function will be 4 units to the right, which is (9, 7). Choice (2).

MATH FACTS

The graph of these four transformations will be congruent to the graph of the original function. The only difference is that it is shifted horizontally, vertically, or some combination of both. The four basic transformations are

1. $f(x) + k$ The graph shifts up by k units.
2. $f(x) - k$ The graph shifts down by k units.
3. $f(x + k)$ The graph shifts left by k units.
4. $f(x - k)$ The graph shifts right by k units.

Check Your Understanding of Section 10.4

A. Multiple-Choice

1. If below is the graph of $y = f(x)$, which is the graph of $y = f(x) - 5$?

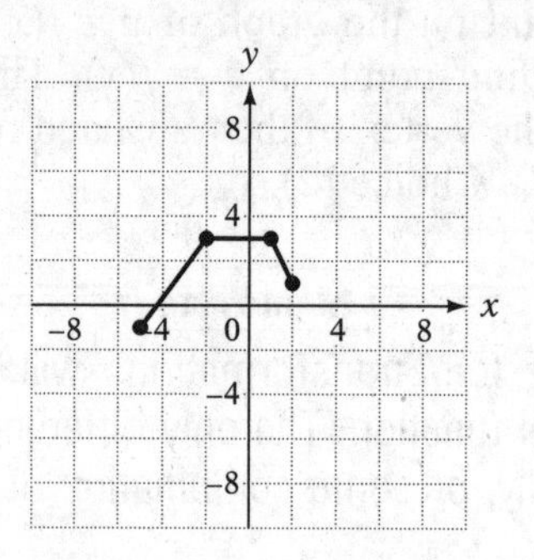

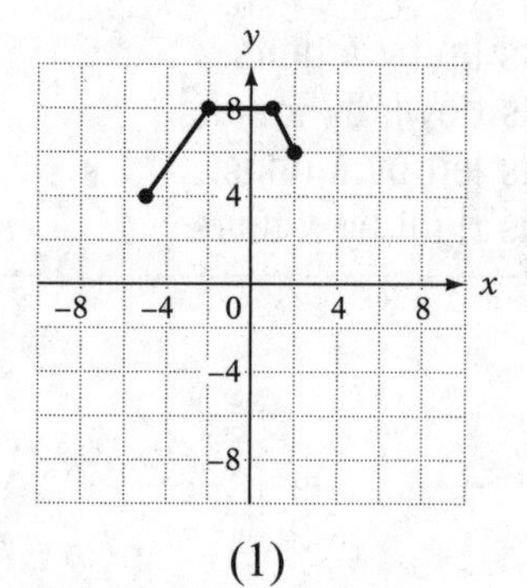

(1)

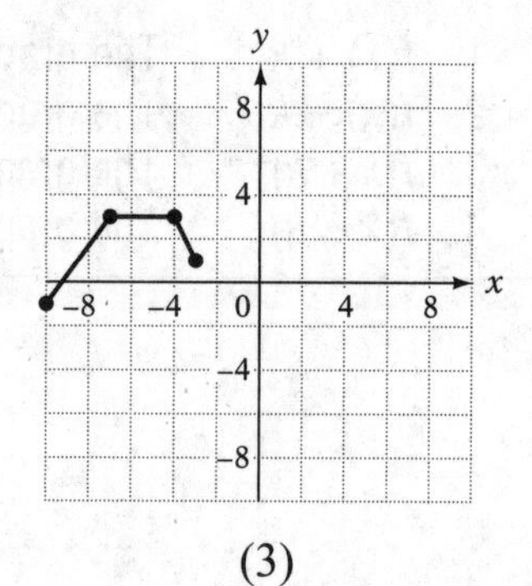

(3)

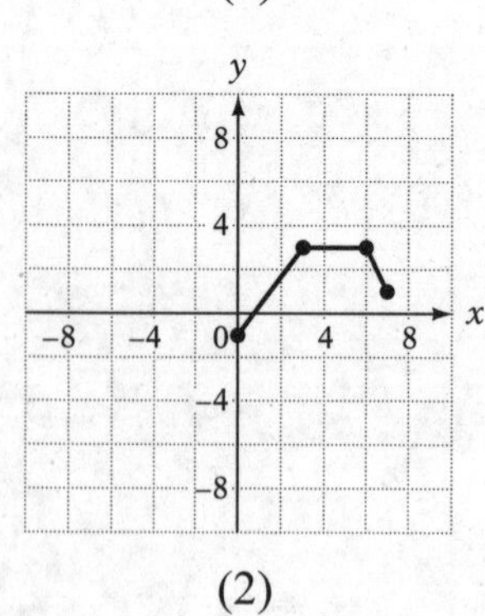

(2)

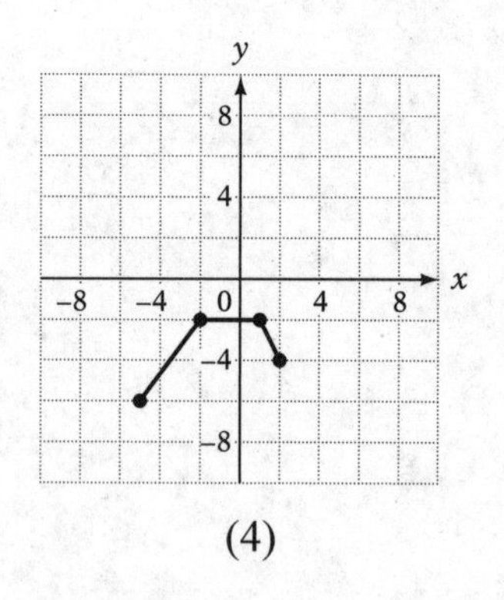

(4)

2. If below is the graph of $y = f(x)$, which is the graph of $y = f(x) + 5$?

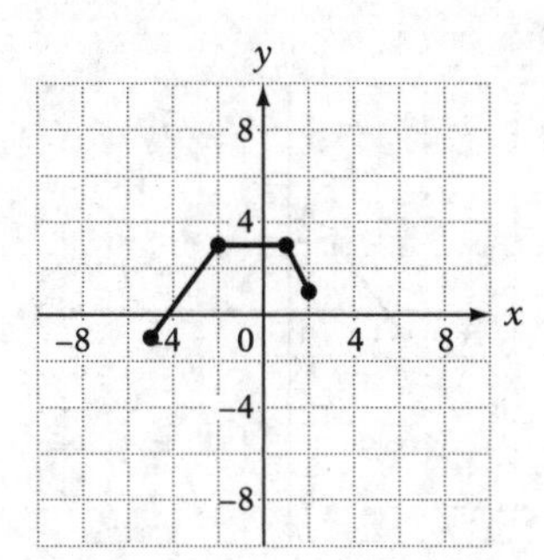

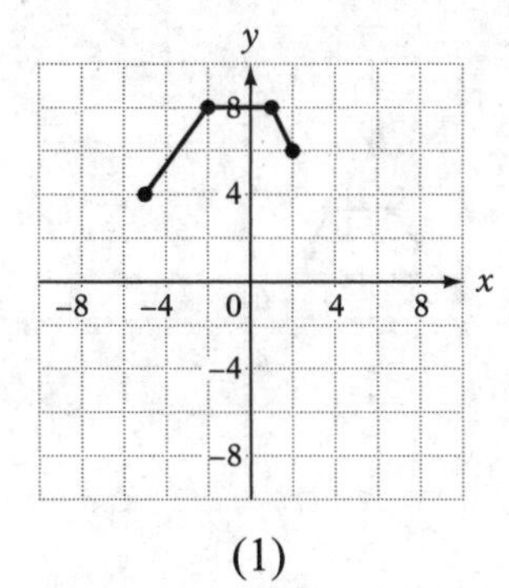

(1)

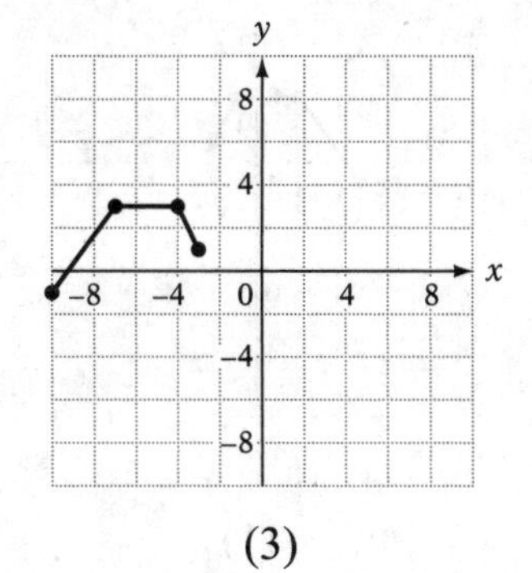

(3)

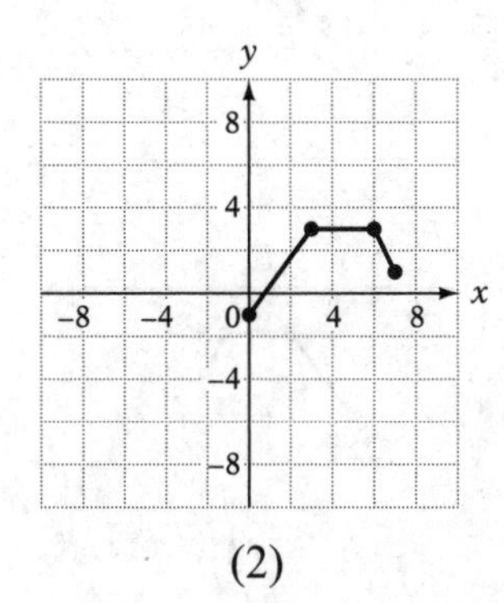

(2)

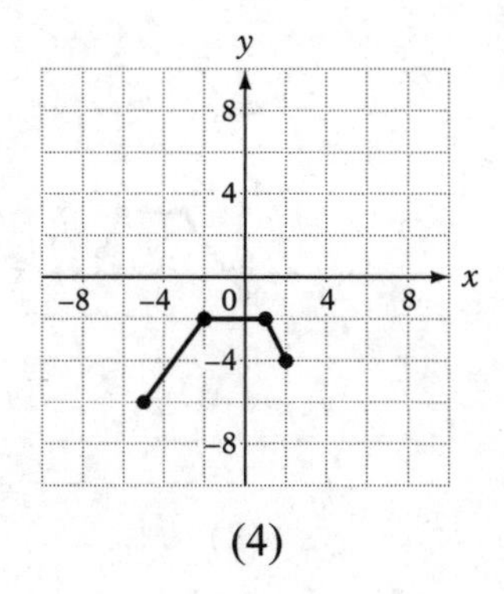

(4)

3. If below is the graph of $y = f(x)$, which is the graph of $y = f(x - 5)$?

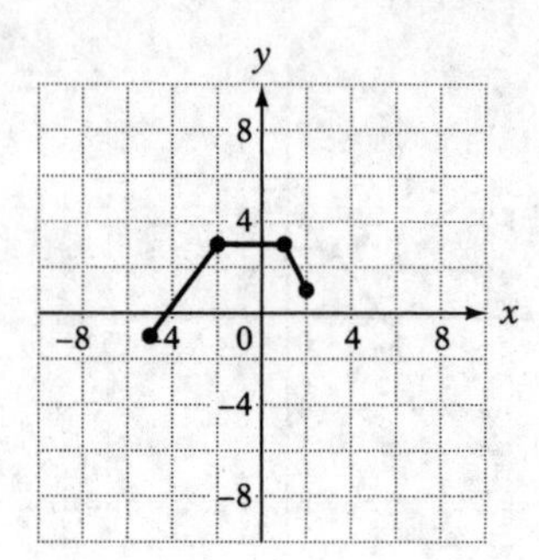

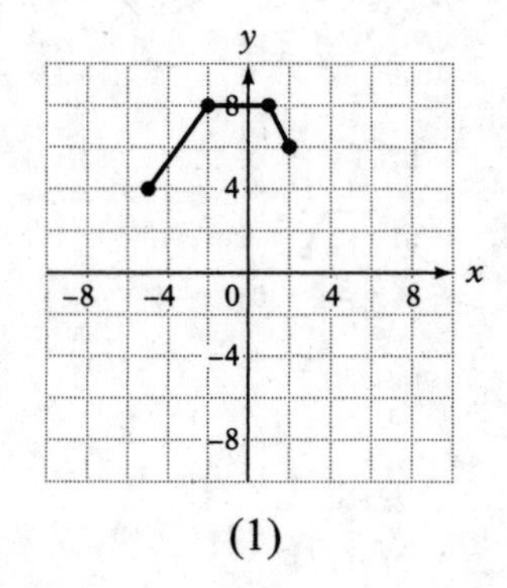

(1)

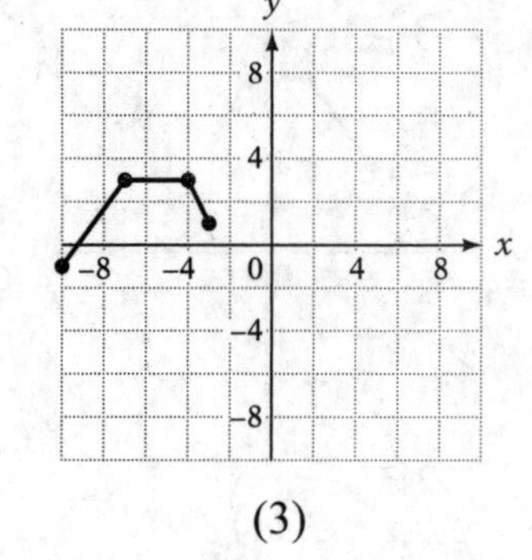

(3)

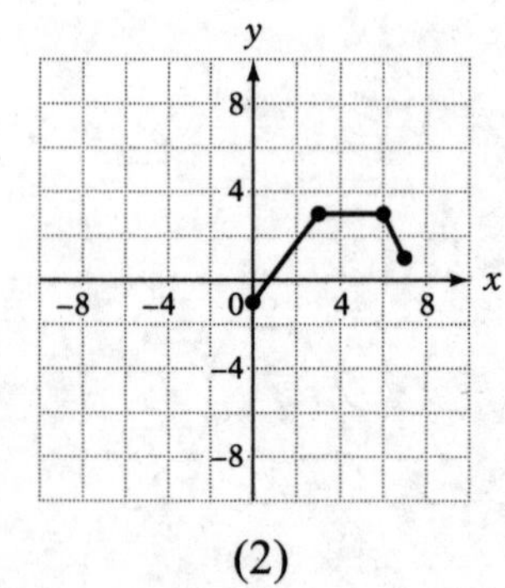

(2)

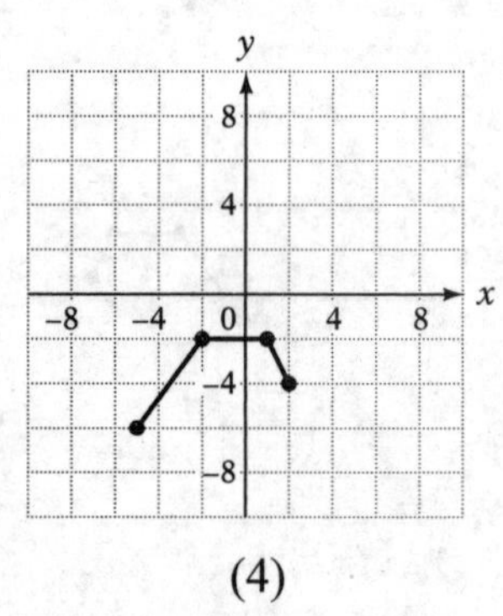

(4)

4. If below is the graph of $y = f(x)$, which is the graph of $y = f(x + 5)$?

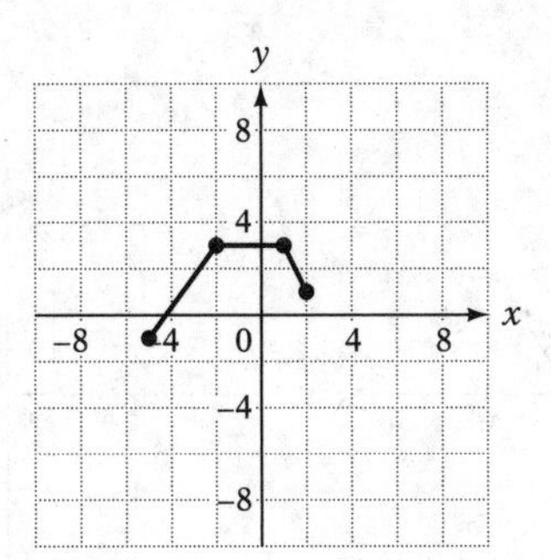

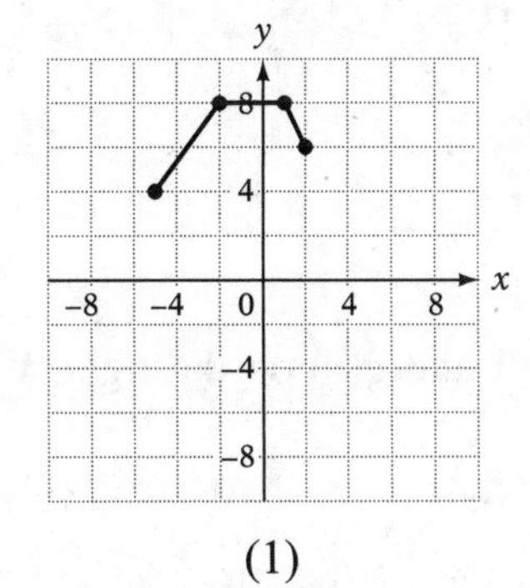

(1)

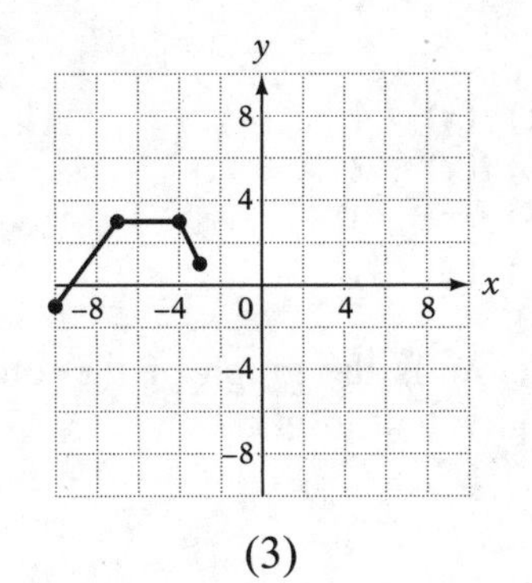

(3)

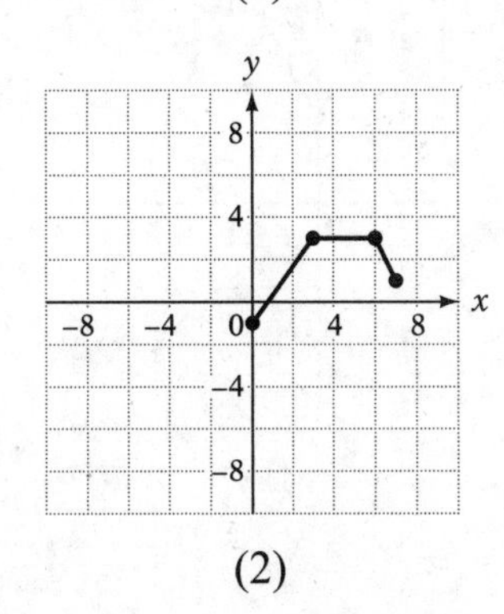

(2)

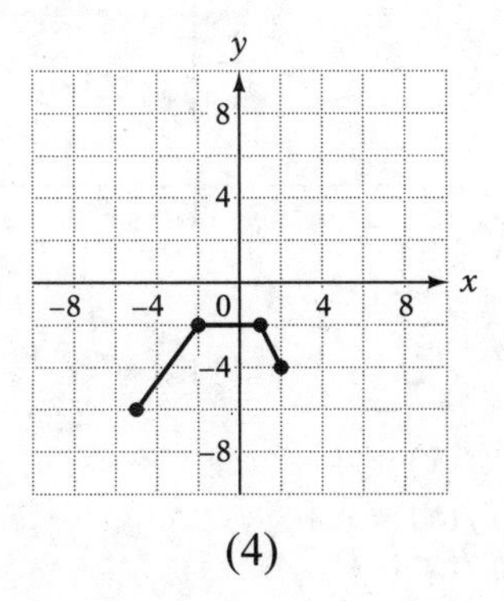

(4)

5. Below is the graph of $f(x)$ on the left and $g(x)$ on the right. Which is equivalent to $g(x)$?

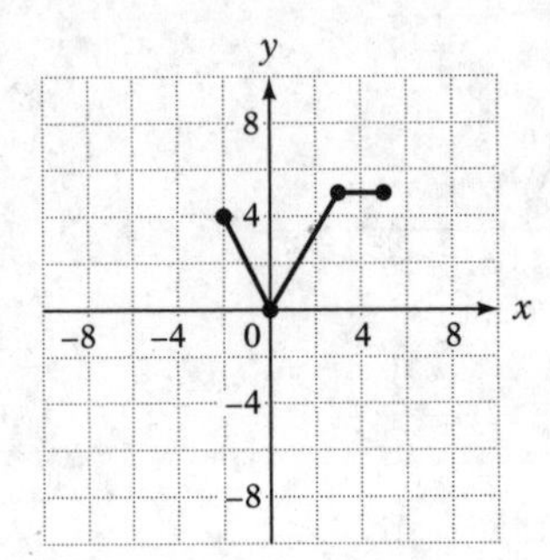

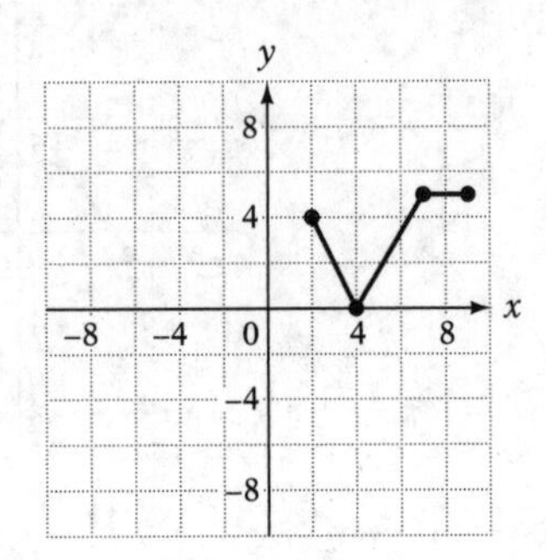

(1) $f(x) + 4$
(2) $f(x) - 4$
(3) $f(x + 4)$
(4) $f(x - 4)$

6. Below is the graph of $f(x)$ on the left and $g(x)$ on the right. Which is equivalent to $g(x)$?

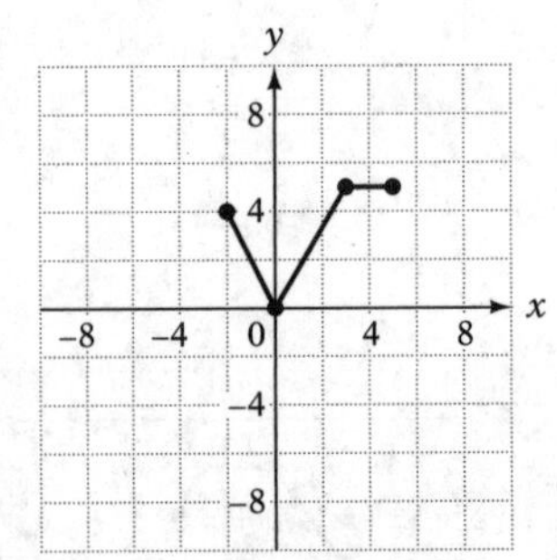

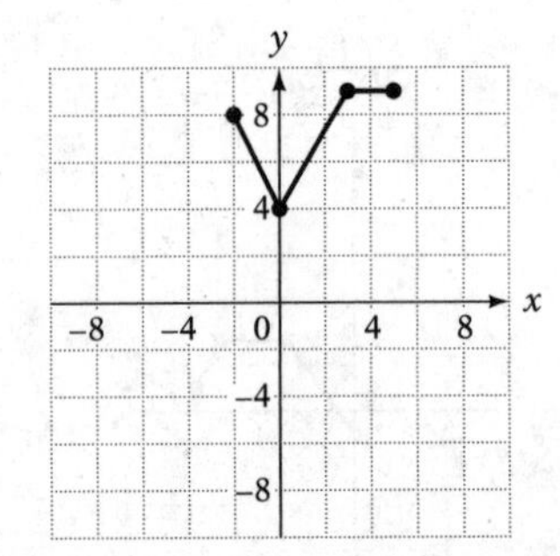

(1) $f(x) + 4$
(2) $f(x) - 4$
(3) $f(x + 4)$
(4) $f(x - 4)$

7. Below is the graph of $f(x)$ on the left and $g(x)$ on the right. Which is equivalent to $g(x)$?

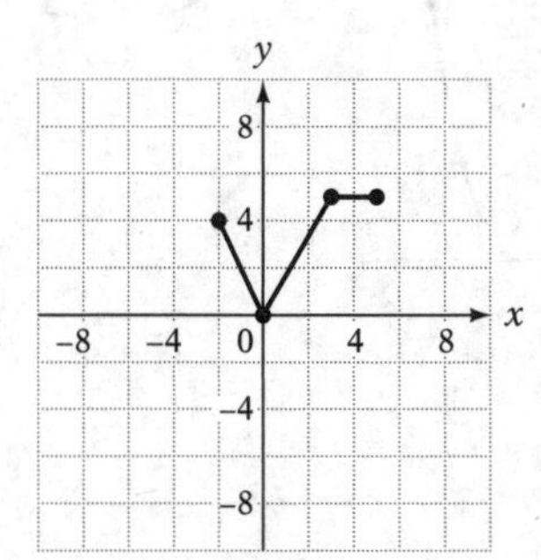

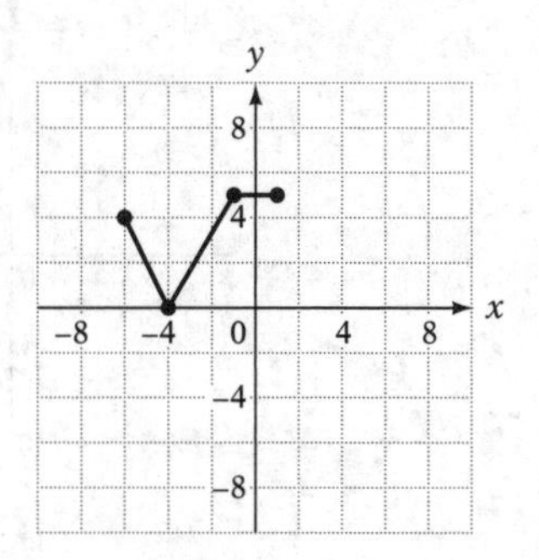

(1) $f(x) + 4$
(2) $f(x) - 4$
(3) $f(x + 4)$
(4) $f(x - 4)$

8. Below is the graph of $f(x)$ on the left and $g(x)$ on the right. Which is equivalent to $g(x)$?

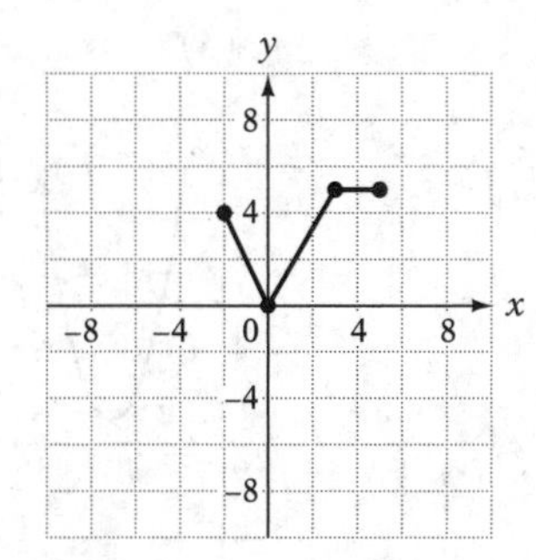

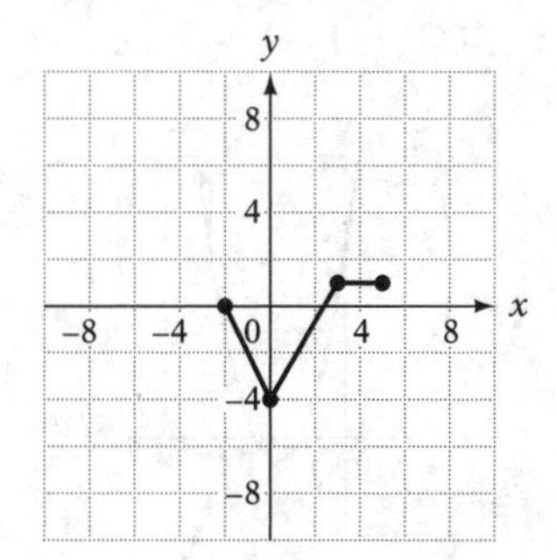

(1) $f(x) + 4$
(2) $f(x) - 4$
(3) $f(x + 4)$
(4) $f(x - 4)$

9. If the graph of $y = f(x)$ is a parabola with the vertex at (5, 1), what is the vertex of the graph of the parabola $y = f(x - 2)$?
(1) (5, 3)
(2) (5, –1)
(3) (7, 1)
(4) (3, 1)

10. If the graph of $y = g(x) + 2$ is below, which is the graph of $y = g(x)$?

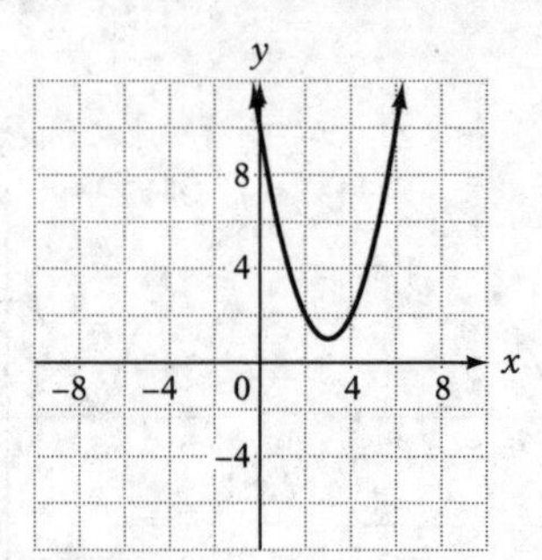

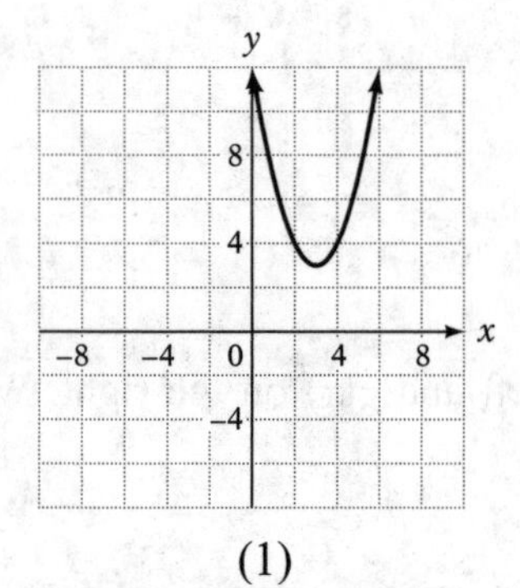

(1)

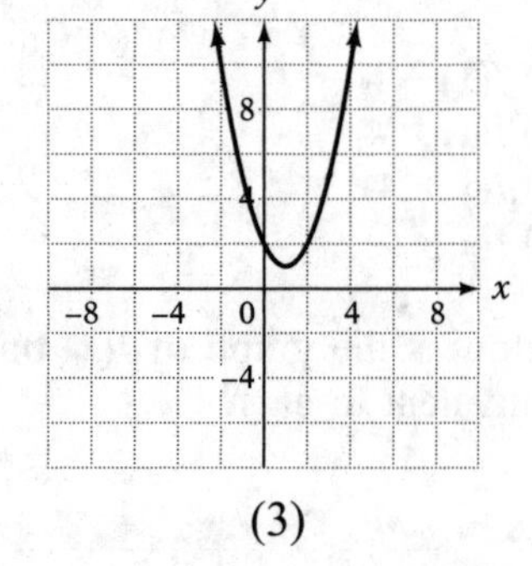

(3)

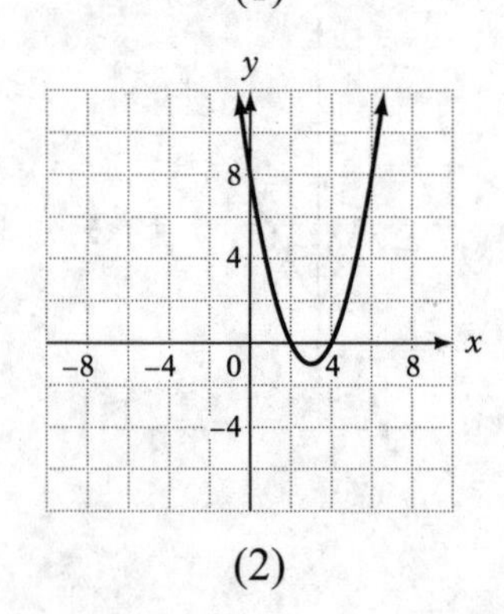

(2)

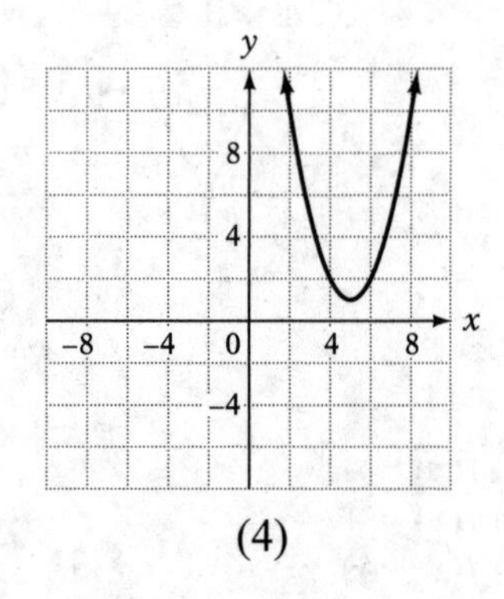

(4)

B. Show how you arrived at your answers.

1. Below is a chart of x values and corresponding $f(x)$ values. (a) If $g(x) = f(x) + 1$, what is the value of $g(3)$? (b) If $h(x) = f(x + 1)$, what is $h(3)$?

x	$f(x)$
1	4
2	3
3	1
4	5
5	2

2. Below is the graph of $y = f(x)$. Make a graph of $y = f(x - 2)$.

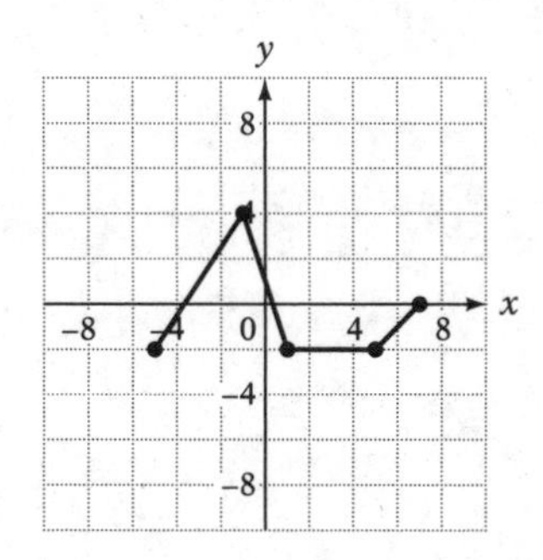

3. The graph of $y = f(x)$ has a domain of $2 \leq x \leq 7$. (a) What is the domain of the graph of $y = f(x) + 4$? (b) What is the domain of the graph of $y = f(x + 4)$?

4. The graph of $y = f(x)$ is below. If $g(x) = f(x + 2)$ and $h(x) = g(x) + 3$, make a graph of $y = h(x)$.

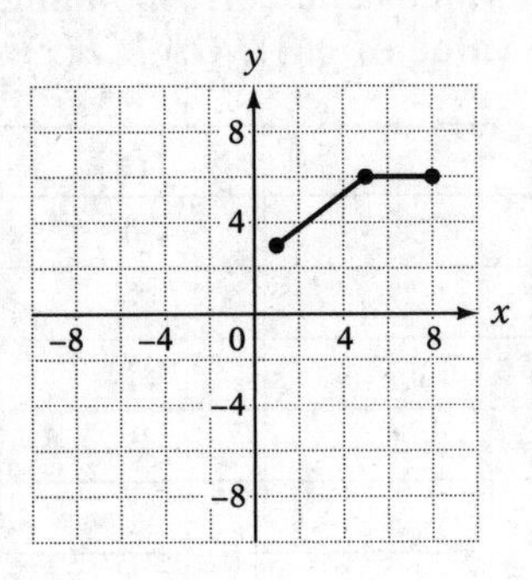

5. If $f(x) = x^2 + 2x - 3$ and $g(x) = x^2 + 6x + 5$, how can $g(x)$ be written as a transformation of $f(x)$? Explain.

Chapter Eleven SEQUENCES

11.1 TYPES OF SEQUENCES

A sequence is a list of numbers. It is sometimes possible to predict the next number on a list by examining the numbers before and detecting a pattern. That pattern could be adding a number to the term before, multiplying a number by the term before, or something even more complicated.

Sequence Notation

An example of a sequence is 3, 7, 11, 15, 19, ...

The sequence is often named with the letter a. The individual elements of the sequence are named by the name of the sequence with a subscript identifying the element's position in the list. For this list, $a_1 = 3$, $a_2 = 7$, $a_3 = 11$, $a_4 = 15$, and $a_5 = 19$. Sometimes instead of a subscript, the term number is put into parentheses, like with functions, like $a(1) = 3$ or $a(2) = 7$.

Arithmetic Sequences

To find the value of a_6, check to see if this is the kind of sequence where each number can be calculated by adding or subtracting the same thing from the previous number. When this happens, the sequence is called an *arithmetic sequence*.

In this example since $3 + 4 = 7$, $7 + 4 = 11$, $11 + 4 = 15$, and $15 + 4 = 19$, this seems to be an arithmetic sequence. To continue the pattern, $a_6 = 19 + 4 = 23$.

The sequence 40, 32, 24, 16 is also an arithmetic sequence. By subtracting 8 from each term, the next term is obtained. When each term is smaller than the previous term, it is a *decreasing* sequence. When each term is larger than the previous term it is an *increasing* sequence.

Geometric Sequences

The sequence 3, 6, 12, 24, 48, … is not an arithmetic sequence. $3 + 3 = 6$, but $6 + 3$ is not 12. In this sequence, each term is gotten by multiplying the previous term by 2. When this happens, the sequence is called a *geometric sequence.* The next term is $48 \cdot 2 = 96$. Since each term is larger than the previous term, this is an increasing geometric sequence. A sequence like 160, 80, 40, 20, ... is a decreasing geometric sequence since each term is 1/2 the previous term.

Check Your Understanding of Section 11.1

A. Multiple-Choice

1. In the sequence a = 1, 3, 6, 10, 15, …, what is the value of a_2?
(1) 1 (2) 3 (3) 6 (4) 10

2. What type of sequence is 3, 7, 11, 15, …?
(1) Increasing arithmetic (3) Increasing geometric
(2) Decreasing arithmetic (4) Decreasing geometric

3. What type of sequence is 25, 18, 11, 4, …?
(1) Increasing arithmetic (3) Increasing geometric
(2) Decreasing arithmetic (4) Decreasing geometric

4. What type of sequence is 4, 8, 16, 32, …?
(1) Increasing arithmetic (3) Increasing geometric
(2) Decreasing arithmetic (4) Decreasing geometric

5. What type of sequence is 1, 215, 405, 135, 45, …?
(1) Increasing arithmetic (3) Increasing geometric
(2) Decreasing arithmetic (4) Decreasing geometric

6. What is the next number in the sequence 8, 13, 18, 23?
(1) 26 (2) 27 (3) 28 (4) 29

7. What is the next number in the sequence 14, 6, –2, –10?
(1) –17 (2) –18 (3) –19 (4) –20

8. What is the next number in the sequence 20, 10, 5, $\frac{5}{2}$?

(1) $\frac{5}{4}$ (2) $\frac{5}{8}$ (3) $\frac{5}{16}$ (4) $\frac{5}{32}$

9. A sequence begins with 2, 6. If this is a geometric series, what is the next number?
(1) 10 (2) 12 (3) 14 (4) 18

10. What type of sequence is $2^1, 2^2, 2^3, 2^4, 2^5, \ldots$?
(1) Increasing arithmetic (3) Increasing geometric
(2) Decreasing arithmetic (4) Decreasing geometric

11.2 RECURSIVELY DEFINED SEQUENCES

Describing the terms of a sequence, including the rule of how to get the next term, requires a shorthand that resembles function notation. This *recursive* description identifies the first number in the sequence and also the rule of how to use any term to obtain the next term.

Listing an Arithmetic Sequence from the Recursive Rule

The sequence $a = 3, 7, 11, 15, 19, \ldots$ can be described by first identifying the first element, $a_1 = 3$. This is sometimes called the *base case*. For this sequence, a_2 can be calculated by adding 4 to a_1 Then a_3 can be calculated by adding 4 to a_2, and so on.

The way to write this rule is $a_n = a_{n-1} + 4$ for $n > 1$. The a_n is the term you are trying to calculate, and the a_{n-1} is the previous term. $a_n = a_{n-1} + 4$ is a fancy way of saying that to get any term, add four to the previous term.

Together these two pieces make up the rule:

$$a_1 = 3$$
$$a_n = a_{n-1} + 4 \text{ for } n > 1$$

When $n = 2$, the bottom formula says $a_2 = a_{2-1} + 4 = a_1 + 4 = 3 + 4 = 7$.
When $n = 3$, the bottom formula says $a_3 = a_{3-1} + 4 = a_2 + 4 = 7 + 4 = 11$.
When $n = 4$, the bottom formula says $a_4 = a_{4-1} + 4 = a_3 + 4 = 11 + 4 = 15$.

Example 1

The recursive rule for a sequence is

$$a_1 = 5$$
$$a_n = a_{n-1} + 6 \text{ for } n > 1$$

Find the values of a_2, a_3, a_4, and a_5.

Solution: The rule says that to get the next term, add 6 to the previous term. From that rule, the next numbers 11, 17, 23, and 29 can be found. This can also be solved by substituting 2, 3, 4, and 5 into the bottom formula.

For $n = 2$, $a_2 = a_{2-1} + 6 = a_1 + 6 = 5 + 6 = 11$.
For $n = 3$, $a_3 = a_{3-1} + 6 = a_2 + 6 = 11 + 6 = 17$.
For $n = 4$, $a_4 = a_{4-1} + 6 = a_3 + 6 = 17 + 6 = 23$.
For $n = 5$, $a_5 = a_{5-1} + 6 = a_4 + 6 = 23 + 6 = 29$.

Listing a Geometric Sequence from the Recursive Rule

The geometric sequence 3, 6, 12, 24, 48, … can also be described recursively. Since the first term is 3, $a_1 = 3$. Then since the next term is calculated by multiplying the previous term by 2, the recursive part of the equation is $a_n = 2 \cdot a_{n-1}$ for $n > 1$. The complete rule is

$$a_1 = 3$$
$$a_n = 2 \cdot a_{n-1} \text{ for } n > 1$$

When given the recursive rule for a geometric sequence, it is possible to list all the terms of it.

Example 2

A sequence is defined by the recursive rule

$$a_1 = 2$$
$$a_n = 3 \cdot a_{n-1} \text{ for } n > 1$$

What are the first four terms of the sequence?

Solution: The first term, a_1, is 2 since it is defined in the first line of the rule. The second term can be calculated by substituting $n = 2$ into the bottom line of the rule.

$$a_2 = 3 \cdot a_{2-1} = 3 \cdot a_1 = 3 \cdot 2 = 6$$

The third term can be calculated by substituting $n = 3$ into the bottom line of the rule.

$$a_3 = 3 \cdot a_{3-1} = 3 \cdot a_2 = 3 \cdot 6 = 18$$

And the fourth term, $a_4 = 3 \cdot a_{4-1} = 3 \cdot a_3 = 3 \cdot 18 = 54$.

So, the sequence is 2, 6, 18, 54, ….

Creating Recursive Rules for Arithmetic and Geometric Sequences

MATH FACTS

The recursive part of the rule for an arithmetic sequence is $a_n = a_{n-1} + d$ where d is what needs to be added (or subtracted if d is negative) to the previous term to get the next term. The recursive part of the rule for a geometric sequence is $a_n = r \cdot a_{n-1}$ where r is what needs to be multiplied by the previous term to get the next term.

Creating the recursive rule for a sequence requires the same analysis as trying to find the next term of a sequence. Examine the first few terms to determine if it is an arithmetic sequence where each term is obtained from the previous term by adding or subtracting or if it is a geometric sequence where each term is obtained from the previous term by multiplying.

If the sequence is 8, 11, 14, 17, …, first decide how to get the *next* term. In this case, each term is 3 more than the one before it, so the next term seems to be $17 + 3 = 20$. Each term is 3 more than the previous term.

So, the recursive rule is

$$a_1 = 8$$
$$a_n = a_{n-1} + 3 \text{ for } n > 1$$

The sequence 4, 20, 100, 500, ... is not an arithmetic sequence since 20 is 16 more than 4 and 100 is 80 more than 20. To see if it is geometric, divide the second term by the first term. 20 divided by 4 is 5. Then divide the third term by the second term. 100 divided by 20 is also 5. This is a geometric series where each term is 5 times the previous term.

So, the recursive rule is

$$a_1 = 4$$
$$a_n = 5 \cdot a_{n-1} \text{ for } n > 1$$

Recursive Definitions of Sequences for Picture Patterns

Instead of providing a list of numbers, there might be a sequence of pictures representing a sequence. A recursive formula can be created for these by either turning the pictures into a number list or by examining the pictures to determine the pattern.

Example 3

Create a recursive formula that describes the number of shaded circles in the *n*th picture of this sequence

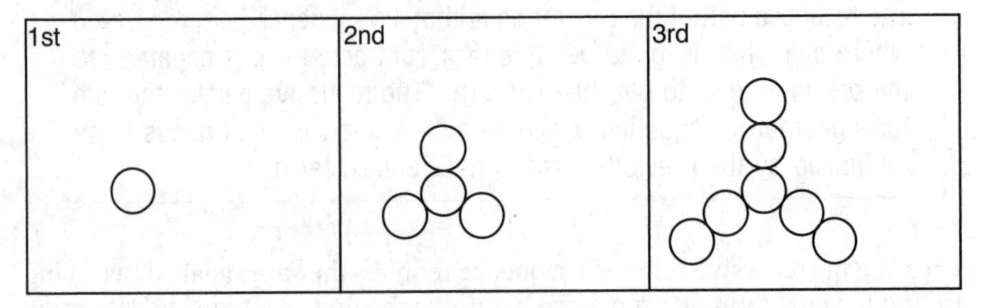

Solution: The numbers follow a pattern 1, 4, 7, …. Since the first picture has 1 circle, $a_1 = 1$. Since each picture has three more circles than the previous picture, the recursive part is $a_n = a_{n-1} + 3$ for $n > 1$.

Recursive Definitions of Sequences for Real-World Scenarios

A sequence of numbers could be related to a real-world scenario. The process of finding the equation is the same as for a list.

Suppose there is a plant that grows according to an arithmetic progression. It is 4 inches tall at the end of the first week, 7 inches tall at the end of the second week, and 10 inches tall at the end of the third week. Finding a recursive rule to describe the height of the plant after *n* weeks is the same as finding the rule for the sequence 4, 7, 10, ….

$$a(1) = 4$$
$$a(n) = a(n-1) + 3 \text{ for } n > 1$$

Check Your Understanding of Section 11.2

A. Multiple-Choice

1. Find the value of a_2 in the sequence defined by

$$a_1 = 4$$
$$a_n = 3 + a_{n-1} \text{ for } n > 1$$

(1) 3 (2) 5 (3) 7 (4) 12

2. Find the value of a_2 in the sequence defined by

$$a_1 = 10$$
$$a_n = a_{n-1} - 6 \text{ for } n > 1$$

(1) 6 (2) 4 (3) 10 (4) 16

3. Find the value of a_2 in the sequence defined by

$$a_1 = 4$$
$$a_n = 3 \cdot a_n - 1 \text{ for } n > 1$$

(1) 7 (2) 10 (3) 12 (4) 16

4. Find the value of a_2 in the sequence defined by

$$a_1 = 6$$
$$a_n = \frac{1}{3} a_{n-1} \text{ for } n > 1$$

(1) 1 (2) 2 (3) 3 (4) 18

5. The sequence 5, 11, 17, 23, ... can be generated by which definition?
(1) $a_1 = 5 \quad a_n = 6a_{n-1}$ for $n > 1$
(2) $a_1 = 5 \quad a_n = \frac{11}{5} a_{n-1}$ for $n > 1$
(3) $a_1 = 5 \quad a_n = -6 + a_{n-1}$ for $n > 1$
(4) $a_1 = 5 \quad a_n = 6 + a_{n-1}$ for $n > 1$

6. The sequence 3, 15, 75, 375, ... can be generated by which definition?
(1) $a_1 = 3 \quad a_n = 5a_{n-1}$ for $n > 1$
(2) $a_1 = 3 \quad a_n = 12 + a_{n-1}$ for $n > 1$
(3) $a_1 = 3 \quad a_n = \frac{1}{5} a_{n-1}$ for $n > 1$
(4) $a_1 = 3 \quad a_n = -12 + a_{n-1}$ for $n > 1$

7. Which definition generates a sequence of numbers to describe the total number of rectangles for each picture?

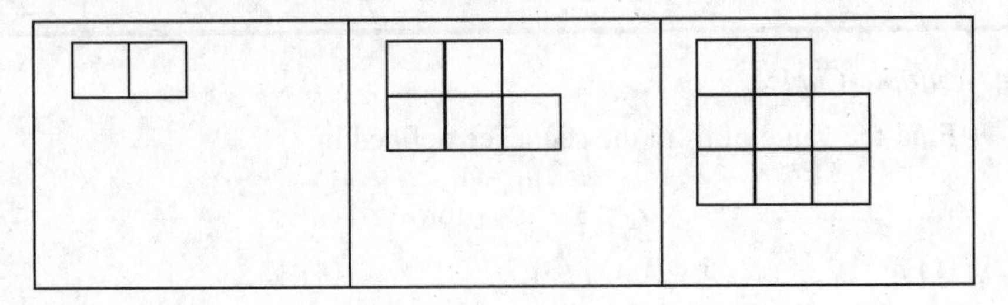

(1) $a_1 = 2$ $\quad a_n = 3 + a_{n-1}$ for $n > 1$

(2) $a_1 = 2$ $\quad a_n = \frac{5}{2} a_{n-1}$ for $n > 1$

(3) $a_1 = 2$ $\quad a_n = a_{n-1} - 3$ for $n > 1$

(4) $a_1 = 2$ $\quad a_n = 5a_{n-1}$ for $n > 1$

8. The side length of the square in the first picture is one unit. What sequence describes the area of each of the pictures in the pattern?

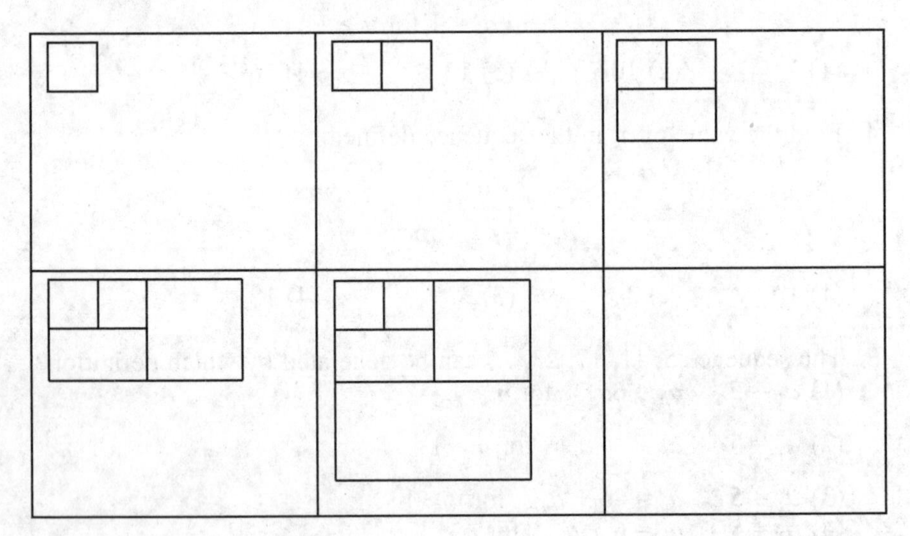

(1) $a_1 = 1$ $\quad a_n = 1 + a_{n-1}$ for $n > 1$

(2) $a_1 = 1$ $\quad a_n = \frac{1}{2} a_{n-1}$ for $n > 1$

(3) $a_1 = 1$ $\quad a_n = 2 + a_{n-1}$ for $n > 1$

(4) $a_1 = 1$ $\quad a_n = 2 \cdot a_{n-1}$ for $n > 1$

9. A person jogs 3 miles on the first day of training and 5 more miles each day after that. Which definition can generate the sequence of the number of miles run each day?

(1) $a_1 = 3 \quad a_n = \frac{8}{3}a_{n-1}$ for $n > 1$

(2) $a_1 = 3 \quad a_n = 5 + a_{n-1}$ for $n > 1$

(3) $a_1 = 3 \quad a_n = 5a_{n-1}$ for $n > 1$

(4) $a_1 = 3 \quad a_n = \frac{8}{3} \cdot a_{n-1}$ for $n > 1$

10. A ball is dropped from a window 50 feet above the ground. Each bounce is $\frac{4}{5}$ the height of the previous bounce. Which definition would generate the height of the bounces?

(1) $a_1 = 50 \quad a_n = -10 + a_{n-1}$ for $n > 1$

(2) $a_1 = 50 \quad a_n = 10 + a_{n-1}$ for $n > 1$

(3) $a_1 = 50 \quad a_n = \frac{4}{5}a_{n-1}$ for $n > 1$

(4) $a_1 = 50 \quad a_n = \frac{5}{4}a_{n-1}$ for $n > 1$

***B.** Show how you arrived at your answers.*

1. A sequence can be generated by the definition

$$a_1 = 5$$
$$a_n = 4 + a_{n-1} \text{ for } n > 1$$

What are the first four terms of this sequence?

2. The recursive part of a sequence definition is $a_n = 3a_{n-1}$. If $a_{17} = 58$, what is a_{18} and a_{19}?

3. Madelyn puts $50 in a bank that offers simple interest. This chart lists the amount in the bank at the beginning of each year. What recursive definition would generate these numbers?

Year	Amount
1	$50
2	$55
3	$60
4	$65
5	$70

4. Archimedes estimated the circumference of a circle by measuring regular polygons with increasing numbers of sides.

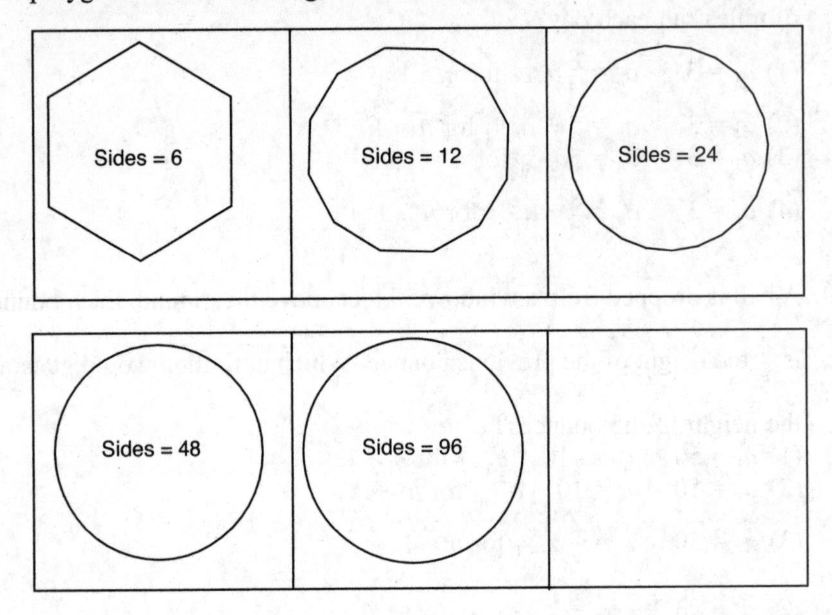

What recursive definition could be used to generate the number of sides for each picture?

5. What are the first five terms of the sequence defined by this recursive definition?

$$a_1 = 3$$
$$a_2 = 4$$
$$a_n = 2a_{n-1} + a_{n-2} \text{ for } n > 2$$

11.3 CLOSED FORM DEFINED SEQUENCES

KEY IDEAS

A recursive rule is useful for finding the next term of a sequence. If, however, you need to find the 100th term of a sequence, the recursive rule is not very convenient as it would require finding all 100 terms. A *closed form* definition of a sequence is a formula that relates the position of the number in the sequence to the number in that position. For arithmetic and geometric sequences, there is a short way to create the closed form definition.

Listing a Sequence from the Closed Form Definition

Like a function, the closed form definition takes the position of the number on the list as an input, and it outputs the number that goes into that position.

If the closed form definition of a sequence is $a_n = 3n + 7$, then the sequence can be created by substituting 1, 2, 3, and so on for n. This formula does not require looking at the previous term.

$$a_1 = 3 \cdot 1 + 7 = 3 + 7 = 10$$
$$a_2 = 3 \cdot 2 + 7 = 6 + 7 = 13$$
$$a_3 = 3 \cdot 3 + 7 = 9 + 7 = 16$$

So the sequence begins 10, 13, 16, …

Example 1

The sequence 10, 13, 16, 19, … has a closed from definition of $a_n = 3n + 7$. What is the 100th term of the sequence?

Solution: To get a_{100}, substitute $n = 100$ into the formula to get $a_{100} = 3 \cdot 100 + 7 = 300 + 7 = 307$. To get this answer with the recursive definition would take a very long time!

Finding the Closed Form Definition of an Arithmetic Sequence

The sequence 8, 11, 14, 17, … is an arithmetic sequence. Since $11 = 8 + 3 \cdot 1$, $14 = 8 + 3 \cdot 2$, and $17 = 8 + 3 \cdot 3$, the closed form definition for this sequence is

$$a_n = 8 + 3(n - 1)$$

MATH FACTS

The closed form formula for describing an arithmetic sequence is $a_n = a_1 + d(n - 1)$ where a_1 is the first term in the sequence and d is the difference between any two consecutive terms.

Example 2

Find the closed form definition to calculate a_n for the sequence 5, 12, 19, 26, … and determine the 50th number in the sequence.

Solution: Since $a_1 = 5$ and $d = 7$, the formula is $a_n = 5 + 7(n - 1)$ which can be simplified to $a_n = 7n - 2$. Substitute 50 into either of these formulas.

$$a_5 = 5 + 7(50 - 1) = 5 + 7 \cdot 49 = 5 + 343 = 348$$

After the closed form definition of an arithmetic sequence is found, it can be used to determine when in a sequence a particular number will appear.

Example 3

What position in the sequence 5, 12, 19, 26, … will the number 651 be in?

Solution: From example 2, the closed form definition of the sequence is $a_n = 5 + 7(n - 1)$. To find the position of 651, replace a_n with 651 and solve for n.

$$656 = 5 + 7(n - 1)$$
$$-5 = -5$$
$$\frac{651}{7} = \frac{7(n-1)}{7}$$
$$93 = n - 1$$
$$+1 = +1$$
$$94 = n$$

Finding the Closed Form Definition of a Geometric Sequence

2, 10, 50, 250, 1250, … is a geometric series since each term is equal to the previous term multiplied by 5. The first three terms of this sequence can also be written as:

$$a_1 = 2 = 2g5^0$$
$$a_2 = 10 = 2g5^1$$
$$a_3 = 50 = 2g5^2$$

In general, the nth term will be $a_n = 2g5^n$

MATH FACTS

The closed from formula for describing a geometric sequence is $a_n = a_1 g r^{n-1}$ where a_1 is the first term and r is what you have to multiply by the previous term to get to the next term.

Example 4

What is the tenth term of the geometric sequence that begins 3, 12, 48, 192, …?

Solution: Since the first term is 3 and each term is 4 times the previous term, $a_1 = 3$ and $r = 4$. The closed form formula for the terms of this sequence, then, is $a_n = 3g4^{n-1}$. For $n = 10$, this becomes $a_{10} = 3g4^{10-1} = 3g4^9 = 3g262144 = 786432$.

Closed Form Definitions of Sequences for Real-World Scenarios

If a real-world scenario has numbers that form an arithmetic sequence, the closed form definition can be used to describe the real-world scenario.

Suppose there is a swimming pool being filled by a hose according to an arithmetic progression. It has 4 inches of water after the first hour, then 7 inches of water at the end of the second hour, and then 10 inches of water at the end of the third hour. Finding a closed form rule to describe the height of the water in the pool after n hours is the same as finding the rule for the sequence 4, 7, 10, 13, ….

Since $a_1 = 4$ and $d = 3$, the formula is $a_n = 4 + 3(n - 1)$, which can be simplified to $a_n = 3n + 1$.

Closed Form Definitions of Sequences for Picture Patterns

If a picture pattern models an arithmetic sequence, the closed form equation will describe that sequence too.

Example 5

Create a closed formula that describes the number of shaded circles in the nth picture of this sequence. Use the formula to determine how many shaded circles would be in the 20th picture.

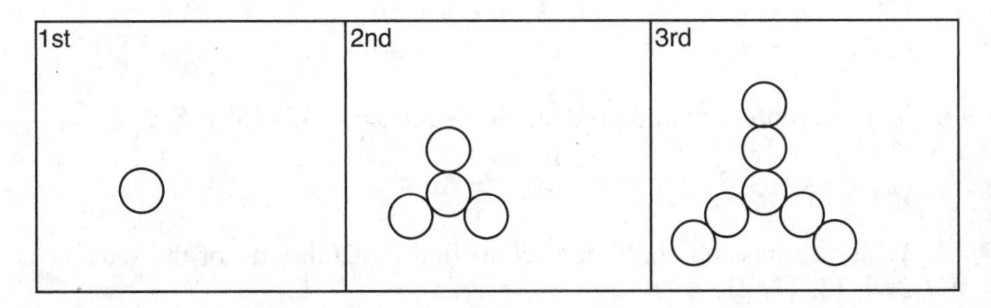

Solution: $a_1 = 1$ and $d = 3$, so the closed form equation is $a_n = 1 + 3(n - 1)$, which can also be simplified to $a_n = 3n - 2$. The value of a_{20} is

$$a_{20} = 1 + 3(20 - 1) = 1 + 3 \cdot 19 = 1 + 57 = 58$$

Check Your Understanding of Section 11.3

A. Multiple-Choice

1. What are the first three terms of the sequence generated by the definition $a_n = 2 + 3(n - 1)$?
(1) 3, 5, 7 (2) 2, 5, 8 (3) 2, 8, 14 (4) 0, 5, 10

2. What is the fifth term of the sequence generated by the definition $a_n = 10 - 4(n - 1)$?
(1) 2 (2) –2 (3) –6 (4) –10

3. What is the third term of the sequence generated by the definition $a_n = 3 \cdot 4^{n-1}$?
(1) 3 (2) 4 (3) 12 (4) 48

4. What is the fifth term of the sequence generated by the definition $a_n = 6 \cdot \left(\frac{2}{3}\right)^{n-1}$?
(1) 96 (2) 48 (3) $\frac{32}{27}$ (4) $\frac{16}{9}$

5. What definition would produce the sequence 4, 13, 22, 31, …?
(1) $a_n = 4 + 9n$
(2) $a_n = 4 + 9(n - 1)$
(3) $a_n = 9 + 4n$
(4) $a_n = 9 + 4(n - 1)$

6. What definition would produce the sequence 4, 12, 36, 108, …?
(1) $a_n = 4 \cdot 3^{n-1}$
(2) $a_n = 4 \cdot 3^n$
(3) $a_n = 3 \cdot 4^{n-1}$
(4) $a_n = 3 \cdot 4^n$

7. Which expression could be used to find the 20th term of the sequence 5, 9, 13, 17, 21, …?
(1) 5 + 4(20)
(2) 5 + 4(19)
(3) 4 + 5(20)
(4) 4 + 5(19)

8. Which expression could be used to find the 20th term of the sequence 5, 15, 45, 135, …?
(1) $3 \cdot 5^{19}$
(2) $3 \cdot 5^{20}$
(3) $5 \cdot 3^{19}$
(4) $5 \cdot 3^{20}$

9. A person has saved \$20. Each week he adds \$7 to the savings. Which definition can be used to generate a sequence of the amount of money saved at the beginning of each week?

(1) $a_n = 7 + 20n$ (3) $a_n = 20 + 7n$
(2) $a_n = 7 + 20(n - 1)$ (4) $a_n = 20 + 7(n - 1)$

10. What definition will generate a sequence of the number of hexagons in each picture?

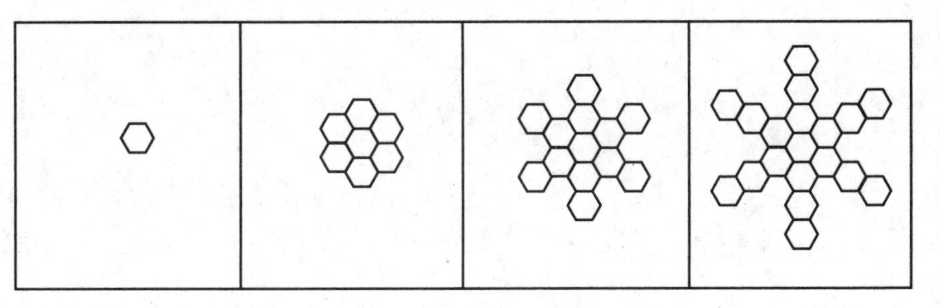

(1) $a_n = 1 + 6(n - 1)$ (3) $a_n = 1 \cdot 6^{n-1}$
(2) $a_n = 1 + 6n$ (4) $a_n = 6 + 1(n - 1)$

B. *Show how you arrived at your answers.*

1. Liam says that the sequence 7, 12, 17, 22, ... can be written as $a_n = 7 + 5(n - 1)$. Cecilia says it can be written as $a_1 = 7$, $a_n = 5 + a_{n-1}$. Who is right? Explain.

2. What is the 100th term of the sequence 26, 19, 12, 5, ...?

3. A folded sheet of paper is $\frac{1}{633,600}$ miles thick. If a sheet of paper is folded the thickness doubles. (a) Create a definition that will generate a sequence where the first number is $\frac{1}{633,600}$ and the next number is always double the previous number. (b) Use the definition to determine the height of the folded paper after 38 folds. Will this reach the Moon (240,000 miles away)?

4. The first terms of sequence a are 1,000, 1,200, 1,400, 1,600, ... The first terms of sequence b are 1, 2, 4, 8, ... Which is greater, a_{20} or b_{20}?

5. To find the 100th term of the sequence 5, 8, 11, 14, ..., would it be faster to use the closed form definition or the recursive definition of the sequence? Explain.

Chapter Twelve

REGRESSION CURVES

12.1 LINE OF BEST FIT

Two points can always be joined with a line. If there are more than two points, however, there may not be one line that passes through all of them. If we need a linear equation for a line that comes close to all the points, there is a feature of the graphing calculator that will calculate the slope and *y*-intercept of this line.

Perfectly straight lines are not common in nature. They also are rare when real-world data is graphed. Below is a chart of some scientific measurements and a graph of the data.

Temperature	Number of people at the beach
30	5
40	21
50	45
60	65
70	77
80	102
90	129

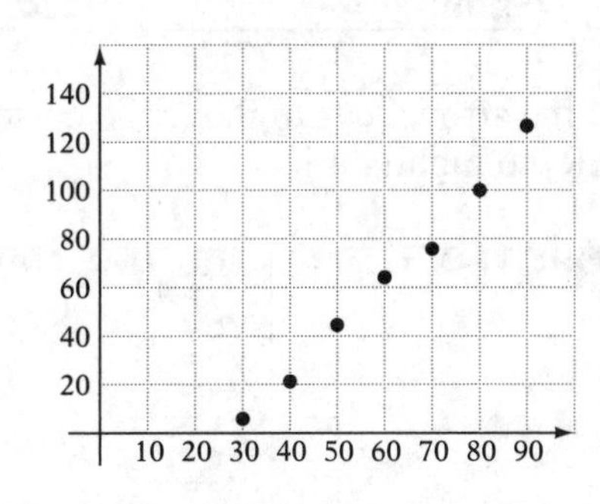

A line drawn through any of the two points will come close but will not pass through any of the other points.

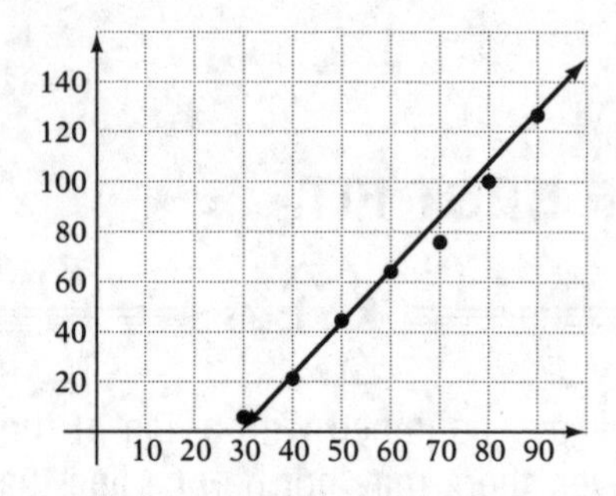

Though there is not one line that passes through all the points, there is a *line of best fit* that will come close to all the points. The equation for the line of best fit will be useful in calculating approximate coordinates for other points on the line.

For the TI-84:

The first step in the process of determining the line of best fit for a data set is to enter the data into a list.

Press [STAT] and then [1] to edit the list. If the lists already have numbers in them, use the arrows to move the cursor to L1 and L2 and press [CLEAR] (not [DEL]!). Enter the temperatures into L1 and the number of people into L2.

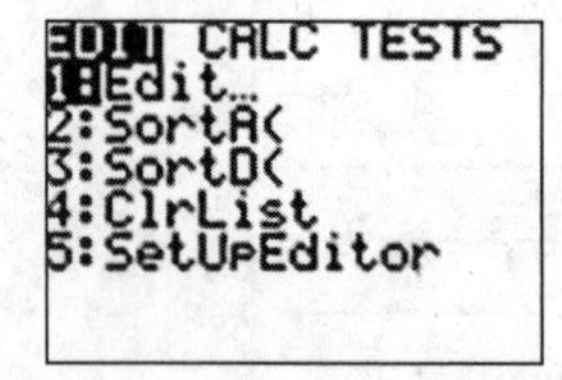

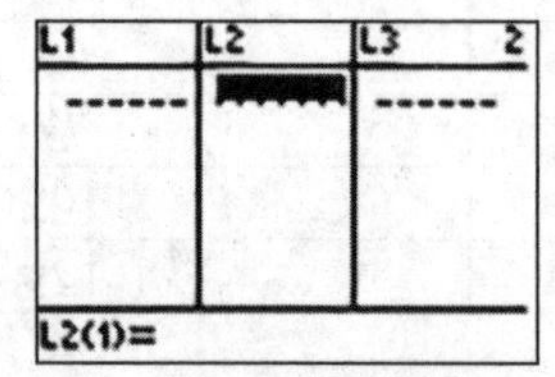

Press [Y=] and use the arrow keys to move the cursor to the Plot1 on the top left. Press [ENTER] to highlight it.

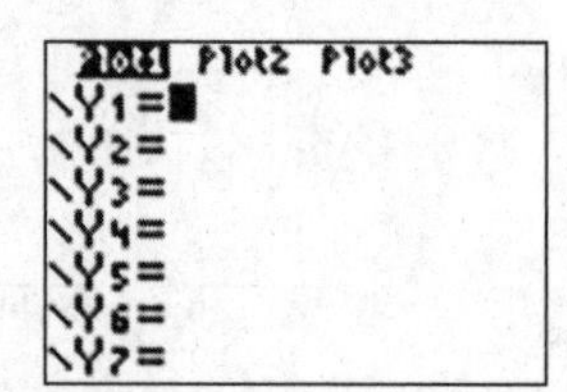

Then press [ZOOM] and [9] to create the scatter plot of the data.

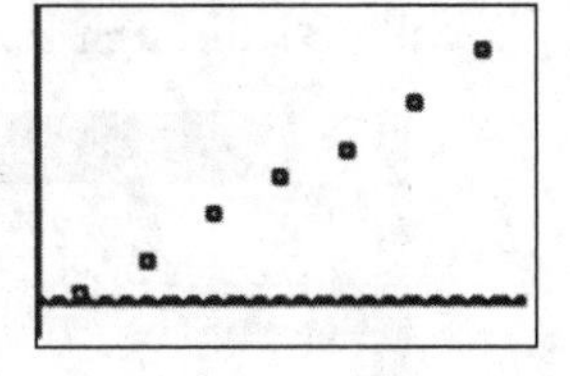

Since the points line up nicely, a line of best fit is appropriate. Press [STAT] and then move to [CALC] and [4]. Then press [ENTER].

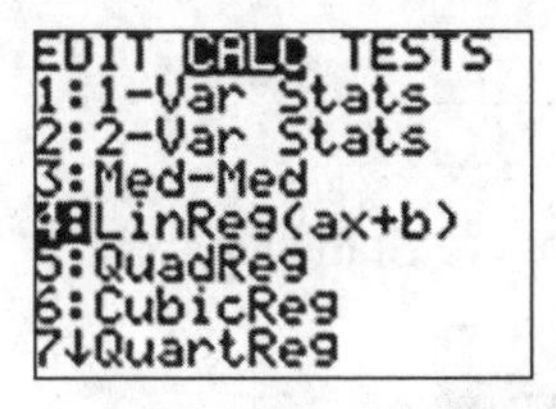

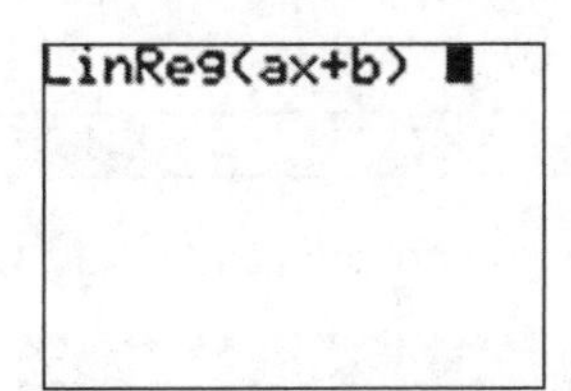

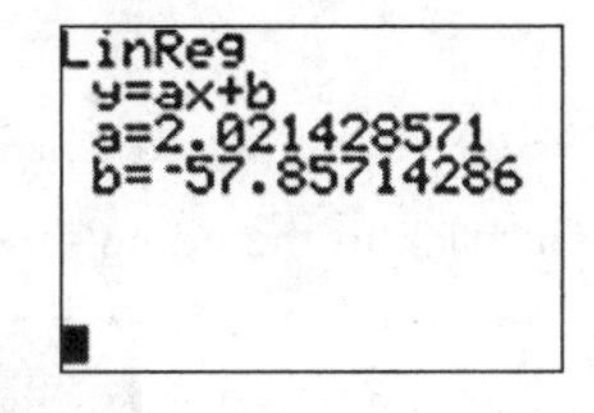

For this example, the line of best fit is approximately $y = 2.02x - 57.86$ or, if using T for temperature and P for people, $P = 2.02T - 57.86$.

This is a graph of the scatter plot and the line of best fit. It does not go through any of the points but comes very close to all of them.

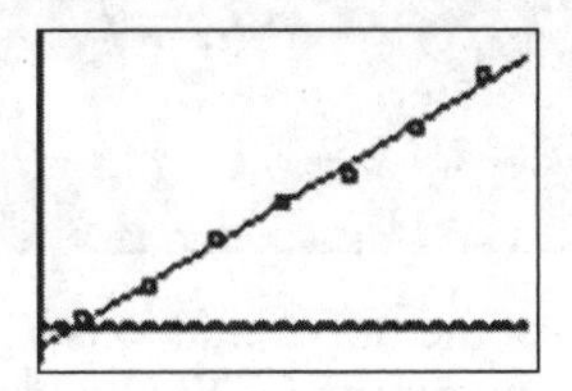

For the TI-Nspire:

From the home screen select the Add Lists & Spreadsheet icon.

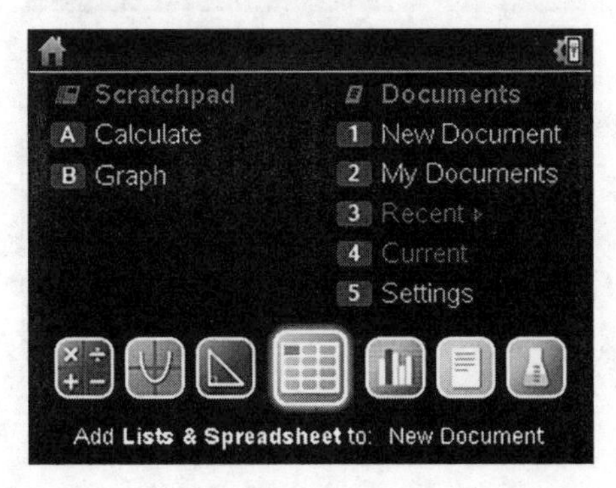

In the first row, label the A column "temp" and the B column "people." Enter the temperature values in cells A1 through A7. Enter the corresponding people values in cells B1 through B7.

	A temp	B people	C	D
◆				
1	30	5		
2	40	21		
3	50	45		
4	60	65		
5	70	77		

C1

Go back to the home screen, and select the Add Data & Statistics icon.

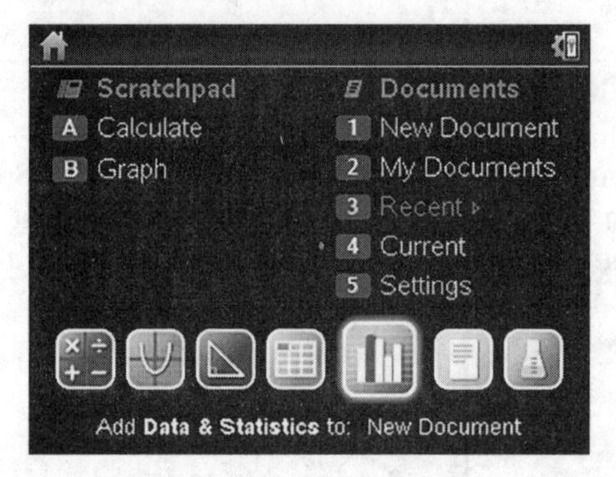

To create the scatter plot, set the horizontal axis to temp and the vertical axis to people. Do this by clicking on the Click to add variable box and selecting the variables temp and people from the list.

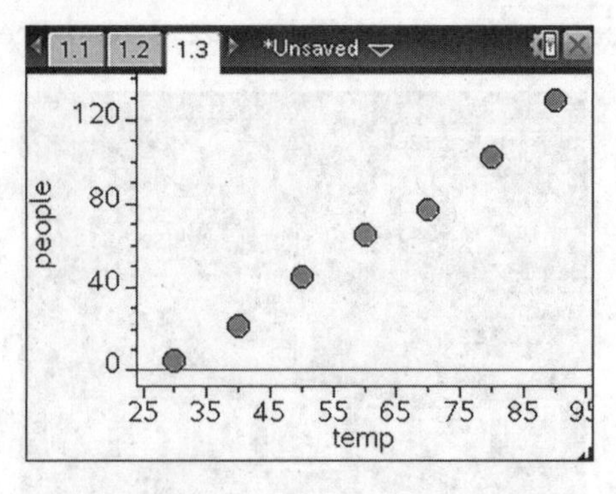

To see the line of best fit and its equation, select [menu], [4], and [6] for Regression. Then select [1] for Show Linear (mx+b) and press [enter].

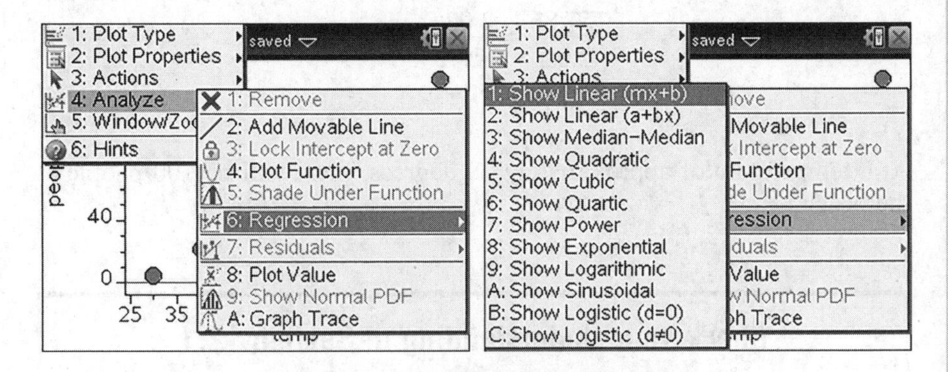

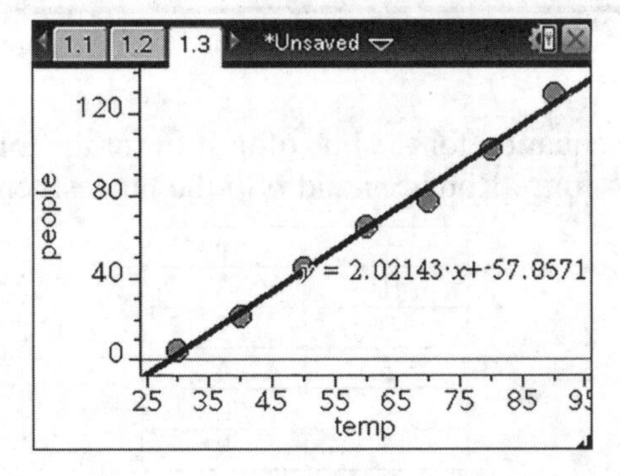

The equation for the line of best fit is $y = 2.02x - 57.86$ or $P = 2.02T - 57.86$.

Using the Line of Best Fit to Answer Questions About the Real-World Scenario

After the equation is determined, it can be used to answer questions. In the temperature/people example above, the equation became $P = 2.02T - 57.86$.

One question that can be answered with this equation is: Approximately how many people will be at the beach when the temperature is 100 degrees?

To solve, substitute 100 for T and solve for P.

$$P = 2.02(100) - 57.86 = 202 - 57.86 = 144.14$$

Approximately 144 people will be there when it is 100 degrees.

Another question that can be answered with this equation is: At approximately what temperature will there be 100 people on the beach?

To solve, substitute 100 for P and solve for T. This will require algebra.

$$100 = 2.02T - 57.86$$
$$+57.86 = \qquad +57.86$$
$$\frac{157.86}{2.02} = \frac{2.02T}{2.02}$$
$$78.15 = T$$

At a temperature of approximately 78 degrees, there will be 100 people at the beach.

Check Your Understanding of Section 12.1

A. Multiple-Choice

1. Calculate the equation for the line of best fit for the following set of data in $y = mx + b$ form. Round m and b to the nearest tenth.

x	y
1	3
2	5
3	4
4	6
5	8

(1) $y = 1.1x + 1.9$
(2) $y = 1.4x + 1.7$
(3) $y = 1.7x + 1.4$
(4) $y = 1.9x + 1.1$

2. Calculate the equation for the line of best fit for the following set of data in $y = mx + b$ form. Round m and b to the nearest tenth.

x	y
1	7
2	6
3	6
4	5
5	4

(1) $y = 7.7x - 0.7$
(2) $y = -0.5x + 9.2$
(3) $y = 9.2x - 0.5$
(4) $y = -0.7x + 7.7$

3. Calculate the equation for the line of best fit for the following set of data in $y = mx + b$ form. Round m and b to the nearest tenth.

x	y
1	1
2	3
3	8
4	7
5	9

(1) $y = 2x - 0.4$
(2) $y = -0.4x + 2$
(3) $y = 2.5x - 0.3$
(4) $y = -0.3x + 2.5$

4. Calculate the equation for the line of best fit for the following set of data in $y = mx + b$ form. Round m and b to the nearest tenth.

x	y
1	4
2	3
3	5
4	4
5	5

(1) $y = 2.8x + 0.4$
(2) $y = 0.4x + 2.8$
(3) $y = 0.3x + 3.3$
(4) $y = 3.3x + 0.3$

5. Calculate the equation for the line of best fit for the following set of data in $y = mx + b$ form. Round m and b to the nearest tenth.

x	y
2	16
4	12
6	2
8	6
10	2

(1) $y = 18.2x - 1.3$
(2) $y = -1.3x + 18.2$
(3) $y = 17.8x - 1.7$
(4) $y = -1.7x + 17.8$

6. Calculate the equation for the line of best fit for the following set of data in $y = mx + b$ form. Round m and b to the nearest tenth.

x	y
10	33
20	20
30	10
40	14
50	6

(1) $y = 34.6x - 0.6$
(2) $y = -0.6x + 34.6$
(3) $y = 31.8x - 0.7$
(4) $y = -0.7x + 31.8$

7. Calculate the equation for the line of best fit for the following set of data in $y = mx + b$ form. Round m and b to the nearest tenth.

x	y
10	250
25	310
37	450
46	560
59	820

(1) $y = 11.1x + 70.3$
(2) $y = 70.3x + 11.1$
(3) $y = 11.5x + 69.8$
(4) $y = 69.8x + 11.5$

8. What is the equation for the line of best fit for the points on this scatter plot?

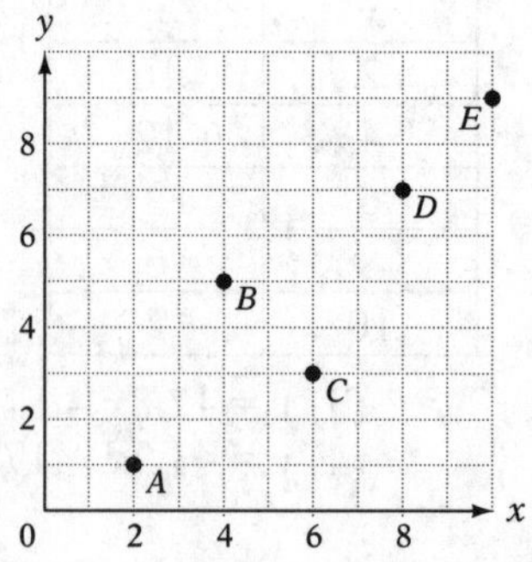

(1) $y = -0.5x + 7$
(2) $y = 0.7x - 0.5$
(3) $y = -0.4x + 0.9$
(4) $y = 0.9x - 0.4$

9. Of these four choices, which line appears to be the best fit for this scatter plot?

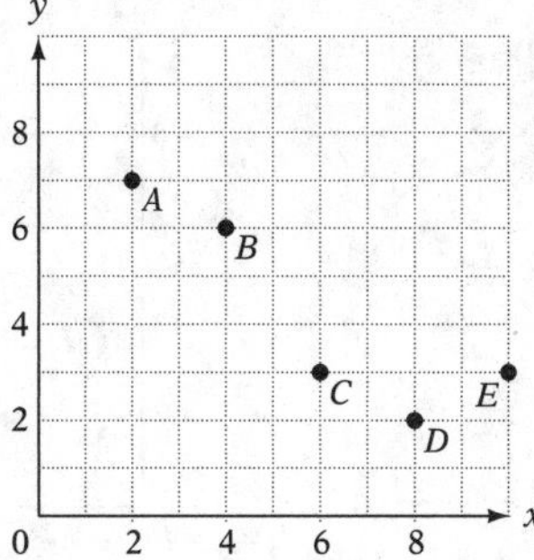

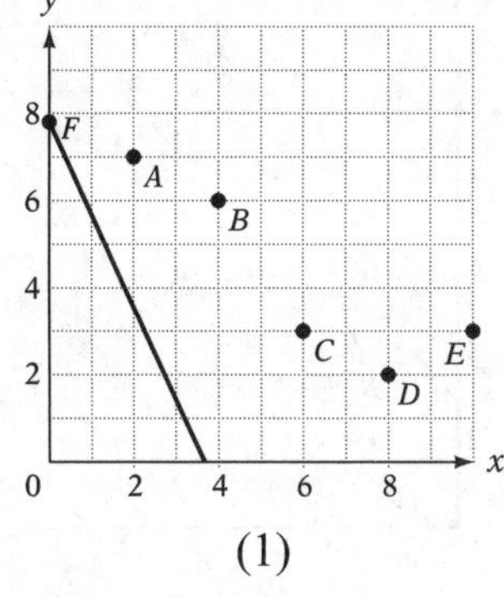

(1)

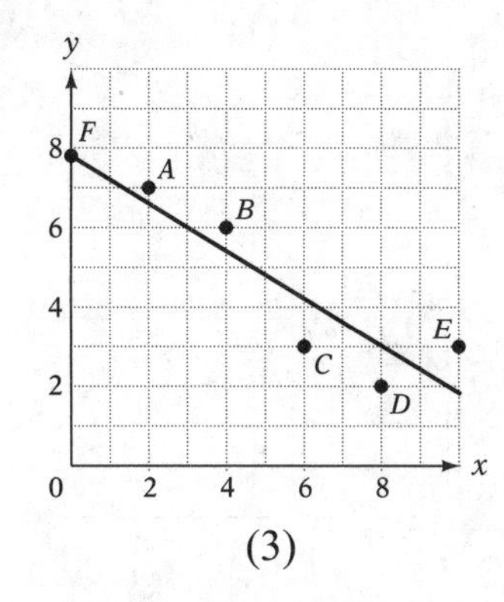

(3)

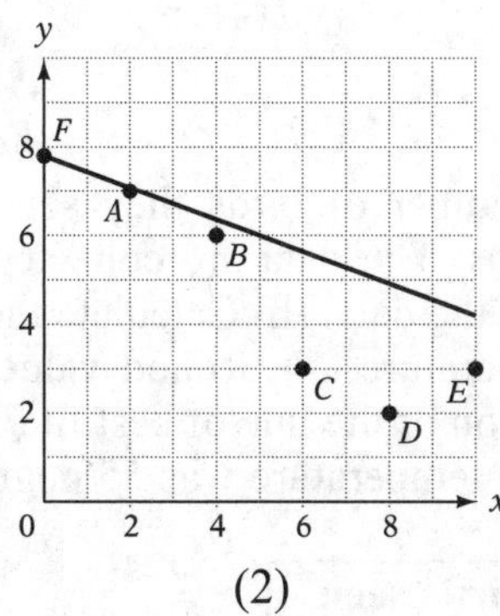

(2)

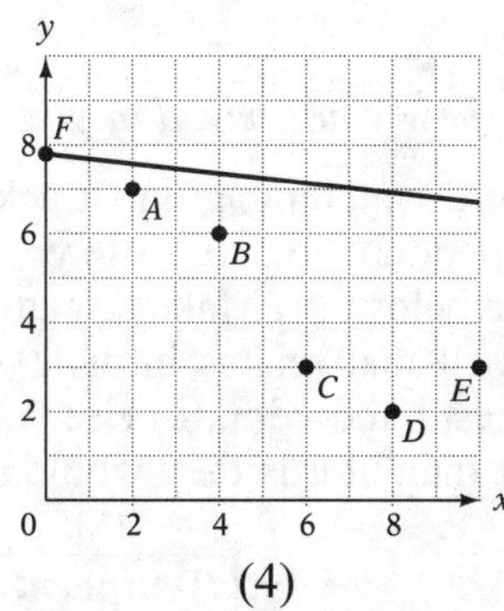

(4)

Chapter Twelve **REGRESSION CURVES**

10. The equation $y = x + 3$ is a line of best fit for which scatter plot?

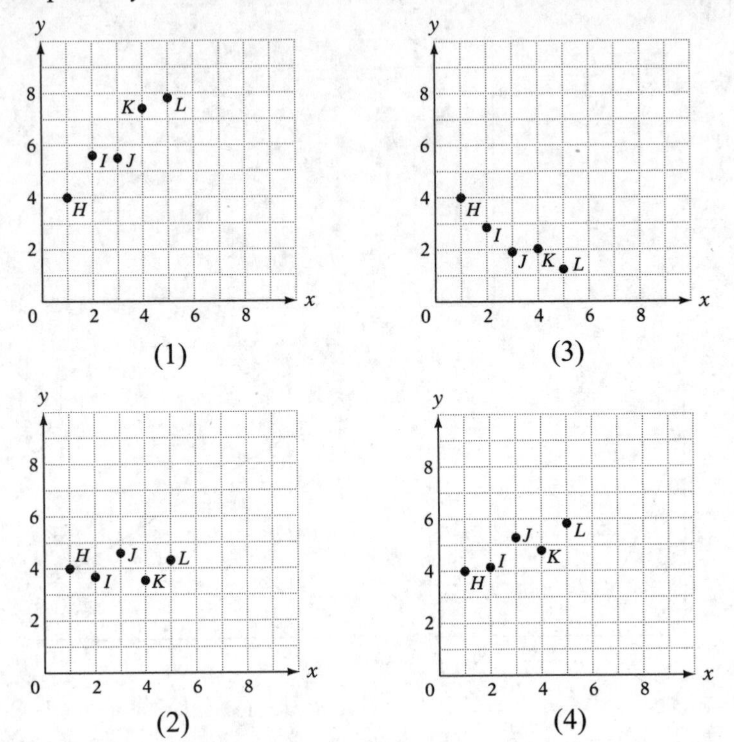

B. Show how you arrived at your answer.

1. A bird-watching group tracks the number of birds they see and the temperature for six different occasions. The data are collected on the table below. (a) Make a scatter plot of the data. (b) Calculate the line of best fit that relates birds (B) to temperature (T). Round values to the nearest hundredth. (c) Use your equation of the line of best fit to predict how many birds they would see if the temperature was 55 degrees.

Temperature	Birds seen
40	30
50	41
60	72
70	91
80	94
90	89

2. A car dealership keeps track of how many cars they sell at different prices. The data are collected on the table below. (a) Make a scatter plot of the data. (b) Calculate the line of best fit that relates cars sold (C) to price (P). Round values to the nearest hundredth. (c) Use your equation of the line of best fit to predict how many cars they would sell at a cost of $31,000.

Price (in thousands of dollars)	Cars sold
25	150
30	127
35	118
40	91
50	65

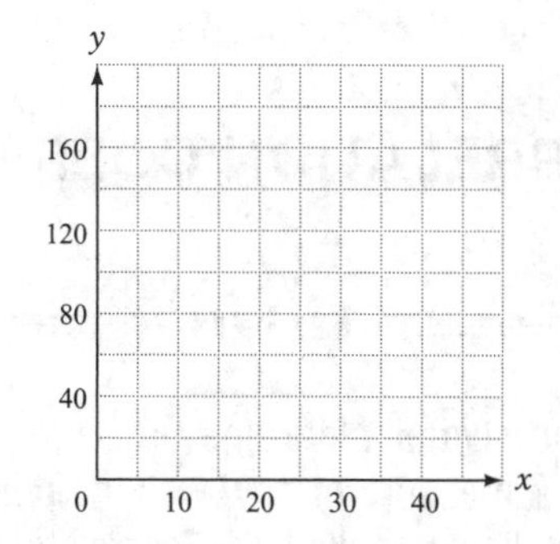

3. How is it possible to find the equation for the line of best fit for the scatter plot below without the use of a calculator?

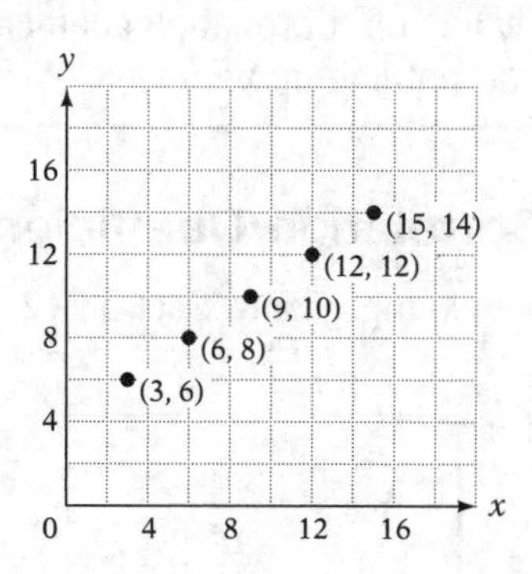

4. The two scatter plots below have all points the same except the one on the left has the point (3, 1) while the one on the right has the point (3, 10). Will they have the same lines of best fit? If not, how will the line of best fit for the scatter plot on the left be different from the line of best fit for the scatter plot on the right?

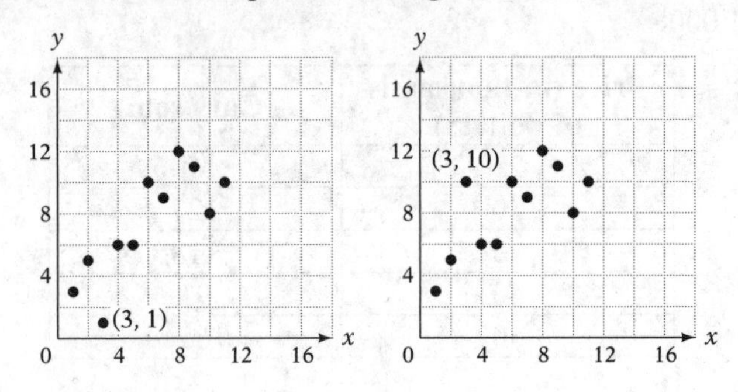

12.2 THE CORRELATION COEFFICIENT

The points on some scatter plots line up better than the points on other scatter plots. A measure of how well the points in a scatter plot resemble a straight line is called the *correlation coefficient* and is denoted by the letter r. The correlation coefficient is a number between −1 and +1. Correlation coefficients very close to −1 or +1 indicate that the points very nearly line up. Correlation coefficients close to 0 indicate that the points do not line up very well.

Calculating the Correlation Coefficient

The temperature / people example from Section 12.1 had a scatter plot that looked like this:

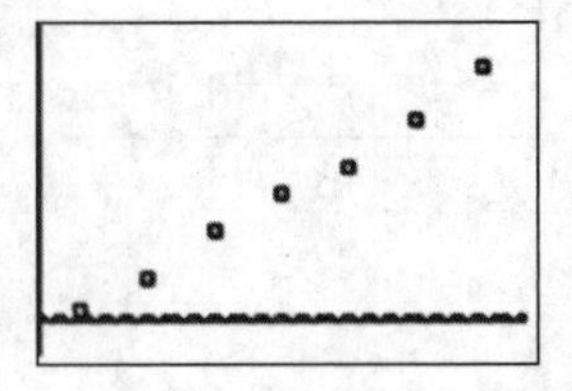

For the TI-84:

To find the correlation coefficient, press [2ND] and then [0] to view the CATALOG. Use the arrows to scroll down to DiagnosticOn and then press [ENTER] and [ENTER] again. This setting will remain until the calculator's memory is reset. If the test proctor has you clear the memory of the calculator, you will have to do this process again.

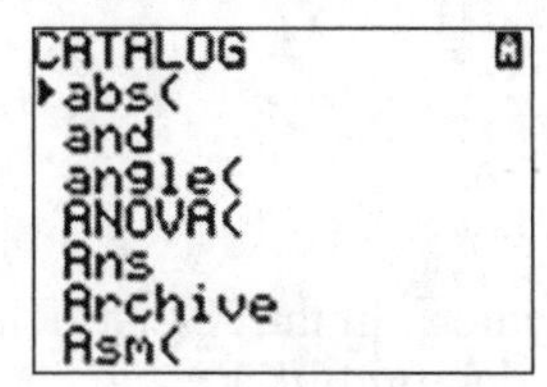

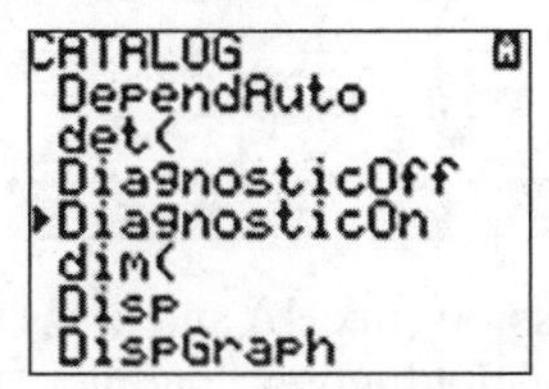

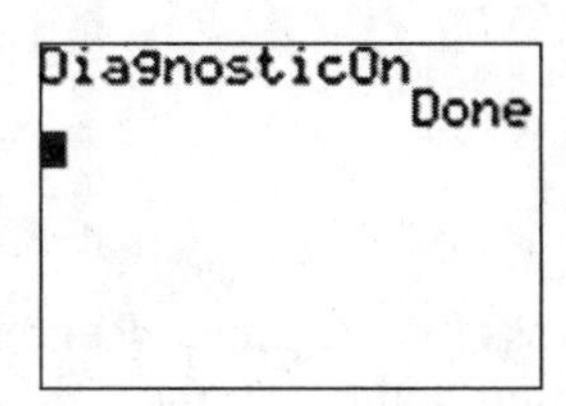

Now when you do the steps from the previous section for determining the line of best fit, the *r* value will be displayed with it.

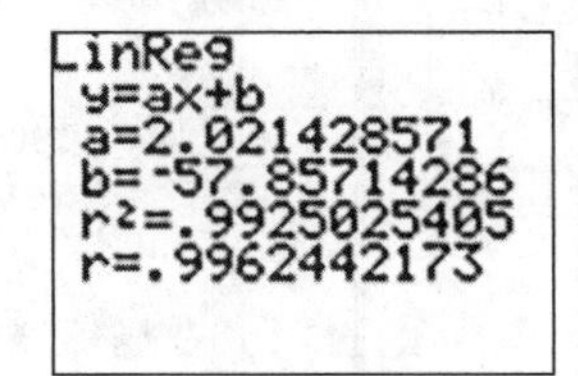

For the TI-Nspire:

From the home screen, select the Add Lists & Spreadsheet icon. Enter the data as you would to determine the line of best fit.

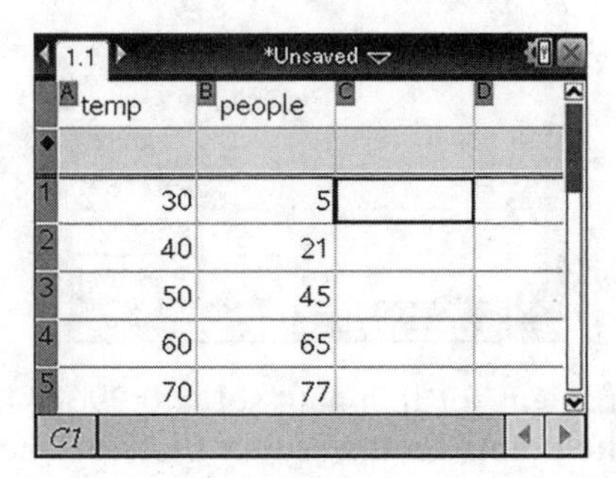

Press [menu], [4], and [1] for Stat Calculations.

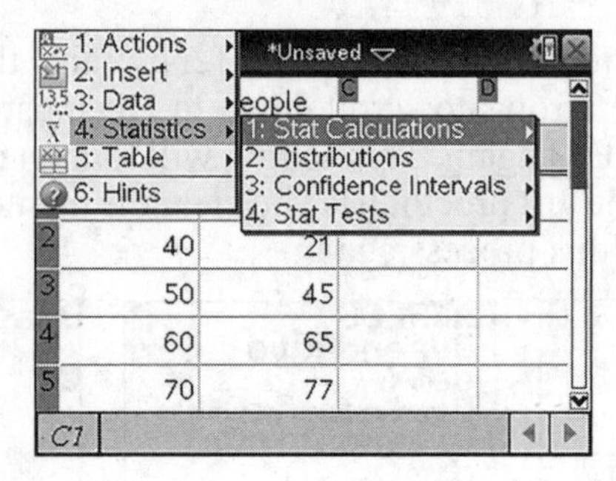

Press [3] for Linear Regression (mx+b) and press [enter]. In the X List field, enter "temp." In the Y List field, enter "people." Click on [OK].

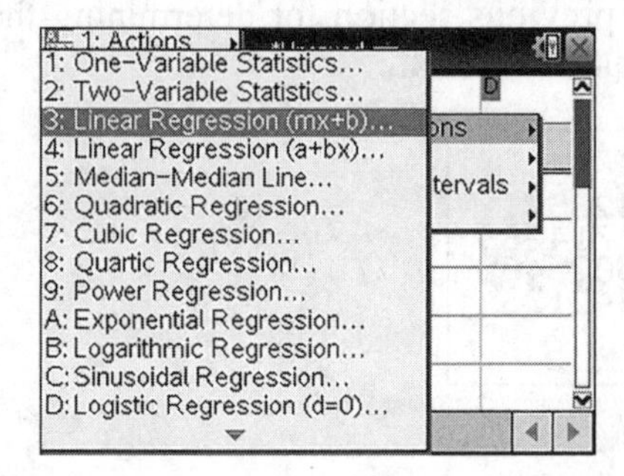

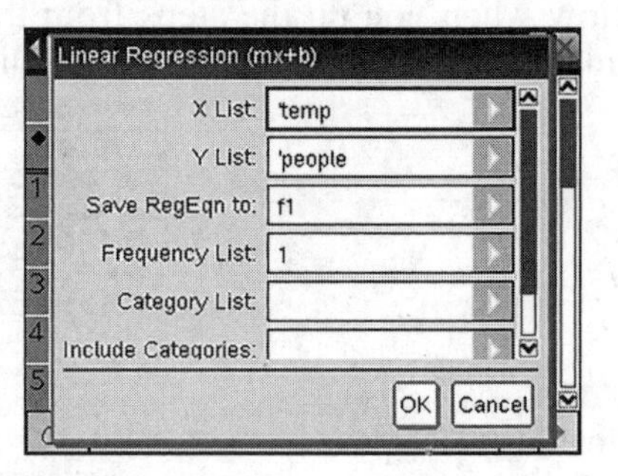

The *r* value of 0.996244 will be in row 6.

	people	C	D	E
◆				=LinRegM
2	21		RegEqn	m*x+b
3	45		m	2.02143
4	65		b	⁻57.8571
5	77		r²	0.992503
6	102		r	0.996244

E6 =0.99624421727882

The correlation coefficient for this data set is 0.9962442173, which is very close to 1 because the points on the scatter plot very nearly line up.

Example 1

What is the correlation coefficient of the line of best fit for the scatter plot below?

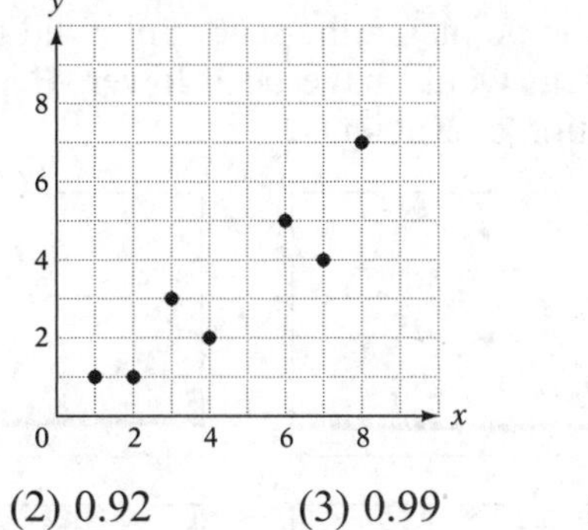

(1) 0.27 (2) 0.92 (3) 0.99 (4) 1

Solution: The r value is approximately 0.92, choice (2).

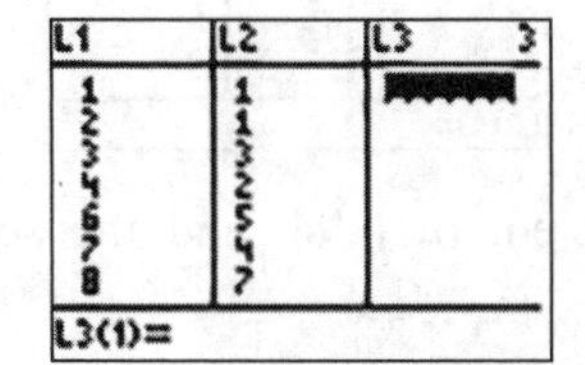

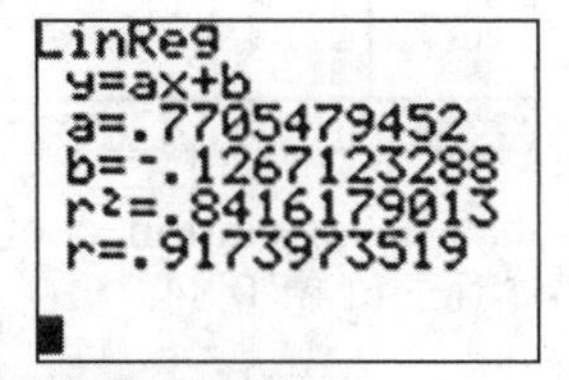

1.1 *Unsaved

	B y		C	D
◆				=LinRegM
2	2	1	RegEqn	m*x+b
3	3	3	m	0.770548
4	4	2	b	-0.126712
5	6	5	r²	0.841618
6	7	4	r	0.917397

D6 =0.91739735192373

Comparing Correlation Coefficients by Looking at the Scatter Plots

The data set from the temperature/people scenario had a correlation coefficient very close to 1. If the points on the scatter plot did not line up so nicely, the correlation coefficient would have been lower. Here are two data sets that have lower correlation coefficients.

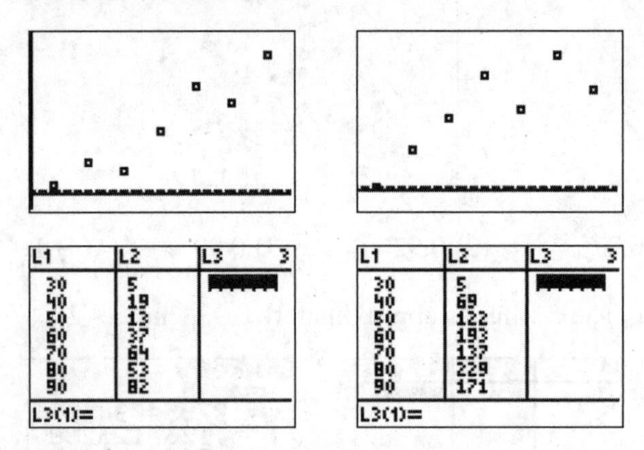

The first set has a correlation coefficient of 0.94, and the second has a correlation coefficient of 0.84.

Example 2

Which scatter plot appears to have the lowest correlation coefficient?

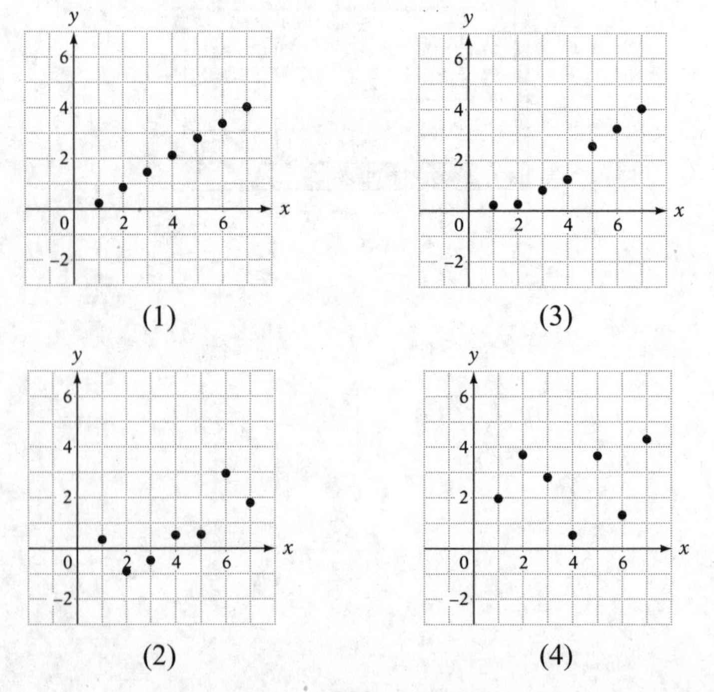

Solution: Since the points on choice (4) do not line up very well, it has the lowest correlation coefficient. The correlation coefficient for choice 4 is about 0.15.

Negative Correlation Coefficients

If the line of best fit has a negative slope, the correlation coefficient will be negative. The more the points line up, the closer the correlation coefficient will be to –1, whereas the less they line up, the closer the correlation coefficient will be to 0.

The scatter plot below has a correlation coefficient of –1.

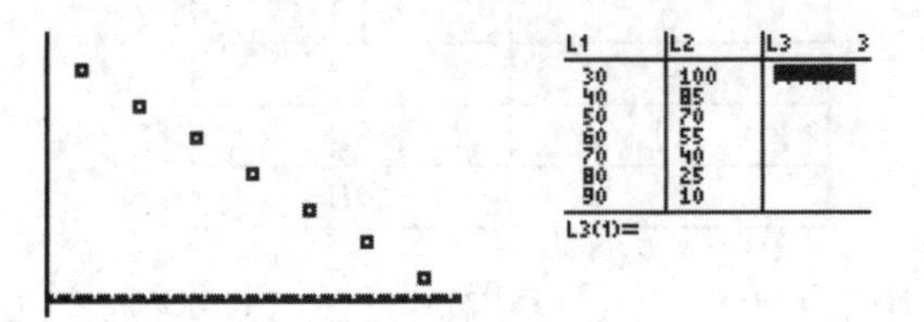

L1	L2	L3 3
30	100	
40	85	
50	70	
60	55	
70	40	
80	25	
90	10	

L3(1)=

The scatter plot below has a correlation coefficient of –0.86.

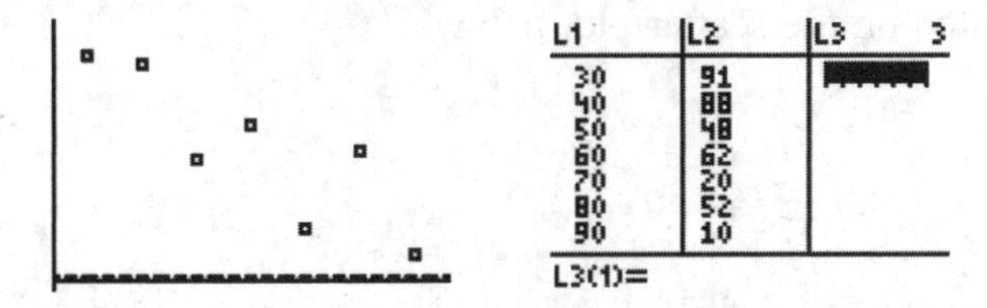

L1	L2	L3 3
30	91	
40	88	
50	48	
60	62	
70	20	
80	52	
90	10	

L3(1)=

In the first example, the correlation coefficient is –1 because all the points land exactly on the line of best fit. In the second example, the correlation coefficient is –0.86, which is closer to 0 because the points are farther away from the line of best fit.

Check Your Understanding of Section 12.2

A. Multiple-Choice

1. What is the correlation coefficient (r), rounded to the nearest hundredth, for the line of best fit for the data on the table below?

x	y
3	10
6	13
9	27
12	38
15	40

(1) 0.97 (2) 0.96 (3) 0.95 (4) 0.94

2. What is the correlation coefficient (r) for the line of best fit for the data represented on the scatter plot below?

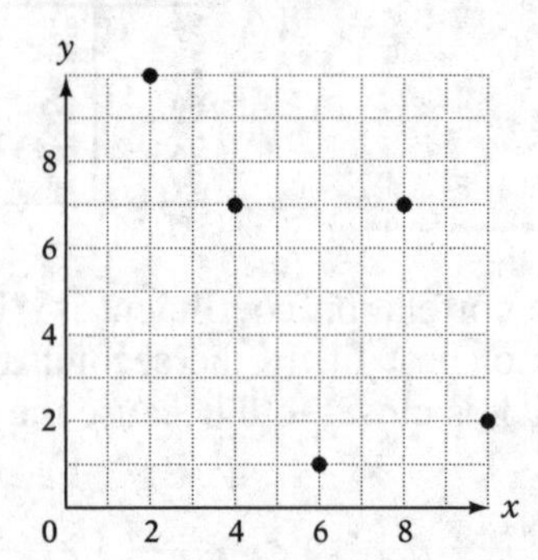

(1) –0.45 (2) –0.52 (3) –0.59 (4) –0.67

3. For which scatter plot is the correlation coefficient closest to 1?

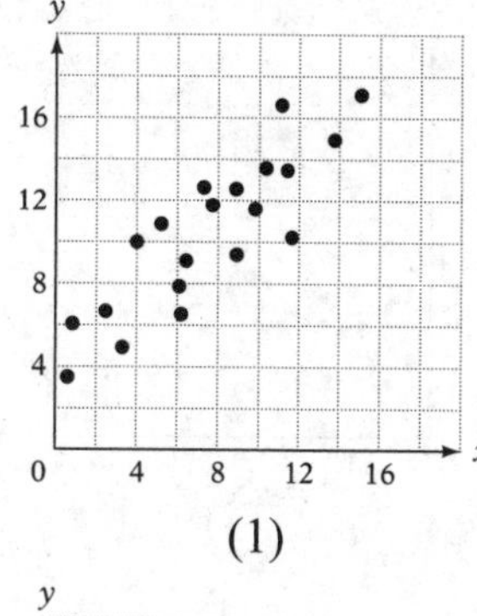

(1)

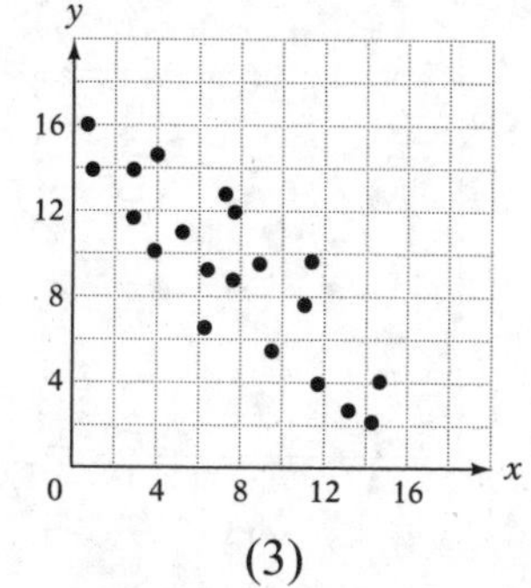

(3)

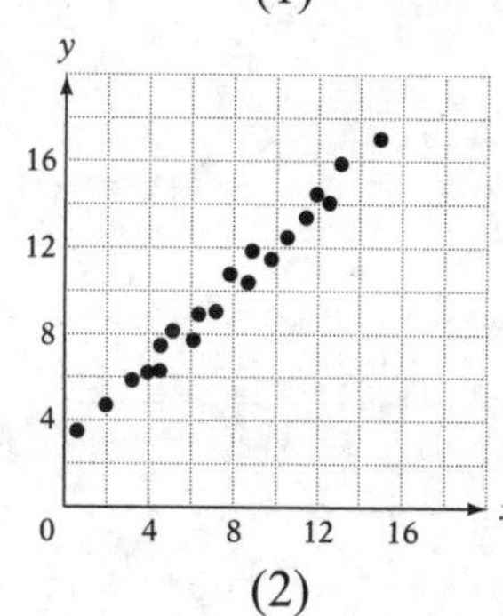

(2)

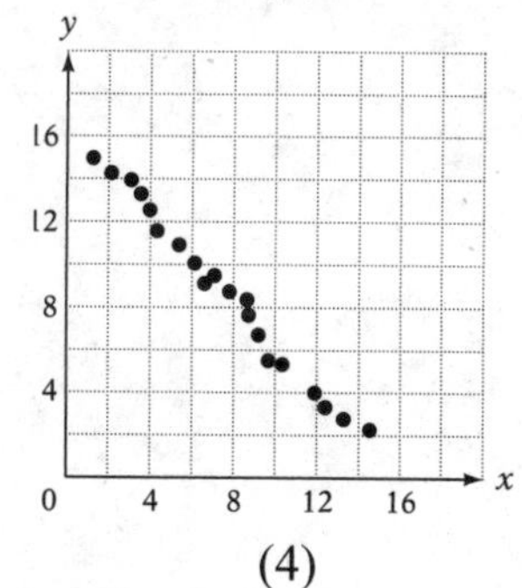

(4)

4. Which scatter plot has a correlation coefficient closest to –1?

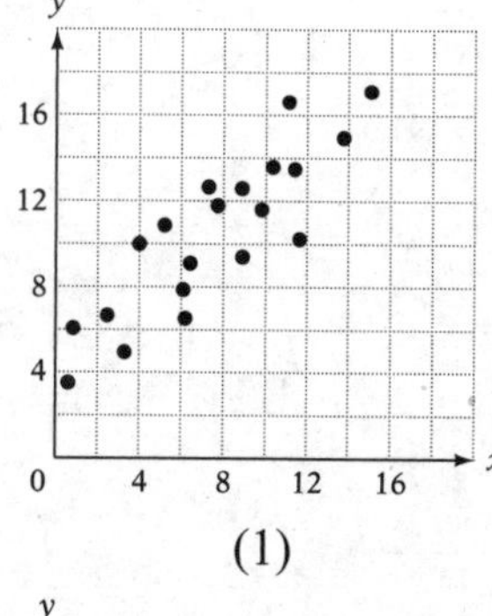

(1)

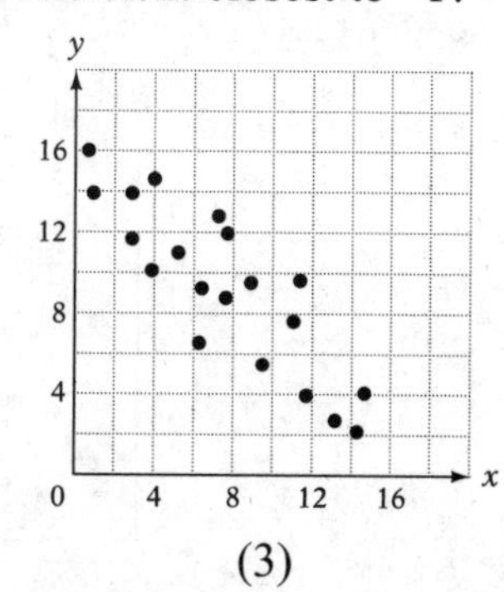

(3)

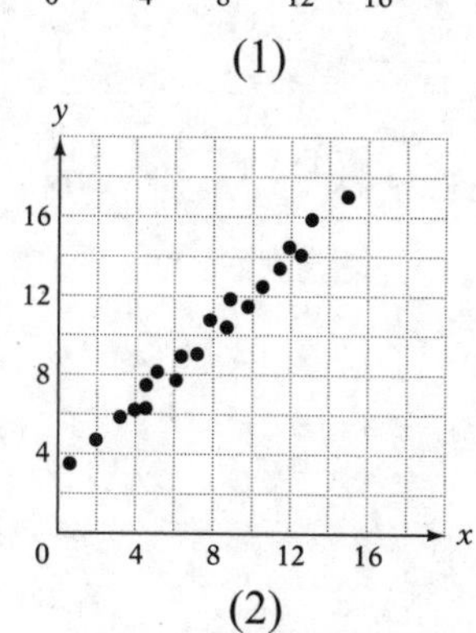

(2)

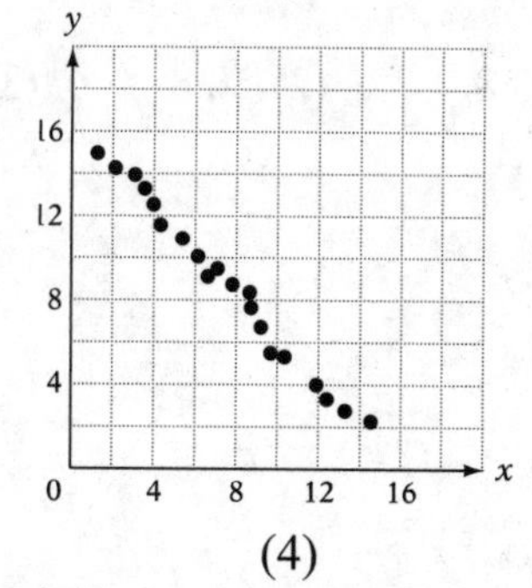

(4)

5. Which scatter plot has a correlation coefficient (r) closest to zero?

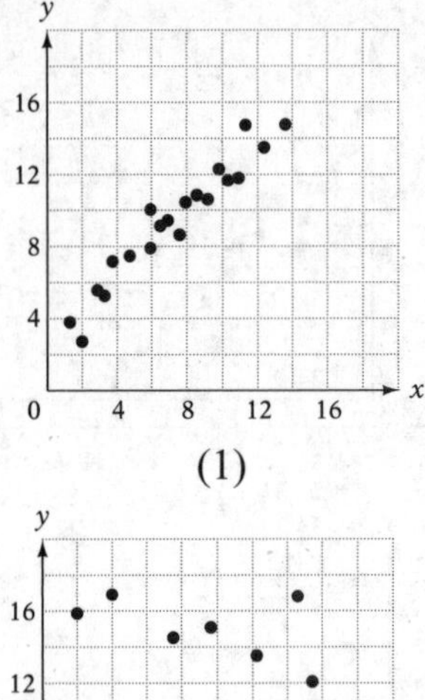

(1)

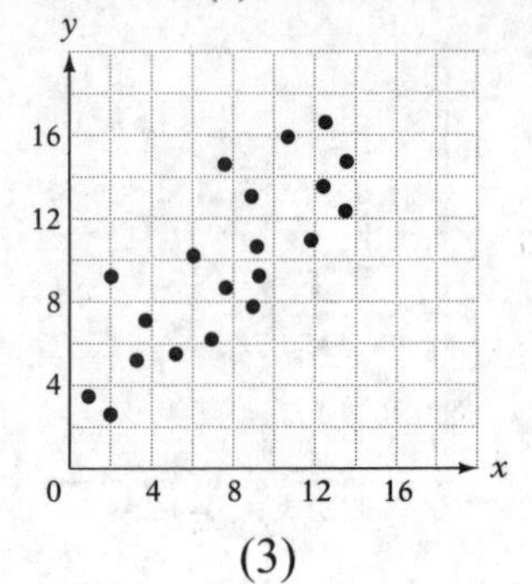

(3)

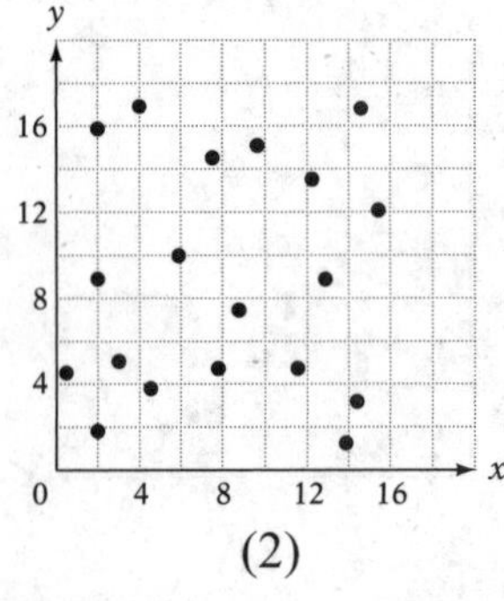

(2)

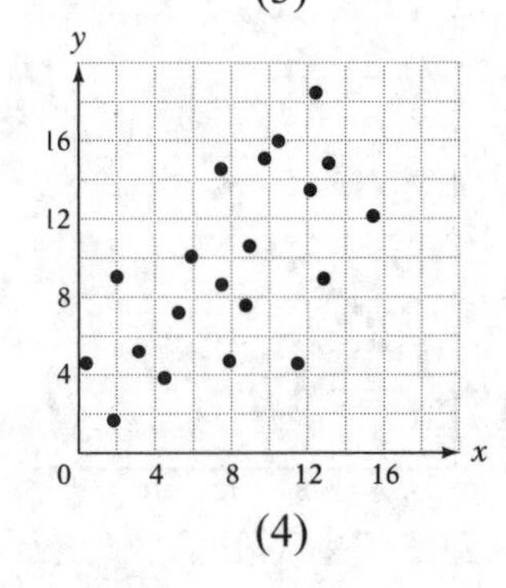

(4)

6. Of the four choices, which is closest to the correlation coefficient for this scatter plot?

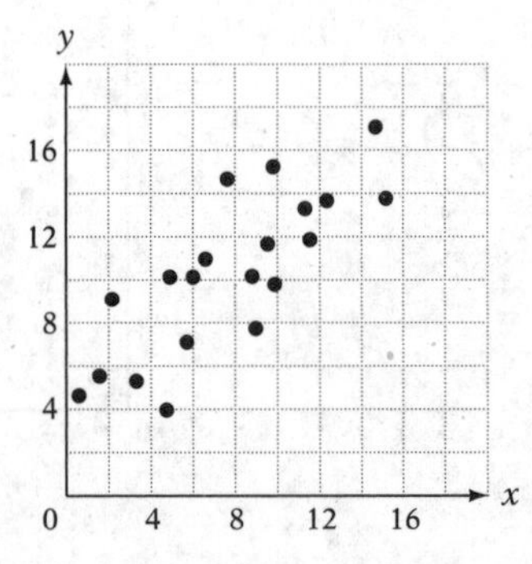

(1) 0.8 (2) –0.8 (3) 1 (4) –1

7. Of the four choices, which is closest to the correlation coefficient for this scatter plot?

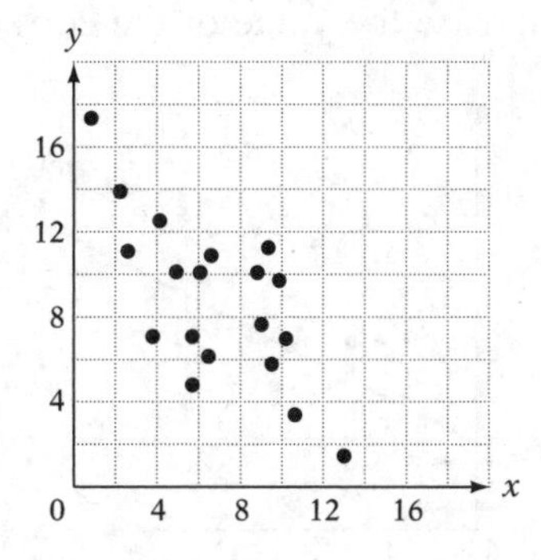

(1) 1 (2) –1 (3) 0.7 (4) –0.7

8. What is the correlation coefficient (r) for the line of best fit for the data represented on the scatter plot below?

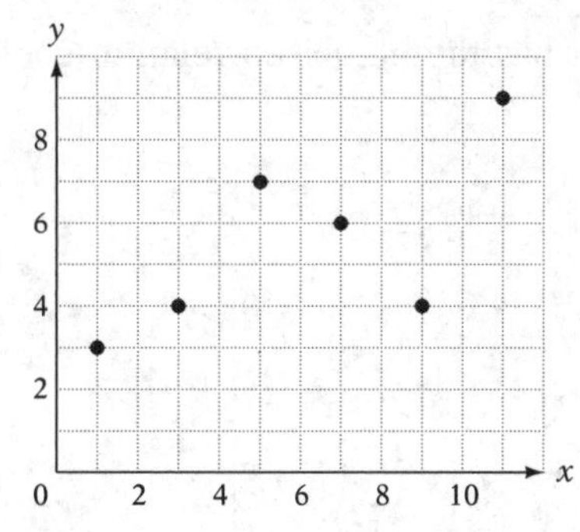

(1) 0.6856 (2) 0.6860 (3) 0.6864 (4) 0.6868

9. What could be the correlation coefficient for the line of best fit for a scatter plot that has only two points?

(1) 0
(2) 1
(3) 0.5
(4) There is not enough information to answer this question.

10. The scatter plot below has a line of best fit with equation $y = 0.5x + 4$ and a correlation coefficient of 0.76. Which point, if added to the scatter plot, would most increase the value of the correlation coefficient?

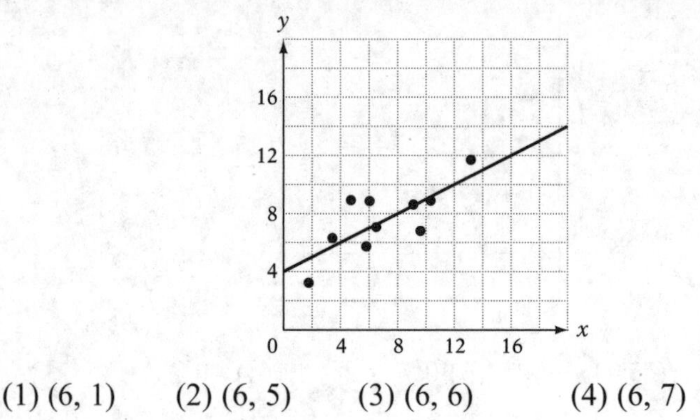

(1) (6, 1) (2) (6, 5) (3) (6, 6) (4) (6, 7)

B. *Show how you arrived at your answers.*

1. Which scatter plot has the higher correlation coefficient? Explain your reasoning.

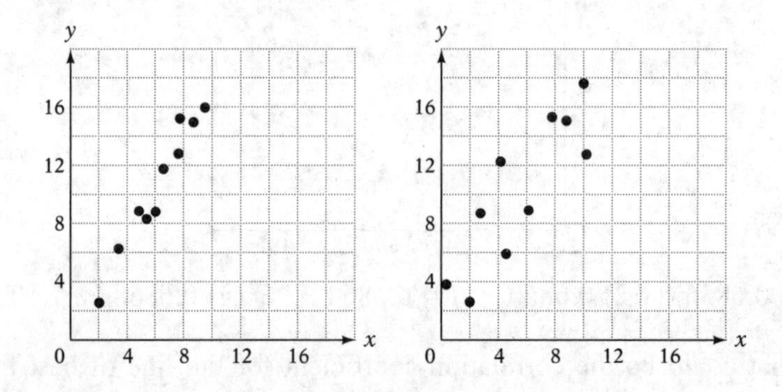

2. Ronald says that a correlation coefficient cannot be negative since completely random placed points have a correlation coefficient of 0, and the points cannot get any more randomly placed. Jett says that it is possible to have a negative correlation coefficient. Who is correct? Explain.

3. The seven points on the scatter plot below all lie on the curve $y = \left(\frac{1}{16}\right)x^2$.

What can be determined about the correlation coefficient of the line of best fit for these seven points without using a calculator?

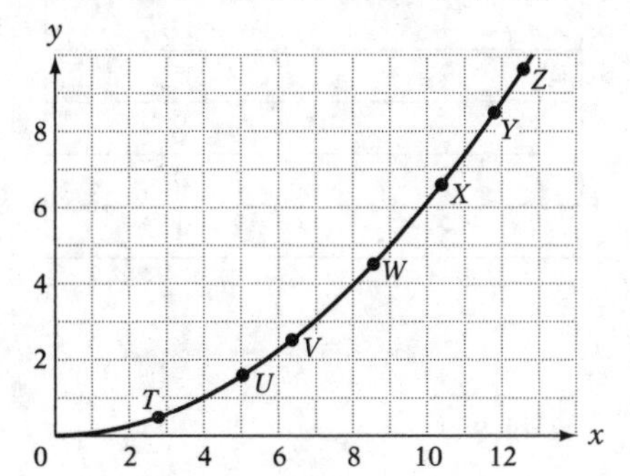

4. Create a five-point scatter plot that has a line of best fit with the equation $y = x + 3$ and a correlation coefficient (r) of $+1$.

12.3 PARABOLAS AND EXPONENTIAL CURVES OF BEST FIT

Not every scatter plot resembles a straight line. When data are plotted from real-world scenarios, sometimes the plot resembles a parabola or an exponential curve. When this happens, the line of best fit is no longer appropriate. Instead the graphing calculator can also be used to find the parabola of best fit or the exponential curve of best fit.

The Parabola of Best Fit

When the scatter plot resembles a parabola, it is possible to find the equation of the parabola of best fit using a process similar to finding the line of best fit. Below is a data set with the scatter plot next to it.

x	y
1	2
2	5
3	11
4	15
5	12
6	8
7	4
8	1

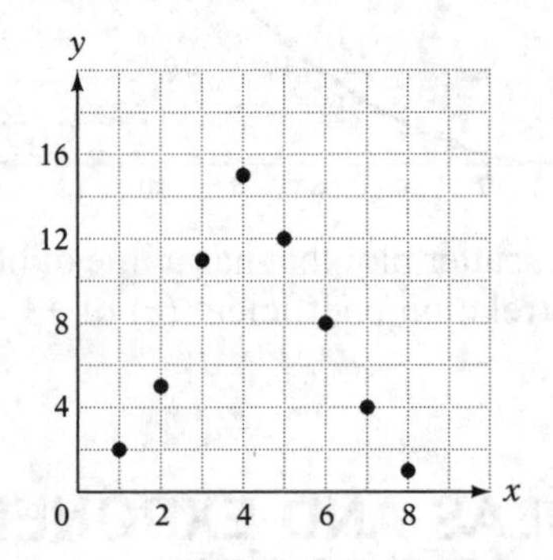

There is a line of best fit for this scatter plot, but it will have a very low correlation coefficient because the points look more like the shape of a parabola than a line. When this happens, it is more appropriate to find the parabola of best fit.

For the TI-84:

Press [STAT] and [1] and enter the x values into L1 and the y values into L2. Press [STAT] and go to the CALC menu. Press [5] for QuadReg.

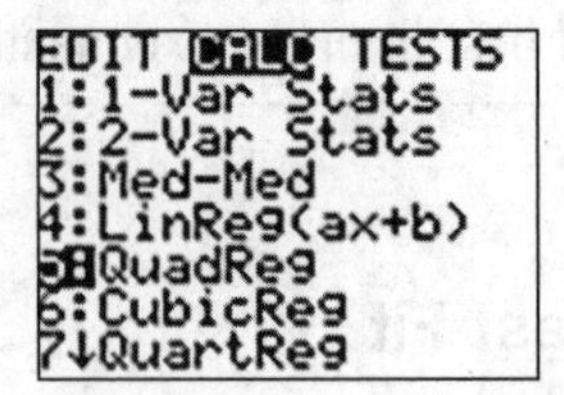

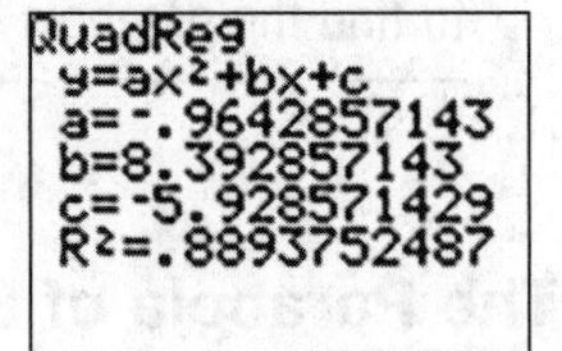

For the TI-Nspire:

From the home screen, select Add Lists & Spreadsheet icon. Enter the data for x into column A and the data for y into column B. Press [menu], [4], [1], and [6] for Quardratic Regression.

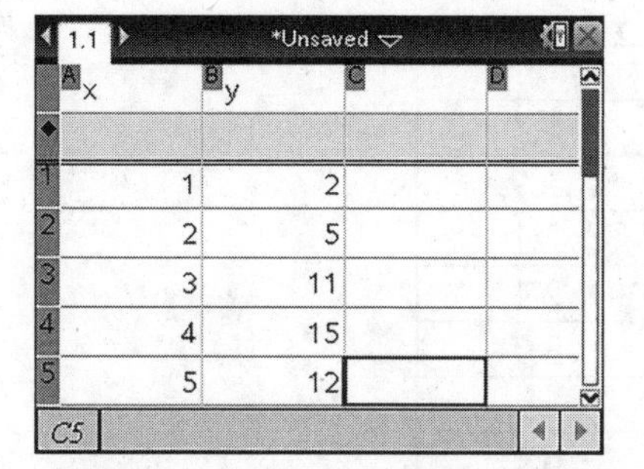

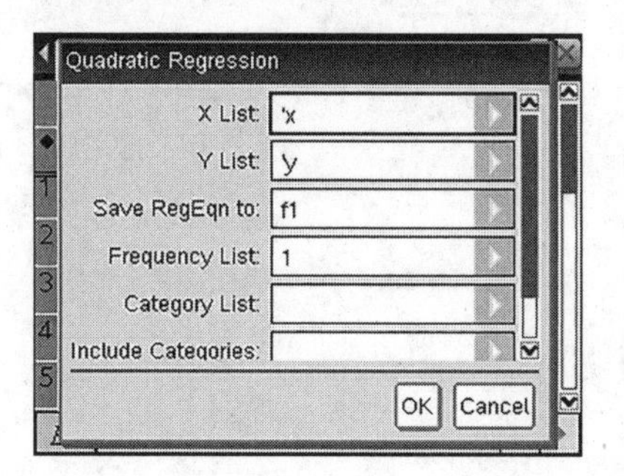

The equation of the parabola of best fit is approximately

$$y = -0.96x^2 + 8.39x - 5.93$$

Here is the graph of the scatter plot and the parabola plot on the same coordinate plane.

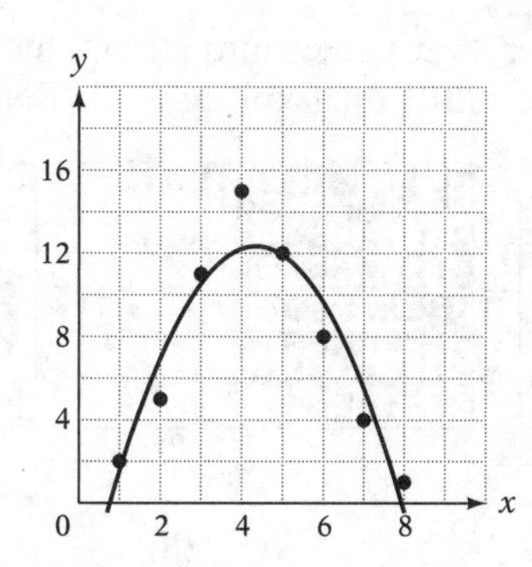

The Exponential Curve of Best Fit

If the scatter plot does not resemble a line or a parabola, it may resemble the playground slide shape of an exponential curve.

Here is a data set with its scatter plot.

x	y
1	1.3
2	1.9
3	2.1
4	2.1
5	3.7
6	4
7	6.7

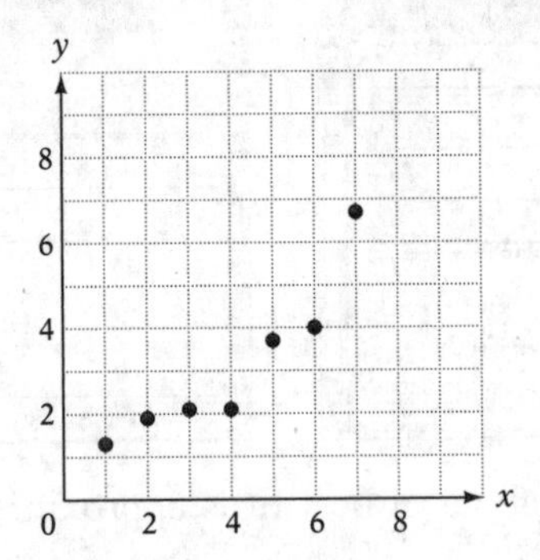

The graphing calculator can quickly find the exponential curve of best fit.

For the TI-84:

Press [STAT], [1] to enter the x values into L1 and the y values into L2. Then press [STAT], [right arrow],[0] for ExpReg, and [ENTER].

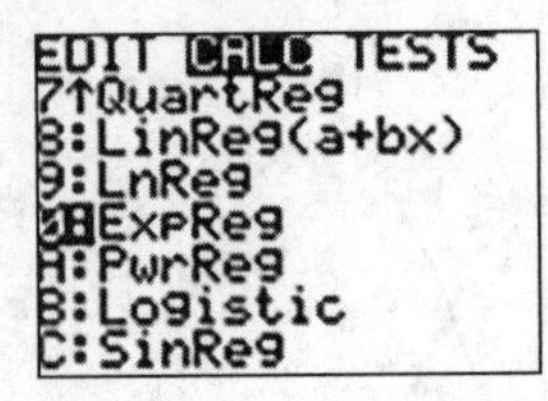

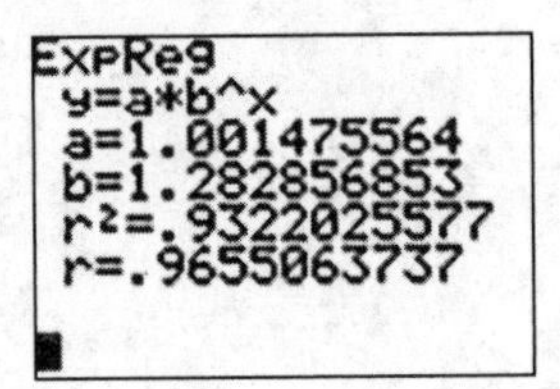

For the TI-Nspire:

From the home screen, select the Add Lists & Spreadsheet icon. Enter the data for x into column A and the data for y into column B. Press [menu], [4], [1], and [A] for Exponential Regression.

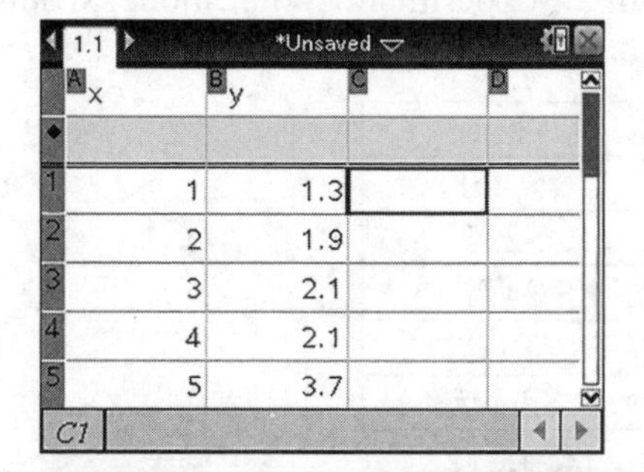

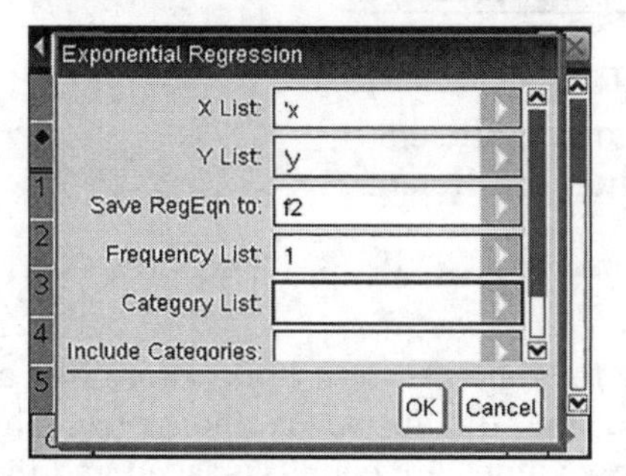

The exponential curve of best fit is approximately $y = 1.0 \cdot 1.28^x$.

Here is the scatter plot with the exponential curve of best fit.

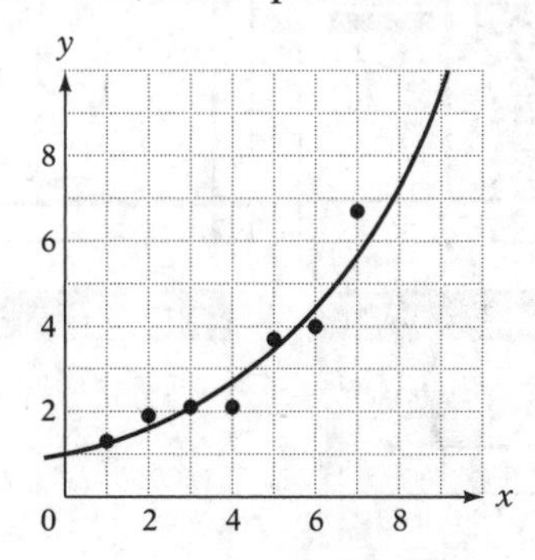

Choosing Which Model Is Most Appropriate

Some questions could indicate which model to use—linear, quadratic (for parabolas), or exponential. Others could ask for you to choose the most appropriate model. By looking at, or by making, the scatter plot, decide if it looks more like a line, a parabola, or an exponential curve. Then choose either linear regression, quadratic regression, or exponential regression.

Example

The following data were collected from an experiment. What model would be most appropriate for determining an equation for this data?

Years	Value
0	\$100
10	\$217
20	\$471
30	\$1,022
40	\$2,217
50	\$4,812
60	\$10,441

(1) Linear with a positive correlation coefficient
(2) Linear with a negative correlation coefficient
(3) Exponential
(4) Quadratic

Solution: Make the scatter plot for the data. Since it looks more like an exponential curve than a parabola or a line, the answer is choice (3). This could also be done by noticing that the x values are equally spaced and that each number in the y column is 2.17 times greater than the number above it.

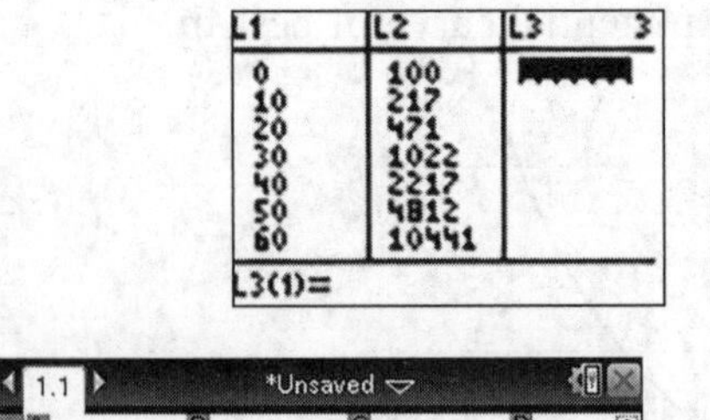

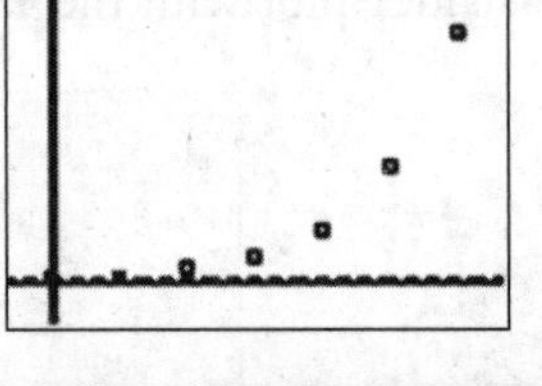

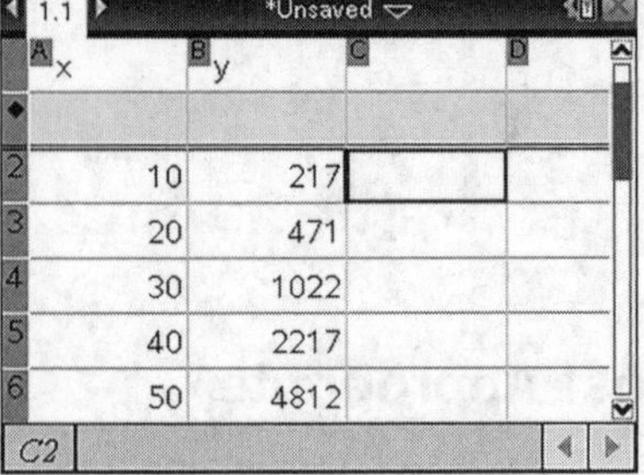

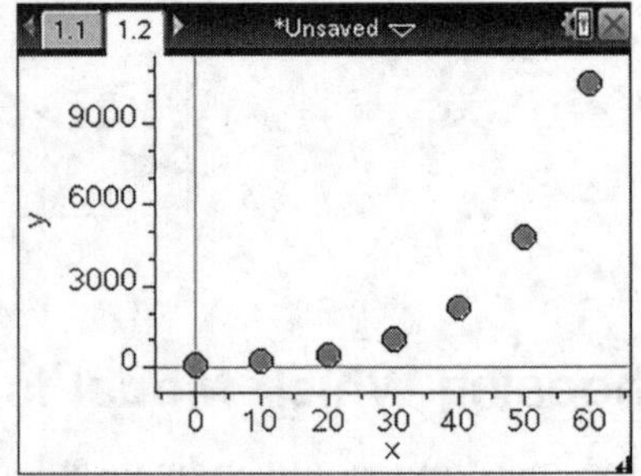

Check Your Understanding of Section 12.3

A. Multiple-Choice

1. Find the equation of the line of best fit for the data on the table below.

x	y
1	2
2	2
3	3
4	4
5	6
6	8
7	10
8	14

(1) $y = 1.27x - 1.73$
(2) $y = -1.73x + 1.27$
(3) $y = 1.68x - 1.43$
(4) $y = -1.43x + 1.68$

2. Find the equation of the parabola of best fit for the data on the table below.

x	y
1	2
2	2
3	3
4	4
5	6
6	8
7	10
8	14

(1) $y = 0.31x^2 - 67x + 3.53$
(2) $y = 0.24x^2 - 0.52x + 2.23$
(3) $y = 0.58x^2 - 0.41x + 4.12$
(4) $y = 0.73x^2 - 0.72x + 6.87$

3. Find the equation of the exponential curve of best fit for the data on the table below.

x	y
1	2
2	2
3	3
4	4
5	6
6	8
7	10
8	14

(1) $y = 1.28 \cdot 1.35^x$
(2) $y = 1.35 \cdot 1.28^x$
(3) $y = 1.47 \cdot 1.39^x$
(4) $y = 1.39 \cdot 1.47^x$

4. For which scatter plot would an exponential curve of best fit be most appropriate?

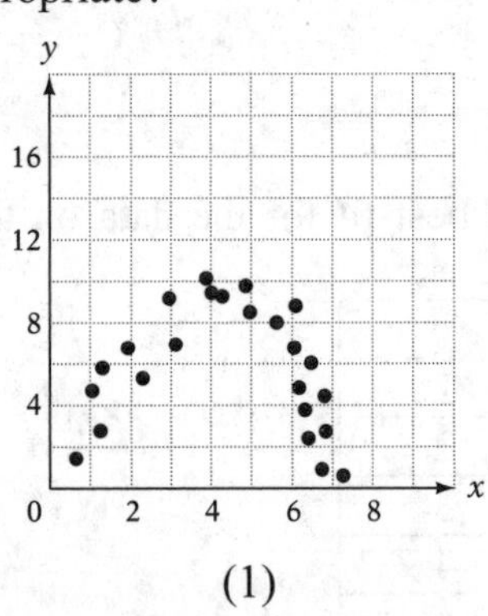

(1)

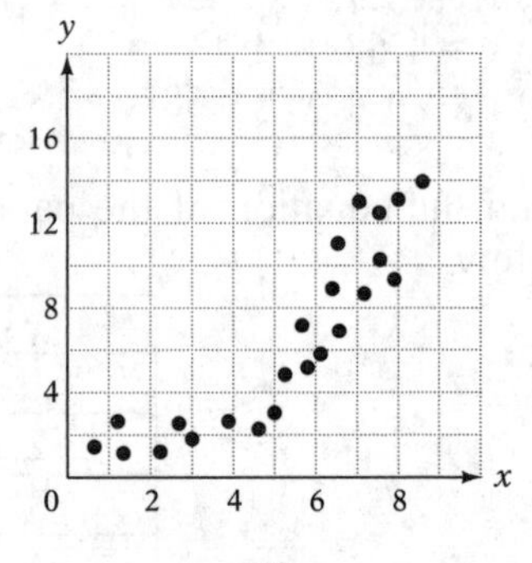

(3)

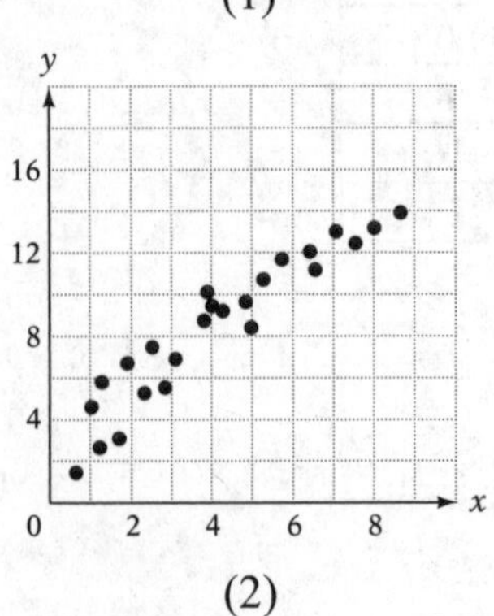

(2)

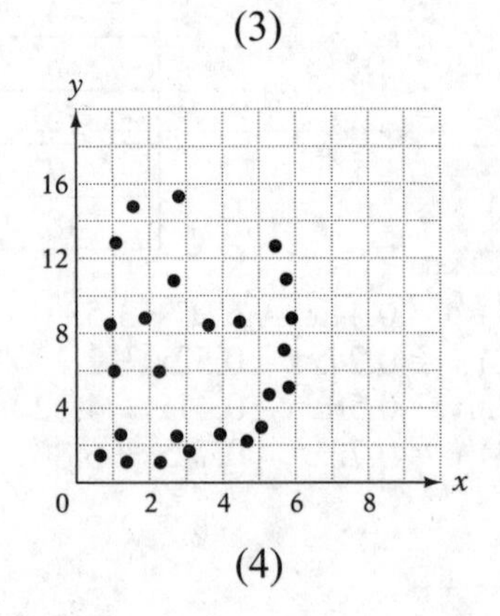

(4)

5. For which scatter plot would a parabola of best fit be most appropriate?

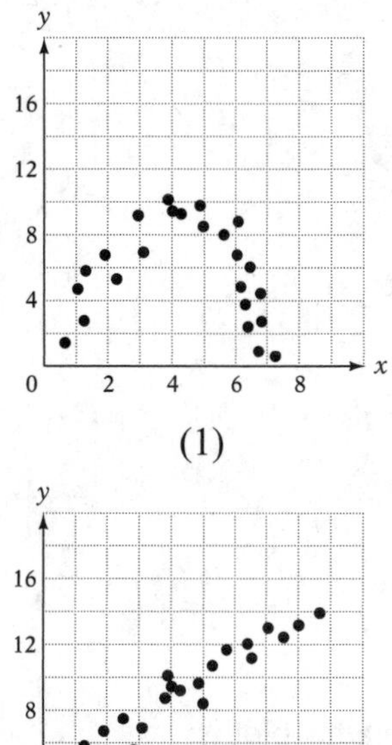

(1)

(2)

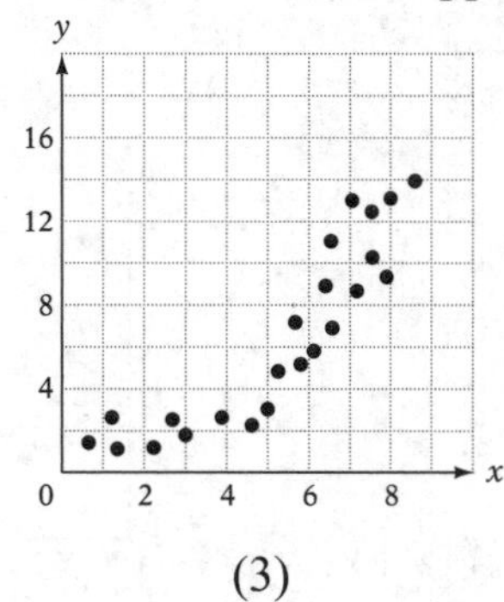

(3)

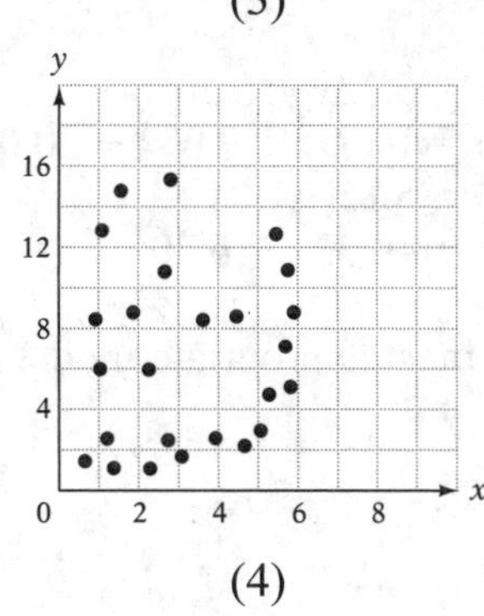

(4)

6. For which scatter plot would a line of best fit be most appropriate?

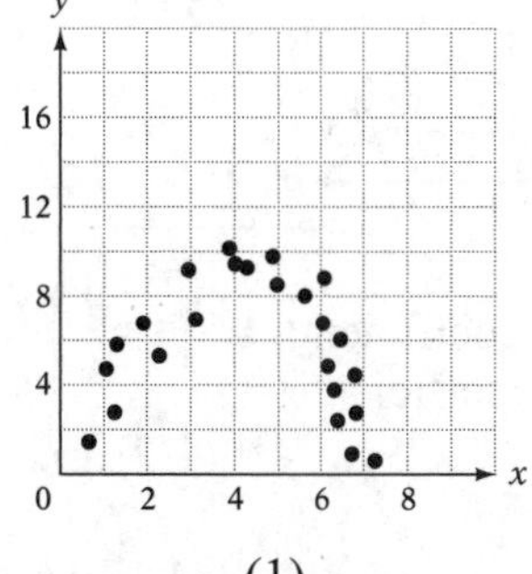

(1)

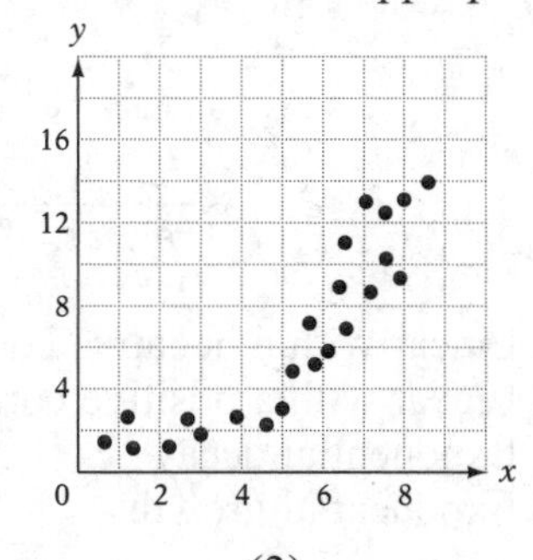

(3)

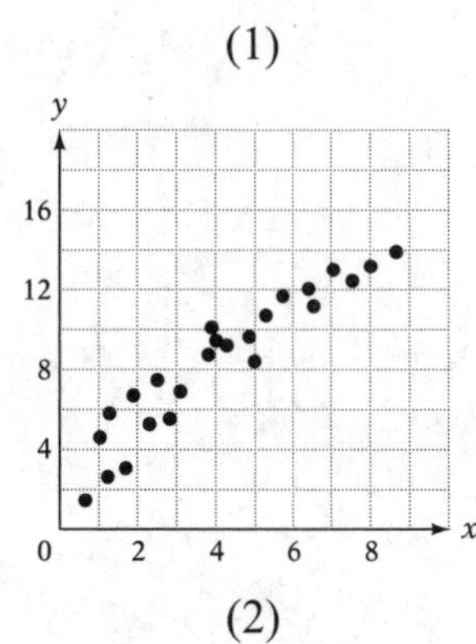

(2)

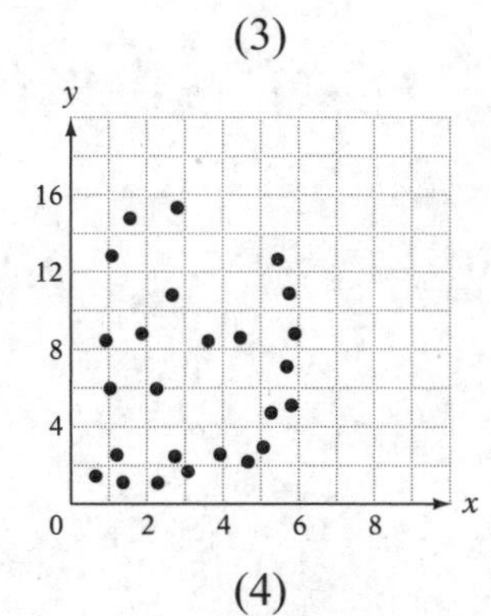

(4)

7. Which equation would be the most accurate model for this scatter plot?

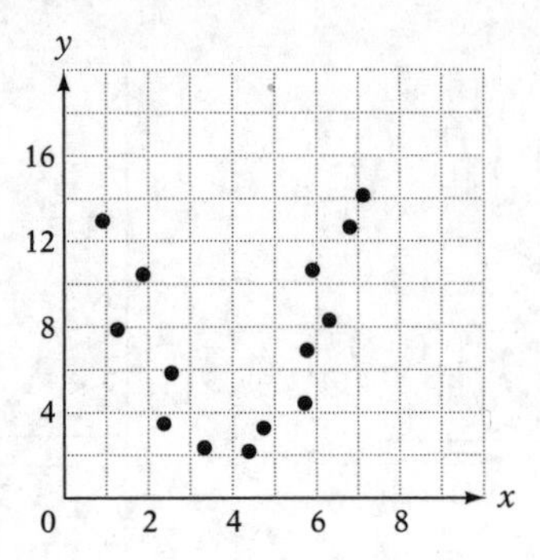

(1) $y = 0.33x + 6.20$
(2) $y = 1.17x^2 - 9.03x + 20.02$
(3) $y = 5.35 \cdot 1.04^x$
(4) $y = -0.33x + 6.20$

8. The most appropriate model for this scatter plot is

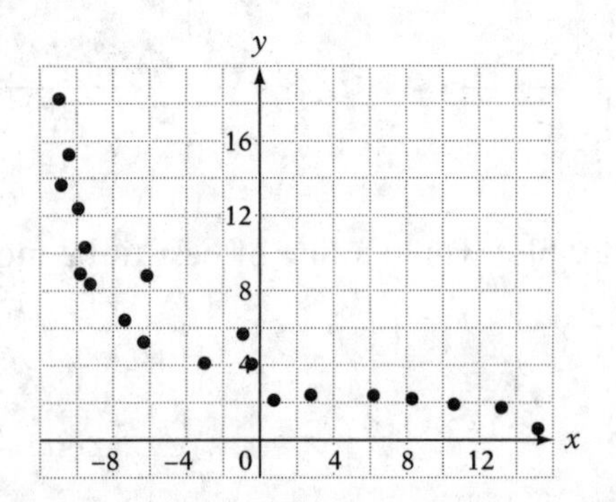

(1) Linear with a negative correlation coefficient
(2) Linear with a positive correlation coefficient
(3) Exponential decay
(4) Exponential growth

9. The height of a ball at different times is graphed on this scatter plot. Which type of equation is most appropriate for modeling this?

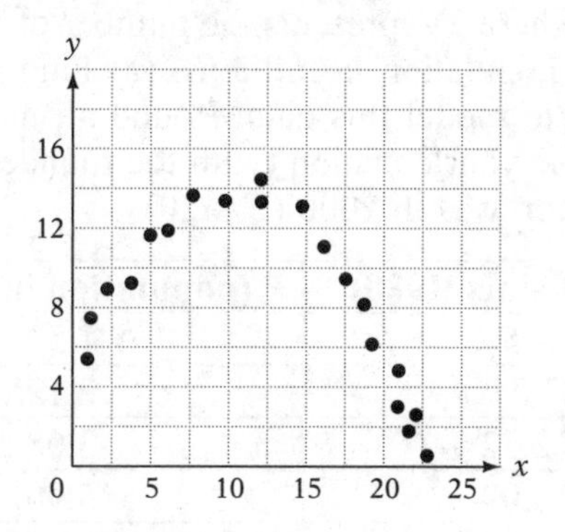

(1) Quadratic with a negative a value
(2) Quadratic with a positive a value
(3) Exponential growth
(4) Exponential decay

10. Which type of equation would be most appropriate for describing the data in this table?

x	y
10	350
20	420
30	490
40	560
50	630
60	700

(1) Decreasing linear function
(2) Increasing linear function
(3) Decreasing exponential function
(4) Increasing exponential function

B. Show how you arrived at your answers.

1. The population of a country at different times was recorded on the following chart where T represents the number of years since 1980 and P represents the population in millions. (a) Find the most appropriate type of equation to model this data. Round all numbers to the nearest hundredth. (b) Use your equation (with the rounded values) to estimate what the population was in 2000 ($T = 20$).

T (years since 1980)	P (population in millions)
1.5	1.59
7.04	2.12
12.32	3.08
16.09	5.90
18.70	8.46
24.17	14.63
27.5	23
30.56	30.32

2. Food is put into a 0 degree freezer. The temperature of the food is recorded at different times after it is put into the freezer on the following chart where M represents the number of minutes since the food was put into the freezer and D represents the temperature of the food in degrees. (a) Find the most appropriate type of equation to model this data. Round all numbers to the nearest hundredth. (b) Use your equation (with the rounded values) to estimate what the temperature of the food will be after 40 minutes.

M (minutes since putting in freezer)	D (temperature in degrees)
3	58
9	31
15	16
20	10
27	5
31	3

3. When a computer company sets the price of a smart phone too low, they don't make a lot of profit. When they set the price too high, few people buy the phone so the company also does not make a lot of profit. Below is a chart showing the amount of profit the company made for six different prices. (a) Decide which model is most appropriate and use it to find an equation to relate the amount of profit they make (F) to the price of the phone (R). Round numbers to the nearest thousandth. (b) Use this equation to determine the price they should charge to make the most amount of profit.

***R* (price of phone in dollars)**	***F* (profit in millions)**
50	2
75	4
100	7
125	8
150	6
175	3

4. A person uses a linear equation to model the data on this scatter plot.

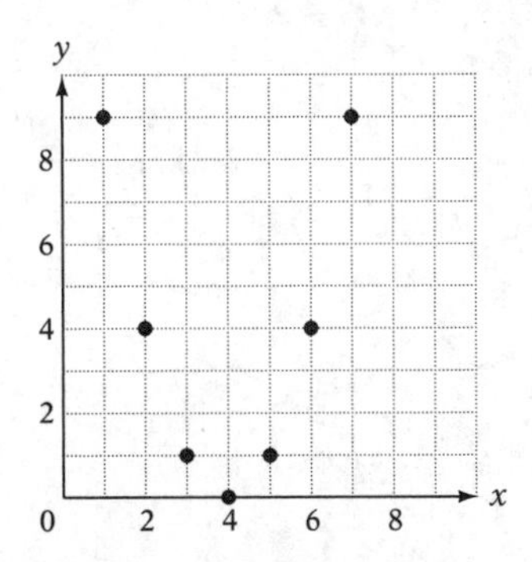

(a) Was a linear model the most appropriate choice? Why or why not?
(b) Find the line of best fit and the correlation coefficient for the data set.

5. Tyler wants to find a curve that best fits the data on this scatter plot and chart.

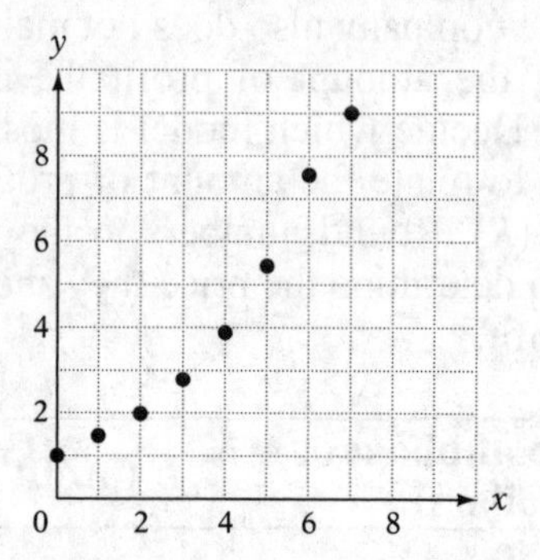

Since it looks like half a parabola and also like an exponential curve, he is not sure whether a quadratic or an exponential model is best. How can he use the calculator to determine which is a more accurate type of equation?

Chapter Thirteen

STATISTICS

13.1 MEASURES OF CENTRAL TENDENCY

A *measure of central tendency* is a number that summarizes a list of numbers. Since one number cannot describe everything someone might want to know about a set of data, there are many different measures of central tendency.

Mean

The *mean* of a data set is also known as the arithmetic mean or the average. To calculate the mean, add up all the numbers in the data set and divide that result by the number of numbers in the set. The symbol for mean is $\bar{x}$.

This is a set of scores on 11 quizzes 70, 80, 73, 91, 75, 83, 85, 96, 85, 79, 85. To calculate the mean, add the 11 numbers and divide the result by 11.

$$\bar{x} = (70 + 80 + 73 + 91 + 75 + 83 + 85 + 96 + 85 + 79 + 85) \div 11 = 902 \div 11 = 82$$

The mean is 82, and it is a single number that gives information about the eleven numbers.

Example 1

On the first four quizzes, Jose gets 85, 88, 90, and 84. What must he get on the fifth quiz to have a mean of 88 for the five quizzes?

Solution: If the unknown quiz grade is called x, the sum of the five quizzes would be $85 + 88 + 90 + 84 + x$. When divided by 5, it needs to equal 88. This becomes the equation

$$\frac{85+88+90+84+x}{5} = 88$$

$$\frac{347+x}{5} = 88$$

$$5\cdot\frac{347+x}{5} = 5\cdot 88$$

$$347 + x = 440$$

$$-347 = -347$$

$$x = 93$$

Mode

The *mode* of a data set is the most "popular" number. Of the eleven numbers, there are three 85s but only one of each other number. The mode of this data set, therefore, is 85.

Median

The median is the "middle" number after the numbers have been arranged from least to greatest. For the data set of the eleven quizzes, the numbers arranged from least to greatest are 70, 73, 75, 79, 80, 83, 85, 85, 85, 91, 96. In this case, 83 is the median because there are five numbers greater than 83 and five numbers less than 83.

One way to locate the median is to cross off the first and the last number of the list to make it

~~70~~, 73, 75, 79, 80, 83, 85, 85, 85, 91, ~~96~~

Then cross off the second and the second to last number of this smaller list to get

~~70~~, ~~73~~, 75, 79, 80, 83, 85, 85, 85, ~~91~~, ~~96~~

Keep doing this until you have either one or two numbers left in the middle.

~~70~~, ~~73~~, ~~75~~, ~~79~~, ~~80~~, 83, ~~85~~, ~~85~~, ~~85~~, ~~91~~, ~~96~~

The median for this data set is 83.

If the list has an even number of numbers, add the two middle numbers and divide by 2. If the data set was six numbers, which when arranged in order were 70, 73, 75, 79, 80, and 83, after crossing off the first and last number twice, the list would look like this

~~70~~, ~~73~~, 75, 79, ~~80~~, ~~83~~

There are two numbers left, 75 and 79. Add them and divide by 2 to get the median.

$$(75 + 79) \div 2 = 154 \div 2 = 77$$

Quartiles

The median is also called the *second quartile* since it is the number that is greater than 50% of the numbers in the set. The *first quartile* is the number that is greater than 25% of the numbers on the list. The *third quartile* is the number that is greater than 75% of the numbers on the list.

To calculate the first and third quartile, find the median of the list. Make a list of all the numbers less than the median and another list of all the numbers greater than the median. The medians of these two lists are the first quartile and the third quartile.

With the 11 quiz scores arranged in order from least to greatest, the median was 83.

70, 73, 75, 79, 80, 83, 85, 85, 85, 91, 96

The five numbers less than 83 are 70, 73, 75, 79, 80.

Since the median of these five numbers is 75, 75 is the first quartile.

The five numbers greater than 83 are 85, 85, 85, 91, 96. Since the median of these five numbers is 85, 85 is the third quartile.

Example 2

Find the mean, first quartile, and third quartile of the fifteen numbers

16, 19, 20, 13, 13, 11, 10, 17, 10, 14, 13, 20, 10, 19, 16

Solution: Arrange the numbers from least to greatest

10, 10, 10, 11, 13, 13, 13, 14, 16, 16, 17, 19, 19, 20, 20

The median of the set is 14.

There are seven numbers less than the median. They are 10, 10, 10, 11, 13, 13, 13. Since the median of those seven numbers is 11, the first quartile is 11.

There are seven numbers greater than the median. They are 16, 16, 17, 19, 19, 20, 20. Since the median of those seven numbers is 19, the third quartile is 19.

Interquartile Range

The difference between the third quartile and the first quartile is called the *interquartile range*. For some data sets this number could be small and for others it could be large. In the example with the eleven quizzes, the difference between the third quartile and the first quartile is 85 – 75 = 10.

Example 3

What is the interquartile range of the data from Example 2?

Solution: Since the third quartile was 19 and the first quartile was 11, the interquartile range is 19 – 11 = 8.

Using the Graphing Calculator to Determine Maximum, Minimum, Median, First Quartile, and Third Quartile

For the TI-84:

The graphing calculator can calculate the five measure of central tendency. First enter all the numbers into L1 by pressing [STAT] and [1] for Edit.

To find the minimum, first quartile, median, third quartile, and maximum of the seven numbers 10, 4, 8, 12, 6, 16, 14, enter them into L1. Then press [STAT] and [1] for 1-Var Stats and press [ENTER].

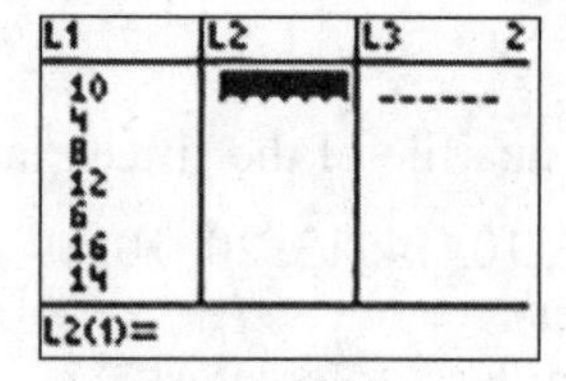

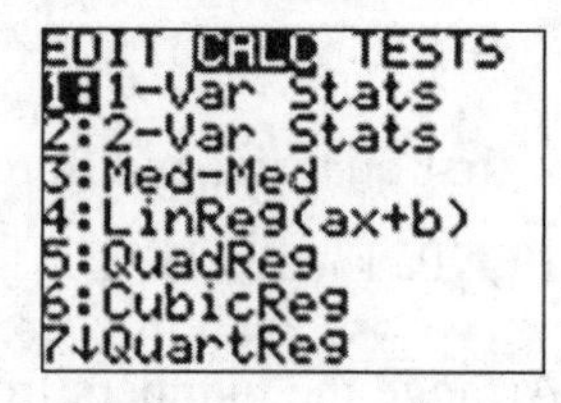

The screen will display

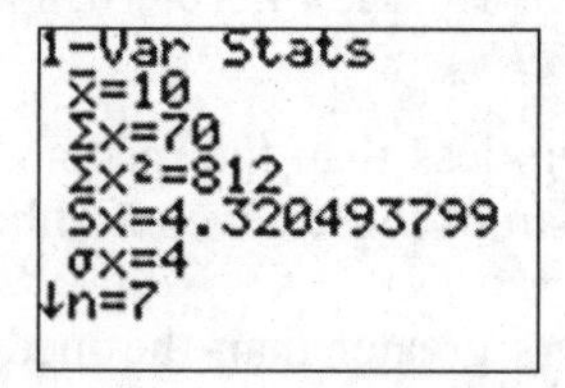

The $\bar{x} = 10$ is for the mean. The $n = 7$ means that there were seven elements in the list. For the minimum, first quartile, median, third quartile, and maximum, press the down arrow five times.

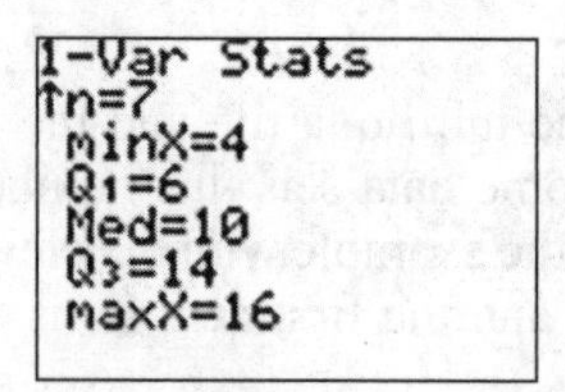

minX is for the minimum, Q1 is for the first quartile, Med is for the median, Q3 is for the third quartile, and maxX is for the maximum element.

For the TI-Nspire:

From the home screen, select the Add Lists & Spreadsheet icon. Name column A *x* and fill in cells A1 through A7 with the numbers 10, 4, 8, 12, 6, 16, 14.

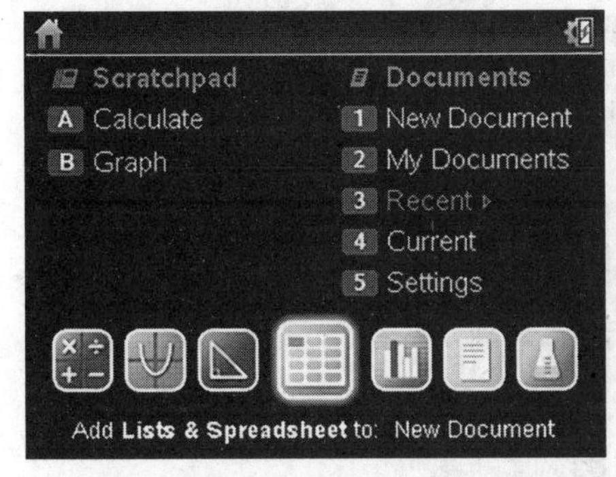

Press [4], [1], [and 1] for One-Variable Statistics.

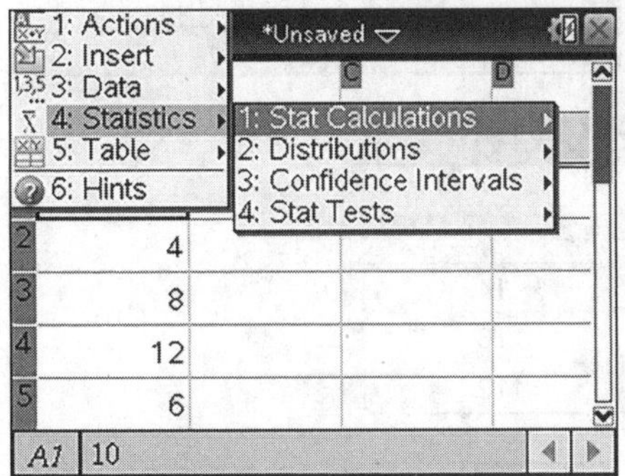

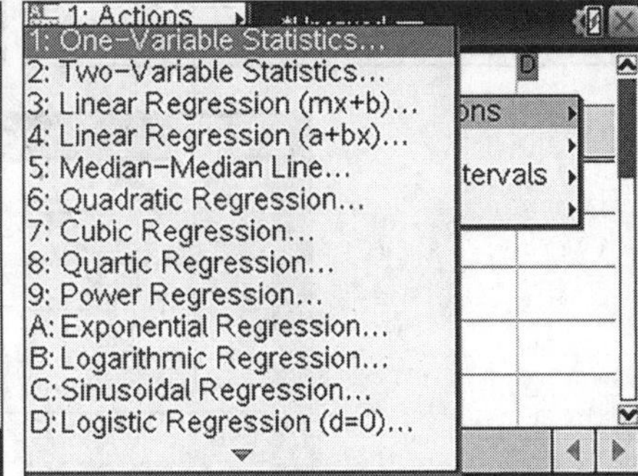

Press [OK] since there is just one list. Set the X1 List to *x* since that was what the column with the data was named in the spreadsheet.

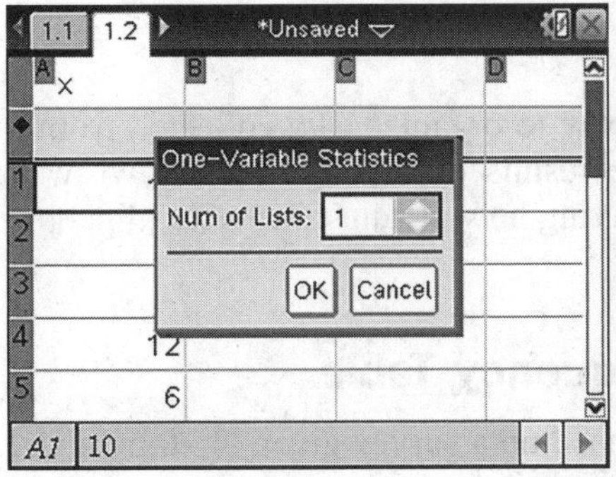

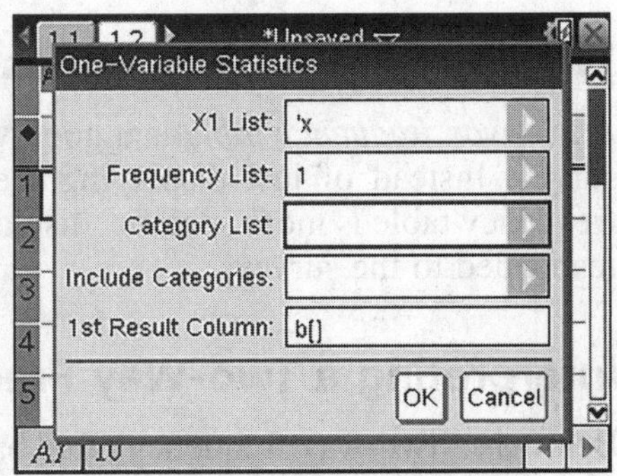

In cells C2 through C13, the one-variable statistics will be displayed. The median is the $\bar{x}$. n is for the number of numbers. MinX is the smallest number. Q1X is the first quartile. MedianX is the median. Q3X is the third quartile. MaxX is the largest number.

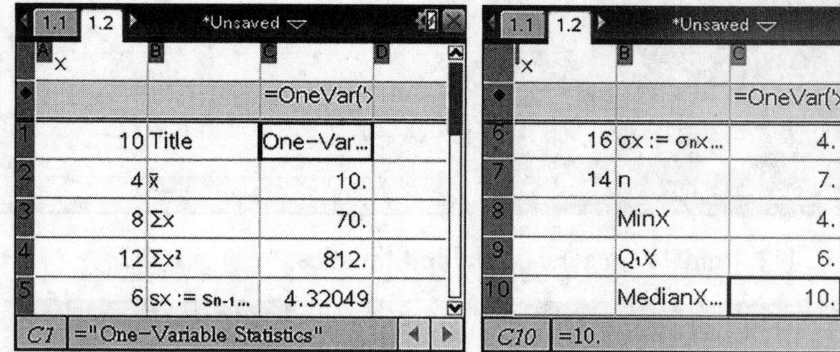

Two-Way Frequency Tables

A *two-way frequency table* is a good way to organize data collected from a survey. Instead of just displaying the results of the survey, the two-way frequency table is more specific, displaying how certain categories of people responded to the survey.

Interpreting a Two-Way Frequency Table

Below is a two-way frequency table based on a survey given to 40 men and 60 women about whether or not math is their favorite subject.

	YES	NO	TOTAL
MEN	32	8	40
WOMEN	54	6	60
TOTAL	86	14	100

From this two-way frequency table, there are nine pieces of information:

100 people took the survey.
Of the 100 people, 40 were men and 60 were women.
Of the 100 people, 86 said "yes" and 14 said "no."
32 men said "yes."
8 men said "no."
54 women said "yes."
6 women said "no."

Creating a Two-Way Frequency Table

If a two-question survey is given to a group of people, the results can be collected on a two-way frequency table. Below are 12 surveys. Create a two-way frequency table based on these 12 surveys.

Circle one answer for each question. Grade: **9**, 10, 11 Course: **Algebra**, Geometry	Circle one answer for each question. Grade: **9**, 10, 11 Course: **Algebra**, Geometry	Circle one answer for each question. Grade: 9, **10**, 11 Course: Algebra, **Geometry**
Circle one answer for each question. Grade: 9, **10**, 11 Course: Algebra, **Geometry**	Circle one answer for each question. Grade: **9**, 10, 11 Course: **Algebra**, Geometry	Circle one answer for each question. Grade: 9, 10, **11** Course: **Algebra**, Geometry
Circle one answer for each question. Grade: 9, **10**, 11 Course: **Algebra**, Geometry	Circle one answer for each question. Grade: 9, 10, **11** Course: Algebra, **Geometry**	Circle one answer for each question. Grade: **9**, 10, 11 Course: **Algebra**, Geometry
Circle one answer for each question. Grade: 9, **10**, 11 Course: Algebra, **Geometry**	Circle one answer for each question. Grade: **9**, 10, 11 Course: Algebra, **Geometry**	Circle one answer for each question. Grade: 9, 10, **11** Course: Algebra, **Geometry**

There needs to be a Total row and column and also a row and column for the labels, so for this data set, make five rows and four columns.

	Algebra	Geometry	Total
9th			
10th			
11th			
Total			

Tally the numbers from each combination 9, 10, 11 with Algebra, Geometry and fill in those numbers first.

	Algebra	Geometry	Total
9th	4	1	
10th	1	3	
11th	1	2	
Total			

Make sure these numbers add up to 12: 4 + 1 + 1 + 3 + 1 + 2 = 12.

Add each row and put the result into the total row. Add each column and put the result into the total column.

	Algebra	Geometry	Total
9th	4	1	5
10th	1	3	4
11th	1	2	3
Total	6	6	

Add up the numbers in the total row and in the total column. They should get the same answer. In this example, 6 + 6 = 12 and 5 + 4 + 3 = 12. Put this nuhmber in the bottom right cell.

	Algebra	Geometry	Total
9th	4	1	5
10th	1	3	4
11th	1	2	3
Total	6	6	12

Creating a Two-Way Relative Frequency Table

A relative two-way frequency table is much like a regular two-way frequency table except that all the numbers are recorded as percentages of the total. To convert the table from the last section into a relative two-way frequency table, divide each number by the total, which is 12.

	Algebra	**Geometry**	**Total**
9th	$\frac{4}{12} = 0.33 = 33\%$	$\frac{1}{12} = 0.08 = 8\%$	$\frac{5}{12} = 0.42 = 42\%$
10th	$\frac{1}{12} = 0.08 = 8\%$	$\frac{3}{12} = 0.25 = 25\%$	$\frac{4}{12} = 0.33 = 33\%$
11th	$\frac{1}{12} = 0.08 = 8\%$	$\frac{2}{12} = 0.17 = 17\%$	$\frac{3}{12} = .25 = 25\%$
Total	$\frac{6}{12} = 0.50 = 50\%$	$\frac{6}{12} = 0.50 = 50\%$	$\frac{12}{12} = 1.00 = 100\%$

Check Your Understanding of Section 13.1

A. Multiple-Choice

1. Find the mean of this set of numbers {4, 5, 8, 8, 8, 10, 10, 13, 15, 17, 23}.
(1) 8 (2) 9 (3) 10 (4) 11

2. Find the median of this set of numbers {4, 5, 8, 8, 8, 10, 10, 13, 15, 17, 23}.
(1) 8 (2) 10 (3) 11 (4) 15

3. Find the mode of this set of numbers {4, 5, 8, 8, 8, 10, 10, 13, 15, 17, 23}.
(1) 23 (2) 11 (3) 10 (4) 8

4. Find the first quartile of this set of numbers {4, 5, 8, 8, 8, 10, 10, 13, 15, 17, 23}.
(1) 8 (2) 10 (3) 11 (4) 15

5. Find the third quartile of this set of numbers {4, 5, 8, 8, 8, 10, 10, 13, 15, 17, 23}.
(1) 10 (2) 15 (3) 17 (4) 23

6. Find the interquartile range of this set of numbers {4, 5, 8, 8, 8, 10, 10, 13, 15, 17, 23}.
(1) 7 (2) 8 (3) 9 (4) 10

7. For the first four days of a five day vacation, the mean temperature was 80 degrees. What must the temperature be on the fifth day in order for the mean temperature to be 82 degrees?
(1) 88 (2) 89 (3) 90 (4) 91

8. What is the median of the set of numbers {12, 4, 8, 3, 1, 4, 9, 5}?
(1) 2 (2) 3 (3) 4 (4) 4.5

9. In a set of seven numbers, the largest number is increased by 10. Which measure of central tendency must increase because of this?
(1) Mean
(2) Mode
(3) First quartile
(4) Interquartile range

10. For which data set is the median greater than the mean?
(1) {4, 7, 10, 13, 16}
(2) {8, 9, 10, 18, 19}
(3) {8, 9, 10, 11, 12}
(4) {1, 2, 10, 11, 12}

B. *Show how you arrived at your answer.*

1. Seven numbers out of a set of nine numbers are 16, 17, 19, 19, 21, 21, and 25. If the mode of the eight numbers is 19 and the mean is 20, what are the other two numbers?

2. Two basketball teams each have a mean height of 6 feet and a median height of 6 feet, but the interquartile range of the first team is much greater than the interquartile range of the second team. How can this be? Explain.

3. Zahra says that on five tests she has a mode of 81 but a mean and median of 90. How is this possible? Explain.

4. What fraction of these 11 numbers is between the first quartile and the third quartile?

3, 17, 4, 8, 4, 9, 9, 15, 14, 19, 8.

5. Lila has taken five algebra tests. Her median score is 90, and her mode score is 90, but her mean score is greater than 90. What is the highest her mean score can be?

13.2 GRAPHICALLY REPRESENTING DATA

When data are listed as just a series of numbers, many aspects of the data set are not clear. With a graphical representation of the data, certain information can be seen, even without doing any calculations. Three ways to represent data graphically are dot plots, histograms, and box plots. Each provides more information than a list of numbers would.

Dot Plots

In a *dot plot*, each piece of data is represented by a circle. When two data points represent the same value, they are stacked in a vertical line. With a dot plot, the mode can be easily determined. It is even possible to estimate the interquartile range from a dot plot.

If a data set for the heights, in inches, of 11 professional men's basketball players are

76, 67, 76, 77, 72, 77, 80, 70, 84, 77, 80

a dot plot is created by putting the numbers from 67 to 84 on a horizontal line and then creating a dot for each number. When a number is repeated, the dot goes on top of the dot, or dots, already representing that number.

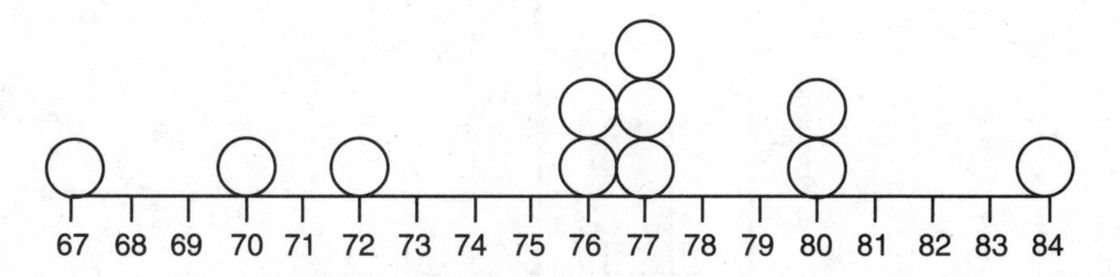

From this dot plot, it can be seen that the mode is 77. Also it can be seen that the data are bunched together so the interquartile range will be small for this data set.

Histograms

The same data set can be represented graphically with a histogram. In a histogram the data is represented by rectangles. The height of the rectangle is the number of times that number appears in the data set. Unlike a bar graph, in a histogram there are no spaces between bars representing consecutive values. In the picture below, there is no space between the bar representing the two 76s and the bar representing the three 77s.

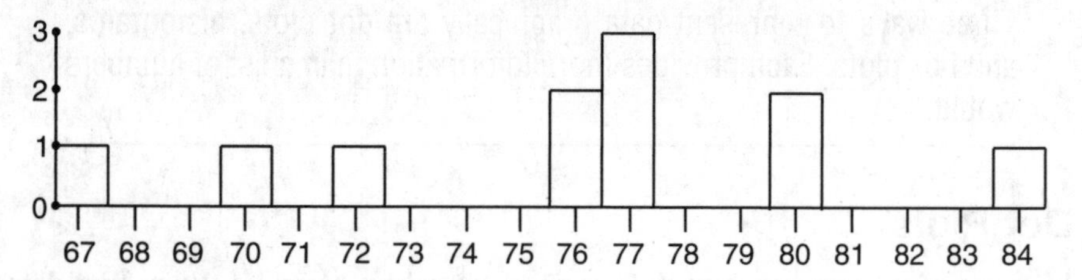

In a histogram, a range of numbers can be grouped together with a larger bin size. Here is the same data on a histogram with bin size of 5. The first bar is for numbers between 66 and 71, including 71 but not 66. The second bar is for numbers between 71 and 76, including 76 but not 71. The third bar is for numbers between 76 and 81, including 81 but not 76. The fourth bar is for numbers between 81 and 85, including 85 but not 81.

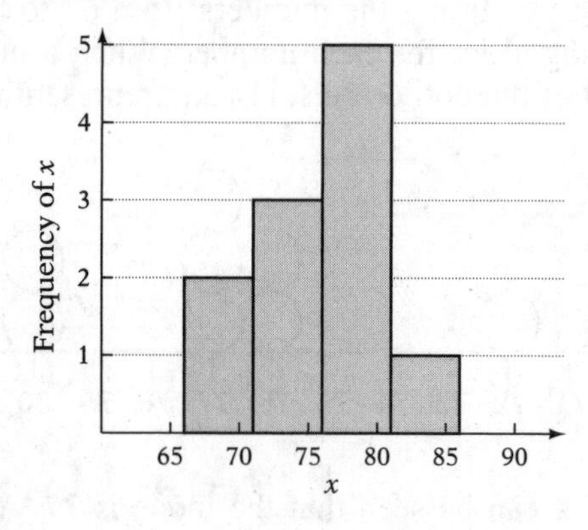

Box Plots

A *box plot* is a way to represent graphically the minimum value, the maximum value, the first quartile, the median, the third quartile, and the maximum value on the same diagram.

To create a box plot, first calculate these five values. For the data set with the 11 heights, the five values are

67	70	72	76	76	77	77	77	80	80	84
min		Q1			med			Q3		max

On a number line, plot these five values.

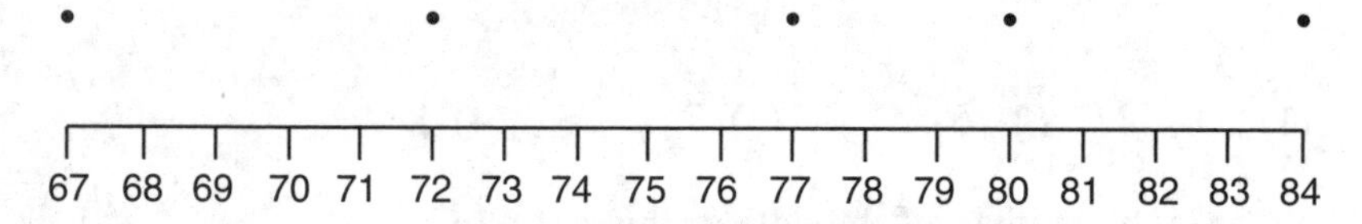

Make short vertical lines through each of these five points.

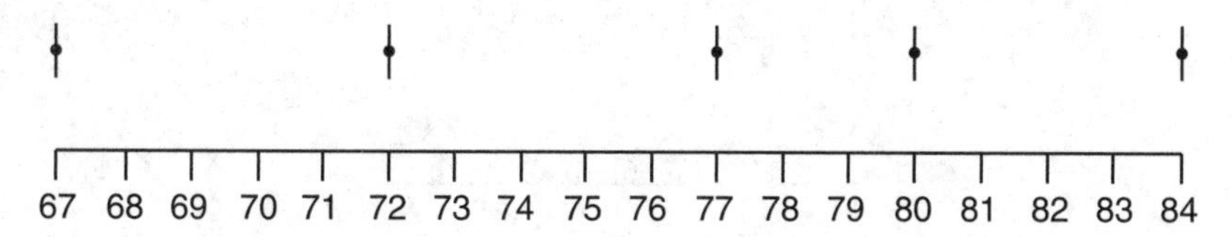

Complete the diagram by making a rectangle with the second and fourth segments as two of its sides and also make a horizontal line segment connecting the first point to the second point and another connecting the fourth point to the fifth point.

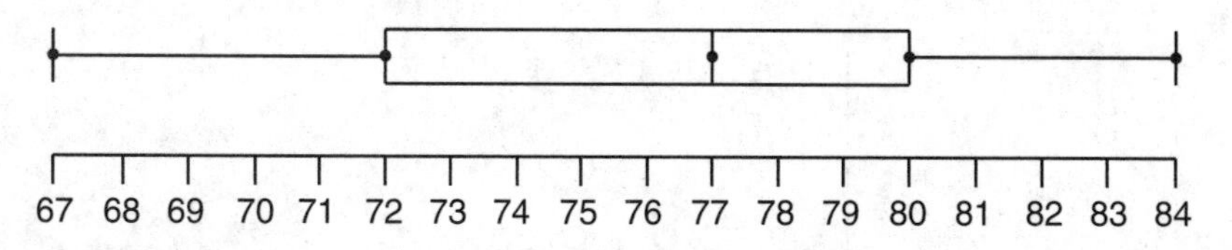

From a box plot (sometimes called a *box and whisker plot*) the interquartile range is the width of the rectangle in the middle. When the numbers in the data set are close together, this rectangle will be very narrow.

Check Your Understanding of Section 13.2

A. Multiple-Choice

1. What is the mode of the data in this dot plot?

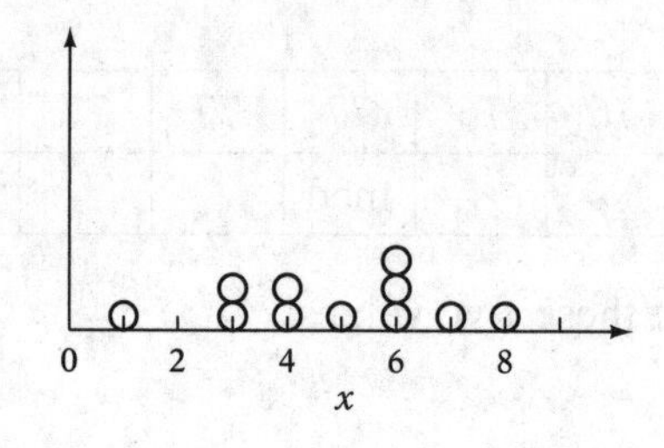

(1) 5 (2) 6 (3) 7 (4) 8

2. What is the median of the data in this dot plot?

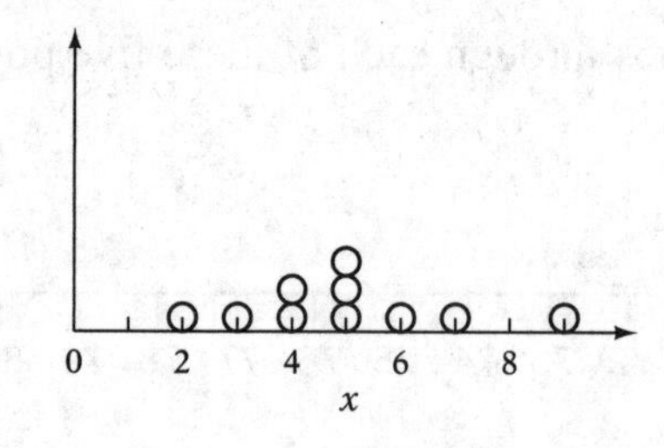

(1) 3 (2) 4 (3) 5 (4) 6

3. What is the mean of the data in this dot plot, rounded to the nearest hundredth?

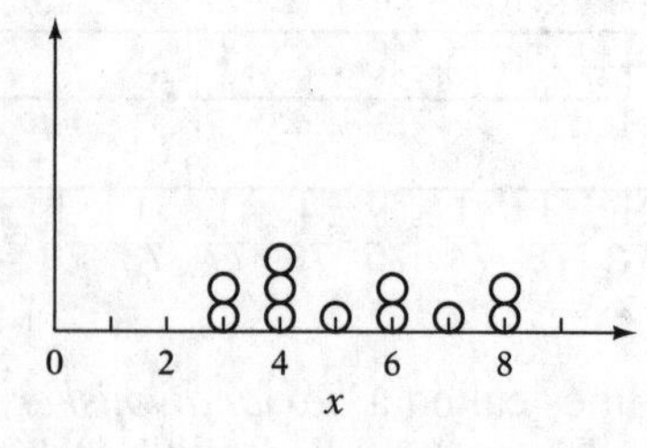

(1) 5.13 (2) 5.27 (3) 5.45 (4) 5.78

4. What is the mode of the data in this histogram?

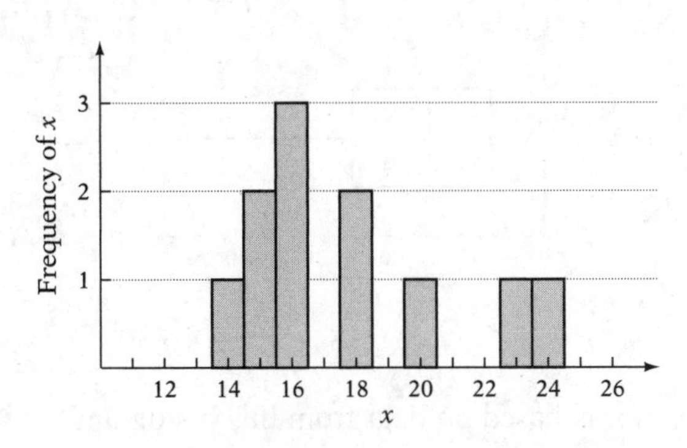

(1) 23 (2) 24 (3) 15 (4) 16

5. What is the median of the data in this histogram?

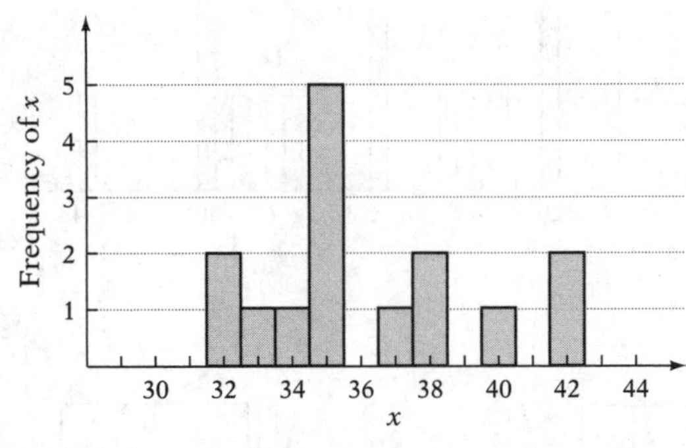

(1) 34 (2) 35 (3) 36 (4) 37

6. What is the median of the data in this box plot?

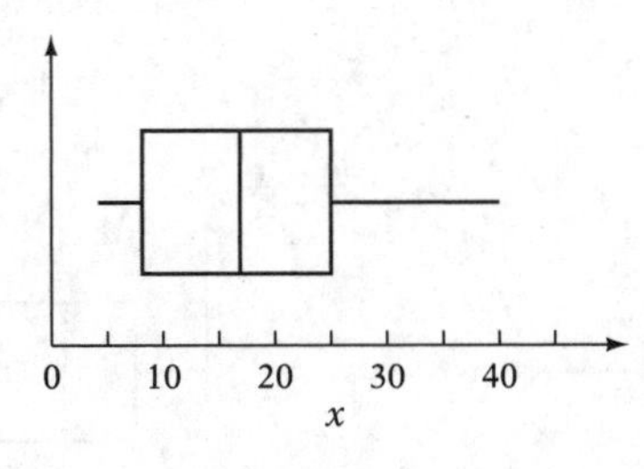

(1) 17 (2) 15 (3) 8 (4) 25

7. What is the interquartile range of the data in this box plot?

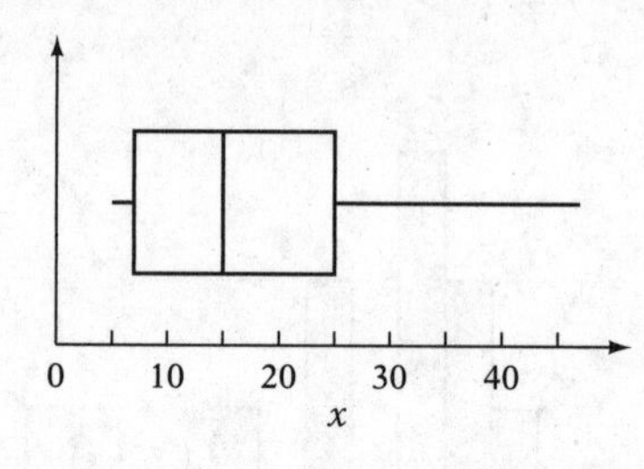

(1) 7 (2) 9 (3) 18 (4) 30

8. Which box plot is based on data from this histogram?

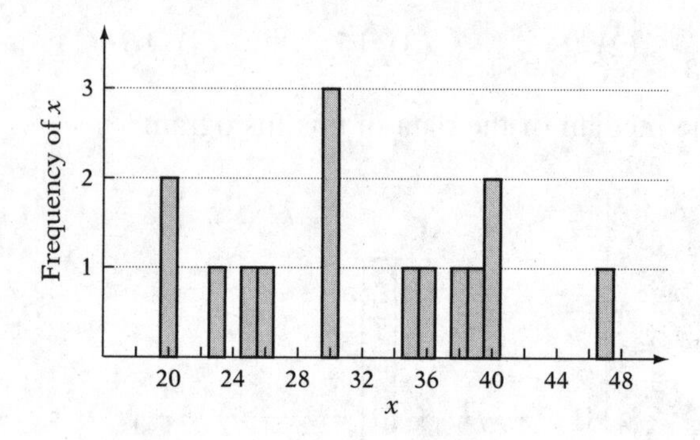

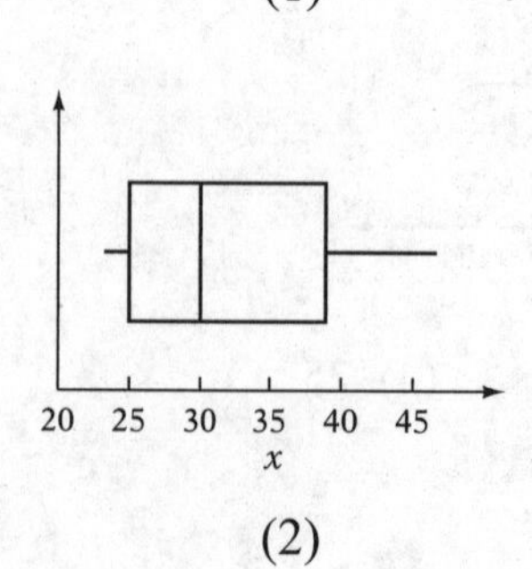

(1)

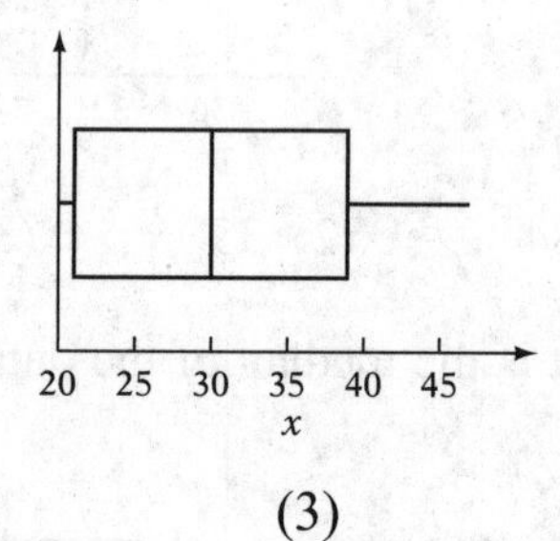

(3)

(2)

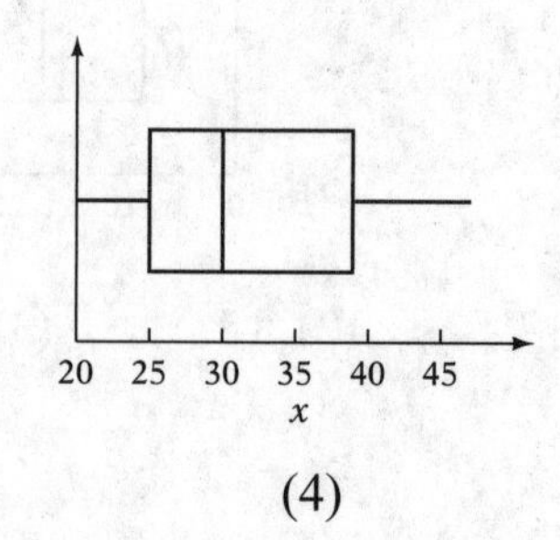

(4)

9. What is true about this data set: 1, 2, 10, 11, 11?
   (1) The median is greater than the mean.
   (2) The median is equal to the mean.
   (3) The median is less than the mean.
   (4) The median is greater than the mode.

10. For which data set is the interquartile range equal to 0?
   (1) 2, 3, 6, 6, 6, 7, 8
   (2) 2, 6, 6, 6, 6, 6, 8
   (3) 1, 2, 3, 6, 7, 8, 9
   (4) 2, 6, 6, 6, 6, 7, 8

***B.** Show how you arrived at your answer.*

1. Create a dot plot for the following set:

   5, 5, 6, 7, 7, 7, 8, 8, 10, 10, 12, 13, 14, 14, 14, 14

2. Create a box plot for the following set:

   20, 12, 4, 8, 4, 8, 12, 17, 2, 12, 9, 3, 2, 6, 15

3. Two basketball teams each have a mean height of 72 inches and median height of 72 inches. The box plots for the two teams are below. How is it that two teams with the same mean and median can have such different box plots?

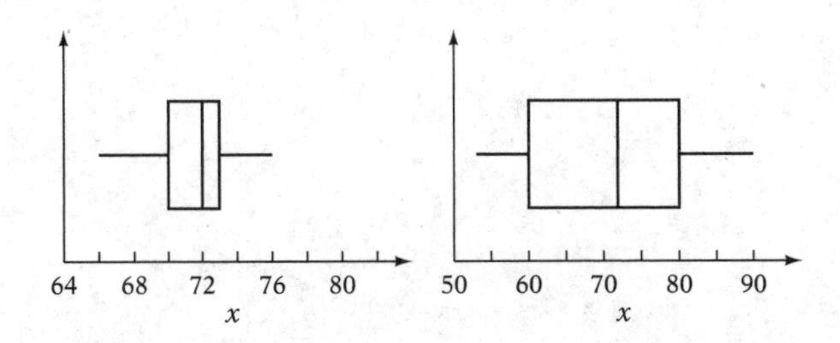

4. The box plot below is based on 11 numbers. Find two different sets of data that would produce this box plot.

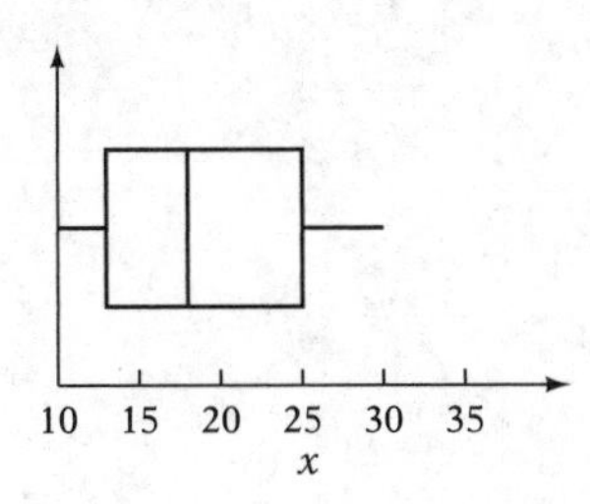

5. Make a histogram for this data set using 6 bars, each with a width of 5, using 0 to 5 including 5 but not 0, 5 to 10 including 10 but not 5, 10 to 15 including 15 but not 10, 15 to 20 including 15 but not 20, 20 to 25 including 25 but not 20, and 25 to 30 including 30 but not 25.

2, 2, 4, 6, 7, 7, 8, 8, 9, 11, 12, 12, 13, 16, 17, 17, 18, 18, 18, 19, 21, 21, 22, 22, 22, 23, 24, 28, 29, 29

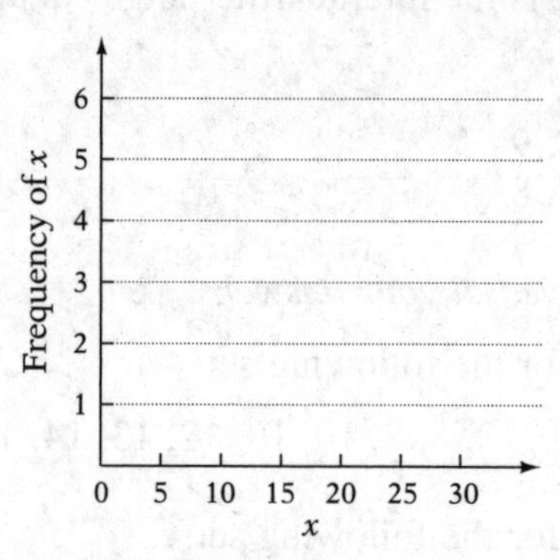

Chapter Fourteen TEST-TAKING STRATEGIES

Knowing the material is only part of the battle in acing the Algebra Regents exam. Things like not managing your time, making careless errors, and struggling with the calculator can cost valuable points. In this chapter some test-taking strategies are explained to help you perform your best on test day.

14.1 TIME MANAGEMENT

Don't Rush

The Algebra Regents exam is three hours long. Even though you are permitted to leave after one and a half hours, to get the best grade possible, stay until the end of the exam. Just as it wouldn't be wise to come to the test an hour late, it is almost the same thing if you leave a test an hour early.

Do the Test Twice

The best way to protect against careless errors is to do the entire test twice and compare the answers you got the first time to the answers you got the second time. For any answers that don't agree, do a "tie breaker" third time. Redoing the test and comparing answers is much more effective than simply "looking over" your work. People tend to skim by careless errors when looking over their work. Redoing the questions, you are less likely to make the same careless error.

Bring a Watch

In some classrooms the clock is broken. Without knowing how much time is left, you might rush and make careless errors. Though the proctor should write the time on the board and update it every so often, it is best if you have your own watch.

The TI–84 graphing calculator has a built in clock. Press the [MODE] to see it. If the time is not right, go to SET CLOCK and set it correctly. The TI–Nspire does not have a built-in clock.

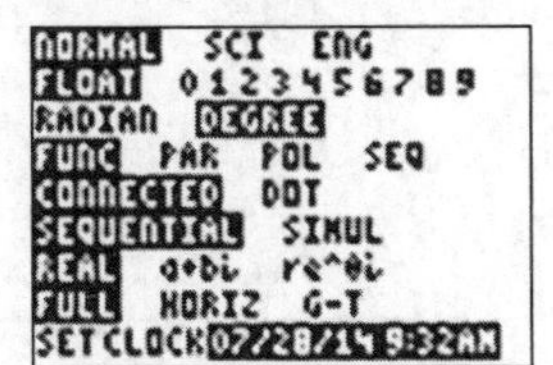

14.2 KNOW HOW TO GET PARTIAL CREDIT

Know the Structure of the Exam

The Algebra Regents exam has 37 questions. The first 24 of those questions are multiple choice worth two points each. There is no partial credit if you make a mistake on one of those questions. Even the smallest careless error, like missing a negative sign, will result in no credit for that question.

Parts Two, Three, and Four are free-response questions with no multiple choice. Besides giving a numerical answer, you may be asked to explain your reasoning.

Part Two has eight free-response questions worth two points each. The smallest careless error will cause you to lose one point, which is half the value of the question.

Part Three has four free-response questions worth four points each. These questions generally have multiple parts.

Part Four has one free-response question worth six points. This question will have multiple parts.

Explaining Your Reasoning

When a free-response question asks to "Justify your answer," "Explain your answer," or "Explain how you determined your answer," the grader is expecting a few clearly written sentences. For these, you don't want to write too little since the grader needs to see that you understand why you did the different steps you did to solve the equation. You also don't want to write too much because if anything you write is not accurate, points can be deducted.

Here is an example followed by two solutions. The first would not get full credit, but the second would.

Example

Use algebra to solve for x in the equation $\frac{2}{3}x + 1 = 11$. Justify your steps.

Solution 1 (part credit):

$\frac{2}{3}x+1=11$ $-1=-1$ $\frac{2}{3}x=10$ $x=15$	I used algebra to get the x by itself. The answer was $x = 15$.

Solution 2 (full credit):

Work	Explanation
$\frac{2}{3}x+1=11$ $-1=-1$ $\frac{2}{3}x=10$ $\frac{3}{2}\cdot\frac{2}{3}x=\frac{3}{2}\cdot 10$ $1x=15$ $x=15$	I used the subtraction property of equality to eliminate the +1 from the left-hand side. Then to make it so the x had a 1 in front of it, I used the multiplication property of equality and multiplied both sides of the equation by the reciprocal of $\frac{2}{3}$, which is $\frac{3}{2}$. Then since $1 \cdot x = x$, the left-hand side of the equation just became x and the right-hand side became 15.

Computational Errors vs. Conceptual Errors

In the Part Three and Part Four questions, the graders are instructed to take off one point for a "computational error" but half credit for a "conceptual error." This is the difference between these two types of errors.

If a four point question was $x - 1 = 2$ and a student did it like this,

$$\begin{aligned} x - 1 &= 2 \\ +1 &= +1 \\ x &= 4 \end{aligned}$$

the student would lose one point out of 4 because there was one computational error since $2 + 1 = 3$ and not 4.

Had the student done it like this,

$$\begin{aligned} x - 1 &= 2 \\ -1 &= -1 \\ x &= 1 \end{aligned}$$

the student would lose half credit, or 2 points, since this error was conceptual. The student thought that to eliminate the –1, he should subtract 1 from both sides of the equation.

Either error might just be careless, but the conceptual error is the one that gets the harsher deduction.

14.3 KNOW YOUR CALCULATOR

Which Calculator Should You Use?

The two calculators used for this book are the TI-84 and the TI-Nspire. Both are very powerful. The TI-84 is somewhat easier to use for the functions needed for this test. The TI-Nspire has more features for courses in the future.

The choice is up to you. This author prefers the TI-84 for the Algebra Regents.

Graphing calculators come with manuals that are as thick as the book you are holding. There are also plenty of video tutorials online for learning how to use advanced features of the calculator. To become an expert user, watch the online tutorials or read the manual.

Clearing the Memory

You may be asked at the beginning of the test to clear the memory of your calculator. When practicing for the test, you should clear the memory too so you are practicing under test-taking conditions.

This is how you clear the memory.

For the TI-84:

Press [2ND] and then [+] to get to the MEMORY menu. Then press [7] for Reset.

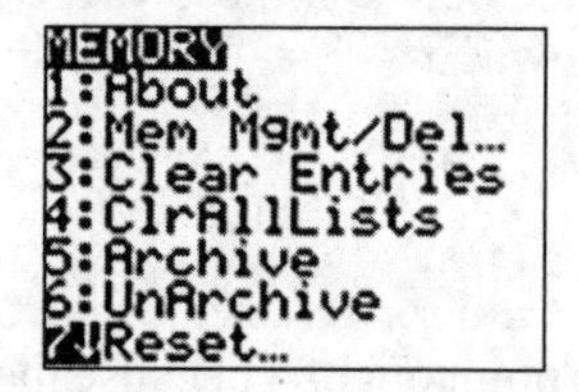

Use the arrows to go to [ALL] for All Memory. Then press [1].

Press [2] for Reset.

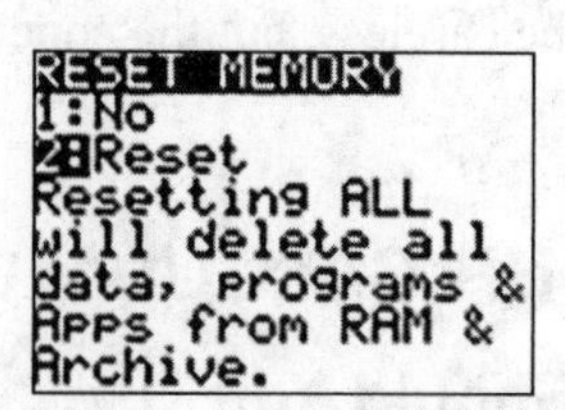

The calculator will be reset as if brand new condition. The one setting that you may need to change is to turn the diagnostics on if you need to calculate the correlation coefficient. (See Chapter 12 for this.)

For the TI-Nspire:

The TI-Nspire must be set to Press-To-Test mode when taking the Algebra Regents. Turn the calculator off by pressing [ctrl] and [home]. Press and hold [esc] and then press [home].

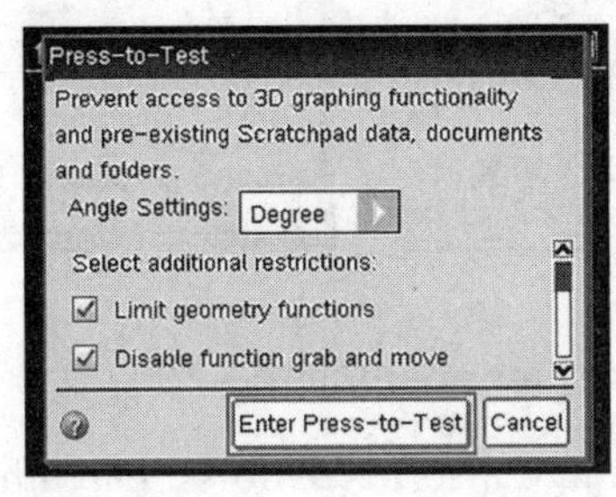

While in Press-to-Test mode, certain features will be deactivated. A small green light will blink on the calculator so a proctor can verify the calculator is in Press-to-Test mode.

To exit Press-to-Test mode, use a USB cable to connect the calculator to another TI-Nspire. Then from the home screen on the calculator in Press-to-Test mode, press [doc], [9] and select Exit Press-to-Test.

Use Parentheses

The calculator always uses the order of operations where multiplication and division happen before addition and subtraction. Sometimes, though, you may want the calculator to do the operations in a different order.

Suppose at the end of a quadratic equation, you have to round $x = \frac{-1+\sqrt{5}}{2}$ to the nearest hundredth. If you enter (–) (1) (+) (2ND) (x^2) (5) (/) (2), it displays

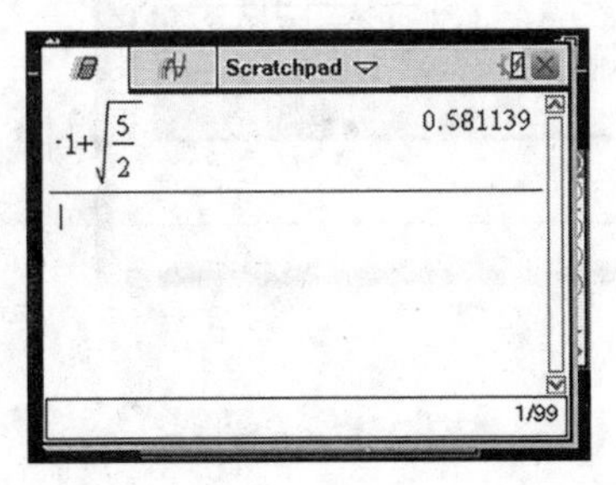

which is not the correct answer.

One reason is that for the TI-84 there needs to be a closing parentheses (or on the TI-Nspire, press [right arrow] to move out from under the radical sign) after the 5 in the square root symbol. Without it, it calculated $-1 + \sqrt{\frac{5}{2}}$.

More needs to be done, though, since

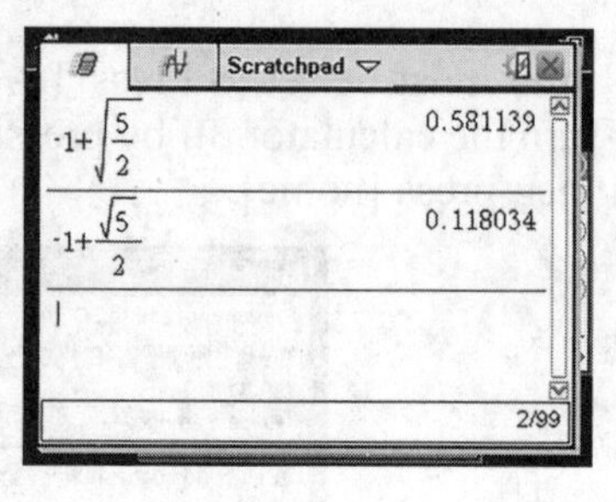

still is not correct. This is the solution to $-1 + \sqrt{\frac{5}{2}}$.

To get this correct, there also needs to be parentheses around the entire numerator, $-1 + \sqrt{5}$.

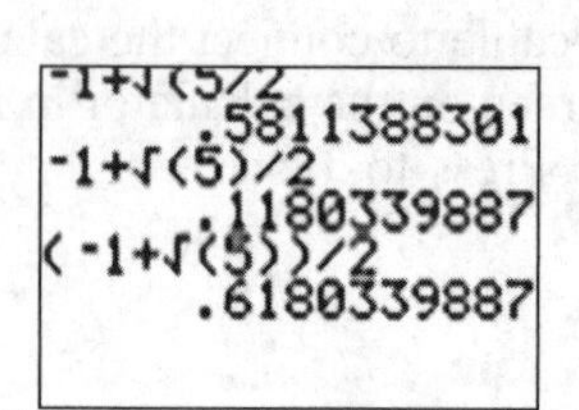

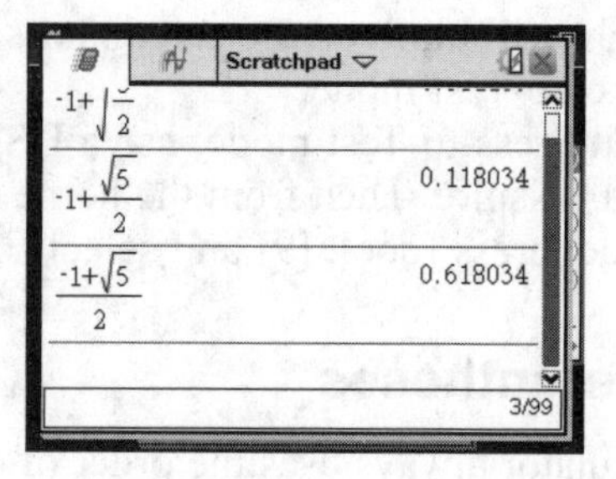

This is the correct answer.

On the TI-Nspire, fractions like this can also be done with [templates].

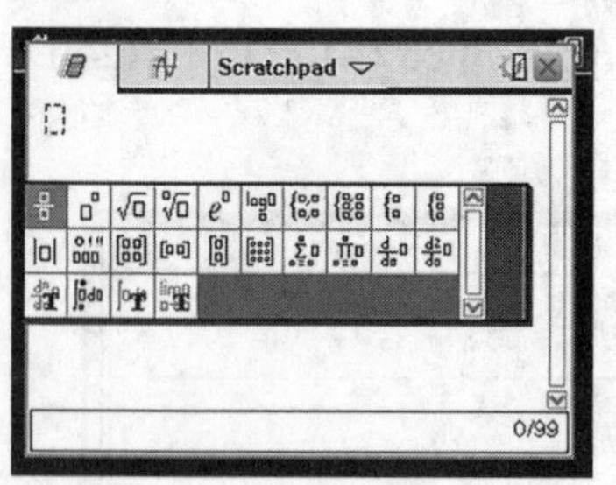

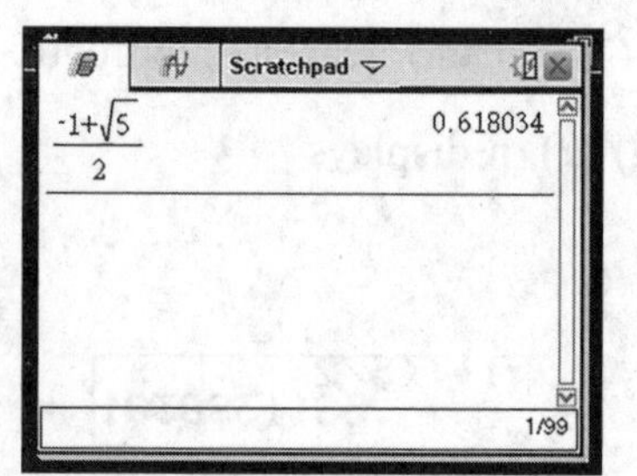

Using the ANS Feature

The last number calculated with the calculator is stored in something called the ANS variable. This ANS variable will appear if you start an expression with a +, –, ×, or ÷. When an answer has a lot of digits in it, this saves time and is also more accurate.

If for some step in a problem you need to calculate the decimal equivalent of $\frac{1}{7}$, it will look like this on the TI-84:

For the TI-Nspire, if you try the same thing, it leaves the answer as $\frac{1}{7}$. To get the decimal approximation, press [ctrl] and [enter] instead of just [enter].

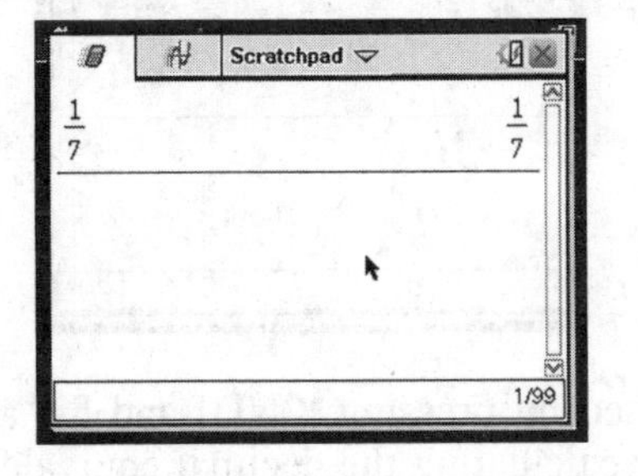

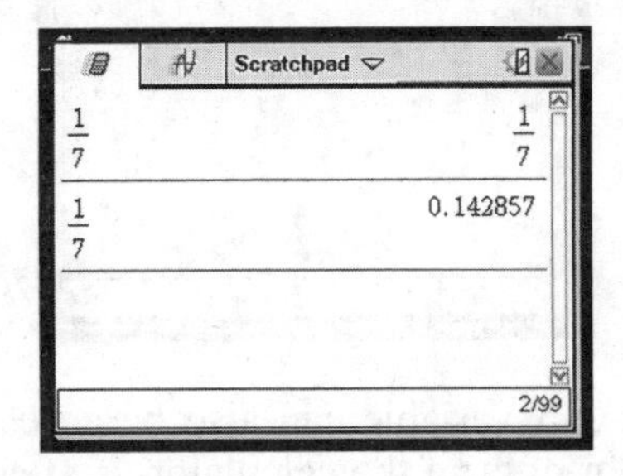

Now if you have to multiply this by 3, just press [×], and the calculator will display "Ans*"; press [3] and [enter].

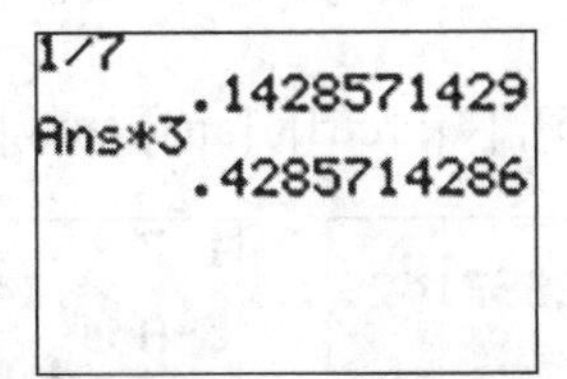

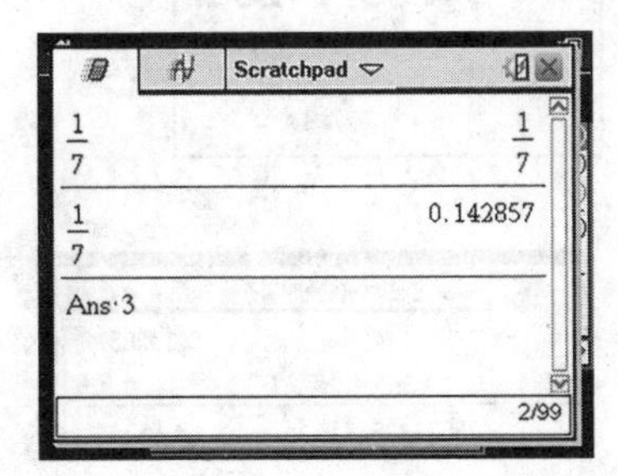

The ANS variable can also help you do calculations in stages. To calculate $x = \dfrac{-1+\sqrt{5}}{2}$ without using so many parentheses as before, it can be done by first calculating $-1 + \sqrt{5}$ and then pressing [÷] and [2] and Ans will appear automatically.

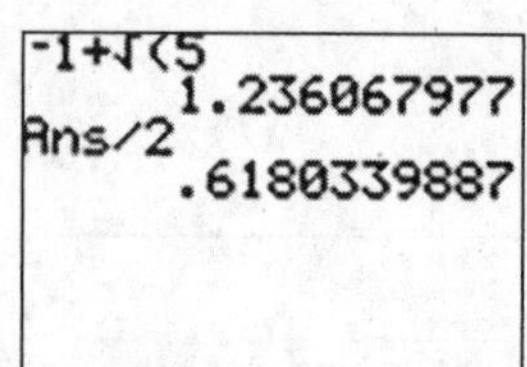

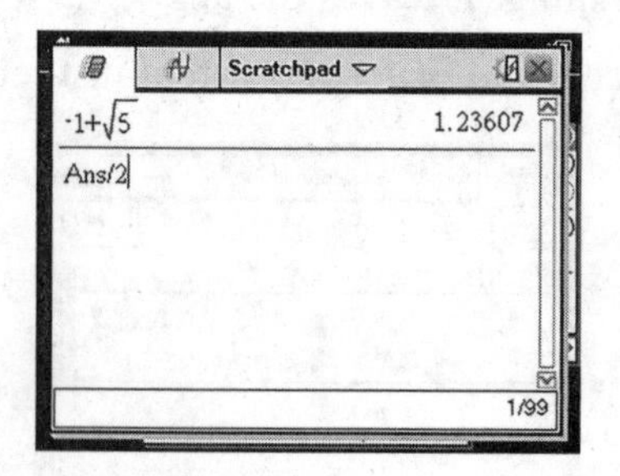

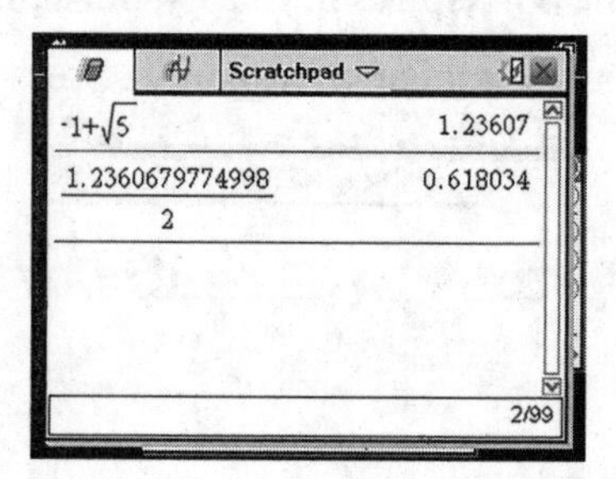

The ANS variable can also be accessed by pressing [2ND] and [–] at the bottom right of the calculator. If after calculating the decimal equivalent of $\frac{1}{7}$ you wanted to subtract $\frac{1}{7}$ from 5, for the TI-84 press [5], [–], [2ND], [ANS], and [ENTER].

For the TI-Nspire press [5], [–], [ctrl], [ans], and [enter].

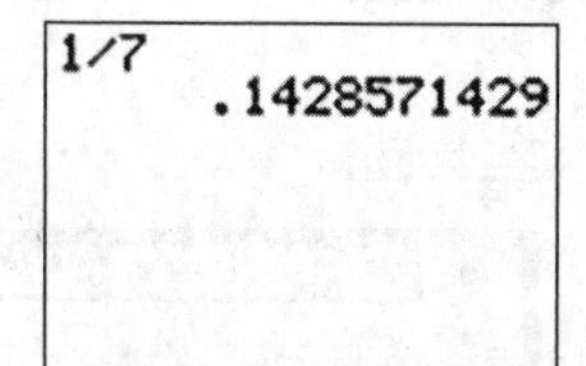

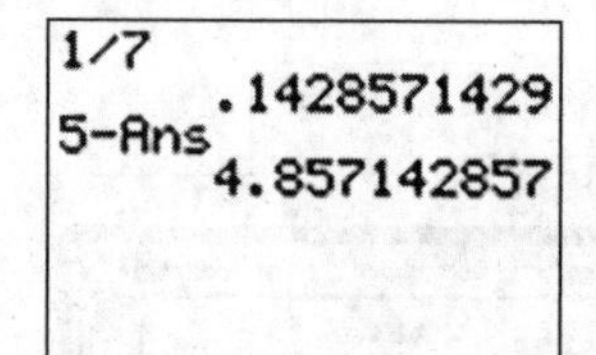

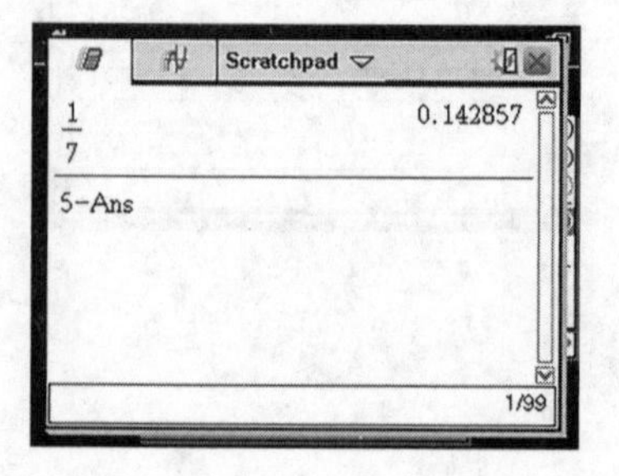

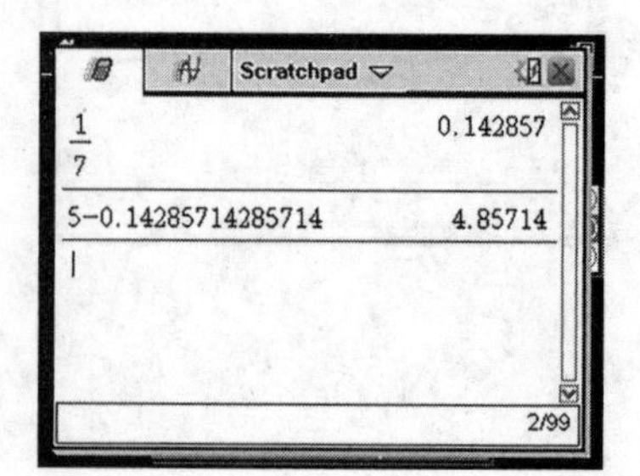

14.4 USE THE REFERENCE SHEET

In the back of the Algebra Regents booklet is a reference sheet with 17 conversion facts, like inches to centimeters and quarts to pints, and also 17 formulas. Many of these conversion facts and formulas will not be needed for an individual test, but the quadratic formula and the arithmetic sequence formula are the two that will come in the handiest.

High School Math Reference Sheet

1 inch = 2.54 centimeters
1 meter = 39.37 inches
1 mile = 5280 feet
1 mile = 1760 yards
1 mile = 1.609 kilometers

1 kilometer = 0.62 mile
1 pound = 16 ounces
1 pound = 0.454 kilogram
1 kilogram = 2.2 pounds
1 ton = 2000 pounds

1 cup = 8 fluid ounces
1 pint = 2 cups
1 quart = 2 pints
1 gallon = 4 quarts
1 gallon = 3.785 liters
1 liter = 0.264 gallon
1 liter = 1000 cubic centimeters

Triangle	$A = \frac{1}{2}bh$
Parallelogram	$A = bh$
Circle	$A = \pi r^2$
Circle	$C = \pi d$ or $C = 2\pi r$
General Prisms	$V = Bh$
Cylinder	$V = \pi r^2 h$
Sphere	$V = \frac{4}{3}\pi r^3$
Cone	$V = \frac{1}{3}\pi r^2 h$
Pyramid	$V = \frac{1}{3}Bh$

Pythagorean Theorem	$a^2 + b^2 = c^2$
Quadratic Formula	$x = \frac{-b \pm \sqrt{b^2 - 4ac}}{2a}$
Arithmetic Sequence	$a_n = a_1 + (n - 1)d$
Geometric Sequence	$a_n = a_1 r^{n-1}$
Geometric Series	$S_n = \frac{a_1 - a_1 r^n}{1 - r}$ where $r \neq 1$
Radians	1 radian $= \frac{180}{\pi}$ degrees
Degrees	1 degree $= \frac{\pi}{180}$ radians
Exponential Growth/Decay	$A = A_0 e^{k(t - t_0)} + B_0$

14.5 HOW MANY POINTS DO YOU NEED TO PASS?

The Algebra Regents exam is scored out of a possible 86 points. Unlike most tests given in the year by your teacher, the score is not then turned into a percent out of 86. Instead each test has a conversion sheet that varies from year to year. For the June 2014 test, the conversion sheet looked like this.

Raw Score	Scale Score	Raw Score	Scale Score	Raw Score	Scale Score
86	100	57	75	28	64
85	99	56	74	27	63
84	97	55	74	26	62
83	96	54	74	25	61
82	95	53	73	24	60
81	94	52	73	23	59
80	92	51	73	22	58
79	91	50	72	21	56
78	90	49	72	20	55
77	89	48	72	19	54
76	88	47	72	18	52
75	87	46	71	17	50
74	86	45	71	16	49
73	85	44	71	15	47
72	84	43	70	14	45
71	83	42	70	13	42
70	82	41	70	12	40
69	82	40	70	11	38
68	81	39	69	10	35
67	80	38	69	9	32
66	79	37	69	8	30
65	79	36	68	7	26
64	78	35	68	6	23
63	78	34	67	5	20
62	77	33	67	4	16
61	77	32	66	3	12
60	76	31	66	2	9
59	76	30	65	1	4
58	75	29	64	0	0

On this test, 30 points became a 65, 57 points became a 75, and 73 points became an 85. This means that for this examination a student who got 30 out of 86, which is just 35% of the possible points, would get a 65 on this exam. 57 out of 86 is 66%, but this scaled to a 75. 73 out of 86, however, is actually 85% and became an 85. So in the past there has been a curve on the exam for lower scores, though the scaling is not released until after the exam.

Answers and Solution Hints to Practice Exercises

CHAPTER 1

Section 1.1

A

1. (4)	**3.** (1)	**5.** (3)	**7.** (4)	**9.** (2)
2. (3)	**4.** (2)	**6.** (1)	**8.** (1)	**10.** (1)

B

1. Malachi would have $6x - 2 = 16$ and Wesley would have $3x - 1 = 8$.
2. Addition property of equality, Distributive property of multiplication over addition, Subtraction property of equality, Division property of equality.
3. The first person is wrong. The proper use of the distributive property is $5(3 + 7) = 5 \cdot 3 + 5 \cdot 7$, not $5 \cdot 3 + 7$.
4. There is no commutative property of subtraction. For example, $5 - 2 = 3$ is not equivalent to $2 - 5 = -3$.
5. 30 is correct. There is no distributive property of multiplication over multiplication so it is not true that $a(b \cdot c) = a \cdot b \cdot a \cdot c$.

Section 1.2

A

1. (3)	**3.** (2)	**5.** (1)	**7.** (3)	**9.** (3)
2. (4)	**4.** (2)	**6.** (3)	**8.** (4)	**10.** (2)

B

1. It will become the equation $5x = 60$, which can then be solved by dividing both sides of the equals sign by 5.
2. If you multiply both sides by $\frac{5}{3}$ or if you divide both sides by $\frac{3}{5}$, you get the same correct answer.
3. By subtracting -2 from both sides, it becomes $x = 5 - (-2) = 5 + 2 = 7$.
4. The addition property of equality. Add 3 to both sides of the equation.
5. There is nothing that when multiplied by 0 becomes 10. If you try to use the division property of equality and divide both sides of the equation by 0, you get $\frac{10}{0}$, which is not defined.

Section 1.3

A

1. (3)	**3.** (3)	**5.** (4)	**7.** (2)	**9.** (1)
2. (1)	**4.** (1)	**6.** (1)	**8.** (1)	**10.** (4)

B

1. $x = 5$. Distributive property, addition property of equality, and then multiplication property or by multiplication property then addition property of equality.
2. $x = 7$. Addition property of equality then division property of equality.
3. $x = 1$. Distributive property, addition property of equality, and then division property of equality or by division property of equality then addition property of equality.
4. They are both right. Either way gets the answer of $x = 5$.
5. $p = 20$. The sneakers sell for $20.

Section 1.4

A

1. (2)	**3.** (3)	**5.** (1)	**7.** (3)	**9.** (4)
2. (3)	**4.** (1)	**6.** (1)	**8.** (3)	**10.** (2)

B

1. $q = \dfrac{p+f}{r}$
2. $w = \dfrac{v}{lh}$
3. $b = \dfrac{2a}{h}$
4. $r = \dfrac{s}{2\pi h}$
5. $n = \dfrac{a_n - a_1}{d} + 1$

CHAPTER 2

Section 2.1

A

1. (2)	**3.** (1)	**5.** (4)	**7.** (3)	**9.** (4)
2. (3)	**4.** (2)	**6.** (3)	**8.** (3)	**10.** (3)

B

1. It is a trinomial with degree 2, also called a quadratic trinomial.
2. It is a binomial with degree 1, also called a linear binomial.
3. Cross off either the $2x^2$, the $-7x$, or the $+9$.
4. Anything of the form ax^4, for example $5x^4$. The variable name can be something other than x.
5. Mark was correct that it was a binomial. It is true that after simplifying it would no longer be a binomial, but before simplifying it is one.

Section 2.2

A

1. (3)	**3.** (4)	**5.** (3)	**7.** (2)	**9.** (4)
2. (1)	**4.** (4)	**6.** (2)	**8.** (3)	**10.** (1)

B

1. She is not correct. The answer is $20x^5$ because $x^2 \cdot x^3 = x^5$, not x^6.

2. When multiplying monomials, multiply the coefficients and add the exponents. So for dividing, do the opposite, divide the coefficients and subtract the exponents to get $4x^4$.
3. Sawyer is right, $x^0 = 1$ provided that $x \neq 0$.
4. 10. There are two ways to do this. One is to multiply the two monomials and get $10x^0$, which is $10 \cdot 1 = 10$. Another way is to rewrite as
$$5x^2 \cdot \frac{2}{x^2} = 10x^0 = 10.$$
5. It will be a fifth degree monomial. For example $3x^2 \cdot 4x^3 = 12x^5$.

Section 2.3

A

1. (2)	**3.** (3)	**5.** (4)	**7.** (2)	**9.** (3)
2. (4)	**4.** (3)	**6.** (1)	**8.** (4)	**10.** (4)

B

1. Yes, this is allowed. The expression $x^2(3 + 5)$ can be simplified to $3x^2 + 5x^2$ and also to $8x^2$. Since each expression is equivalent to $x^2(3 + 5)$, $3x^2 + 5x^2$ must equal $8x^2$.
2. Josie is correct. Since $5y^2x$ is equivalent to $5xy^2$ by the commutative property of multiplication, these are like terms and can be combined to become $8xy^2$.
3. The variable a must be equivalent to $2xy^2z$ so that it can be combined with the $5xy^2z$.
4. It becomes $x^3 + 3x^2y + 3xy^2 + y^3$
5. $6 - i - i^2$ or $- i_2 - i + 6$

Section 2.4

A

1. (2)	**3.** (3)	**5.** (4)	**7.** (2)	**9.** (1)
2. (4)	**4.** (2)	**6.** (4)	**8.** (3)	**10.** (2)

B

1. Alexander distributes the 5 through the terms in the parentheses to get $10x + 15x = 25x$, which is the same answer as Tucker.
2. Jack is correct. Because of the commutative property of multiplication, it is the same thing as $5x(3x^2 + 2x + 1)$.
3. The a represents $2x$ since $2x \cdot 5x^2 = 10x^3$, $2x \cdot (-3x) = -6x^2$, and $2x \cdot 7 = 14x$.
4. First it becomes $8x^2 + 12x - 4x - 6$. Then after combining like terms becomes $8x^2 + 8x - 6$.
5. First it becomes $6x^2 - 10x - 18x + 30$. Then after combining like terms becomes $6x^2 - 28x + 30$.

Section 2.5

A

1. (1)	**3.** (2)	**5.** (4)	**7.** (1)	**9.** (1)
2. (1)	**4.** (3)	**6.** (2)	**8.** (3)	**10.** (3)

B

1. He did not do it correctly. He should have distributed the negative through the parentheses to get $2x + 3 - 4x - 2$, which would lead to the correct answer of $-2x + 1$.
2. $3x + 5 - 3(2x - 5) = 3x + 5 - 6x + 15 = -3x + 20$
3. The rules for combining polynomials are the same as for combining numbers so this process will still get the right answer.
4. $2(-2x + 5) - 5(3x - 7) = -4x + 10 - 15x + 35 = -19x + 45$
5. $(x^2 + 5x + 2) + (x^2 - 3x + 4) - (x^2 - 2x + 5) =$
   $x^2 + 5x + 2 + x^2 - 3x + 4 - x^2 + 2x - 5 = x^2 + 4x + 1$

Section 2.6

A

1. (1)	**3.** (3)	**5.** (4)	**7.** (2)	**9.** (1)
2. (3)	**4.** (1)	**6.** (4)	**8.** (3)	**10.** (1)

B

1. The equation can be $x + 20 = 2x + 3$ with a solution of $x = 17$.
2. One reason for choosing to subtract $2x$ from both sides is to avoid negatives. Also fine to write that the methods are the same degree of difficulty so you would not choose either way over the other.
3. $x = 4$
4. $30x$ is the distance the first train is from its starting point. The second train begins 300 miles from the first train's starting point and then every hour gets 20 miles closer so in x hours it will be $300 - 20x$ miles from the first train's starting point. When both expressions are equal, the trains are the same distance from the first train's starting point so they are passing. The solution to the equation is $x = 6$.
5. The equation is $x + (2x - 5) + (3x + 10) = 125$, which becomes $6x + 5 = 125$ or $x = 20$. So the store sold 20 items on Monday, 35 items on Tuesday, and 70 items on Wednesday.

Section 2.7

A

1. (3)	**3.** (3)	**5.** (4)	**7.** (1)	**9.** (3)
2. (1)	**4.** (4)	**6.** (2)	**8.** (2)	**10.** (2)

B

1. $(x + 4)(x - 1) + 2(x + 4) = x^2 + 3x - 4 + 2x + 8 = x^2 + 5x + 4$
2. $4 + 12i + 9i^2 + 1 + 2i = 5 + 14i + 9i^2$
3. She is not correct. There is not a distributive property for exponents. Instead $(x + 4)(x + 4) = x^2 + 4x + 4x + 16 = x^2 + 8x + 16$.

4. Since this is not a binomial multiplied by a binomial, FOIL can't be used. Instead, use the distributive property.
$(x + 3)x^2 + (x + 3)7x + (x + 3)10 = x^3 + 3x^2 + 7x^2 + 21x + 10x + 30 =$
$x^3 + 10x^2 + 31x + 30$
5. $(50 - 2)(50 + 2)$ has the $(a - b)(a + b)$ pattern so the solution is $50^2 - 2^2 = 2500 - 4 = 2496$.

Section 2.8
Solutions
A

1. (4)	**3.** (1)	**5.** (2)	**7.** (1)	**9.** (4)
2. (2)	**4.** (2)	**6.** (1)	**8.** (1)	**10.** (2)

B
1. Ariana is correct. $(x + 5)^2 = x^2 + 10x + 25$, not $x^2 + 25$.
2. First factor out the greatest common factor of x. $x(x^2 + 4x - 21)$. Then factor the quadratic trinomial in the parentheses $x(x + 7)(x - 3)$.
3. First factor out the greatest common factor of 5. $5(x^2 + 4x + 4)$. Then factor the quadratic trinomial in the parentheses $5(x + 2)(x + 2)$ or $5(x + 2)^2$.
4. He is not correct. $(x - 6)(x - 1)$ is $x^2 - 7x + 6$, not $x^2 - 5x + 6$.
5. The other factor is $(x + 5)$. To get the 15 as the last term there needs to be a +5 in the other factor. Also to get $2x^2$ in the first term there needs to be a $1x$ in the other factor. To check that this is correct, multiply $(2x + 3)(x + 5) = 2x^2 + 10x + 3x + 15 = 2x^2 + 13x + 15$.

Section 2.9
A

1. (3)	**3.** (3)	**5.** (2)	**7.** (3)	**9.** (1)
2. (1)	**4.** (1)	**6.** (3)	**8.** (2)	**10.** (2)

B
1. $x^4 - 81 = (x^2)^2 - 81 = (x^2 - 9)(x^2 + 9)$ Then $x^2 - 9$ can be factored into $(x - 3)(x + 3)$. The solution is $(x - 3)(x + 3)(x^2 + 9)$.
Complete answer is $(x - 3)(x + 3)(x^2 + 9)$.
2. $x^4 + 4x^2 - 5 = (x^2)^2 + 4(x^2) - 5 = (x^2 + 5)(x^2 - 1) = (x^2 + 5)(x - 1)(x + 1)$
3. $x^4 - 13x^2 + 36 = (x^2)^2 - 13x^2 + 36 = (x^2 - 4)(x^2 - 9) = (x - 2)(x + 2)(x - 3)$
$(x + 3)$
4. This is like $x^2 + 2x - 3 = (x + 3)(x - 1)$ but instead of an x there is a $(3x - 5)$. The solution is $(3x - 5 + 3)(3x - 5 - 1)$ which can also be simplified further to $(3x - 2)(3x - 6)$.
5. $x^6 - 8 = (x^2)^3 - 2^3$ so a is x^2 and b is 2. Solution is $(x^2 - 2)(x^4 + 2x^2 + 4)$.

CHAPTER 3

Section 3.1

A

1. (4)	**3.** (3)	**5.** (3)	**7.** (4)	**9.** (3)
2. (1)	**4.** (2)	**6.** (4)	**8.** (3)	**10.** (1)

B

1. The solutions are $\sqrt{15}$ and $-\sqrt{15}$. Rounded to the nearest hundredth, 3.87 and –3.87.
2. $(x + 5)^2 = 64$ becomes $x + 5 = \pm 8$ with solutions $x = 3$, $x = -13$.
3. $(x + 2)^2 = 49$ becomes $x + 2 = 7$ so $x = 5$.
4. $(x + 4)^2$ does not equal $x^2 + 16$. The correct answer is $x = -4 \pm \sqrt{97}$.
5. Taking the cube root of both sides of the equation, it becomes $x + 1 = 2$ (not ± 2 since $(-2)^3 = -8$), or $x = 1$.

Section 3.2

A

1. (1)	**3.** (2)	**5.** (2)	**7.** (2)	**9.** (2)
2. (4)	**4.** (3)	**6.** (4)	**8.** (1)	**10.** (4)

Section 3.3

A

1. (1)	**3.** (1)	**5.** (2)	**7.** (2)	**9.** (1)
2. (2)	**4.** (4)	**6.** (3)	**8.** (1)	**10.** (4)

B

1. $b = 16, p = 8$ and $b = -16, p = -8$
2. $x^2 - 16x = 7$, $x^2 - 16x + 64 = 7 + 64 = 71$, $(x - 8)^2 = 71$, $x - 8 = \pm\sqrt{71}$, $x = 8 \pm \sqrt{71}$
3. $\frac{22}{7}$ is just an approximation for π. $\frac{22}{7}$ is rational since it is a fraction with an integer numerator and denominator.
4. $x^2 - 5x + (\frac{5}{2})^2 = 14 + (\frac{5}{2})^2$, $(x - \frac{5}{2})^2 = 14 + \frac{25}{4}$, $x - \frac{5}{2} = \pm\sqrt{\frac{81}{4}}$, $x = \frac{5}{2} \pm \frac{9}{2}$, $x = -2, 7$.
5. $x^2 + 2ax + a^2 = b + a^2$, $(x + a)^2 = b + a^2$, $x + a = \pm\sqrt{b + a^2}$, $x = -a \pm \sqrt{b + a^2}$.

Section 3.4

A

1. (1)	**3.** (4)	**5.** (2)	**7.** (1)	**9.** (2)
2. (1)	**4.** (2)	**6.** (3)	**8.** (3)	**10.** (3)

B

1. No, she is not correct. By factoring the left-hand side, it becomes $x(x-3)=0$ so either $x=0$ or $x-3=0$, which means there are two solutions 0 and 3.
2. $(x-1)(x-2)(x-3)=0$ so $x-1=0$ or $x-2=0$ or $x-3=0$ with three solutions 1, 2, and 3.
3. Either $a=0$ or $b=0$ or both. Since $a \neq 0$, b must equal 0.
4. It can be factored into $(x^2-9)(x^2-4)=0$ and then into $(x-3)(x+3)(x-2)(x+2)=0$ so $x-3=0$ or $x+3=0$ or $x-2=0$ or $x+2=0$ so the four solutions are 3, –3, 2, and –2.
5. When two things have a product of 40, it does not follow that one of them must be 40, the way it is when the product is 0. The most common way to do this question is to start by subtracting 40 from both sides and factoring.

$$x^2+6x-40=0$$
$$(x+10)(x-4)=0$$
$$x=10 \text{ or } x=4$$

It actually can be done from $x(x+6)=40$ through guess and check. Find two numbers where one is 6 larger than the other so that the product is 40. The numbers are 4 and 10, so $x=4$ is one solution.

Section 3.5

A

1. (2)	**3.** (2)	**5.** (1)	**7.** (4)	**9.** (3)
2. (3)	**4.** (3)	**6.** (1)	**8.** (2)	**10.** (2)

B

1. $(x-2)(x+2)(x-4)=0$
2. $(x+4)(x+7)=0$ or $x^2+11x+28=0$
3. No. The root is a number, whereas the factor is an expression, like $x+2$. So changing the sign would make it $-(x+2)=-x-2$, which is not the same as the number –2.
4. $2x+3=0, x=-\frac{3}{2}$ and $2x+5=0, x=-\frac{5}{2}$. When there is a coefficient in front of the x, the roots are not just the opposite of the constant term.
5. The roots are $2+\sqrt{7}$ and $2-\sqrt{7}$ so the factors can be written as $(x-(2+\sqrt{7}))$ and $(x-(2-\sqrt{7}))$.

Section 3.6

A

1. (1)	**3.** (2)	**5.** (1)	**7.** (3)	**9.** (1)
2. (2)	**4.** (4)	**6.** (4)	**8.** (1)	**10.** (2)

B

1. $x=5 \pm \sqrt{3}$, $x=6.73, 3.27$

2. $2a = 2$ so $a = 1$. $-b = -5$ so $b = 5$. $b^2 - 4ac = 17$ so $25 - 4c = 17$ and $c = 2$.

3. $\frac{-4 \pm \sqrt{(16-4c)}}{2}$, which can be optionally simplified to

$$\frac{-4 \pm \sqrt{4 \cdot (4-c)}}{2} = -2 \pm \sqrt{4-c}.$$

4. The solutions are both irrational. $\sqrt{17}$ is irrational so $-3 \pm \sqrt{17}$ are both irrational, and $\frac{-3 \pm \sqrt{17}}{2}$ are both irrational.

5. First, get into the form $ax^2 + bx + c = 0$ by subtracting $4x$ and 6 from both sides to get $x^2 - 4x - 3 = 0$. Then use the quadratic formula to get $x = 2 \pm \sqrt{7}$.

Section 3.7

A

1. (4)	**3.** (3)	**5.** (4)	**7.** (4)	**9.** (1)
2. (1)	**4.** (2)	**6.** (2)	**8.** (3)	**10.** (4)

B

1. $-16t^2 + 128t + 144 = 384$ becomes $-16t^2 + 128t - 240 = 0$. By factoring or by using the quadratic formula, the solution is $t = 3$ and $t = 5$.

2. The equation is $(4 + 2x)(6 + 2x) = 80$. It becomes $4x^2 + 20x + 24 = 80$ and then $4x^2 + 20x - 56 = 0$ with a solution of $x = 2$.

3. The equation is $(0.99 + 0.10x)(200{,}000 - 10{,}000x) = 223{,}500$. It becomes $-1{,}000x^2 + 10{,}100x - 25{,}500 = 0$ with a solution of $x = 5$.

4. The equation is $x(x - 35) = 884$ becomes $x^2 - 35x - 884 = 0$ with a solution of $x = 52$.

5. Since the perimeter is 44, $2x + 2w = 44$ or $2w = 44 - 2x$ or $w = 22 - x$. The area of the rectangle is $120 = lw = x(22 - x)$. This becomes the equation $120 = x(22 - x)$ or $120 = 22x - x^2$ or $x^2 - 22x + 120 = 0$. This can be factored as $(x - 12)(x - 10) = 0$ to get the solution $x = 12$ or $x = 10$. (Or it can be solved with the quadratic formula.)

CHAPTER 4

Section 4.1

A

1. (2)	**3.** (4)	**5.** (1)	**7.** (1)	**9.** (4)
2. (1)	**4.** (1)	**6.** (4)	**8.** (3)	**10.** (2)

B

1. (4, 1) and (1, 3)

2. $7 + 1 = 8$, $7 - 1 = 6$ so the solution is (7, 1).

3. Any pair that satisfies the first equation cannot also satisfy the second equation since if the left-hand side evaluates to 25, then it cannot equal 26.
4. $c = 25$ since $4 + 3 \cdot 7 = 25$
5. There are many, but (6, –2) is the quickest. Since the second equation is obtained by multiplying both sides of the original equation by 3, it has the same solution set, including (6, –2).

Section 4.2

A

1. (4)	**3.** (1)	**5.** (3)	**7.** (4)	**9.** (2)
2. (3)	**4.** (4)	**6.** (3)	**8.** (1)	**10.** (1)

B

1. There are many answers to this. One is the equation $5x + 10y = 25$ obtained by adding the two equations together. Any equation that has (1, 2) as its solution is acceptable.
2. By adding $2x$ to both sides of the top equation, it becomes $y = 2x + 4$. Then the y in the second equation can be replaced with $2x + 4$ to solve for the answer (3, 10).
3. By subtracting the two equations, it becomes $0 = 5x - 25$, which becomes $5x = 25$ or $x = 5$. By substituting the y in the second equation with $2x - 9$, it becomes $2x - 9 = -3x + 16$, which also becomes $5x = 25$. The solution is (5, 1).
4. The system is $x + y = 17$, $x - y = 11$. Add them together to get $2x = 28$, $x = 14$, $y = 3$.
5. They are both correct. Adding makes the equation $4y = -24$, or $y = -6$. Subtracting makes $10x = 40$, or $x = 4$. Either way the solution (4, –6) can be obtained.

Section 4.3

A

1. (3)	**3.** (2)	**5.** (3)	**7.** (1)	**9.** (1)
2. (2)	**4.** (4)	**6.** (4)	**8.** (4)	**10.** (4)

B

1. Eliminating the x requires less work because it can be done by just changing the first equation, multiplying both sides of the first equation by –3; however, to eliminate the y requires changing both equations because 7 is not a multiple of 5.
2. Multiply both sides of the first equation by –2. It becomes $-2x - 2y = -80$. Add to the second equation to eliminate the x, $2y = 56$ so $y = 28$ and $x = 12$.

3. It is not accurate. Most ordered pairs do not satisfy either equation, for instance (0, 0). But every ordered pair that satisfies the first equation, like (0, 7) will also satisfy the second equation and so will satisfy the system of equations. There still are an infinite number of ordered pairs, but not all ordered pairs satisfy the system.
4. There are no solutions. The top equation is equivalent to $10x - 4y = 10$ so any ordered pair that satisfies this equation will not satisfy the equation $10x - 4y = 11$. When using elimination, it becomes $0 = -1$, which is never true.
5. Victoria will continue by adding the two equations together to eliminate the x. Porter will continue by subtracting the changed second equation from the first equation to eliminate the x. Either way, the answer (6, 2) will be correct.

Section 4.4

A

1. (4)	**3.** (1)	**5.** (2)	**7.** (1)	**9.** (1)
2. (1)	**4.** (3)	**6.** (2)	**8.** (3)	**10.** (2)

B

1. x is the number of red roses, and y is the number of pink roses. The equations are $x + y = 24$, $3x + 2y = 68$. The solution is $x = 20$, $y = 4$ so there are 20 red roses and 4 pink roses.
2. (a) \$60; (b) \$40; (c) \$50; (d) 4 cones and 6 shakes. This could be done with the system of equations $x + y = 10$, $4x + 6y = 52$, or by noticing that if it were 5 and 5, it would cost \$50 so to make it cost \$2 more, switch one of the cones with a shake to become 4 cones and 6 shakes.
3. x is the number of ounces of marshmallows, and y is the number of ounces of cookies. The cost equation is $2x + 3y = 29$. The calorie equation is $500x + 400y = 4{,}800$. The solution is 4 ounces of marshmallows and 7 ounces of cookies.
4. x is the cost of a burger, and y is the cost of an order of French fries. $3x + 2y = 24$, $5x + 1y = 33$, solution is $x = 6$ and $y = 3$ so a burger costs \$6 and an order of French fries costs \$3.
5. x is the number of bicycles, and y is the number of tricycles. $x + y = 58$, $2x + 3y = 134$. The solution is 40 bicycles and 18 tricycles.

CHAPTER 5

Section 5.1

A

1. (4)	**3.** (1)	**5.** (2)	**7.** (1)	**9.** (3)
2. (1)	**4.** (3)	**6.** (2)	**8.** (4)	**10.** (2)

B

1. The table could look like

x	y
0	8
1	6
2	4

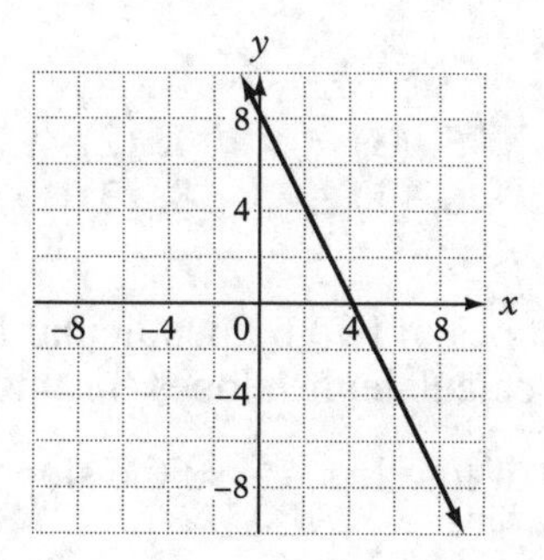

2. For the x-intercept, $4x - 6(0) = 24$ has a solution of $x = 6$. For the y-intercept $4(0) - 6y = 24$ has a solution of $y = -4$. The x-intercept is $(6, 0)$ and the y-intercept is $(0, -4)$

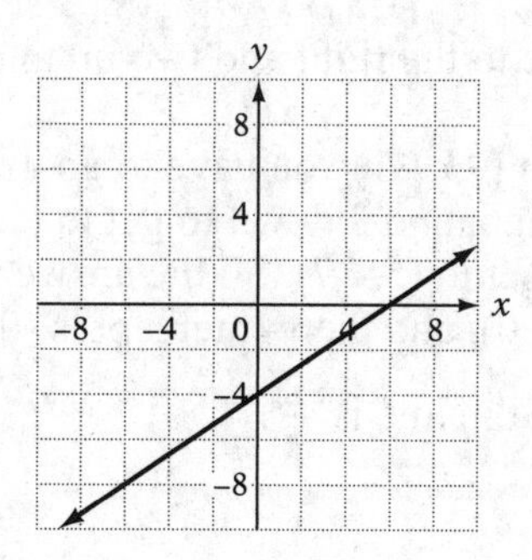

3. It is $y = 3$ since it contains points $(0, 3)$, $(1, 3)$, and all points that have 3 as the y-coordinate.

4. Using a chart or the intercept method, graph the two lines. The intersection is located at (5, 1).

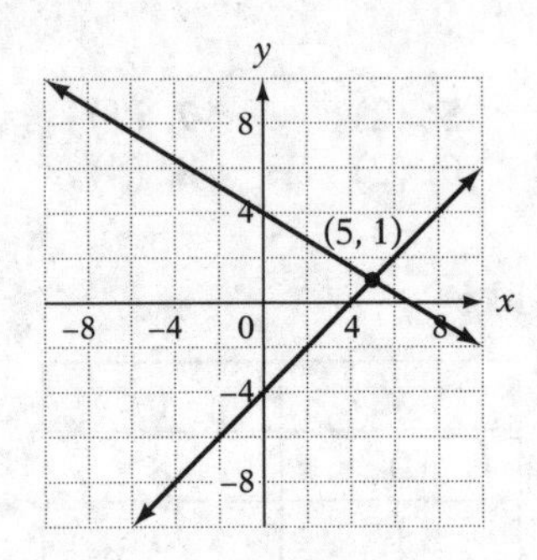

5. Substituting either ordered pair into the equation will lead to $k = 8$.

Section 5.2

A

1. (2)	**3.** (1)	**5.** (3)	**7.** (2)	**9.** (4)
2. (4)	**4.** (1)	**6.** (1)	**8.** (3)	**10.** (3)

B

1. The product of the slopes of two perpendicular lines is equal to −1. It is not necessary to calculate the slopes for this question, but it is permitted. The slope of line 1 is $-\frac{3}{5}$ so the slope of line 2 is $\frac{5}{3}$ and $-\frac{3}{5} \cdot \frac{5}{3} = -1$.

2. Side *AB* has the greatest slope. *BC* has a slope of 0, *AB* has a slope of $\frac{5}{2}$, and *AC* has a slope of $\frac{5}{4}$.

3. From (−5, 2) go three to the right and two up to get to (−2, 4), then to (1, 6), then to (4, 8).

4. To get from (−7, 2) to (−3, −1) you have to go 4 to the right and 3 down. From (−3, −1) go 4 right and 3 down to get to (1, −4). Again go 4 to the right and 3 down to get to (5, −7). So the answer is 5.

5. The *x*-intercept is (3, 0), and the *y*-intercept is (0, −2). Using the slope formula, the slope of the line is $\frac{2}{3}$.

Section 5.3

A

1. (1)	**3.** (2)	**5.** (1)	**7.** (4)	**9.** (3)
2. (2)	**4.** (2)	**6.** (4)	**8.** (3)	**10.** (3)

B

1. Slope is 3, and the y-intercept is at (0, –9).

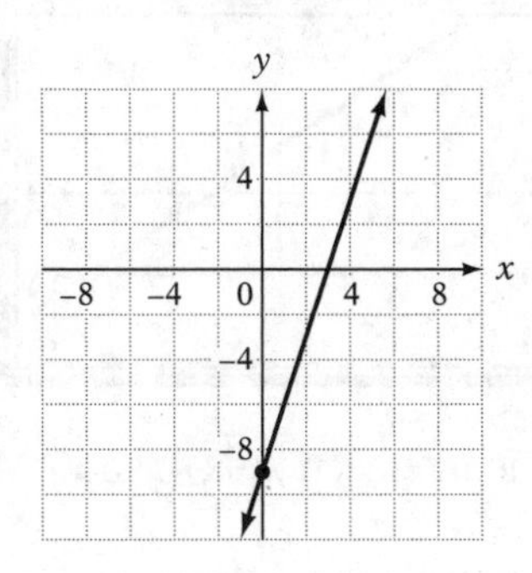

2. Edwin is correct. After dividing both sides of the equation by 2 to get it into slope-intercept form, it becomes $y = \frac{3}{2}x - 4$.
3. (0, 3). To get from (4, 5) to (2, 4) go down one and left two. Then go down one and left two from (2, 4) and get to (0, 3), which is the y-intercept.
4. The solution is (3, 2).

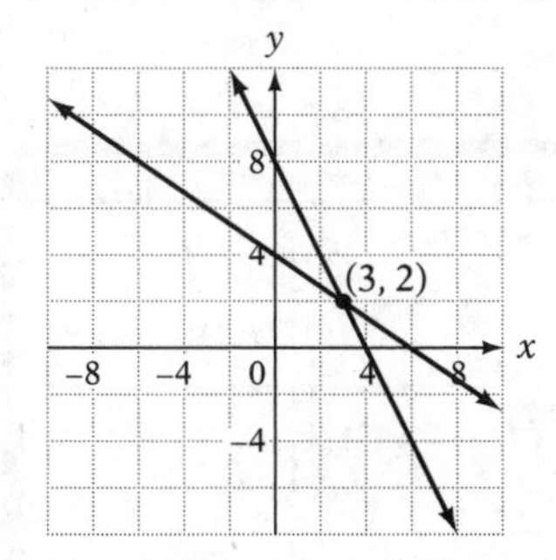

5. They are all of the form $y = mx + 2$ where m is positive.

Section 5.4

B

1.

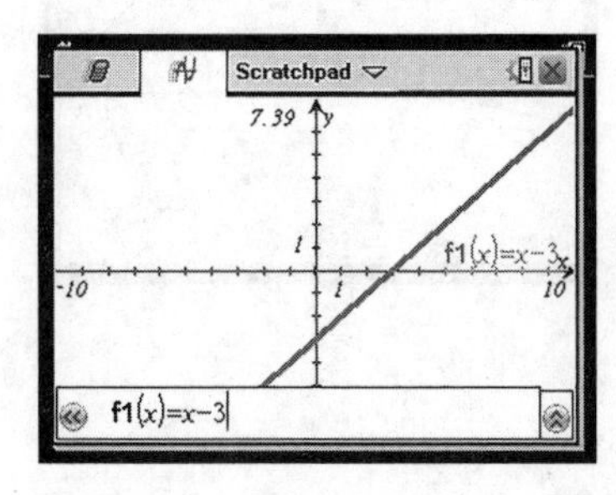

2.

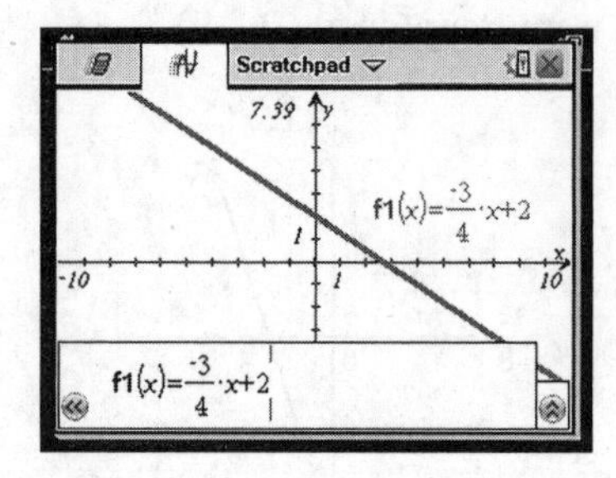

3. The x-coordinate of the intersection is 4, so $x = 4$ is the solution to the equation.

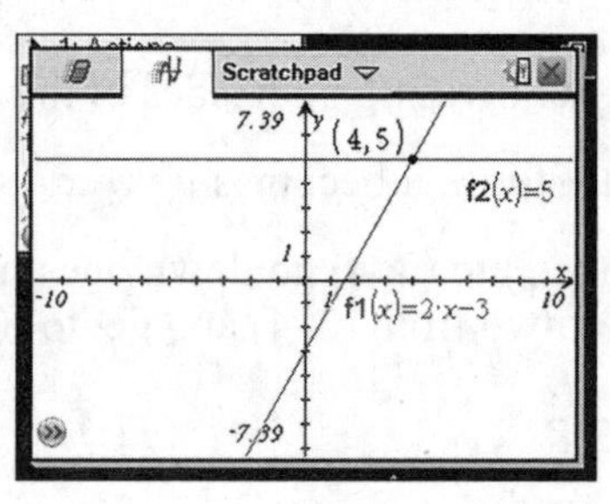

4.

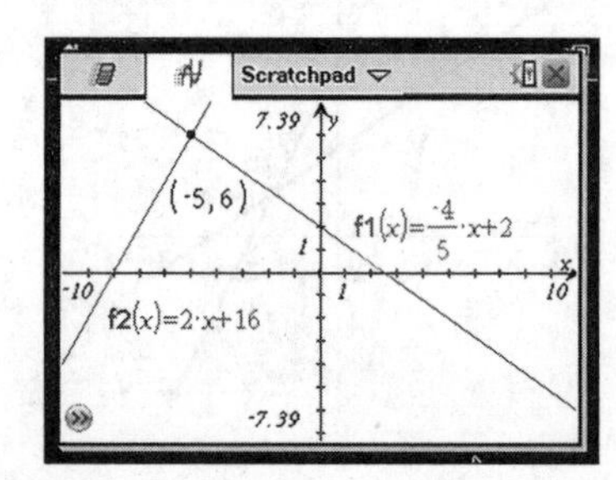

5.

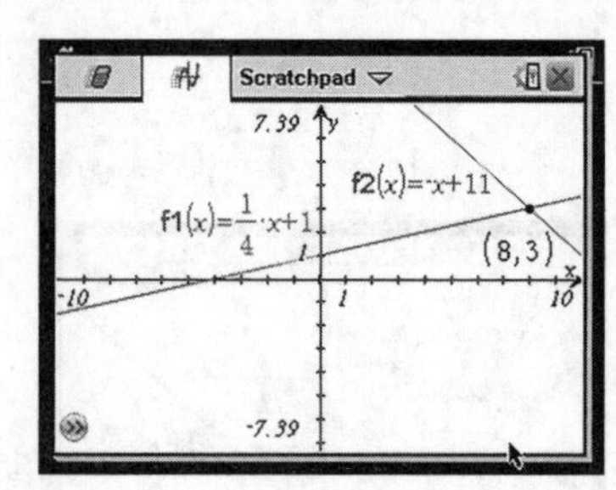

Section 5.5

A

1. (2)	**3.** (3)	**5.** (4)	**7.** (2)	**9.** (2)
2. (3)	**4.** (1)	**6.** (3)	**8.** (4)	**10.** (2)

B

1. The equation is $y = \frac{2}{3}x + 2$, which passes through (5, 5.3) since $\frac{2}{3}(5) + 2 = 5.3$.

2. The equation is $y = -\frac{1}{4}x + 8$, which passes through (18, 3.5) since the equation $3.5 = -\frac{1}{4}a + 8$ becomes $-4.5 = -\frac{1}{4}a$ becomes $18 = a$.

3. The equation is $y = -\frac{4}{5}x + 10$. When $x = 0$, $y = 10$ so the y-intercept is (0, 10). When $y = 0$, $0 = -\frac{4}{5}x + 10$ becomes $-10 = -\frac{4}{5}x$ becomes $x = \frac{50}{4} = 12.5$ so the x-intercept is (12.5, 0).

4. One way to do this is to use the slope formula. The slope is known to be $\frac{1}{3}$ so $\frac{1}{3} = \frac{a-7}{12-6} = \frac{a-7}{6}$. If $\frac{1}{3} = \frac{a-7}{6}$, cross multiply to get $6 = 3(a - 7) = 3a - 21$ or $27 = 3a$ or $a = 9$. Another way to do this would be to substitute 12 for x and a for y into the equation to get $a = \frac{1}{3}(12) + 5 = 4 + 5 = 9$.

5. The equation of the line is $y = \frac{1}{5}x + 3$. The line passes through (6, 4.2), which is 0.8 away from (6, 5). The line passes through (7, 4.4), which is 0.6 away from (7, 5). The line passes through (9. 4.8) which is 0.8 away from (9, 4). The sum of the three vertical segments is $0.8 + 0.6 + 0.8 = 2.2$.

Section 5.6

A

1. (4)	**3.** (2)	**5.** (4)	**7.** (3)	**9.** (4)
2. (3)	**4.** (1)	**6.** (4)	**8.** (1)	**10.** (2)

B

1. (a) $H = 4Y + 6$; (b) 46 feet tall

2. (a) $P = \frac{3}{2}W + 3$; (b) 3; (c) 14 weeks

3. (a) $V = 300D + 100$; (b) 4,300; (c) 9,400

4. (a) $P = 10T + 230$; (b) 2,050
5. (a) $E = -4M + 200$; (b) 140; (c) 8:20; (d) 8:50

Chapter 6

Section 6.1

A

1. (4)	**3.** (4)	**5.** (3)	**7.** (2)	**9.** (2)
2. (2)	**4.** (4)	**6.** (1)	**8.** (1)	**10.** (3)

B

1. $b = 6$ and c can be anything.
2. The equation can be $y = (x - 1)(x + 4) = x^2 + 3x - 4$ so one answer is $b = 3$ and $c = -4$.
3. Because the axis of symmetry passes through the vertex, reflect the point (6, 0) over the line $x = 4$ to get (2, 0). Also reflect (1, 5) over the line $x = 4$ to get (7, 5).
4. The vertex is (3, 0) and the one x-intercept is also (3, 0).
5. $x = 3$. One way to do this is to expand $y = (x - 1)^2 - 4(x - 1) - 2 = x^2 - 2x + 1 - 4x + 4 - 2 = x^2 - 6x + 3$ and then use

$$x = -\frac{b}{2a} = -\frac{-6}{2 \cdot 1} = 3$$

Section 6.2

A

1. (2)	**3.** (3)	**5.** (2)	**7.** (3)	**9.** (2)
2. (3)	**4.** (2)	**6.** (1)	**8.** (3)	**10.** (4)

B

1. Approximately 3.5 and 6.5. More accurate answers are 3.3 and 6.7.
2. It becomes the equation $x^2 - x - 1 = 0$ with solutions 1.62 and –0.62.
3. The horizontal line is $y = 5$. The parabola has x-intercepts (1, 0) and (4, 0), so its equation is $y = (x - 1)(x - 4) = x^2 - 5x + 4$. The intersection points solve the equation $x^2 - 5x + 4 = 5$.
4. Graph $y = -16x^2 + 96x + 112$ and find x–intercept at $x = 7$.
5. All three are correct since the equations are all equivalent. One can be turned into the other using the addition property of equality.

Section 6.3

A

1. (2)	**3.** (3)	**5.** (4)	**7.** (2)	**9.** (3)
2. (2)	**4.** (2)	**6.** (4)	**8.** (1)	**10.** (1)

B

1. (–2, 2) and (4, 8)
2. (0, 5) and (3, 8)
3. (6, 4) and (2.67, 5.11)
4. Jimena is correct. The second intersection occurs off the screen. The two solutions are (–1, 3) and (3, 27).
5. (a) Any number greater than –1; (b) any number less than –1.

Section 6.4

B

1. At $t = 1$ and $t = 5$. $336 = -16t^2 + 96t + 256$, which becomes $16t^2 - 96t + 80 = 0$. Then $16(t - 1)(t - 5) = 0$.
2. The graphs intersect at (5, 5), which means that for 5 days it costs $5,000 to use either construction company.
3. The graphs intersect at the point (1, 34) so after 1 second the stunt man and the elevator are both 34 feet high and the stunt man has landed on top of the elevator.
4. The solution is (2, 36) so after 2 seconds the superhero catches the car 36 feet in the air.
5. The solution to the system of equations is (6, 27) so after 6 seconds the iguanas will crash into each other.

CHAPTER 7

Section 7.1

A

1. (3)	**3.** (3)	**5.** (2)	**7.** (3)	**9.** (4)
2. (1)	**4.** (3)	**6.** (2)	**8.** (2)	**10.** (1)

B

1. The solution set is $x \geq 5$. On a number line, graph a ray with initial point at 5 and going to the right.

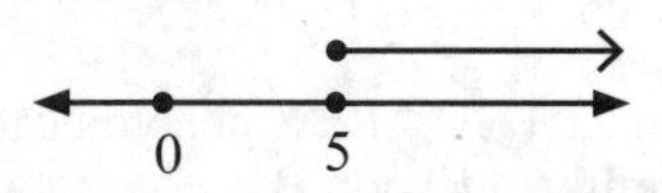

2. Lorenzo is correct. When dividing both sides of an inequality by a negative, the direction of the inequality must be switched to keep it true.
3. It is valid. Rather than divide by a negative and switch the direction of the inequality sign, adding or subtracting the same thing from both sides of an inequality does not require changing the direction of the sign.
4. Subtract $2x$ from both sides and add 2 to both sides to transform inequality into $3x \geq 12$ with solution set $x \geq 4$.
5. $1.20x \leq 60$ or $x \leq 50$. So the most the meal can cost before the tip is $50.

Section 7.2

A

1. (4) **3.** (2) **5.** (2) **7.** (2) **9.** (3)
2. (1) **4.** (1) **6.** (2) **8.** (4) **10.** (2)

B

1. It is like the graph of $x + y <= 4$

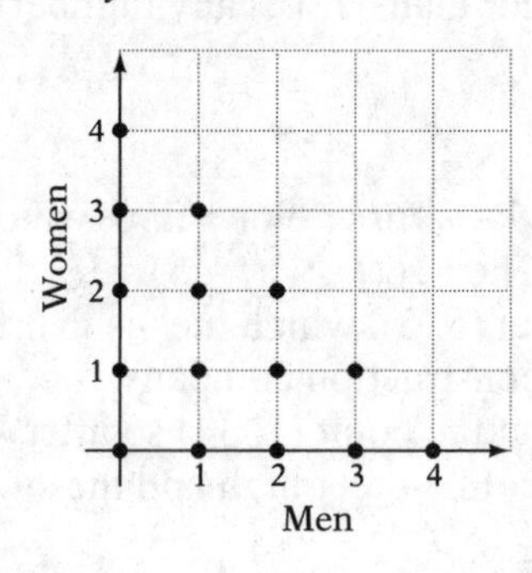

2. $y < \frac{1}{2}x + 3$

3. He does not have enough information. (0, 0) is on the line but he needs to check a point that is not on the line.

4. Divide both sides by –1 and switch the direction of the inequality to get $y > -2x + 4$

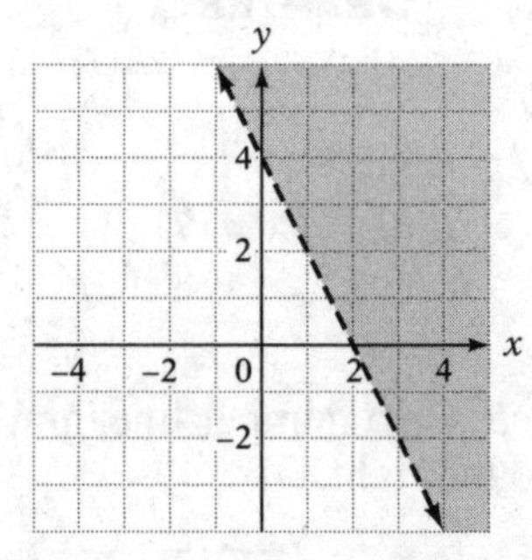

5. It is like the equation $2x + 3y \leq 12$. Since you can buy fractions of a pint, this is shaded unlike question B1.

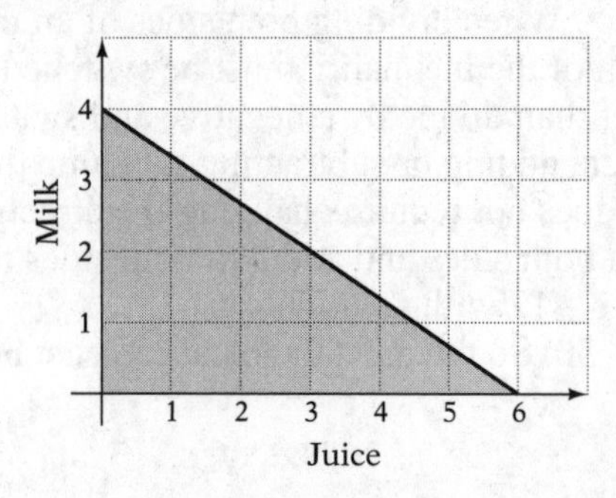

Section 7.3

A

1. (1)	**3.** (1)	**5.** (1)	**7.** (2)	**9.** (2)
2. (2)	**4.** (3)	**6.** (2)	**8.** (3)	**10.** (4)

B

1.

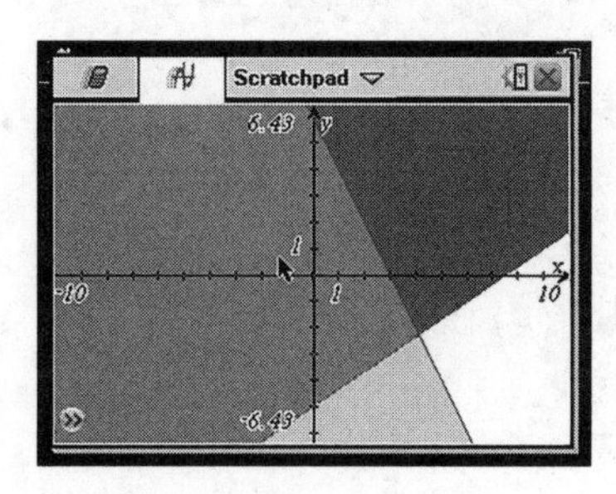

2. $y \leq x + 1, y \leq \frac{1}{2}x + 2$

3. $x + y \leq 50, .75x + 2.50y \leq 90$

4. $x + y \leq 300, 8x + 12y \geq 3{,}200$

5.

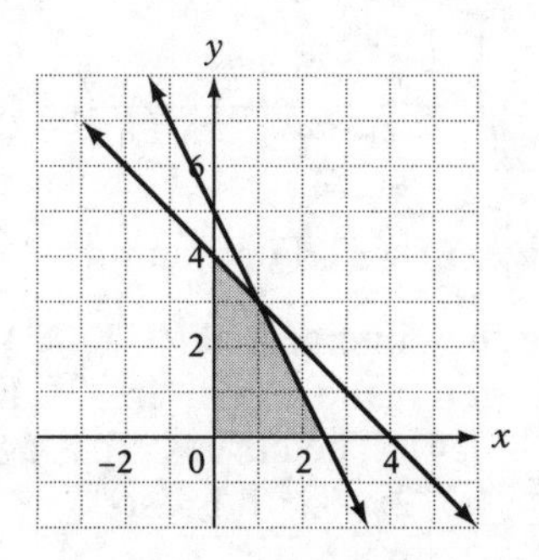

CHAPTER 8

Section 8.1

A

1. (2)	**3.** (3)	**5.** (3)	**7.** (1)	**9.** (4)
2. (3)	**4.** (4)	**6.** (4)	**8.** (2)	**10.** (1)

B

1. \$607.75

2. A) 12,553 after 10 years. B) Using guess-and-check, 14,065 after 15 years.

3. 6.25 mg

4. At $h = 10$, $T = 32.25$. At $h = 11$, $T = 31.58$. So between 10 and 11 hours.

5. Julia is correct. When x is negative, 5^x is greater than 6^x. For example, $5^{-2} = \frac{1}{25}$, which is greater than $6^{-2} = \frac{1}{36}$.

Section 8.2

A

1. (4)	**3.** (3)	**5.** (4)	**7.** (4)	**9.** (1)
2. (3)	**4.** (1)	**6.** (1)	**8.** (2)	**10.** (2)

B

1. $b = 3$ since $3^5 = 243$

2.

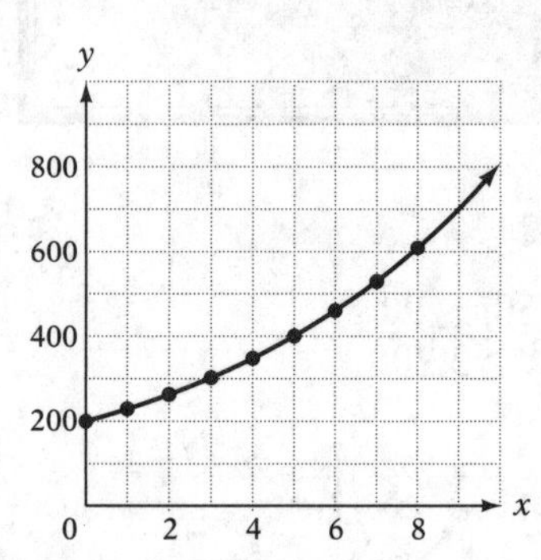

3. No. When x is negative, $2^x > 3^x$. For example $2^{-2} = \frac{1}{4}$ while $3^{-2} = \frac{1}{9}$.

4. (0,1). Graph both and they intersect at (0, 1).

5. No. For numbers between 0 and 1, $x^2 + 1 < 2^x$. For example, $.5^2 + 1 = 1.25$ but $2^{.5} = 1.41$. The curves cross again at approximately $x = 4.3$, so for any $x > 4.3$, $2x > x^2 + 1$ also.

Section 8.3

A

1. (1)	**3.** (2)	**5.** (2)	**7.** (1)	**9.** (4)
2. (1)	**4.** (3)	**6.** (1)	**8.** (3)	**10.** (3)

B

1. It resembles an exponential equation since it is somewhat flat at the beginning and then gets much steeper very quickly.

2. There is not enough information. Helpful would be some points with negative x-coordinates to see if they are close to the x-axis like an exponential curve or if they form a picture with symmetry like a parabola for a quadratic equation.

3. It is an exponential equation. Because the curve is always decreasing, it is exponential decay. The number being raised to the exponent will be between 0 and 1.

4. It is an exponential equation. Since $\frac{85}{50} = 1.7$, each number in the y column is 1.7 multiplied by the number above it. The missing number is $85 \cdot 1.7 = 144.5$.

5. A chart would have the y-values going from 30 to 24 to 19.2 to 15.36, which has the shape of curve with exponential decay.

Section 8.4

A

1. (2)	**3.** (4)	**5.** (2)	**7.** (3)	**9.** (1)
2. (3)	**4.** (1)	**6.** (4)	**8.** (3)	**10.** (3)

B

1. A) 901 million. B) By guess and check or with the graphing calculator, when $t = 8$, which is year 2018.

2. A) It will be approximately 52 degrees. B) By guess and check or with the graphing calculator, approximately 17 minutes.

3. A) For 10 years bank A will have $500 while bank B will have $179.08. B) For 54 years bank A will have $2,260 while bank B will have $2,325.50.

4. 1024 line segments

5. A) 147 pounds. B) Using guess and check, after 20 weeks he will be able to bench press about 250 pounds.

CHAPTER 9

Section 9.1

A

1. (1)	**3.** (4)	**5.** (2)	**7.** (1)	**9.** (3)
2. (4)	**4.** (2)	**6.** (3)	**8.** (2)	**10.** (4)

B

1. $S = 5{,}000Y + 40{,}000$

2. The 2 means that it is $2 for each mile. The 4 means that it is $4 to get into the taxi.

3. $H = 3Y + 5$ or $H = 5 + 3Y$

4. $W = 260 - 2M$

5. Many correct answers. One answer is the price of getting into a carnival is $20 while each ride costs $6. C is the cost of getting into the carnival and going on N rides.

Section 9.2

A

1. (1) **3.** (4) **5.** (4) **7.** (3) **9.** (3)
2. (1) **4.** (2) **6.** (1) **8.** (1) **10.** (4)

B

1. $P = 300{,}000(1.04)^t$
2. $S = 64(.75)^t$
3. $H = 40(1.15)^t$
4. 200,000 represents the $200,000 profits in the company's first year. 0.36 is the growth rate, so 1.36 is the growth factor or 1 plus the growth rate.
5. 300 means that when the training begins it takes her 300 minutes to run the marathon. 0.05 is the decay rate, so 0.95 is the growth (or decay) factor or 1 minus the decay rate.

CHAPTER 10

Section 10.1

A

1. (3) **3.** (3) **5.** (1) **7.** (4) **9.** (1)
2. (4) **4.** (3) **6.** (3) **8.** (2) **10.** (2)

B

1. Yes, this is a function. For each input value, there is only one output value. There are no repeats in the x-coordinates.
2. (a) $4 + 1 = 5$; (b) $f(3) = 2$; (c) $f(4) = 3$
3. The only value is $a = 6$.
4. No, it is not. $f(3)$ cannot be both 4 and 5.
5. Mia is correct. If there are no repeated y-values, the number will be the same. If there are repeated y-values, there will be fewer numbers in the range. There is no way to have more numbers in the range since that would require having repeated x values, which is not permitted for a function.

Section 10.2

A

1. (3) **3.** (4) **5.** (4) **7.** (4) **9.** (2)
2. (1) **4.** (1) **6.** (2) **8.** (1) **10.** (3)

B

1. Tyson is correct. It does pass the vertical line test. Two x values may correspond to the same y value.
2. Domain is $0 \le x \le 6$; range is $0 \le y \le 3$.
3. $f(g(3)) = f(5) = 2$
4. 2

5. $a = 2$ and $a = -2$ since the points $(2, 4)$ and $(-2, 4)$ are both on the graph.

Section 10.3

A

1. (4)	**3.** (3)	**5.** (2)	**7.** (2)	**9.** (1)
2. (1)	**4.** (4)	**6.** (4)	**8.** (3)	**10.** (1)

B

1. (a) 40; (b) $(x + 1)^2 + 8(x + 1) + 7 = x^2 + 10x + 16$

2. (a) $f(11) = 19$; (b) $g(3) = 11$.

3. Elizabeth is right. $h(g(f(x))) = [2(x + 2)]^2$. The correct expression is $g(h(f(x)))$.

4. 

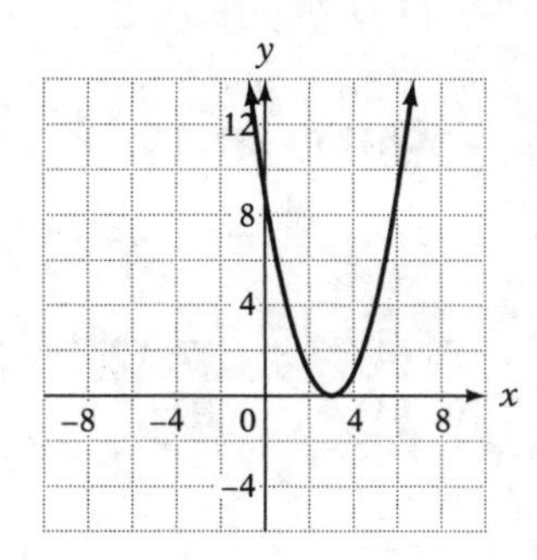

5. Since $g(7) = 15$, $2f(3) + 1 = 15$, so $f(3)$ must be 7.

Section 10.4

A

1. (4)	**3.** (2)	**5.** (4)	**7.** (3)	**9.** (3)
2. (1)	**4.** (3)	**6.** (1)	**8.** (2)	**10.** (2)

B

1. (a) $g(3) = f(3) + 1 = 1 + 1 = 2$; (b) $h(3) = f(3 + 1) = f(4) = 5$.

2.

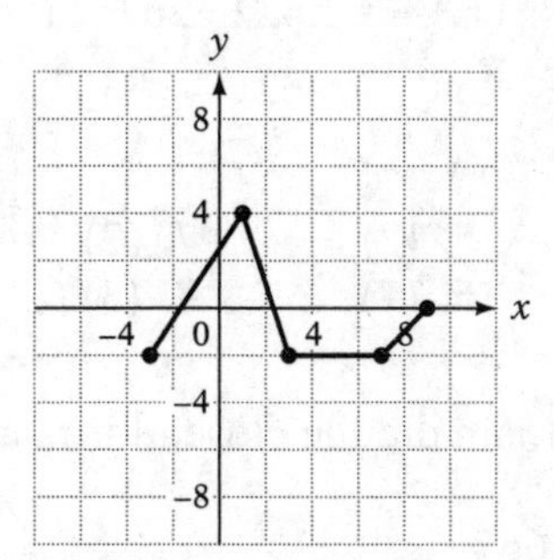

3. (a) The domain is $2 \le x \le 7$ because shifting up does not affect the domain. (b) The domain is $-2 \le x \le 3$ since the graph is shifted four units to the left.

4.

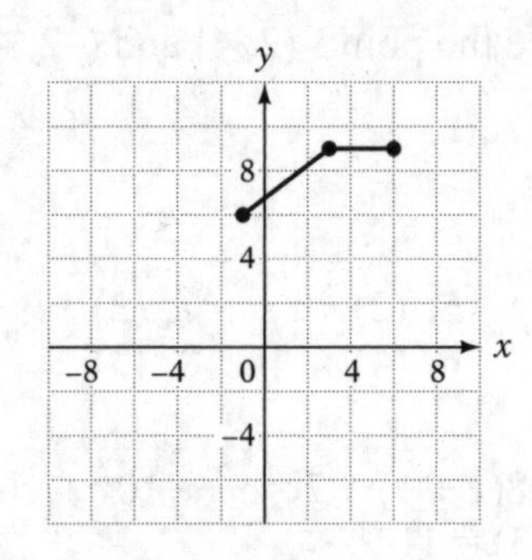

5. Since the graph of $g(x)$ is the graph of $f(x)$ shifted to the left two units, it can be expressed as $g(x) = f(x + 2)$.

CHAPTER 11

Section 11.1

A

1. (2)	**3.** (2)	**5.** (4)	**7.** (2)	**9.** (4)
2. (1)	**4.** (3)	**6.** (3)	**8.** (1)	**10.** (3)

Section 11.2

A

1. (3)	**3.** (3)	**5.** (4)	**7.** (1)	**9.** (2)
2. (2)	**4.** (2)	**6.** (1)	**8.** (4)	**10.** (3)

B

1. 5, 9, 13, 17
2. $a_{18} = 3 \cdot 58 = 174$, $a_{19} = 3 \cdot 174 = 522$
3. $a_1 = 50$, $a_n = 5 + a_{n-1}$ for $n > 1$
4. $a_1 = 6$, $a_n = 2a_{n-1}$ for $n > 1$
5. 3, 4, $2 \cdot 4 + 3 = 11$, $2 \cdot 11 + 4 = 26$, $2 \cdot 26 + 11 = 63$

Section 11.3

A

1. (2)	**3.** (4)	**5.** (2)	**7.** (2)	**9.** (4)
2. (3)	**4.** (3)	**6.** (1)	**8.** (3)	**10.** (1)

B

1. They are both right. Liam did the closed form, and Cecilia did the recursive form.
2. $a_{100} = 26 - 7(99) = -667$
3. (a) $a_n = \dfrac{1}{633{,}600} \cdot 2^n$ (b) $a_{38} = \dfrac{1}{633{,}600}\ 1 \cdot 2^{38} = 433{,}835$ miles. Yes, it will reach the moon.

4. $b_n = 1 \cdot 2^{n-1}$ so $b_{20} = 1 \cdot 2^{19} = 524{,}288$. $a_n = 1{,}000 + 200(n - 1)$ so $a_{20} = 1{,}000 + 200 \cdot 19 = 4{,}800$. So, $b_{20} > a_{20}$.

5. The closed form will be faster. The recursive form would require calculating all 100 values of the sequence.

CHAPTER 12

Section 12.1

A

1. (1)	**3.** (1)	**5.** (4)	**7.** (3)	**9.** (3)
2. (4)	**4.** (3)	**6.** (2)	**8.** (4)	**10.** (1)

B

1.

(a)

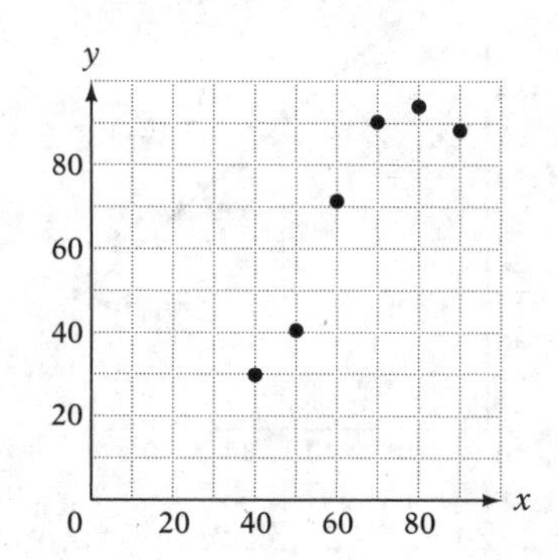

(b) $B = 1.35T - 18.34$

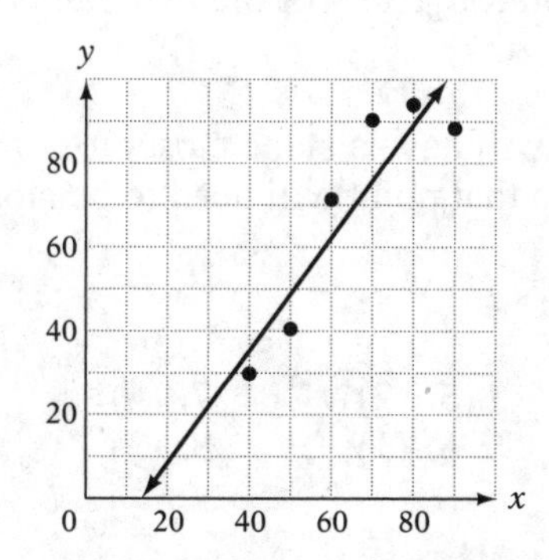

(c) If $T = 55$, $B = 1.35 \cdot 55 - 18.34 = 55.91$ or approximately 56 birds.

2.

(a)

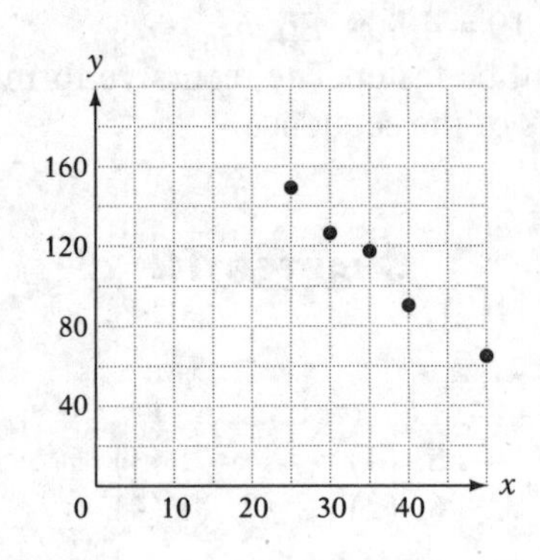

(b) $C = -3.39P + 232.41$

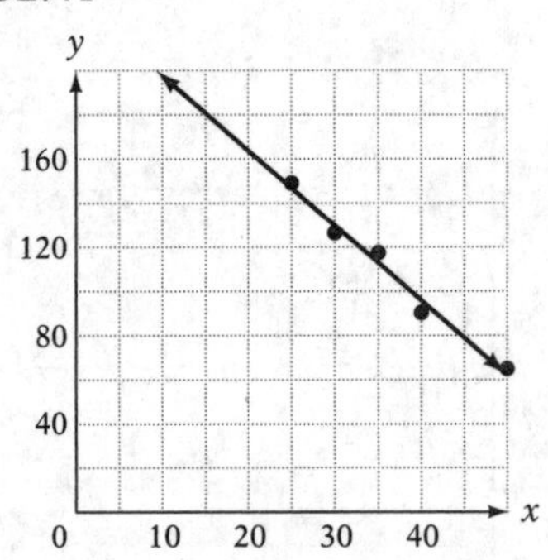

(c) $C = -3.39 \cdot 31 + 232.41 = 127.32$ or approximately 127 cars sold.

3. Because the five points all lie on a line, the line of best fit is just the equation of the line through any of the two points. The equation is $y = \frac{2}{3}x + 4$.

4. The one with (3, 1) will have a slope closer to zero since the line of best fit for the graph with that point will need to be closer to that point.

Section 12.2

A

1. (1)	**3.** (2)	**5.** (2)	**7.** (4)	**9.** (2)
2. (4)	**4.** (4)	**6.** (1)	**8.** (3)	**10.** (4)

B

1. The scatter plot on the left has the higher correlation coefficient since the points more closely resemble a straight line.

2. Jett is correct. If the line of best fit has a negative slope, the correlation coefficient will be negative.

3. The correlation coefficient will be positive since the line of best fit will have a positive slope. Also it will be less than 1 since the points do not all lie on the same line.
4. Pick any five x-coordinates and calculate the y-coordinates with the equation $y = x + 3$. That way all five points will be on the line of best fit so the correlation coefficient will be $+1$.

Section 12.3

A

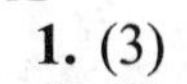

1. (3)	**3.** (1)	**5.** (1)	**7.** (2)	**9.** (1)
2. (2)	**4.** (3)	**6.** (2)	**8.** (3)	**10.** (2)

B

1.

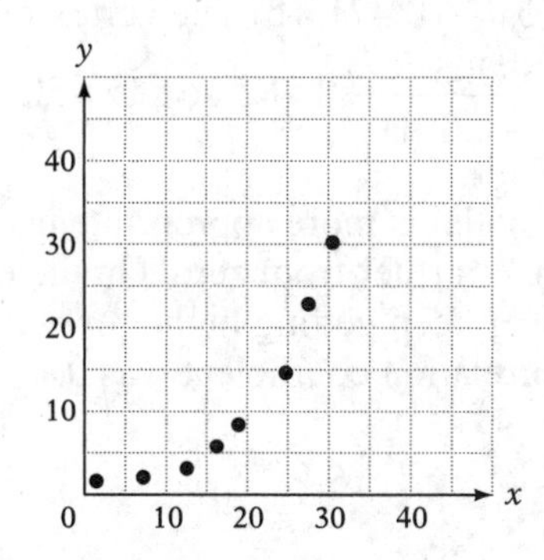

(a) The exponential model is most appropriate. The equation is $P = 1.07 \cdot 1.11^T$. (b) When $T = 20$, $P = 8.63$ or 8.63 million people.

2.

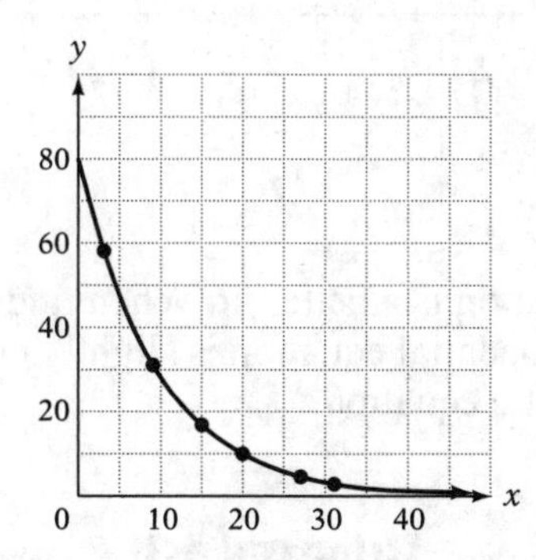

(a) The exponential decay model is most appropriate. The equation is $D = 78.99 \cdot 0.90^M$. (b) When $M = 40$, $D = 1.17$ degrees.

3.

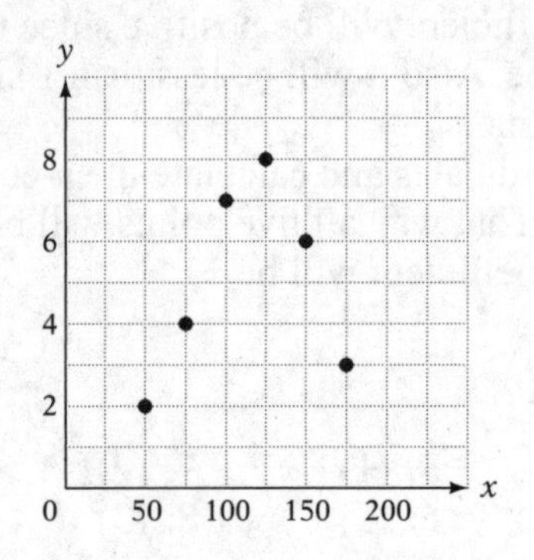

(a) The quadratic model is most appropriate. The equation is $F = -0.001x^2 + 0.303x - 10.471$. (b) The vertex of the parabola has an x-coordinate of $-\dfrac{0.303}{2\cdot(-0.001)} = 151.5$ so \$151.50 is the price that will generate the most profit.

4. (a) No. A quadratic model is more appropriate because the points get closer to zero and then further from zero. On the other hand, in a linear model the points either keep getting higher or lower. (b) The line of best fit was $y = 4$. The correlation coefficient was 0.

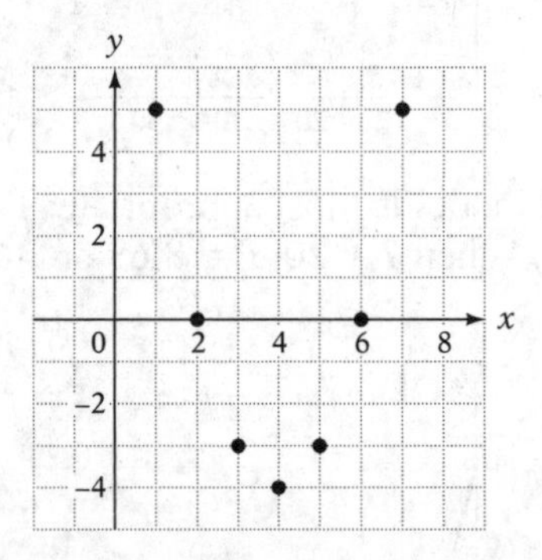

5. An exponential equation is a better fit. When finding the equation, the r^2 value for the exponential equation is slightly closer to +1 than the r^2 value for the quadratic equation.

CHAPTER 13

Section 13.1

A

1. (4)	**3.** (4)	**5.** (2)	**7.** (3)	**9.** (1)
2. (2)	**4.** (1)	**6.** (1)	**8.** (4)	**10.** (4)

B

1. One number is 19 so there will be three 19s making it the mode. The sum of all eight numbers is 160 so the other missing number is 23.
2. The second team must have many players that are close in height to six feet while the first team has some people much shorter than six feet and others much taller than six feet.
3. There can be two 81s, and then a 90 as the first three grades. To get a mean of 90, the sum of all five grades must be 450. The 81, 81, and 90 have a sum of 252 so the other two grades must have a sum of 198. They can't be 99 and 99 since that would change the mode, but they can be 98 and 100.
4. Five out of eleven. The first quartile is 4 and the third quartile is 15 so the numbers 8, 8, 9, 9, and 14 are between them.
5. 94. Her grades can be 90, 90, 90, 100, 100 and still meet the conditions of both median and mode being 90.

Section 13.2

A

1. (2)	**3.** (2)	**5.** (2)	**7.** (3)	**9.** (1)
2. (3)	**4.** (4)	**6.** (1)	**8.** (4)	**10.** (2)

B

1.

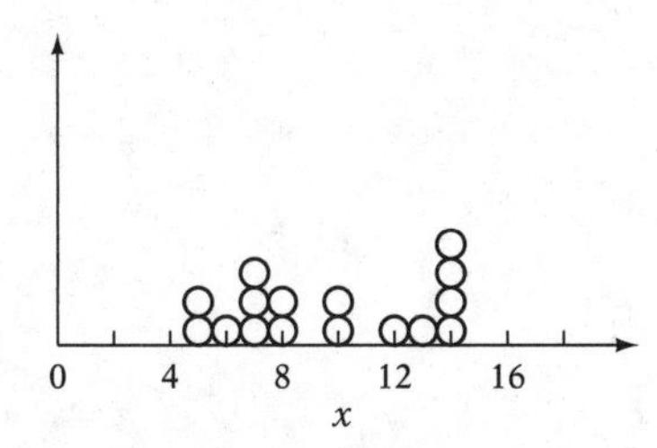

2. min = 2, Q1 = 4, med = 8, Q3 = 12, max = 20

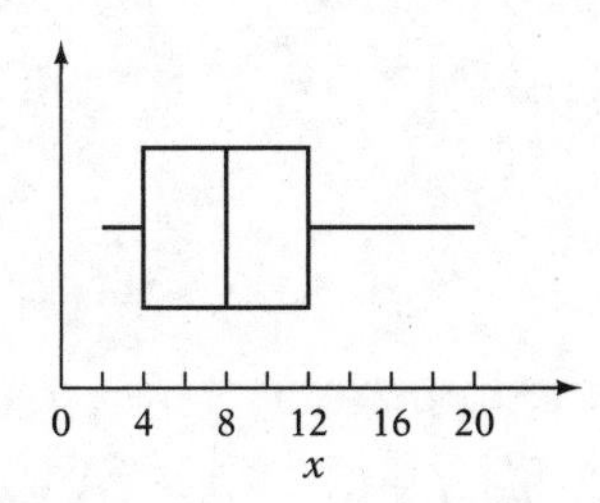

3. The players on the first team all have heights very close to 72 inches, whereas on the other team there are players that are much shorter and players that are much taller than 72 inches.

4. Any data set that when arranged from smallest to largest has 10, x, 13, x, x, 18, x, x, 25, x, 30 will produce this box plot.

5.

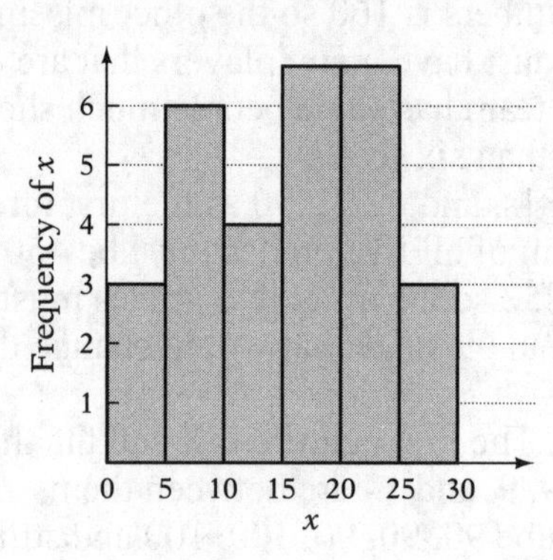

Glossary of Algebra I Terms

A

Addition property of equality A property of algebra that states that when equal values are added to both sides of a true equation, the equation continues to be true. To solve the equation $x - 2 = 5$, add 2 to both sides of the equation by using the addition property of equality.

Arithmetic sequence A number sequence in which the difference between two consecutive terms is a constant. The sequence 2, 5, 8, 11, 14, ... is an arithmetic sequence because the difference between consecutive terms is always 3.

Axis of symmetry An imaginary vertical line that passes through the vertex of a parabola. The equation for the axis of symmetry of a parabola is defined by $y = ax^2 + bx + c$ is $x = -\frac{b}{2a}$.

B

Base The number being raised to a power in an exponential expression. In the expression $2 \cdot 3^x$, the 3 is the base.

Binomial A polynomial with only two terms. $3x + 5$ is a binomial.

Box plot A graphical way to summarize data. The five numbers represented by the minimum, first quartile, median, third quartile, and maximum are graphed on a number line. A line segment connects the minimum to the first quartile. A rectangle is drawn around the first quartile and third quartile with a vertical line at the median. A line segment connects the third quartile to the maximum.

C

Closed form defined sequence A formula that defines the nth term of a sequence. The formula $n = 3 + 2(n - 1)$ is a closed form definition of a sequence. To get the 50th term of the sequence, substitute 50 for n in the definition.

Coefficient A number multiplied by a variable expression. In the expression $5x + 2$, 5 is the coefficient of x.

Common difference In an arithmetic sequence, the difference between consecutive terms. The common difference in the sequence 2, 5, 8, 11, 14, ... is 3.

Common ratio In a geometric sequence, the ratio between consecutive terms. The common ratio in the sequence 2, 6, 18, 54, 162, ... is 3 since $162/54 = 54/18 = 18/6 = 6/2 = 3$.

Commutative property of addition or multiplication The law from arithmetic that states that the order in which two numbers are added or multiplied does not matter. Because of the commutative property of addition, $5 + 2 = 2 + 5$.

Completing the square A method of solving a quadratic equation that involves turning one side of the equation into a perfect square trinomial.

Constant A number that does not have a variable part. In the expression $5x + 2$, 2 is a constant.

Correlation coefficient A number represented by r that measures how well a curve of best fit matches the points in a scatter plot. When the correlation coefficient is very close to 1 or to –1, the curve is a very good fit.

D

Degree of a polynomial The exponent on the highest power of a polynomial. In the polynomial $x^3 - 2x^2 + 5x - 2$, the degree is 3 since the highest power is a 3.

Difference of perfect squares Factoring a quadratic binomial in the form $x^2 - a^2$ into $(x - a)(x + a)$. For example, $x^2 - 9 = (x - 3)(x + 3)$.

Distributive property of multiplication over addition The rule that allows expressions of the form $a(b + c)$ to become $a \cdot b + a \cdot c$. For example, $2(3x + 5) = 6x + 10$.

Division property of equality A property of algebra that states that when both sides of a true equation are divided by the same non-zero number, the equation continues to be true. To solve the equation $2x = 8$, divide both sides of the equation by 2 using the division property of equality.

Domain The numbers that can be input into a function. When the function is defined as a set of ordered pairs or as a graph, the domain is the set of x-coordinates.

Dot plot A graphical way of representing a data set where each piece of data is represented with a dot.

E

Elimination method A way of solving a system of linear equations by combining the two equations in such a way as to eliminate one of the variables. In the set of equations,

$$x + 2y = 12$$
$$3x - 2y = -4$$

the y variable is eliminated by adding the two equations together.

Equation Two mathematical expressions with an equals sign between them. $3x + 2 = 8$ is an equation.

Exponential equation An equation in which the variable is an exponent. $2 \cdot 3^x = 18$ is an exponential equation.

Exponential function A function in which the variable is an exponent. $f(x) = 2 \cdot 3^x$ is an exponential function.

Expression Numbers and variables that are combined with the operations from math.

F

Factoring a polynomial Finding two polynomials that can be multiplied to become another polynomial. The polynomial $x^2 + 5x + 6$ can be factored into $(x + 2)(x + 3)$.

Factors The polynomials that evenly divide into a polynomial. The factors of $x^2 + 5x + 6$ are $(x + 2)$ and $(x + 3)$.

First quartile The number in a data set that is bigger than just 25% of the numbers in the set.

FOIL A way of multiplying two binomials of the form $(a + b)(c + d)$ where (F)irst the a and c are multiplied, then the (O)uters a and d are multiplied, then the (I)nners b and c are multiplied, and finally the (L)asts b and d are multiplied. Then the four results are added together. The product of $(x + 2)$ and $(x + 3)$ is $x^2 + 3x + 2x + 6 = x^2 + 5x + 6$ by this method.

Function Something that takes numbers as inputs and outputs numbers. Functions are often labeled with the letters f or g. The notation $f(2) = 7$ means that when the number 2 is input into function f, it outputs the number 7.

G

Geometric sequence A number sequence in which the ratio between two consecutive terms (what you get when you divide one term by the term before it) is a constant. The sequence 2, 6, 18, 54, 162, … is a geometric sequence because the ratio between consecutive terms is always 3.

Graph A visual way to describe the solution set to an equation. Each solution to the equation corresponds to an ordered pair that is graphed as a point on the coordinate plane. Each of the ordered pairs that satisfy an equation produce a point on the coordinate plane and the collection of all the points is the graph of the equation.

Greatest common factor The largest expression that divides evenly into two or more monomials. The greatest common factor of $6x^2$ and $8x^3$ is $2x^2$.

Growth rate In an exponential expression $a \cdot b^x$, the b is the growth rate. For example, in the equation $y = 500 \cdot 1.05^x$, the growth rate is 1.05.

H

Histogram A way of representing data with repeated values. Each value is represented by a bar whose height corresponds to the number of times that value is repeated. There are no spaces between the bars.

I

Increasing A function is increasing on an interval if making the input value larger also makes the y output larger. On a graph, an increasing function "goes up" from left to right.

Inequality Like an equation, but there is a $<$, $>$, $\leq$, or $\geq$ sign between the two expressions. $x + 2 > 5$ is a one-variable inequality. $y \leq 2x + 6$ is a two-variable inequality.

Interquartile range The difference between the number that is the third quartile of a data set and the number that is the first quartile of a data set.

Isolating a variable A variable is isolated when it is by itself on one side of an equation. In the equation $x + 2 = 5$, the x is not yet isolated. Subtracting 2 from both sides of the equation transforms the original equation into $x = 3$ with the x now isolated.

L

Like terms Terms that have the same variable part. They can be combined by adding or subtracting. $2x^2$ and $3x^2$ are like terms. $2x^2$ and $3x^3$ are not like terms.

Line of best fit A line that comes closest to the set of points in a scatter plot.

Linear equation An equation in which the greatest exponent is a 1. The equation $2x + 3 = 7$ is a linear equation.

Linear function A function in which the greatest exponent is a 1. The function $f(x) = 2x + 3$ is a linear function.

M

Mean The average of the numbers in a data set. Calculate the mean by adding all the numbers and dividing the total by the number of numbers in the set.

Median The middle number in a data set after it has been arranged from least to greatest. If there are an even number of numbers in the data set, the median is found by adding the two middle numbers and dividing by 2.

Mode The most frequent number in a data set.

Monomial A mathematical expression that has a coefficient and/or a variable part. $3x^2$ is a monomial. $3 + x^2$ is not a monomial.

Multiplication property of equality A property of algebra that states that when equal values are multiplied by both sides of a true equation, the equation continues to be true. To solve the equation $(1/2)x = 5$, multiply both sides of the equation by 2, using the multiplication property of equality.

O

Ordered pair Two numbers written in the form (x, y). An ordered pair can be a solution to a two-variable equation. For example, $(2, 5)$ is one solution to the equation $y = 2x + 1$. Ordered pairs can be graphed on the coordinate axes by locating the point with the x-coordinate equal to the x value and the y-coordinate equal to the y value.

P

Parabola A "U"-shaped curve that is the graph of the solution set of a quadratic equation.

Perfect square trinomial A quadratic trinomial of the form $x^2 + bx + (b/2)^2$, which can be factored into $(x + b/2)^2$. For example, $x^2 + 6x + 9 = (x + 3)^2$.

Piecewise function A function that has multiple rules for determining output values from input values, depending on what the input values are. If the function $f(x)$ is defined as

$$f(x) = \begin{cases} 2x + 1 \text{ if } x < 0 \\ x^2 \text{ if } x \geq 0 \end{cases}$$

then $f(-3) = 2(-3) + 1 = -5$ and $f(5) = 5^2 = 25$.

Polynomial The sum of one or more monomials. Each term of the polynomial has the form ax^n. $3x^4 + 2x^3 - 3x^2 + 6x - 1$ is a polynomial.

Q

Quadratic equation An equation in which the highest power on a variable is a 2. $x^2 + 5x + 6 = 0$ is a quadratic equation.

Quadratic formula The formula $x = \dfrac{\left(-b \pm \sqrt{(b^2 - 4ac)}\right)}{2a}$. This formula can be used to find the two solutions to the quadratic equation $ax^2 + bx + c = 0$.

Quadratic function A function in which the highest power on a variable is 2. $f(x) = x^2 + 5x + 6$ is a quadratic function.

Quadratic polynomial A polynomial in which the highest power on a variable is 2. $x^2 + 5x + 6$ is a quadratic polynomial.

R

Range The set of values that can be output from a function is the range of that function.

Recursively defined sequence A way to define a sequence in which the first term or terms of the sequence are given and a formula is given for calculating the next term based on the previous term or terms.

$a_1 = 5$

$a_n = 3 + a_{n-1}$ for $n > 1$

is a recursive definition for the sequence 5, 8, 11, 14, ….

Regression Finding a curve that best fits a scatter plot. Three types of regression are linear, quadratic, and exponential.

Residual plot A set of points that represent how far points on a graph deviate from a curve of best fit.

Roots The roots of an equation are the values that solve that equation. The roots of $x^2 + 5x + 6 = 0$ are -3 and -2.

S

Sequence A list of numbers that usually has some kind of pattern.

Slope The slope of a line is a measure of the line's steepness. The equation for the slope of a line that passes through the two points (x_1, y_1) and (x_2, y_2) is $m = \dfrac{y_2 - y_1}{x_2 - x_1}$.

Slope-Intercept form An equation in the form $y = mx + b$ where m and b are numbers in slope intercept form. $y = 2x - 1$ is in slope intercept form. When a two-variable equation is in slope intercept form, the graph of the equation has a y-intercept of (0, b) and a slope of m.

Solution set The set of numbers or ordered pairs that satisfies an equation. The solution set of $x + 2 = 5$ is $\{3\}$. The solution set of $x + y = 10$ has an infinite number of ordered pairs in its solution set, including (2, 8), (3, 7), and (4, 6).

Substitution method A method for solving a system of equations in which one variable is isolated in one of the equations, and the expression equal to that variable is substituted for it in the other equation.

Subtraction property of equality A property of algebra that states that when equal values are subtracted from both sides of a true equation, the equation continues to be true. To solve the equation $x + 2 = 5$, subtract 2 from both sides of the equation by using the subtraction property of equality.

System of equations Two or more equations with two or more unknowns to solve for. An example of a system of equations with a solution of (8, 2) is
$x + y = 10$
$x - y = 6$

T

Third quartile The number in a data set that is greater than just 75% of the numbers in the set.

Translation A translation of a graph is when the points on it are each shifted the same amount in the same direction. Examples of translations are vertical translations, horizontal translations, and combinations of vertical and horizontal translations.

Trinomial A polynomial with three terms. The polynomial $x^2 + 5x + 6$ is a trinomial.

V

Variable A letter, often an x, y, or z, that represents a value in a mathematical expression. In an algebraic equation, the variable is often the unknown that needs to be solved for.

Vertex The turning point of a parabola is its vertex. If the parabola opens upward, the vertex is the minimum point. If the parabola opens downward, the vertex is the maximum point.

Vertical line test A way of testing to see if a graph can represent a function. If at least one vertical line can pass through at least two points on the graph, the graph fails the vertical line test and cannot represent a function. If there are no vertical lines that can pass through at least two points, then the graph can be the graph of a function.

X

***x*-Intercept** The location where a curve crosses the x-axis. The y-coordinate of the x-intercept is 0.

Y

***y*-Intercept** The location where a curve crosses the y-axis. The x-coordinate of the y-intercept is 0. In slope-intercept form, $y = mx + b$, the y-intercept is located at $(0, b)$.

Z

Zeros The zeros of a function f are the numbers that can be input into the function so that 0 is output from the function. For example, the function $f(x) = 2x - 6$ has the number 3 as its only zero since $f(3) = 2(3) - 6 = 6 - 6 = 0$.

THE ALGEBRA I REGENTS EXAMINATION

The Regents Examination in Algebra I is a three-hour exam that is divided into four parts with a total of 37 questions. All 37 questions must be answered. Part I consists entirely of regular multiple-choice questions. Parts II, III, and IV each contain a set of questions that must be answered directly in the question booklet. You are required to show how you arrived at the answers for the questions in Parts II, III, and IV. The accompanying table shows how the exam breaks down.

Question Type	**Number of Questions**	**Credit Value**
Part I: Multiple choice	24	$24 \times 2 = 48$
Part II: 2-credit open ended	8	$8 \times 2 = 16$
Part III: 4-credit open ended	4	$4 \times 4 = 16$
Part IV: 6-credit open ended	1	$1 \times 6 = 6$
	Total = 37 questions	Total = 86 points

How Is the Exam Scored?

- Each of your answers to the 24 multiple-choice questions in Part I will be scored as either right or wrong.
- Solutions to questions in Parts II, III, and IV that are not completely correct may receive partial credit according to a special rating guide that is provided by the New York State Education Department. In order to receive full credit for a correct answer to a question in Parts II, III, or IV, you must show or explain how you arrived at your answer by indicating the key steps taken, including appropriate formula substitutions, diagrams, graphs, and charts. A correct numerical answer with no work shown will receive only 1 credit.
- The raw scores for the four parts of the test are added together. The maximum total raw score for the Algebra I Regents Examination is 86 points. Using a special conversion chart that is provided by the New York State Education Department, your total raw score will be equated to a final test score that falls within the usual 0 to 100 scale.

What Type of Calculator Is Required?

Graphing calculators are *required* for the Algebra I Regents Examination. During the administration of the Regents exam, schools are required to make a graphing calculator available for the exclusive use of each student. You will need to use your calculator to work with trigonometric functions of angles, find roots of numbers, and perform routine calculations.

Knowing how to use a graphing calculator gives you more options when deciding how to solve a problem. Rather than solving a problem algebraically with pen and paper, it may be easier to solve the same problem using a graph or table created by a graphing calculator. A graphical or numerical solution using a calculator can also be used to help confirm an answer obtained by solving the problem algebraically.

Are Any Formulas Provided?

The Algebra I Regents Examination test booklet will include a reference sheet containing the formulas in the accompanying table. Keep in mind that you may be required to know other formulas that are not included in this sheet. Please see pages 450 and 451 for these formulas.

What Else Should I Know?

- Do not omit any questions from Part I. Since there is no penalty for guessing, make certain that you record an answer for each of the 24 multiple-choice questions.
- If the method of solution is not stated in the problem, choose an appropriate method (numerical, graphical, or algebraic) with which you are most comfortable.
- If you solve a problem in Parts II, III, or IV using a trial-and-error approach, show the work for at least three guesses with appropriate checks. Should the correct answer be reached on the first trial, you must further illustrate your method by showing that guesses below and above the correct guess do not work.
- Avoid rounding errors when using a calculator. Unless otherwise directed, the (pi) key on a calculator should be used in computations involving the constant π rather than the common rational approximation of 3.14 or $\frac{22}{7}$. When performing a sequence of calculations in which the result of one calculation is used in a second calculation, do not round off. Instead, use the full power/display of the calculator by performing a "chain" calculation, saving intermediate results in the calculator's memory. Unless otherwise specified, rounding, if required, should be done only when the *final* answer is reached.
- Check that each answer is in the requested form. If a specific form is not required, answers may be left in any equivalent form, such as $\sqrt{75}$, $5\sqrt{3}$, or 8.660254038 (the full power/display of the calculator).
- If a problem requires using a formula that is not provided in the question, check the formula reference sheet in the test booklet to see if it is listed. Clearly write any formula you use before making any appropriate substitutions. Then evaluate the formula in step-by-step fashion.
- For any problem solved in Parts II, III, and IV using a graphing calculator, you must indicate how the calculator was used to obtain the answer such as by copying graphs or tables created by your calculator together with the equations used to produce them. When copying graphs, label each graph with its equation, state the dimensions of the viewing window, and identify the intercepts and any points of intersection with their coordinates. Whenever appropriate, indicate the rationale of your approach.

Examination June 2018

Algebra I

HIGH SCHOOL MATH REFERENCE SHEET

Conversions

1 inch = 2.54 centimeters	1 cup = 8 fluid ounces
1 meter = 39.37 inches	1 pint = 2 cups
1 mile = 5280 feet	1 quart = 2 pints
1 mile = 1760 yards	1 gallon = 4 quarts
1 mile = 1.609 kilometers	1 gallon = 3.785 liters
	1 liter = 0.264 gallon
1 kilometer = 0.62 mile	1 liter = 1000 cubic centimeters
1 pound = 16 ounces	
1 pound = 0.454 kilogram	
1 kilogram = 2.2 pounds	
1 ton = 2000 pounds	

Formulas

Triangle	$A = \frac{1}{2}bh$
Parallelogram	$A = bh$
Circle	$A = \pi r^2$
Circle	$C = \pi d$ or $C = 2\pi r$
General Prisms	$V = Bh$
Cylinder	$V = \pi r^2 h$
Sphere	$V = \frac{4}{3}\pi r^3$

Formulas (continued)

Cone	$V = \frac{1}{3}\pi r^2 h$
Pyramid	$V = \frac{1}{3}Bh$
Pythagorean Theorem	$a^2 + b^2 = c^2$
Quadratic Formula	$x = \frac{-b \pm \sqrt{b^2 - 4ac}}{2a}$
Arithmetic Sequence	$a_n = a_1 + (n - 1)d$
Geometric Sequence	$a_n = a_1 r^{n-1}$
Geometric Series	$S_n = \frac{a_1 - a_1 r^n}{1 - r}$ where $r \neq 1$
Radians	1 radian $= \frac{180}{\pi}$ degrees
Degrees	1 degree $= \frac{\pi}{180}$ radians
Exponential Growth/Decay	$A = A_0 e^{k(t - t_0)} + B_0$

PART I

Answer all 24 questions in this part. Each correct answer will receive 2 credits. No partial credit will be allowed. For each statement or question, write in the space provided the numeral preceding the word or expression that best completes the statement or answers the question. [48 credits]

1 The solution to $4p + 2 < 2(p + 5)$ is

(1) $p > -6$ (3) $p > 4$
(2) $p < -6$ (4) $p < 4$ 1 ____

2 If $k(x) = 2x^2 - 3\sqrt{x}$, then $k(9)$ is

(1) 315 (3) 159
(2) 307 (4) 153 2 ____

3 The expression $3(x^2 + 2x - 3) - 4(4x^2 - 7x + 5)$ is equivalent to

(1) $-13x - 22x + 11$ (3) $19x^2 - 22x + 11$
(2) $-13x^2 + 34x - 29$ (4) $19x^2 + 34x - 29$ 3 ____

4 The zeros of the function $p(x) = x^2 - 2x - 24$ are

(1) -8 and 3 (3) -4 and 6
(2) -6 and 4 (4) -3 and 8 4 ____

5 The box plot below summarizes the data for the average monthly high temperatures in degrees Fahrenheit for Orlando, Florida.

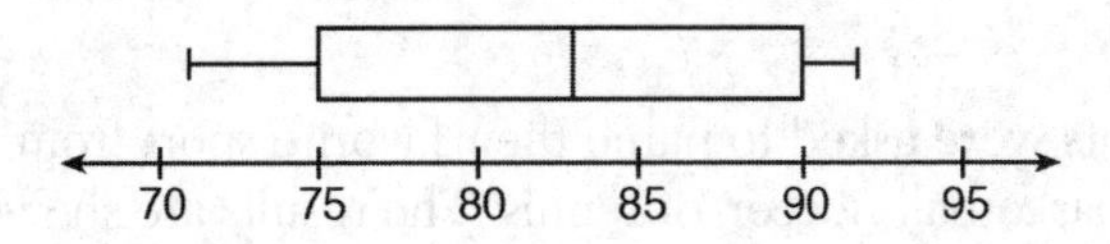

The third quartile is

(1) 92 (3) 83
(2) 90 (4) 71 5____

6 Joy wants to buy strawberries and raspberries to bring to a party. Strawberries cost $1.60 per pound and raspberries cost $1.75 per pound. If she only has $10 to spend on berries, which inequality represents the situation where she buys x pounds of strawberries and y pounds of raspberries?

(1) $1.60x + 1.75y \leq 10$ (3) $1.75x + 1.60y \leq 10$
(2) $1.60x + 1.75y \geq 10$ (4) $1.75x + 1.60y \geq 10$ 6____

7 On the main floor of the Kodak Hall at the Eastman Theater, the number of seats per row increases at a constant rate. Steven counts 31 seats in row 3 and 37 seats in row 6. How many seats are there in row 20?

(1) 65 (3) 69
(2) 67 (4) 71 7____

8 Which ordered pair below is *not* a solution to $f(x) = x^2 - 3x + 4$?

(1) (0, 4)
(2) (1.5, 1.75)
(3) (5, 14)
(4) (–1, 6)

8 _____

9 Students were asked to name their favorite sport from a list of basketball, soccer, or tennis. The results are shown in the table below.

	Basketball	Soccer	Tennis
Girls	42	58	20
Boys	84	41	5

What percentage of the students chose soccer as their favorite sport?

(1) 39.6%
(2) 41.4%
(3) 50.4%
(4) 58.6%

9 _____

10 The trinomial $x^2 - 14x + 49$ can be expressed as

(1) $(x - 7)^2$
(2) $(x + 7)^2$
(3) $(x - 7)(x + 7)$
(4) $(x - 7)(x + 2)$

10 _____

11 A function is defined as {(0, 1), (2, 3), (5, 8), (7, 2)}. Isaac is asked to create one more ordered pair for the function. Which ordered pair can he add to the set to keep it a function?

(1) (0, 2)
(2) (5, 3)
(3) (7, 0)
(4) (1, 3)

11 _____

12 The quadratic equation $x^2 - 6x = 12$ is rewritten in the form $(x + p)^2 = q$, where q is a constant. What is the value of p?

(1) –12 (3) –3
(2) –9 (4) 9 12 ____

13 Which of the quadratic functions below has the *smallest* minimum value?

$h(x) = x^2 + 2x - 6$
(1)

$k(x) = (x + 5)(x + 2)$
(3)

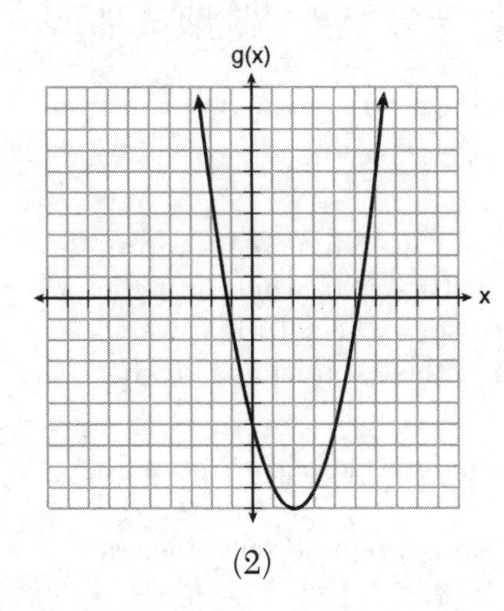

(2)

x	f(x)
–1	–2
0	–5
1	–6
2	–5
3	–2

(4) 13 ____

14 Which situation is *not* a linear function?

(1) A gym charges a membership fee of \$10.00 down and \$10.00 per month.
(2) A cab company charges \$2.50 initially and \$3.00 per mile.
(3) A restaurant employee earns \$12.50 per hour.
(4) A \$12,000 car depreciates 15% per year. 14 ____

15 The Utica Boilermaker is a 15-kilometer road race. Sara is signed up to run this race and has done the following training runs:

I. 10 miles
II. 44,880 feet
III. 15,560 yards

Which run(s) are at least 15 kilometers?

(1) I, only
(2) II, only
(3) I and III
(4) II and III

15 ____

16 If $f(x) = x^2 + 2$, which interval describes the range of this function?

(1) $(-\infty, \infty)$
(2) $[0, \infty)$
(3) $[2, \infty)$
(4) $(-\infty, 2]$

16 ____

17 The amount Mike gets paid weekly can be represented by the expression $2.50a + 290$, where a is the number of cell phone accessories he sells that week. What is the constant term in this expression and what does it represent?

(1) $2.50a$, the amount he is guaranteed to be paid each week
(2) $2.50a$, the amount he earns when he sells a accessories
(3) 290, the amount he is guaranteed to be paid each week
(4) 290, the amount he earns when he sells a accessories

17 ____

18 A cubic function is graphed on the set of axes below.

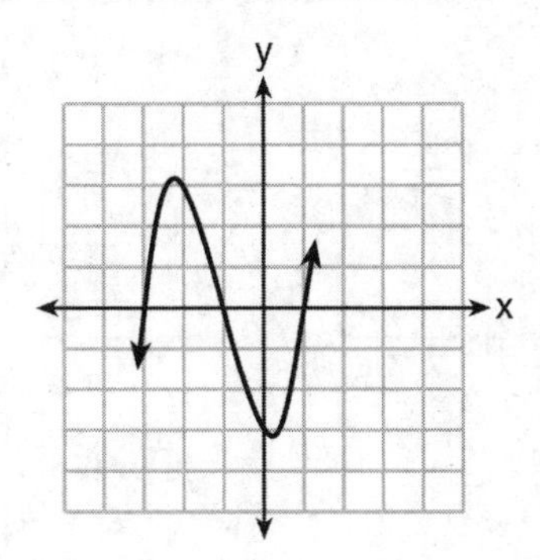

Which function could represent this graph?

(1) $f(x) = (x - 3)(x - 1)(x + 1)$
(2) $g(x) = (x + 3)(x + 1)(x - 1)$
(3) $h(x) = (x - 3)(x - 1)(x + 3)$
(4) $k(x) = (x + 3)(x + 1)(x - 3)$

18 ____

19 Mrs. Allard asked her students to identify which of the polynomials below are in standard form and explain why.

I. $15x^4 - 6x + 3x^2 - 1$
II. $12x^3 + 8x + 4$
III. $2x^5 + 8x^2 + 10x$

Which student's response is correct?

(1) Tyler said I and II because the coefficients are decreasing.
(2) Susan said only II because all the numbers are decreasing.
(3) Fred said II and III because the exponents are decreasing.
(4) Alyssa said II and III because they each have three terms.

19 ____

20 Which graph does *not* represent a function that is always increasing over the entire interval $-2 < x < 2$?

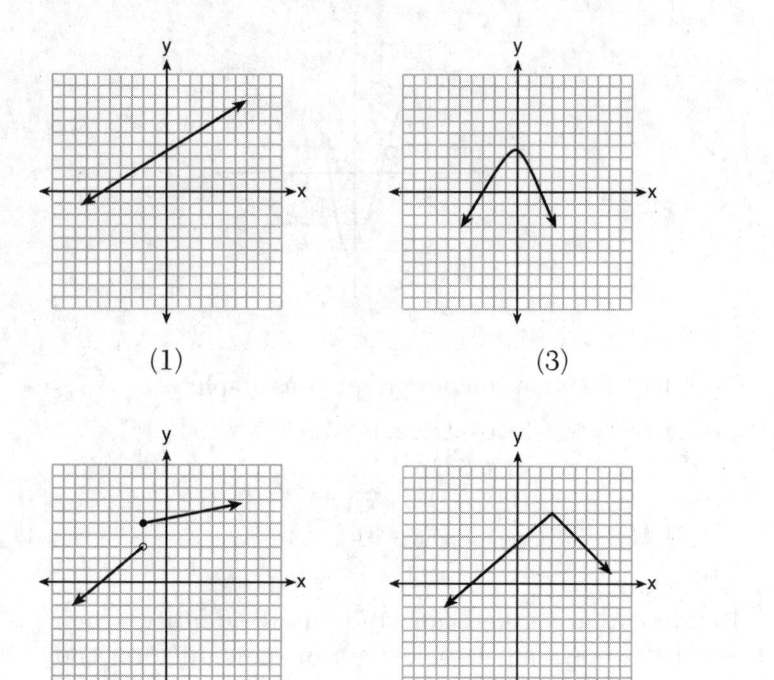

20 _____

21 At an ice cream shop, the profit, $P(c)$, is modeled by the function $P(c) = 0.87c$, where c represents the number of ice cream cones sold. An appropriate domain for this function is

(1) an integer ≤ 0
(2) an integer ≥ 0
(3) a rational number ≤ 0
(4) a rational number ≥ 0

21 _____

22 How many real-number solutions does $4x^2 + 2x + 5 = 0$ have?

(1) one
(2) two
(3) zero
(4) infinitely many

22 ____

23 Students were asked to write a formula for the length of a rectangle by using the formula for its perimeter, $p = 2\ell + 2w$. Three of their responses are shown below.

I. $\ell = \frac{1}{2}p - w$

II. $\ell = \frac{1}{2}(p - 2w)$

III. $\ell = \frac{p - 2w}{2}$

Which responses are correct?

(1) I and II, only
(2) II and III, only
(3) I and III, only
(4) I, II, and III

23 ____

24 If $a_n = n(a_{n-1})$ and $a_1 = 1$, what is the value of a_5?

(1) 5
(2) 20
(3) 120
(4) 720

24 ____

PART II

Answer all 8 questions in this part. Each correct answer will receive 2 credits. Clearly indicate the necessary steps, including appropriate formula substitutions, diagrams, graphs, charts, etc. For all questions in this part, a correct numerical answer with no work shown will receive only 1 credit. [16 credits]

25 Graph $f(x) = \sqrt{x+2}$ over the domain $-2 \leq x \leq 7$.

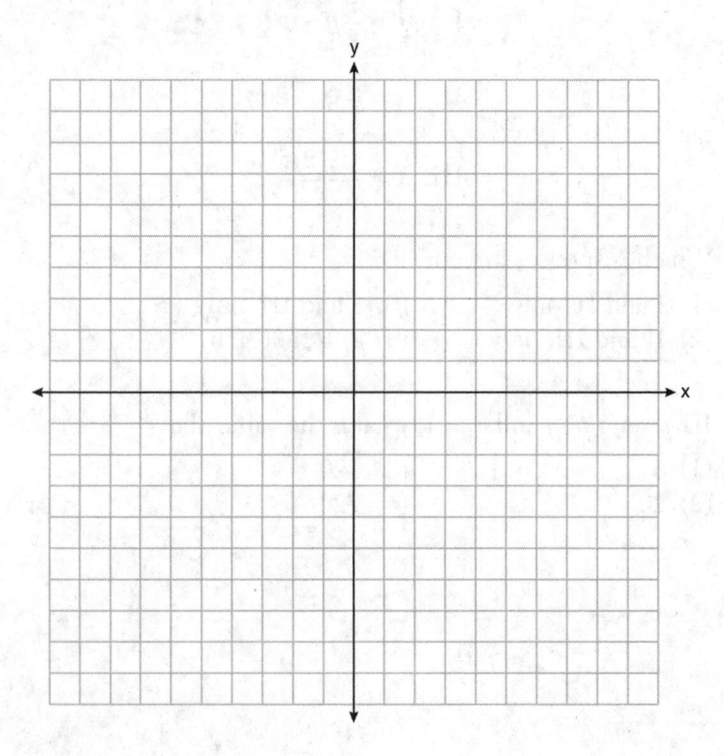

26 Caleb claims that the ordered pairs shown in the table below are from a nonlinear function.

x	f(x)
0	2
1	4
2	8
3	16

State if Caleb is correct. Explain your reasoning.

27 Solve for x to the *nearest tenth*: $x^2 + x - 5 = 0$.

28 The graph of the function $p(x)$ is represented below. On the same set of axes, sketch the function $p(x + 2)$.

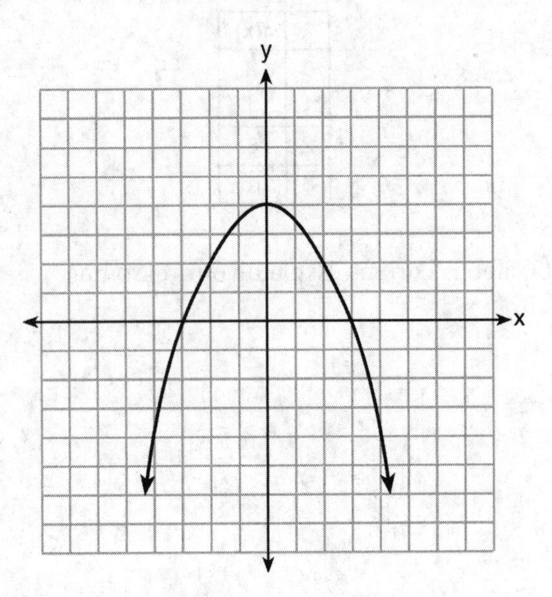

29 When an apple is dropped from a tower 256 feet high, the function $h(t) = -16t^2 + 256$ models the height of the apple, in feet, after t seconds. Determine, algebraically, the number of seconds it takes the apple to hit the ground.

30 Solve the equation below algebraically for the exact value of x.

$$6 - \frac{2}{3}(x + 5) = 4x$$

31 Is the product of $\sqrt{16}$ and $\frac{4}{7}$ rational or irrational?

Explain your reasoning.

32 On the set of axes below, graph the piecewise function:

$$f(x) = \begin{cases} -\frac{1}{2}x, & x < 2 \\ x, & x \geq 2 \end{cases}$$

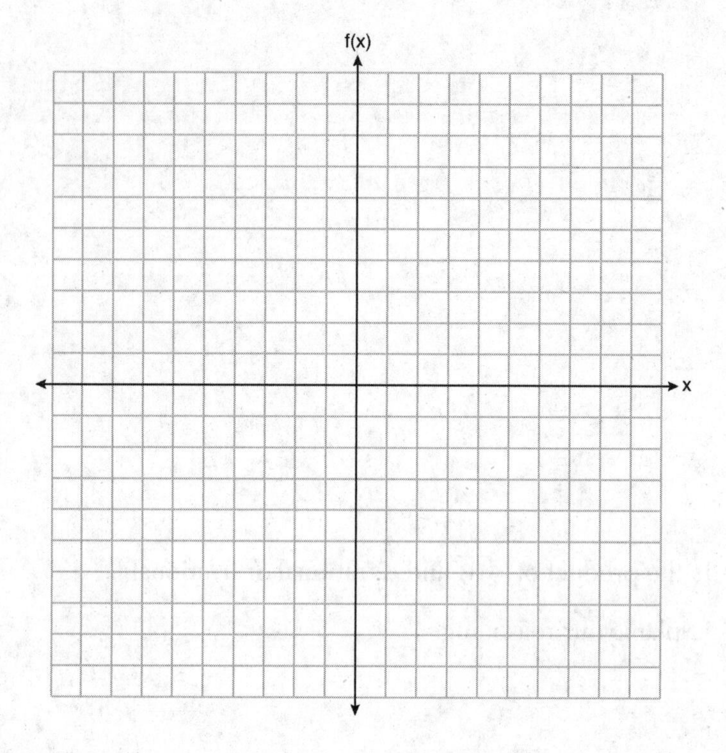

PART III

Answer all 4 questions in this part. Each correct answer will receive 4 credits. Clearly indicate the necessary steps, including appropriate formula substitutions, diagrams, graphs, charts, etc. For all questions in this part, a correct numerical answer with no work shown will receive only 1 credit. [16 credits]

33 A population of rabbits in a lab, $p(x)$, can be modeled by the function $p(x) = 20(1.014)^x$, where x represents the number of days since the population was first counted.

Explain what 20 and 1.014 represent in the context of the problem.

Determine, to the *nearest tenth*, the average rate of change from day 50 to day 100.

34 There are two parking garages in Beacon Falls. Garage A charges \$7.00 to park for the first 2 hours, and each additional hour costs \$3.00. Garage B charges \$3.25 per hour to park.

When a person parks for at least 2 hours, write equations to model the cost of parking for a total of x hours in Garage A and Garage B.

Determine algebraically the number of hours when the cost of parking at both garages will be the same.

35 On the set of axes below, graph the following system of inequalities:

$$2y + 3x \leq 14$$
$$4x - y < 2$$

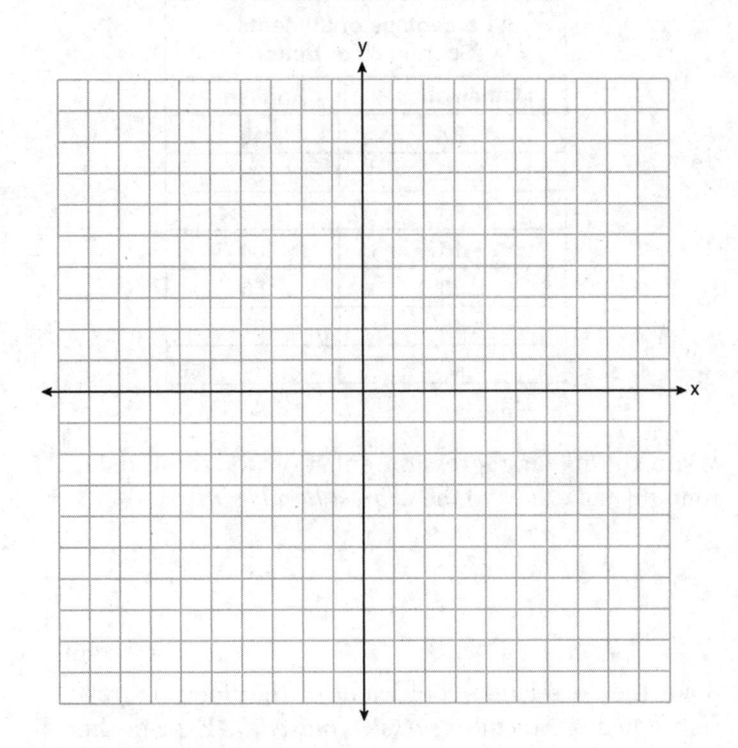

Determine if the point (1, 2) is in the solution set. Explain your answer.

36 The percentage of students scoring 85 or better on a mathematics final exam and an English final exam during a recent school year for seven schools is shown in the table below.

Percentage of Students Scoring 85 or Better	
Mathematics, x	**English, y**
27	46
12	28
13	45
10	34
30	56
45	67
20	42

Write the linear regression equation for these data, rounding all values to the *nearest hundredth*.

State the correlation coefficient of the linear regression equation, to the *nearest hundredth*. Explain the meaning of this value in the context of these data.

PART IV

Answer the question in this part. A correct answer will receive 6 credits. Clearly indicate the necessary steps, including appropriate formula substitutions, diagrams, graphs, charts, etc. A correct numerical answer with no work shown will receive only 1 credit. [6 credits]

37 Dylan has a bank that sorts coins as they are dropped into it. A panel on the front displays the total number of coins inside as well as the total value of these coins. The panel shows 90 coins with a value of $17.55 inside of the bank.

If Dylan only collects dimes and quarters, write a system of equations in two variables or an equation in one variable that could be used to model this situation.

Using your equation or system of equations, algebraically determine the number of quarters Dylan has in his bank.

Question 37 is continued on the next page.

Question 37 continued.

Dylan's mom told him that she would replace each one of his dimes with a quarter. If he uses all of his coins, determine if Dylan would then have enough money to buy a game priced at $20.98 if he must also pay an 8% sales tax. Justify your answer.

Answers June 2018

Algebra I

Answer Key

PART I

1. (4)	**5.** (2)	**9.** (1)	**13.** (2)	**17.** (3)	**21.** (2)
2. (4)	**6.** (1)	**10.** (1)	**14.** (4)	**18.** (2)	**22.** (3)
3. (2)	**7.** (1)	**11.** (4)	**15.** (1)	**19.** (3)	**23.** (4)
4. (3)	**8.** (4)	**12.** (3)	**16.** (3)	**20.** (3)	**24.** (3)

PART II

25.

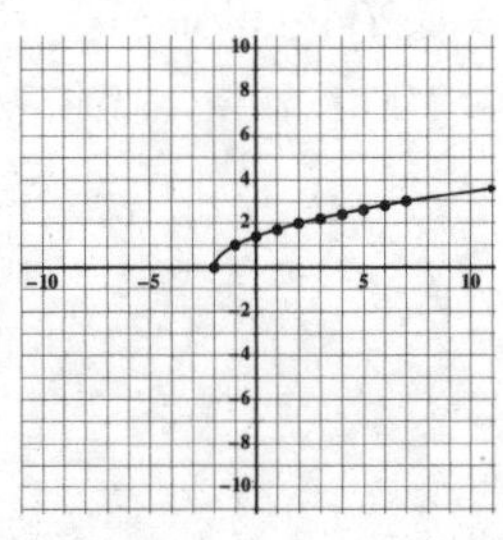

26. Caleb is correct. It is a nonlinear function.

27. $x = 1.8$ and $x = -2.8$

28. See graph

29. 4 seconds

30. $\frac{4}{7}$

31. The product is rational.

32. See graph

PART III

33. 20 is the starting population. 1.014 is the growth factor. The average rate of change is 0.8.

34. For Garage A, $A = 7 + 3(x - 2)$. For Garage B, $B = 3.25x$. Both garages cost the same when you park for 4 hours.

35. (1, 2) is not in the solution set.

36. $y = 0.96x + 23.95$ The correlation coefficient is 0.92, which is a strong positive correlation.

PART IV

37. $0.10d + 0.25q = 17.55$

$d + q = 90$

Dylan has 57 quarters. No, he would not be able to buy the game.

In **Parts II–IV**, you are required to show how you arrived at your answers. For sample methods of solutions, see Barron's *Regents Exams and Answers* book for Algebra I.

Examination June 2019

Algebra I

HIGH SCHOOL MATH REFERENCE SHEET

Conversions

1 inch = 2.54 centimeters	1 cup = 8 fluid ounces
1 meter = 39.37 inches	1 pint = 2 cups
1 mile = 5280 feet	1 quart = 2 pints
1 mile = 1760 yards	1 gallon = 4 quarts
1 mile = 1.609 kilometers	1 gallon = 3.785 liters
	1 liter = 0.264 gallon
1 kilometer = 0.62 mile	1 liter = 1000 cubic centimeters
1 pound = 16 ounces	
1 pound = 0.454 kilogram	
1 kilogram = 2.2 pounds	
1 ton = 2000 pounds	

Formulas

Triangle	$A = \frac{1}{2}bh$
Parallelogram	$A = bh$
Circle	$A = \pi r^2$
Circle	$C = \pi d$ or $C = 2\pi r$
General Prisms	$V = Bh$
Cylinder	$V = \pi r^2 h$
Sphere	$V = \frac{4}{3}\pi r^3$

Formulas (continued)

Cone	$V = \frac{1}{3}\pi r^2 h$
Pyramid	$V = \frac{1}{3}Bh$
Pythagorean Theorem	$a^2 + b^2 = c^2$
Quadratic Formula	$x = \frac{-b \pm \sqrt{b^2 - 4ac}}{2a}$
Arithmetic Sequence	$a_n = a_1 + (n - 1)d$
Geometric Sequence	$a_n = a_1 r^{n-1}$
Geometric Series	$S_n = \frac{a_1 - a_1 r^n}{1 - r}$ where $r \neq 1$
Radians	1 radian $= \frac{180}{\pi}$ degrees
Degrees	1 degree $= \frac{\pi}{180}$ radians
Exponential Growth/Decay	$A = A_0 e^{k(t-t_0)} + B_0$

PART I

Answer all 24 questions in this part. Each correct answer will receive 2 credits. No partial credit will be allowed. For each statement or question, write in the space provided the numeral preceding the word or expression that best completes the statement or answers the question. [48 credits]

1 The expression $w^4 - 36$ is equivalent to

(1) $(w^2 - 18)(w^2 - 18)$ (3) $(w^2 - 6)(w^2 - 6)$

(2) $(w^2 + 18)(w^2 - 18)$ (4) $(w^2 + 6)(w^2 - 6)$ 1 ____

2 If $f(x) = 4x + 5$, what is the value of $f(-3)$?

(1) -2 (3) 17

(2) -7 (4) 4 2 ____

3 Which relation is *not* a function?

x	y
–10	–2
–6	2
–2	6
1	9
5	13

(1)

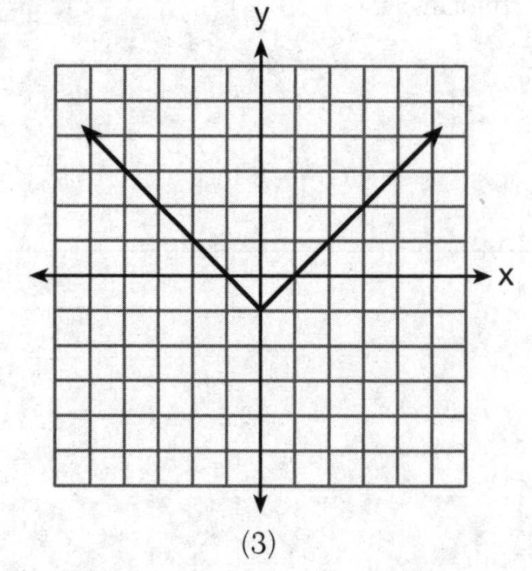

(3)

$3x + 2y = 4$

(2)

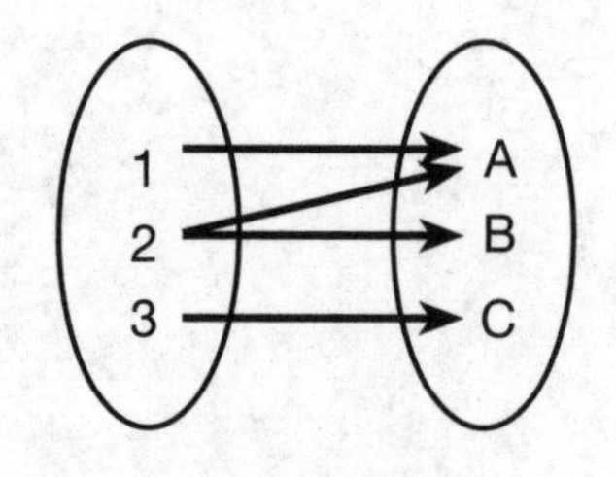

(4) 3 ____

4 Given: $f(x) = (x - 2)^2 + 4$
$g(x) = (x - 5)^2 + 4$

When compared to the graph of $f(x)$, the graph of $g(x)$ is

(1) shifted 3 units to the left
(2) shifted 3 units to the right
(3) shifted 5 units to the left
(4) shifted 5 units to the right

4 _____

5 Students were asked to write $6x^5 + 8x - 3x^3 + 7x^7$ in standard form.

Shown below are four student responses.

Anne: $7x^7 + 6x^5 - 3x^3 + 8x$
Bob: $-3x^3 + 6x^5 + 7x^7 + 8x$
Carrie: $8x + 7x^7 + 6x^5 - 3x^3$
Dylan: $8x - 3x^3 + 6x^5 + 7x^7$

Which student is correct?

(1) Anne
(2) Bob
(3) Carrie
(4) Dylan

5 _____

6 The function f is shown in the table below.

x	f(x)
0	1
1	3
2	9
3	27

Which type of function best models the given data?

(1) exponential growth function
(2) exponential decay function
(3) linear function with positive rate of change
(4) linear function with negative rate of change

6 _____

7 Which expression results in a rational number?

(1) $\sqrt{2} \bullet \sqrt{18}$
(2) $5 \bullet \sqrt{5}$
(3) $\sqrt{2} + \sqrt{2}$
(4) $3\sqrt{2} + 2\sqrt{3}$

7 ____

8 A polynomial function is graphed below.

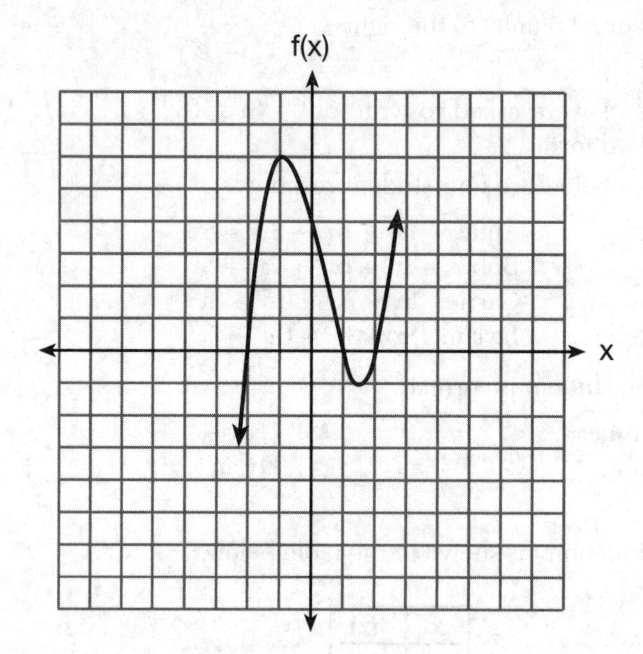

Which function could represent this graph?

(1) $f(x) = (x + 1)(x^2 + 2)$
(2) $f(x) = (x - 1)(x^2 - 2)$
(3) $f(x) = (x - 1)(x^2 - 4)$
(4) $f(x) = (x + 1)(x^2 + 4)$

8 ____

9 When solving $p^2 + 5 = 8p - 7$, Kate wrote $p^2 + 12 = 8p$. The property she used is

(1) the associative property
(2) the commutative property
(3) the distributive property
(4) the addition property of equality

9 ____

10 David wanted to go on an amusement park ride. A sign posted at the entrance read "You must be greater than 42 inches tall and no more than 57 inches tall for this ride." Which inequality would model the height, x, required for this amusement park ride?

(1) $42 < x \leq 57$
(2) $42 > x \geq 57$
(3) $42 < x$ or $x \leq 57$
(4) $42 > x$ or $x \geq 57$

10 _____

11 Which situation can be modeled by a linear function?

(1) The population of bacteria triples every day.
(2) The value of a cell phone depreciates at a rate of 3.5% each year.
(3) An amusement park allows 50 people to enter every 30 minutes.
(4) A baseball tournament eliminates half of the teams after each round.

11 _____

12 Jenna took a survey of her senior class to see whether they preferred pizza or burgers. The results are summarized in the table below.

	Pizza	Burgers
Male	23	42
Female	31	26

Of the people who preferred burgers, approximately what percentage were female?

(1) 21.3
(2) 38.2
(3) 45.6
(4) 61.9

12 _____

13 When $3a + 7b > 2a - 8b$ is solved for a, the result is

(1) $a > -b$
(2) $a < -b$
(3) $a < -15b$
(4) $a > -15b$

13 _____

14 Three functions are shown below.

A: $g(x) = -\frac{3}{2}x + 4$

B: $f(x) = (x + 2)(x + 6)$

C:

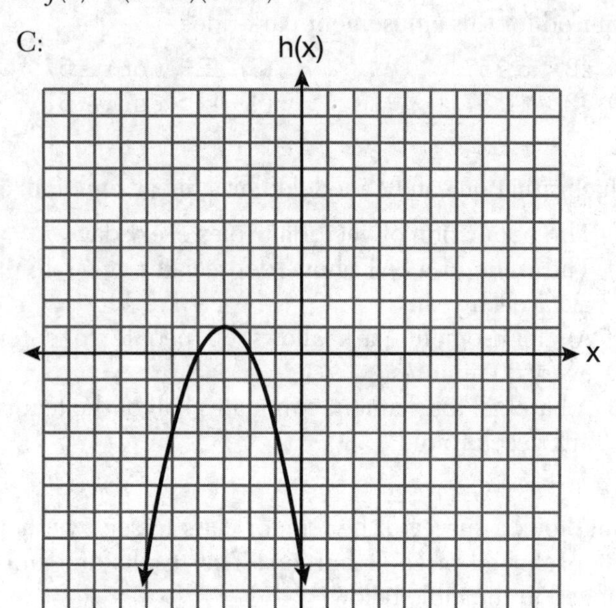

Which statement is true?

(1) B and C have the same zeros.
(2) A and B have the same y-intercept.
(3) B has a minimum and C has a maximum.
(4) C has a maximum and A has a minimum. 14 ____

15 Nicci's sister is 7 years less than twice Nicci's age, a. The sum of Nicci's age and her sister's age is 41. Which equation represents this relationship?

(1) $a + (7 - 2a) = 41$
(2) $a + (2a - 7) = 41$
(3) $2a - 7 = 41$
(4) $a = 2a - 7$ 15 ____

16 The population of a small town over four years is recorded in the chart below, where 2013 is represented by $x = 0$. [Population is rounded to the nearest person]

Year	2013	2014	2015	2016
Population	3810	3943	4081	4224

The population, $P(x)$, for these years can be modeled by the function $P(x) = ab^x$, where b is rounded to the nearest thousandth. Which statements about this function are true?

I. $a = 3810$
II. $a = 4224$
III. $b = 0.035$
IV. $b = 1.035$

(1) I and III
(2) I and IV
(3) II and III
(4) II and IV

16 ____

17 When written in factored form, $4w^2 - 11w - 3$ is equivalent to

(1) $(2w + 1)(2w - 3)$
(2) $(2w - 1)(2w + 3)$
(3) $(4w + 1)(w - 3)$
(4) $(4w - 1)(w + 3)$

17 ____

18 Which ordered pair does *not* represent a point on the graph of $y = 3x^2 - x + 7$?

(1) $(-1.5, 15.25)$
(2) $(0.5, 7.25)$
(3) $(1.25, 10.25)$
(4) $(2.5, 23.25)$

18 ____

19 Given the following three sequences:

I. 2, 4, 6, 8, 10...
II. 2, 4, 8, 16, 32...
III. $a, a + 2, a + 4, a + 6, a + 8...$

Which ones are arithmetic sequences?

(1) I and II, only
(2) I and III, only
(3) II and III, only
(4) I, II, and III

19 ____

20 A grocery store sells packages of beef. The function $C(w)$ represents the cost, in dollars, of a package of beef weighing w pounds. The most appropriate domain for this function would be

(1) integers
(2) rational numbers
(3) positive integers
(4) positive rational numbers

20 ____

21 The roots of $x^2 - 5x - 4 = 0$ are

(1) 1 and 4
(2) $\frac{5 \pm \sqrt{41}}{2}$
(3) –1 and –4
(4) $\frac{-5 \pm \sqrt{41}}{2}$

21 ____

22 The following table shows the heights, in inches, of the players on the opening-night roster of the 2015–2016 New York Knicks.

84	80	87	75	77	79	80	74	76	80	80	82	82

The population standard deviation of these data is approximately

(1) 3.5
(2) 13
(3) 79.7
(4) 80

22 ____

23 A population of bacteria can be modeled by the function $f(t) = 1000(0.98)^t$, where t represents the time since the population started decaying, and $f(t)$ represents the population of the remaining bacteria at time t. What is the rate of decay for this population?

(1) 98%
(2) 2%
(3) 0.98%
(4) 0.02%

23 ____

24 Bamboo plants can grow 91 centimeters per day. What is the approximate growth of the plant, in inches per hour?

(1) 1.49
(2) 3.79
(3) 9.63
(4) 35.83

24 ____

PART II

Answer all 8 questions in this part. Each correct answer will receive 2 credits. Clearly indicate the necessary steps, including appropriate formula substitutions, diagrams, graphs, charts, etc. For all questions in this part, a correct numerical answer with no work shown will receive only 1 credit. [16 credits]

25 Solve algebraically for x: $-\frac{2}{3}(x + 12) + \frac{2}{3}x = -\frac{5}{4}x + 2$

26 If $C = G - 3F$, find the trinomial that represents C when $F = 2x^2 + 6x - 5$ and $G = 3x^2 + 4$.

27 Graph the following piecewise function on the set of axes below.

$$f(x) = \begin{cases} |x|, & -5 \le x < 2 \\ -2x + 10, & 2 \le x \le 6 \end{cases}$$

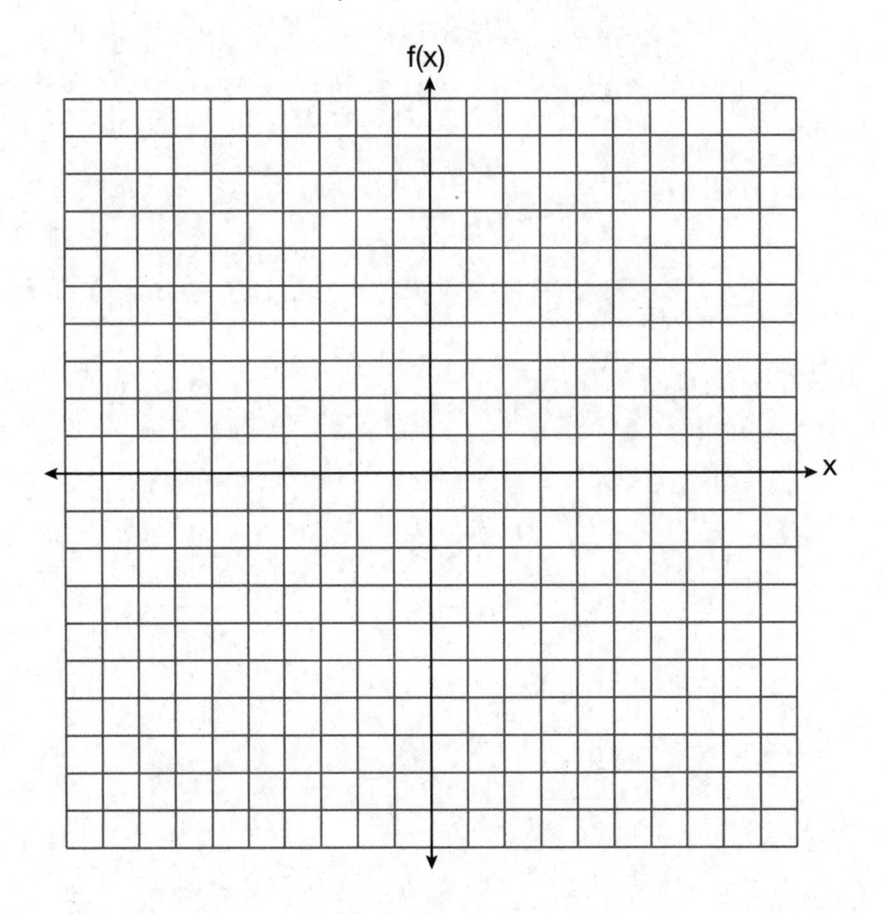

28 Solve $5x^2 = 180$ algebraically.

29 A blizzard occurred on the East Coast during January, 2016. Snowfall totals from the storm were recorded for Washington, D.C. and are shown in the table below.

Washington, D.C.	
Time	**Snow** (inches)
1 A.M.	1
3 A.M.	5
6 A.M.	11
12 noon	33
3 P.M.	36

Which interval, 1 A.M. to 12 noon or 6 A.M. to 3 P.M., has the greatest rate of snowfall, in inches per hour? Justify your answer.

30 The formula for the volume of a cone is $V = \frac{1}{3}\pi r^2 h$.
Solve the equation for h in terms of V, r, and π.

31 Given the recursive formula:

$$a_1 = 3$$

$$a_n = 2(a_{n-1} + 1)$$

State the values of a_2, a_3, and a_4 for the given recursive formula.

32 Determine and state the vertex of $f(x) = x^2 - 2x - 8$ using the method of completing the square.

PART III

Answer all 4 questions in this part. Each correct answer will receive 4 credits. Clearly indicate the necessary steps, including appropriate formula substitutions, diagrams, graphs, charts, etc. For all questions in this part, a correct numerical answer with no work shown will receive only 1 credit. [16 credits]

33 A school plans to have a fundraiser before basketball games selling shirts with their school logo. The school contacted two companies to find out how much it would cost to have the shirts made. Company A charges a \$50 set-up fee and \$5 per shirt. Company B charges a \$25 set-up fee and \$6 per shirt.

Write an equation for Company A that could be used to determine the total cost, A, when x shirts are ordered. Write a second equation for Company B that could be used to determine the total cost, B, when x shirts are ordered.

Determine algebraically and state the *minimum* number of shirts that must be ordered for it to be cheaper to use Company A.

34 Graph $y = f(x)$ and $y = g(x)$ on the set of axes below.

$$f(x) = 2x^2 - 8x + 3$$
$$g(x) = -2x + 3$$

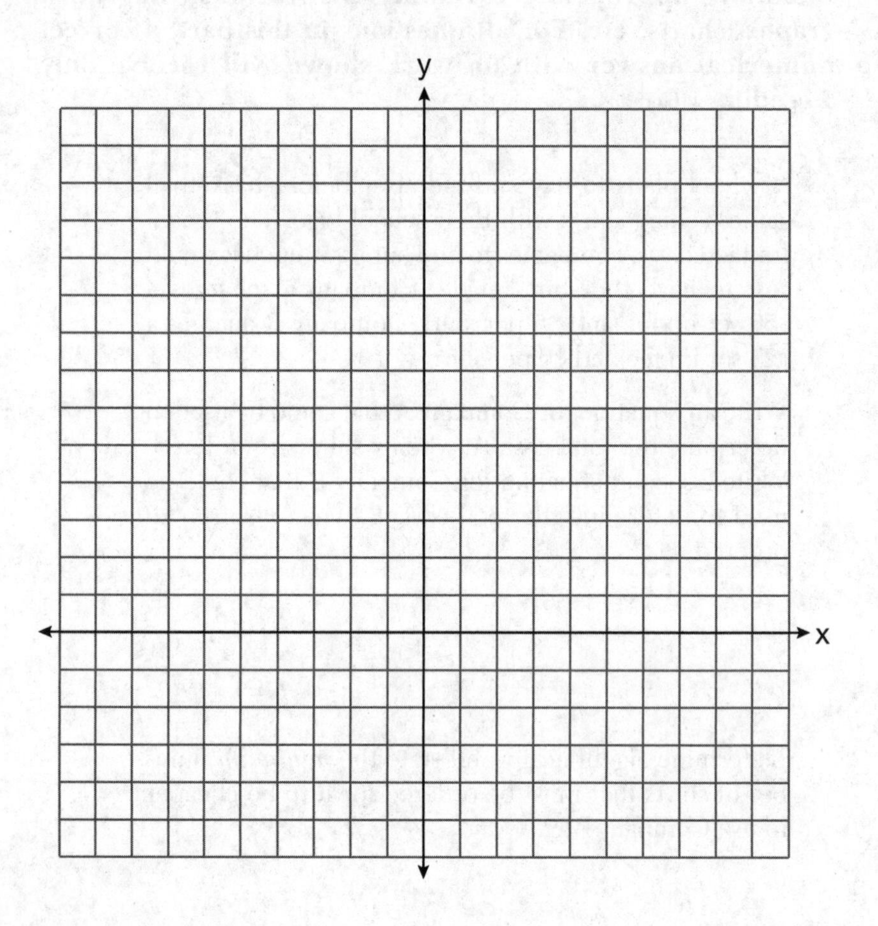

Determine and state all values of x for which $f(x) = g(x)$.

35 The table below shows the number of hours ten students spent studying for a test and their scores.

Hours Spent Studying (x)	0	1	2	4	4	4	6	6	7	8
Test Scores (y)	35	40	46	65	67	70	82	88	82	95

Write the linear regression equation for this data set. Round all values to the *nearest hundredth*.

State the correlation coefficient of this line, to the *nearest hundredth*.

Explain what the correlation coefficient suggests in the context of the problem.

36 A system of inequalities is graphed on the set of axes below.

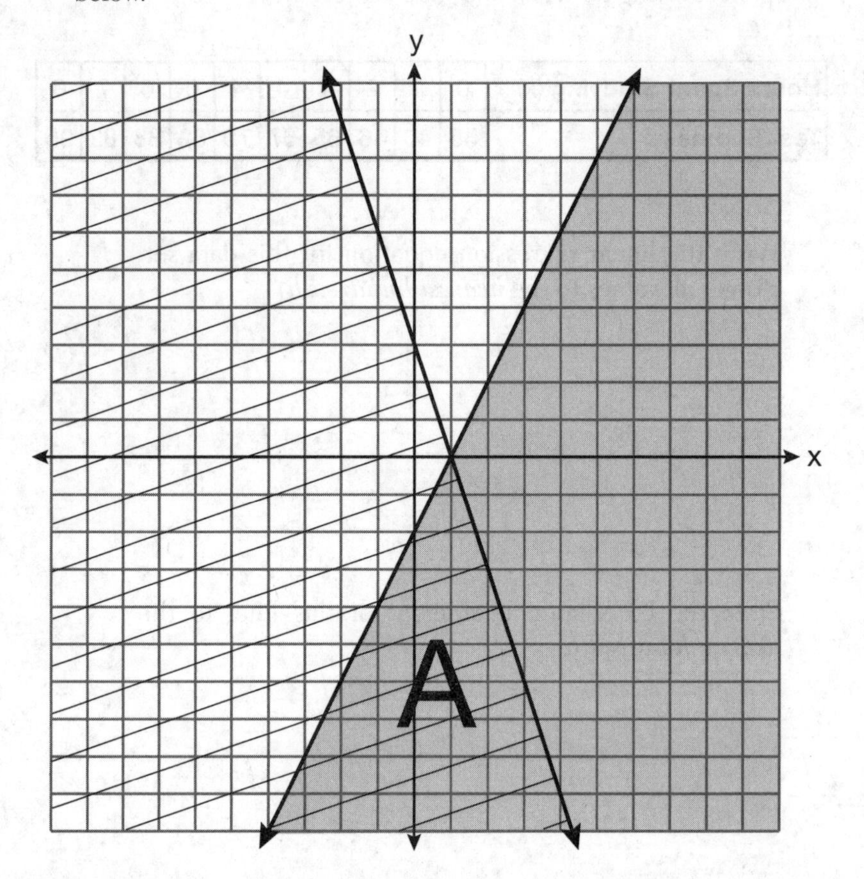

State the system of inequalities represented by the graph.

State what region A represents.

State what the entire gray region represents.

PART IV

Answer the question in this part. A correct answer will receive 6 credits. Clearly indicate the necessary steps, including appropriate formula substitutions, diagrams, graphs, charts, etc. A correct numerical answer with no work shown will receive only 1 credit. [6 credits]

37 When visiting friends in a state that has no sales tax, two families went to a fast-food restaurant for lunch. The Browns bought 4 cheeseburgers and 3 medium fries for $16.53. The Greens bought 5 cheeseburgers and 4 medium fries for $21.11.

Using c for the cost of a cheeseburger and f for the cost of medium fries, write a system of equations that models this situation.

The Greens said that since their bill was $21.11, each cheeseburger must cost $2.49 and each order of medium fries must cost $2.87 each. Are they correct? Justify your answer.

Using your equations, algebraically determine both the cost of one cheeseburger and the cost of one order of medium fries.

Answers June 2019

Algebra I

Answer Key

PART I

1. (4)	**5.** (1)	**9.** (4)	**13.** (4)	**17.** (3)	**21.** (2)
2. (2)	**6.** (1)	**10.** (1)	**14.** (3)	**18.** (3)	**22.** (1)
3. (4)	**7.** (1)	**11.** (3)	**15.** (2)	**19.** (2)	**23.** (2)
4. (2)	**8.** (3)	**12.** (2)	**16.** (2)	**20.** (4)	**24.** (1)

PART II

25. $x = 8$

26. $-3x^2 - 18x + 19$

27.

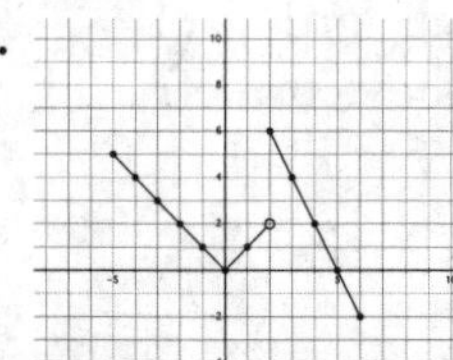

28. $\{-6, 6\}$

29. 1 A.M. to 12 noon because

$$2\frac{10}{11} > 2\frac{7}{9}$$

30. $h = \dfrac{3v}{\pi r^2}$

31. 8, 18, 38

32. (1, −9)

PART III

33. $A = 5x + 50$, $B = 6x + 25$; $x = 26$

34.

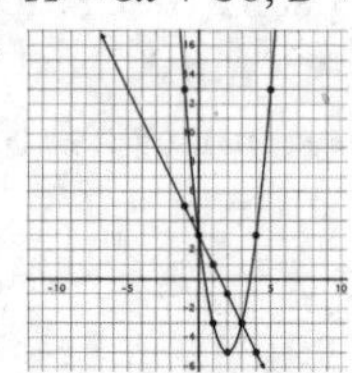

$x = 0, x = 3$

35. $y = 7.79x + 34.27$, $r = .98$

36. $y \leq 2x - 2$, $y < -3x + 3$, A represents the solution set of the system of inequalities. The gray region is the solution set to $y \leq 2x - 2$.

PART IV

37. $4c + 3f = 16.53$, $5c + 4f = 21.11$; they are not correct because it would be \$23.93, not \$21.11; one cheeseburger is \$2.79, and one order of fries is \$1.79

In **Parts II–IV**, you are required to show how you arrived at your answers. For sample methods of solutions, see *Barron's Regents Exams and Answers:* book for Algebra I.

Index

A
$(a - b)(a + b)$ pattern, multiplying of, 42–43
Absolute value, 123–124, 148–150
Addition, of polynomials, 32–33
Addition property of equality, 1–2
Algebra
  equations used in. *See* Equations
  properties of, 1–7
Algebra Reagents exam
  computational errors, 403
  conceptual errors, 403
  credits distribution by topic area, 449
  duration of, 401, 449
  formulas provided, 450–451
  partial credit on, 402–403
  points needed to pass, 410
  reasoning, 402–403
  sample examinations, 455–537
  scoring of, 450
  structure of, 402, 449
  time management for, 401
  tips for, 451–452
ANS feature, of graphing calculator, 406–408
Arithmetic sequences, 329, 331–333
Axis of symmetry, 174

B
Base case, 331
Best fit
  exponential curve of, 371–372
  line of. *See* Line of best fit
  parabola of, 368–370
Binomials
  definition of, 19
  factoring a trinomial into the product of, 46–47
  monomials multiplied by, 29
  multiplying of
    with distributive property, 40
    with FOIL shortcut, 40–41
  squaring, 42
Box plots, 395

C
Calculator. *See* Graphing calculator
Closed form defined sequences, 338–342
Coefficients
  definition of, 8
  negative, applying the quadratic formula with, 80–82
  two-step quadratic equations with, 58
Common factor, 45
Commutative property of addition or multiplication, 3–4
Completing the square, solving quadratic equations by, 66–71
Computational errors, 403
Conceptual errors, 403
Constant polynomials, 20
Constants
  description of, 7
  two-step quadratic equations with, 56–58
Correlation coefficient
  calculation of, 358–360
  definition of, 358
  negative, 362
  scatter plots used to compare, 361–362
Cubic polynomials, 20

D
Data representation
  box plots, 395
  dot plots, 393
  histograms, 394
Difference, 261
Distributive property
  of multiplication over addition, 3
  multiplying binomials with, 40
Division, of monomials, 24
Division property of equality, 2
Domain of function, 288–289, 297–298

E
Equations
  exponential. *See* Exponential equations
  for functions, 304–306
  identifying the type of
    by looking at graph of solution set, 254–256
    from table without graphing, 261–262
  isolating variables in, 14–15
  linear. *See* Linear equations
  multistep, 36–38
  one-step, 7–8
  quadratic. *See* Quadratic equations
  simplifying one side of, 36

solving of, with variables on both sides of the equals sign, 37
three-step, 12
two-step, 11–12
Exponential curve of best fit, 371–372
Exponential decay, 249, 283
Exponential equations
distinguishing characteristics of, 255
interpreting of, 282–283
real-world problems involving, 268–272, 279–285
solutions to, 243
solving for exponent in, 243–245
table of values for identifying, 256–257
two-variable, graphing solution sets to, 246–254
Exponential growth, 249, 282
Exponents
in exponential equations, solving for, 243–245
multiplying expressions involving, 22

F

Factoring
greatest common factor, 45–46
multiple-choice questions, 49–50
perfect squares, 48–49
perfect square trinomial into the square of a binomial, 48
polynomials, 45–50
trinomial into the product of two binomials, 46–47
Factors, 76–78
First quartile, 384–386
FOIL shortcut, multiplying binomials with, 40–41
Formulas, 450–451
Functions
defining functions based on other functions, 305
definition of, 287
domain of, 288–289, 297–298
equations for, 304–306
graphing calculator with
determining function values, 306–310
piecewise functions, 312–313
graphing of
description of, 292–298
horizontal translations, 317–319
from its equation, 305–306
piecewise functions, 312–313
vertical translations, 316
horizontal translations, 317–319
as list of ordered pairs, 287–289
negative, determining intervals for, 313–314
piecewise, 310–313
positive, determining intervals for, 313–314
range of, 288–289, 297–298
transformations, 316–319
vertical translations, 316

G

Geometric sequences, 330
Graphing calculator
ANS feature, 406–408
description of, 450
for determining function values, 306–310
for graphing piecewise functions, 312–313
for graphing solution sets
to exponential equations, 247–249
to linear equations, 146–157, 160–162
to quadratic equations, 179–183
to square root equations, 183
for graphing systems of inequalities, 233
for graphing table of values, 257–260
for graphing two-variable inequalities, 223–224
measures of central tendency, 386–388
memory of, 404–405
parentheses, 405–406
for solving linear-quadratic systems of equations, 199–201
for solving quadratic equations, 188–192
types of, 403–404
Graphs/graphing
of the equation, 120
equations for
description of, 158–162
horizontal or vertical lines, 122–123
of functions
description of, 292–298
horizontal translations, 317–319
from its equation, 305–306
piecewise functions, 312–313
vertical translations, 316
linear equations involving absolute value, 123–124
quadratic equations, 205–210
solution sets to exponential equations, 246–254
solution sets to linear equations
with graphing calculator, 146–157

in slope-intercept form, 138–141
table of values for, 117–121
two-intercept method for, 121–122
solution sets to quadratic equations
graphing calculator for, 179–183
table of values for, 171–173
solution sets to square root equations
graphing calculator for, 183
table of values for, 173
solving systems of linear equations by, 124–125
systems of linear inequalities, 230–241
two-variable linear inequalities
involving $>$ or $<$ signs, 221–222
involving $\leq$ or $\geq$ signs, 218–220
Greatest common factor factoring, 45–46

H

Histograms, 394

I

Interquartile range, 385
Intersect feature, graphing systems of linear equations using, 150–156
Irrational numbers, 68–69

L

Like terms, 26–27, 37
Line
slope of. *See* Slope
word problems finding equation of a, 165–167
Linear equations
creating of, 273–278
distinguishing characteristics of, 254
interpreting of, 273–278
from real-world scenarios, 273–278
systems of. *See* Systems of linear equations
Linear inequalities
one-variable, 213–217
systems of, graphing of, 230–241
two-variable, graphing of
on graphing calculator, 223–224
involving $>$ or $<$ signs, 221–222
involving $\leq$ or $\geq$ signs, 218–220
Linear polynomials, 20
Linear-quadratic systems of equations
solving with algebra, 197–198
solving with graphing calculator, 199–201
solving with table of values, 197
Line of best fit
description of, 343–347
real-world scenario questions answered with, 347–348
residual plots, 348–351

M

Maximum vertex of parabola, 178
Mean, 383
Measures of central tendency
definition of, 383
graphing calculator for determining, 386–388
interquartile range, 385
mean, 383
median, 384, 386
mode, 384
quartiles, 384–386
Median, 384, 386
Minimum vertex of parabola, 178
Mode, 384
Monomials
definition of, 19
dividing of, 24
multiplying of
by binomials, 29
with coefficients and variable parts, 22–23
by polynomials with more than two terms, 29–30
Multiplication
of $(a - b)(a + b)$ pattern, 42–43
of monomials, 22–23
Multiplication property of equality, 2–3
Multiplicative inverse, 3, 8

N

Negative correlation coefficients, 362
Negative functions, 313–314
Negative reciprocal, 131

O

One-step algebra equations
one-step solutions, 7–8
zero-step solutions, 7
One-variable linear inequalities, 213–217
Ordered pairs
description of, 89–90
function represented as list of, 287–289
list of, function represented as, 287–289

P

Parabola
definition of, 171
vertex of, 174–178

Parabola of best fit, 368–370
Parallel lines, 131
Parentheses, with graphing calculator, 405–406
Partial credit, 402–403
Percent decrease problems, 280–282
Percent increase problems, 268–269, 279–281
Perfect square pattern, 51–52
Perfect squares factoring, 48–49
Perfect square trinomial
  factoring of, into the square of a binomial, 48
  recognizing the pattern of, 53
Picture patterns, sequences for, 334, 340
Piecewise functions, 310–313
Polynomials
  addition of, 32–33
  constant, 20
  cubic, 20
  definition of, 19
  degree of, 19–20
  determining of, by number of terms, 19
  factoring, 45–50
  linear, 20
  monomials multiplied using, 29–30
  multiplying of, by polynomials, 40–43
  quadratic, 20
  subtraction of, 33–34
Positive functions, 313–314

Q
Quadratic equations
  distinguishing characteristics of, 255
  graphing of, for real-world applications, 205–210
  solving of
    by completing the square, 66–71
    by factoring, 71–75
    graphing calculator for, 188–192
    by guess and check, 63–65
    with quadratic formula, 79–82
    by taking the square root of both sides of the equation, 55–62
  word problems involving, 83–88
Quadratic formula, 79–82
Quadratic polynomials, 20
Quadratic trinomial, solving equations by factoring, 72–74
Quartiles, 384–386
Quotient, 261

R
Range of function, 288–289, 297–298
Rational numbers, 68–69
Real-world problems
  domain and range of function, 289
  exponential equations, 268–272, 279–285
  linear equations, 273–278
  line of best fit used to answer questions about, 347–348
  quadratic equations, 205–210
  sequences, 334, 340
Reasoning, 402–403
Reciprocal
  definition of, 3, 8
  negative, 131
Recursively defined sequences, 331–337
Reference sheet, 409, 453–454
Regression curves
  correlation coefficient. *See* Correlation coefficient
  line of best fit, 343–348
Relative two-way frequency table, 391
Residual plots, 348–351
Roots
  factors and, 76–78
  nature of, 68–69

S
Scatter plots, for comparing correlation coefficients, 361–362
Second quartile, 384
Sequence notation, 329
Sequences
  arithmetic, 329, 331–333
  closed form defined, 338–342
  definition of, 329
  geometric, 330, 332–333
  for picture patterns, 334, 340
  real-world scenarios, 334, 340
  recursively defined, 331–337
Slope
  calculating of, 130–131
  definition of, 129
  negative, 129
  positive, 129–131
  in time/distance graph, 131–132
Slope-intercept form
  definition of, 138
  finding the equation in, when *y*-intercept and another point are known, 141–142

graphing solution set of a linear equation in, 138–141
Square root equations, 60, 183
Squaring binomials, 42
Statistics
data representation methods, 393–398
measures of central tendency. *See* Measures of central tendency
Subtraction, of polynomials, 33–34
Subtraction property of equality, 2
Symmetry, axis of, 174
Systems of linear equations
combining two equations to form new equation, 95–96
with intersect feature, 150–156
setting up, 111–112
solving of
by changing both equations, 104–107
by changing one of the equations, 102–104
by graphing, 124–125
with guess and check, 89–94
solving word problems with, 111–112
substitution method for solving, 98–99
subtracting two equations to eliminate a variable, 98
without solutions or infinite number of solutions, 107
Systems of linear inequalities, 230–241

T

Table of values
calculator used to create, 148
exponential equations identified by looking at, 256–257
graphing calculator used to graph, 257–260
for graphing the solution set
of a linear equation, 117–121
to quadratic equation, 171–173
to square root equations, 173
for solving linear-quadratic systems of equations, 197
Test-taking strategies
graphing calculator, 403–408
partial credit, 402–403
time management, 401
Third quartile, 384–386
Three-step algebra equations, 12
TI-84, 403–404
Time/distance graphs
for changing speeds, 132–133
interpreting slope in, 131–132
Time management, 401
TI-Nspire, 403
Trinomials
definition of, 19
factoring of
into the product of two binomials, 46–47
recognizing of, into two binomials pattern, 52–53
perfect square
factoring of, into the square of a binomial, 48
recognizing the pattern of, 53
x term of, 32
Two-intercept method, for graphing the solution set of a linear equation, 121–122
Two-step algebra equations, 11–12
Two-step quadratic equations
with coefficients, 58
with constants, 56–58
with multiple variables, 59
Two-way frequency table, 388–390

V

Variables
description of, 1
isolating, in equations with multiple variables, 14–15
subtracting two equations to eliminate, 98
two-step quadratic equations with, 59
Vertex of parabola, 174–178
Vertical line test, 295

W

Word problems
combining like terms, 37–38
finding equation of a line, 165–167
quadratic equations, 83–88
solving, with systems of equations, 111–112

X

x-coordinate, 89–90
x-intercept, 121
x term, 32

Y

y-coordinate, 89–90
y-intercept, 121, 140, 142, 158

Z

Zeros feature, 188–191